Harmonic Modeling of Voltage Source Converters using Basic Numerical Methods

Harmonic Modeling of Voltage Source Converters using Basic Numerical Methods

Ryan Kuo-Lung Lian
National Taiwan University of Science and Technology, Taipei, Taiwan

Ramadhani Kurniawan Subroto
Technical University of Denmark, Kgs. Lyngby, Denmark

Victor Andrean
HedgeDesk, CA, USA

Bing Hao Lin
Industrial Technology Research Institute, Hsinchu, Taiwan

WILEY

IEEE PRESS

This edition first published 2022

Registered Offices
John Wiley & Sons, Inc., 111 River Street, Hoboken, NJ 07030, USA
John Wiley & Sons Ltd, The Atrium, Southern Gate, Chichester, West Sussex, PO19 8SQ, UK

Editorial Office
The Atrium, Southern Gate, Chichester, West Sussex, PO19 8SQ, UK

For details of our global editorial offices, customer services, and more information about Wiley products visit us at www.wiley.com.

Wiley also publishes its books in a variety of electronic formats and by print-on-demand. Some content that appears in standard print versions of this book may not be available in other formats.

Library of Congress Cataloging-in-Publication Data

Names: Lian, Ryan Kuo-Lung, author. | Subroto, Ramadhani Kurniawan, author. | Andrean, Victor, author. | Lin, Bing Hao, author.
Title: Harmonic modeling of voltage source converters using basic numerical methods / Ryan Kuo-Lung Lian, National Taiwan University of Science and Technology, Taipei, Taiwan, Ramadhani Kurniawan Subroto, Technical University of Denmark, Kgs. Lyngby, Denmark, Victor Andrean, HedgeDesk, CA, USA, Bing Hao Lin, Industrial Technology Research Institute, Hsinchu, Taiwan.
Description: Hoboken, NJ, USA : Wiley, 2022. | Includes bibliographical references and index.
Identifiers: LCCN 2021015328 (print) | LCCN 2021015329 (ebook) | ISBN 9781119527138 (hardback) | ISBN 9781119527145 (adobe pdf) | ISBN 9781119527152 (epub)
Subjects: LCSH: Electric current converters–Mathematical models. | Harmonics (Electric waves)–Mathematical models. | Electromagnetic interference–Mathematical models. | Numerical analysis.
Classification: LCC TK2799 .L53 2022 (print) | LCC TK2799 (ebook) | DDC 621.3815/322–dc23
LC record available at https://lccn.loc.gov/2021015328
LC ebook record available at https://lccn.loc.gov/2021015329

Cover Design: Wiley
Cover Images: © dowell/Moment/Getty Images, Olivier Le Moal/iStock/Getty Images

Set in 9.5/12.5pt STIXTwoText by Straive, Chennai, India
Printed and bound by CPI Group (UK) Ltd, Croydon, CR0 4YY

C9781119527138_191021

In Memory of Corrine Su.

"No two things have been combined better than knowledge and patience."

– Prophet Muhammad

Contents

Preface

The purpose of this book is threefold. One is to help students to see how some of the basic numerical analysis methods can be applied for harmonic analysis and modeling of power converter systems. Electrical engineering students usually do not have an undergraduate course specifically devoted to numerical techniques. Thus, some of them may have difficulty in seeing that some seemingly hard problems can be easily solved by some simple numerical techniques. This book covers some basic numerical techniques and shows how they can be effectively used for power system analysis and modeling.

The second purpose is that the deployment of voltage source converters (VSCs) has been gradually increased over the past decade due to their greater controllability and versatility, compared to other counterparts in power conversion, conditioning, compensation, and active filtering. The harmonics generated by the VSCs and what adverse effects they may have on the power system need to be fully studied. Nevertheless, a book that specifically addresses harmonic modeling of a VSC with various topologies is currently lacking. This book is to fill in this gap in the research of harmonic modeling.

The third objective is to unite the time-domain and frequency-domain modeling approaches for harmonic modeling. Frequency domain modeling of power electronic converter has been well covered in books such as *Power System Harmonic Analysis* by J. Arrillaga, B. C. Smith, N. R. Watson, and A. R. Wood and *Power Systems Harmonics: Computer Modelling and Analysis* by E. Acha and M. Madrigal. These two books have addressed models of various power system components including power converters. The reader is encouraged to read these two books, though it is not mandatory for understanding this book. This book mainly focuses on harmonic modeling of a VSC. The modeling process starts from deriving the state-space models of VSCs. Then it shows how a frequency coupling matrix (FCM) can be constructed in the time domain. Finally, to model a closed-loop VSC or other power converters, the monograph shows that one does not need to limit oneself to doing modeling in a certain domain as long as one follows the loop-breaking procedure laid out in the book. Moreover, although this book primarily focuses on VSC modeling, the presented techniques in general can be applied to current source converters, and other types of converters such as matrix converters.

This book is organized in the following fashion: Chapter 1 covers some of the fundamental concepts such as the definition of harmonics, the evaluation of Fourier series, and the concept of large-signal, small-signal, steady-state, and dynamic modeling. Chapter 2 briefly covers some of the commonly used semiconductor switches and highlights the importance of the VSC. Chapter 3 covers some of the basic numerical iterative techniques, which will be used throughout the book. Chapter 4 briefly reviews some of the commonly used numerical techniques for evaluating large exponential matrices, which play an essential role in multilevel converters, where the size of their

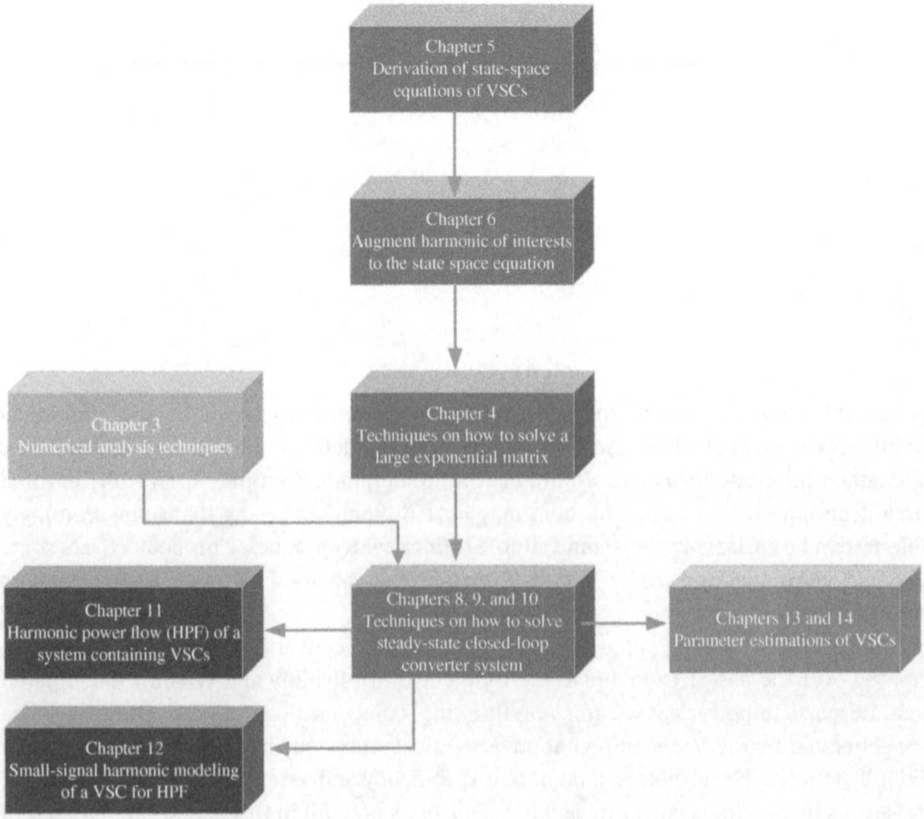

Figure 1 Connection between the main chapters.

state matrices can drastically increase for high-power applications. Chapter 5 derives the state space equations of various types of VSCs. Chapter 6 shows how an FCM of a VSC can be constructed in the time domain. The constructed FCM can be directly used in a harmonic power flow program, which relies on $I = YV$, where I is the harmonic current vector, V is the harmonic voltage vector, and Y is the system admittance matrix. Chapter 7 briefly describes the reference frames including a-b-c, α-β, and d-q frames, and some of the commonly used control configurations. Chapter 8 outlines the solution procedure for obtaining steady-state model of a closed-loop VSC. Chapters 9 and 10 show how the solution procedure presented in Chapter 8 can be used to model a single VSC with full control loops. Chapter 11 shows how the developed models in Chapters 9 and 10 can be interfaced with OpenDSS, which is a comprehensive electrical simulator, developed by Electric Power Research Institute (EPRI) for studies of distribution systems. Chapter 12 shows how a small-signal harmonic model of a VSC is derived. Such a model can be used to analytically predict the AC current and dc voltage harmonics, provided that the change of the input stimuli is small. Moreover, such a small-signal model can also be used in a harmonic power flow program where an analytical Jacobian matrix is required. Finally, Chapters 13 and 14 demonstrate one new application for the harmonic models of VSCs. Essentially, the VSC harmonic model, which is a white-box model, can be combined with a (deep) neural network model, a black-box model, to form a cost-effective and highly accurate parameter estimator of a VSC.

The order of the chapters in the book is exactly how the authors' graduate course is taught. Thus, instructors can follow the same order for conducting lectures. Alternatively, for readers who want to have a quick grasp of what the core of the book is, they may follow the flow of the chapters presented in Figure 1.

It is worth noting that program code is provided in snippets. That is, the code is provided to carry out only a particular function. Therefore, the reader needs to piece these codes together, and perhaps add a few more lines in order to obtain the same simulation results presented in the book. In this way, the reader does not need to read pages of code and can quickly comprehend how each snippet is written to carry out a certain function or task. Nevertheless, for some very simple example cases, the whole code is provided in the book.

Acknowledgments

While teaching and revising the drafts of this book, I have received invaluable help and feedback from my graduate students and I was very grateful. In particular, I would like to thank Miss Yi-Chun Chen, Mr. Jing-Tong Tsai, Mr. Jason Chang, and Mr. Perry Chang for their help on some of the code development and graph drawing. Prof. P. W. Lehn from University of Toronto is greatly appreciated for introducing harmonic modeling to me and providing insightful guidance, patient supervision, and invaluable comments during my Ph.D. studies. In fact, writing a book on harmonic modeling for voltage source converters was first suggested by him during one of the weekly research meetings at the University of Toronto.

I would also like to sincerely thank the Wiley editorial team for being super patient with me and keeping me on track and Prof. Chia-Chi Chu from National Tsing Hua University for his encouragement on writing this book. Finally, I would like to thank my family, especially my wife Zujun for taking care of our sons, Konrad and Jeff, while I was busy writing this book.

K. L. Lian, Taipei

Symbols

$\otimes$: convolution operator
$\Delta\mathbf{F}$: small-signal frequency coupling matrix
$\zeta^{(k)}$: the unknown variable at the k-th iteration
θ_w: weights of the network W,
θ_t: weights of the network h_t,
θ_d: weights of the network h_d
$\mathbf{\Phi}$: state transition of system matrix of $\mathbf{A}$
$\bar{\mathbf{\Phi}}$: state transition of system matrix of $\bar{\mathbf{A}}$
$\hat{\mathbf{\Phi}}$: state transition of system matrix of $\hat{\mathbf{A}}$
$\check{\mathbf{\Phi}}$: state transition of system matrix of $\check{\mathbf{A}}$
$\breve{\mathbf{\Phi}}$: state transition of system matrix of $\breve{\mathbf{A}}$
$\mathbf{\Omega}$: input dynamics matrix
$\mathbf{A}$: system matrix
$\mathbf{A}_r$: reduced system matrix
$\bar{\mathbf{A}}$: system matrix with input dynamics augmented
$\hat{\mathbf{A}}$: system matrix with input and harmonic state dynamics augmented
$\check{\mathbf{A}}$: system matrix with input, control, and reference state dynamics augmented
$\breve{\mathbf{A}}$: system matrix with input, control, reference, and harmonic state dynamics augmented
$\mathbf{B}$: input matrix
$\mathbf{B}_r$: reduced input matrix
C: capacitor
$C_c(s)$: transfer function of the current controller
$\mathbf{C}_{\mathbf{con}}$: connection matrix
$C_v(s)$: transfer function of the dc voltage controller
$\mathbf{C}_{\mathbf{x}}$: Clarke transformation matrix
$\mathcal{D}$: domain
$\mathcal{D}_S$: source domain
$\mathcal{D}_T$: target domain
$\mathbb{E}$: statistical expectation
h: hypothesis
h_t: task hypothesis
h_d: discrepancy hypothesis
i: state variable for current
i_{td}: d-frame current
i_{tdref}: d-frame current reference
i_{tq}: q-frame current
i_{tqref}: q-frame current reference

$\mathbb{I}$: identity matrix
I_i: i-th current harmonic component
$\mathbf{I}$: vector of current harmonics
$\mathbf{J}$: Jacobian matrix
L: inductor
$L_{\mathcal{D}}$: risk over $\mathcal{D}$
$\mathcal{L}_C$: cross-entropy
$\mathcal{L}_{CCSA}$: classification and contrastive semantic alignment loss
$\mathcal{L}_S$: separation loss
$\mathcal{L}_{SA}$: semantic alignment loss
m_a: modulation index
m_f: frequency modulation index
$\mathbf{M}$: vector of mismatch equations
$\mathbf{P}_M$: Park transformation matrix
$P_s(\boldsymbol{X})$: marginal probability distribution of source
$P_s(\boldsymbol{X}, Y)$: joint probability distribution of source
$P_s(Y|\boldsymbol{X})$: conditional probability distribution of source
$P_t(\boldsymbol{X})$: marginal probability distribution of target
$P_t(\boldsymbol{X}, Y)$: joint probability distribution of target
$P_t(Y|\boldsymbol{X})$: conditional probability distribution of target
R: resistor
SS_{res}: the sum of squares of errors between the true value and the estimated value of a model
SS_{tot}: the total variation
$\mathbf{u}$: vector of input variable
v: state variable for voltage
v_{dc}: dc voltage
v_{dcref}: dc voltage reference
v_{ma}: modulating signal of phase-a
v_{mb}: modulating signal of phase-b
v_{mc}: modulating signal of phase-c
$v_{m\alpha}$: modulating signal of α frame
$v_{m\beta}$: modulating signal of β frame
v_{md}: modulating signal of d frame
v_{mq}: modulating signal of q frame
V_i: i-th voltage harmonic component
$\mathbf{V}$: vector of voltage harmonics
$\mathbf{V_m}$: vector of modulating voltage harmonics
W: weighting network
$\mathbf{x}$: vector of the state variable
$\bar{\mathbf{x}}$: vector of the system and input state variable augmented
$\hat{\mathbf{x}}$: vector of the system with input and harmonic state variable augmented
x_d: integral of $i_{tdref} - i_{td}$
x_{dd}: integral of $v_{dcref} - v_{dc}$
$\mathbf{x_h}$: vector of the harmonic state of interest
x_q: integral of $i_{tqref} - i_{tq}$
$\mathbf{x}_r$: reduced state variable vector
$\boldsymbol{X}$: input space
Y: output space
$\boldsymbol{Z}$: embedding space

1
Fundamental Theory

1.1 Background

AC electric power systems are essentially designed to operate with sinusoidal voltages and currents at frequencies of 50 or 60 Hz. However, certain types of power components or loads produce currents and voltages with frequencies that are integer multiples of these frequencies (i.e. the fundamental frequencies). These higher frequencies are a form of electrical pollution known as power system harmonics. Power system harmonics are not a new phenomenon, and it is as old as the distribution of alternating current, which began in 1895–1896 [1]. It is reported that in 1893 Charles Proteus Steinmetz worked on the problem of motor heating while working at Thomson-Houston [2]. After rigorous calculations and experimental validation, Steinmetz concluded that the problem was due to the resonance in the transmission circuit feeding the plant and a generator with a substantial amount of waveform distortion. Consequently, Steinmetz proposed two solutions to overcome this harmonic problem. The first was to reduce the system frequency to one half of its original value. That is, to reduce the original frequency value of 125 Hz to a new value of 62.5 Hz. Note that at that time, most single-phase generators were operated at 125 Hz, 140 Hz or $133\frac{1}{3}$ Hz [3]. By 1892 there were a large number of Westinghouse-designed 60 Hz central stations in existence, and 60 Hz had taken over a share of the AC business from the higher frequencies [3] in North America. The second option was to restack iron laminations in the motor to withstand the higher operating voltage.

In 1895, both General Electric (GE) and Westinghouse started to use distributed armature windings for their manufactured generator to substantially improve their waveforms, making them more sinusoidal. In fact, the older-type design limited the transmission line lengths to 5 to 10 miles due to the resonance problems. In 1916, a text published by Steinmetz devotes considerable attention to the study of harmonics in three-phase power systems. The main concern was third harmonic currents caused by saturated iron in transformers and machines [4]. He was the first to propose delta connections for blocking third harmonic currents. In about the same year, a new problem, called the telephone interference factor, arose as a result of harmonic interference in communication circuits. Open-wire telecommunications were developed in the 19th and early 20th centuries without any consideration for other transmission lines because they were about five decades older than those of power transmission. Problems were caused by the introduction of the coexistence of power lines and telephone lines. Transformers and rectifiers in power systems produced harmonic currents that inductively coupled into adjacent open-wire telephone circuits and produced audible telephone interference. These problems were gradually alleviated by filtering and

Harmonic Modeling of Voltage Source Converters using Basic Numerical Methods, First Edition.
Ryan Kuo-Lung Lian, Ramadhani Kurniawan Subroto, Victor Andrean, and Bing Hao Lin.

by minimizing transformer core magnetizing currents [4]. Isolated telephone interference problems still occur nowadays, but these problems are infrequent because open-wire telephone circuits have been replaced with twisted pair, buried cables, and fiber optics [4].

Today, power electronic converters are the largest nonlinear loads and are used in industry [5] for a variety of purposes, such as electrochemical power supplies, adjustable speed drives, uninterruptible power supplies, flexible AC transmission system devices [6], interfaces for renewable energy sources with electric power systems, etc. These devices are useful because they can convert ac to dc, dc to dc, dc to ac, and ac to ac. According to an estimate by the Electric Power Research Institute, roughly 70% of electrical energy in the USA now flows through power converters, which will eventually grow to 100%. We expect to see the tremendous impact of power electronics not only in global industrialization and general energy systems, but also in energy saving, renewable energy systems and electric/hybrid vehicles [7]. Nevertheless, this comes at the expense of the generation of harmonics. Power electronic converters use static switches such as diodes, silicon-controlled rectifiers (SCRs), transistors, and other electronic switches to convert frequency, adjust voltage levels, and change power flow. This is done through chopping sinusoidal or dc waveforms, which leads to the generation of harmonics. The current emphasis on power quality has reinforced the need for harmonic studies as an important practice of power system analysis and design activities. Harmonic studies essentially involve quantifying the distortions in voltage and current waveforms at various points in a power system and determining whether they are harmful to the system and how these adverse conditions might be mitigated. This led to the availability of fairly general techniques and software for the formulations and solutions of harmonic propagation problems. This book reveals some of the numerical methods and techniques to model a power converter for harmonic analysis.

1.2 Definition of Harmonics

The term "harmonics" originates from the field of acoustics, where it refers to the vibration of a string or column of air at a frequency that is a multiple of the basic repetition frequency. Similarly, for electrical signals, a harmonic is defined as a sinusoidal component of a periodic waveform or quantity having a frequency that is an integral multiple of the fundamental frequency. For a 60 Hz power system, a harmonic wave is a sinusoid having a frequency expressed by:

$$f_k = k \times 60Hz, \tag{1.1}$$

where k is an integer.

Fig. 1.1 illustrates the fundamental frequency sine wave and its second, third, and fifth harmonics.

1.3 Fourier Series

The French mathematician Joseph Fourier found that any periodic waveform, $x(t)$, can be expressed as

$$x(t) = \frac{1}{2}a_0 + a_1 \cos(\omega t) + a_2 \cos(2\omega t) + \cdots + b_1 \sin(\omega t) + b_2 \sin(2\omega t) + \cdots, \tag{1.2}$$

where a_0, a_1, b_1, a_2, b_2, and so on are constants.

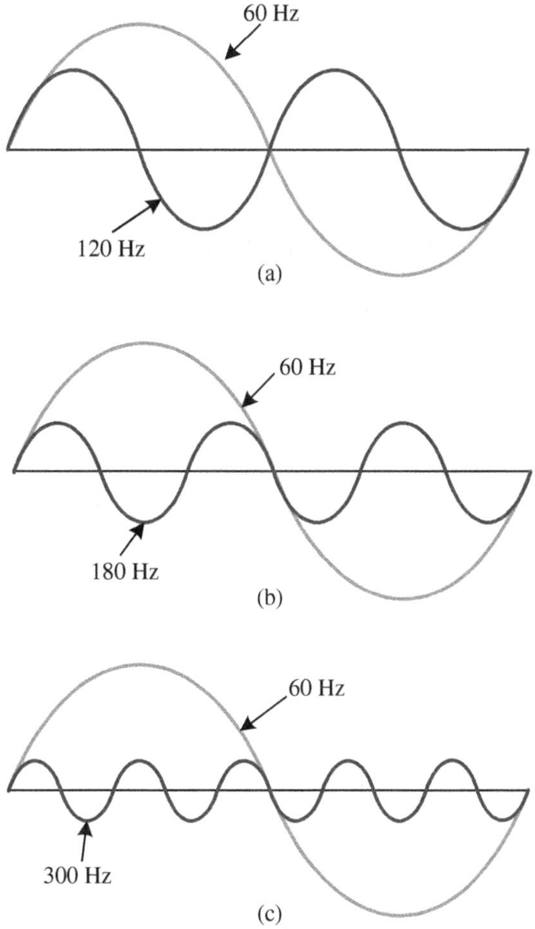

Figure 1.1 Fundamental frequency sine wave (60 Hz) and harmonics: (a) 2nd harmonic (120 Hz); (b) 3rd harmonic (180 Hz); (c) 5th harmonic (300 Hz).

1.3.1 Trigonometric Form

Equation (1.2) can be compactly written as

$$x(t) = \frac{1}{2}a_0 + \sum_{k=1}^{\infty} a_k \cos(k\omega t) + b_k \sin(k\omega t), \tag{1.3}$$

which is generally referred as the trigonometric form of the Fourier series. The first term $a_0/2$ represents the dc (average) component of $x(t)$. The terms with coefficient a_1 and b_1 together represent the fundamental frequency component ω. Similarly, the coefficients a_2 and b_2 together represent the second harmonic component 2ω.

Evaluation of a_k and b_k coefficients of (1.3) is not a difficult task because the sines and cosines are orthogonal functions, which means that their inner products are zero. Note that the inner product of two complex functions, f and g is defined as

$$\int_0^T f(t)g(t)^* dt. \tag{1.4}$$

If $f(t)$ and $g(t)$ are real-valued functions, (1.4) is then simply

$$\int_0^T f(t)g(t)\, dt. \tag{1.5}$$

The coefficients of the Fourier series can consequently be found by setting $f(t) = x(t)$ and $g(t) = \sin(k\omega t)$ or $g(t) = \cos(k\omega t)$. Thus,

$$a_k = \frac{2}{T}\int_0^T x(t)\cos(k\omega t)\,dt \tag{1.6}$$

$$b_k = \frac{2}{T}\int_0^T x(t)\sin(k\omega t)\,dt \tag{1.7}$$

Since the dc term is the average component of $x(t)$,

$$a_0 = \frac{1}{T}\int_0^T x(t)dt. \tag{1.8}$$

1.3.2 Phasor Form

An alternative form to (1.3) is usually preferred. The sine and cosine terms of the same frequency are grouped together, and the sum is combined to a single term, either sine or cosine. This is usually referred as the phasor form of the Fourier series. Equation (1.9) is the phasor form of the Fourier series in terms of cosines

$$x(t) = c_0 + \sum_{k=1}^{\infty} c_k \cos\left(k\omega t + \theta_k\right), \tag{1.9}$$

where

$$c_0 = \frac{1}{2}a_0 \tag{1.10}$$

$$c_k = \sqrt{a_k^2 + b_k^2} \tag{1.11}$$

$$\theta_k = -\arctan\left(\frac{b_k}{a_k}\right) \tag{1.12}$$

The derivations of (1.11) and (1.12) can be observable from the triangle of Fig. 1.2.

1.3.3 Exponential Form

According to Euler's formula, as stated in (1.13) and (1.14), a sinusoidal function can be expressed in terms of sums of exponentials.

$$\cos(\omega t) = \frac{e^{j\omega t} + e^{-j\omega t}}{2} \tag{1.13}$$

$$\sin(\omega t) = \frac{e^{j\omega t} - e^{-j\omega t}}{2j} \tag{1.14}$$

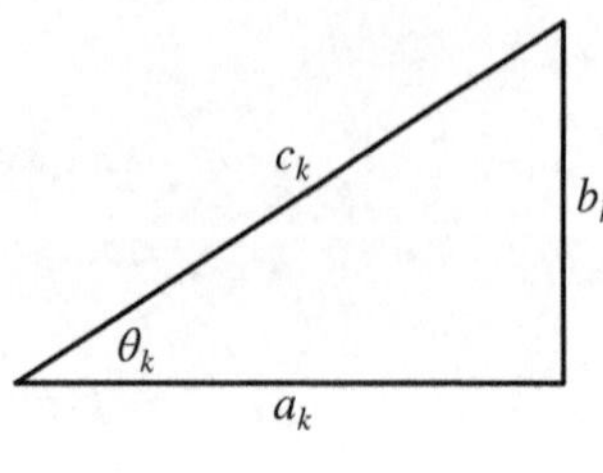

Figure 1.2 The triangle for deriving the phasor form of the Fourier series.

Hence, the Fourier series can also be expressed in an exponential form. Substituting (1.13) and (1.14) into (1.2), we get

$$x(t) = \frac{1}{2}a_0 + a_1\left(\frac{e^{j\omega t} + e^{-j\omega t}}{2}\right) + a_2\left(\frac{e^{j2\omega t} + e^{-j2\omega t}}{2}\right) + b_1\left(\frac{e^{j\omega t} - e^{-j\omega t}}{2j}\right) + b_2\left(\frac{e^{j2\omega t} - e^{-j2\omega t}}{2j}\right) + \cdots \tag{1.15}$$

Grouping terms with the same exponents, we get

$$x(t) = \cdots + \left(\frac{a_2}{2} - \frac{b_2}{2j}\right)e^{-2j\omega t} + \left(\frac{a_1}{2} - \frac{b_1}{2j}\right)e^{-j\omega t} + \frac{a_0}{2} + \left(\frac{a_1}{2} + \frac{b_1}{2j}\right)e^{j\omega t} + \left(\frac{a_2}{2} + \frac{b_2}{2j}\right)e^{2j\omega t} + \cdots \tag{1.16}$$

Compactly written, we get

$$x(t) = \sum_{k=-\infty}^{\infty} D_k e^{jk\omega t}, \tag{1.17}$$

where

$$D_0 = \frac{a_0}{2} \tag{1.18}$$

$$D_k = \frac{a_k}{2} - j\frac{b_k}{2} \tag{1.19}$$

$$D_{-k} = \frac{a_k}{2} + j\frac{b_k}{2} \tag{1.20}$$

Thus, the coefficients, except D_0, are complex and occur in complex conjugate pairs; that is,

$$D_k = D^*_{-k} \tag{1.21}$$

The Fourier series can be found by utilizing (1.4) with $g(t) = e^{jk\omega t}$. Thus,

$$D_k = \frac{1}{T}\int_0^T x(t)e^{-jk\omega t}dt. \tag{1.22}$$

1.4 Waveform Symmetry

Waveform symmetry can greatly simplify the integration effort required to develop Fourier coefficients. In general, symmetry arguments should be applied to the waveform after the average value has been removed [4].

1.4.1 Even Symmetry

A periodic function $f(t)$ has even symmetry if $f(t) = f(-t)$. That is, if in an even function, t is replaced with $-t$, the function $f(t)$ does not change. The cosine function is a good example of an even function because it can be written as the power series

$$\cos(t) = 1 - \frac{t^2}{2!} + \frac{t^4}{4!} - \frac{t^6}{6!} + \cdots \tag{1.23}$$

A typical example of an even function, observable in power system, is shown in Fig. 1.3. Such a waveform is a typical dc voltage ripple of an ac/dc converter. Since even functions have no sine terms, the b_k coefficients in (1.19) are zero. Therefore, D_k and D_{-k} in (1.19) and (1.20) are the same and real.

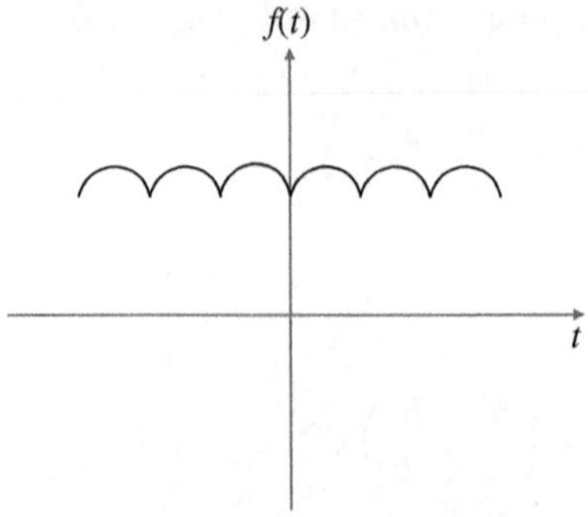

Figure 1.3 An example of an even function, $f(t)$.

1.4.2 Odd Symmetry

A periodic function $f(t)$ has odd symmetry if $f(t) = -f(-t)$. That is, if in an odd function, t is replaced with $-t$, we obtain the negative of the function $f(t)$. The sine function is a good example of an odd function because it can be written as the power series

$$\sin(t) = 1 - \frac{t^3}{3!} + \frac{t^5}{5!} - \frac{t^7}{7!} + \cdots \tag{1.24}$$

Since odd functions have no cosine terms, a_k coefficients in (1.19) are zero. Therefore, both D_k and D_{-k} in (1.19) and (1.20) are imaginary.

1.4.3 Half-Wave Symmetry

A periodic waveform with period T, has half-wave symmetry if

$$-f(t + T/2) = f(t). \tag{1.25}$$

That is, the shape of the negative half-cycle of the waveform is the same as that of the positive half-cycle, but inverted. Signal with this type of symmetry only have odd harmonics and all the even harmonics are zero. To see this, the following derivations are applied. Based on (1.6),

$$a_k = \frac{2}{T}\int_0^T f(t)\cos(k\omega t)\,dt \tag{1.26}$$

$$= \frac{2}{T}\left[\int_0^{T/2} f(t)\cos(k\omega t)\,dt + \int_{T/2}^{T} f(t)\cos(k\omega t)\,dt\right] \tag{1.27}$$

$$= \frac{2}{T}\left(In_1 + In_2\right), \tag{1.28}$$

where $In_1 = \int_0^{T/2} f(t)\cos(k\omega t)\,dt$ and $In_2 = \int_{T/2}^{T} f(t)\cos(k\omega t)\,dt$.

Making the substitution $\tau = t - T/2$ in In_2 gives

$$In_2 = \int_0^{T/2} f\left(\tau + \frac{T}{2}\right)\cos\left(k\omega\left(\tau + \frac{T}{2}\right)\right)d\tau \tag{1.29}$$

According to (1.25), (1.29) can be restated as

$$In_2 = -\int_0^{T/2} f(\tau)\cos(k\omega\tau + k\pi)\,d\tau. \tag{1.30}$$

Equation (1.30) can be simplified by using the trigonometric identity

$$\cos(u \pm v) = \cos(u)\cos(v) \mp \sin(u)\sin(v). \tag{1.31}$$

Hence,

$$\cos(k\omega\tau + k\pi) = \cos(k\omega\tau)\cos(k\pi) - \sin(k\omega\tau)\sin(k\pi) \tag{1.32}$$

$$= (-1)^k \cos(k\omega\tau), \tag{1.33}$$

and

$$In_2 = -(-1)^k \int_0^{T/2} f(\tau)\cos(k\omega\tau)\,d\tau. \tag{1.34}$$

Combining (1.28) and (1.34), we have

$$a_k = \frac{2}{T}\left(1-(-1)^k\right)\int_0^{T/2} f(t)\cos(k\omega t)\,dt, \tag{1.35}$$

and

$$a_k = \begin{cases} \frac{4}{T}\int_0^{T/2} f(t)\cos(k\omega t)\,dt, & k \textit{ is odd} \\ 0, & k \textit{ is even} \end{cases} \tag{1.36}$$

Similarly,

$$b_k = \begin{cases} \frac{4}{T}\int_0^{T/2} f(t)\sin(k\omega t)\,dt, & k \textit{ is odd} \\ 0, & k \textit{ is even} \end{cases} \tag{1.37}$$

Therefore, as long as a waveform possesses half-wave symmetry, regardless of how non-sinusoidal it looks like, only odd harmonics will be present. A waveform as shown in Fig. 1.4 only contains odd harmonics due to its half-wave symmetry.

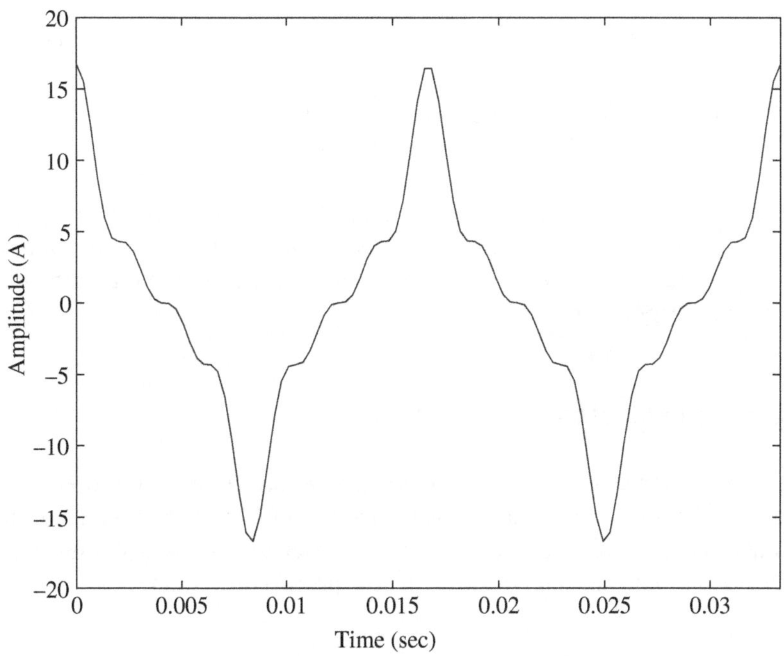

Figure 1.4 An example of current waveform with half-wave symmetry.

Table 1.1 The phase sequences for different order of harmonics.

Harmonics	Phase Sequence
1	+
2	−
3	0
4	+
5	−
6	0
⋮	⋮

1.5 Phase Sequence of Harmonics

In a balanced three-phase power system, the currents and voltage are shifted in time by 120° of the fundamental component. Therefore, phase *a* current can be expressed as

$$i_a(t) = \sum_{k=1}^{\infty} I_k \cos\left(k\omega_0 t + \theta_k\right). \tag{1.38}$$

For phases *b* and *c* current, their expressions are expressed as (1.39) and (1.40), respectively.

$$i_b(t) = \sum_{k=1}^{\infty} I_k \cos\left(k\omega_0 t + \theta_k - k\frac{2\pi}{3}\right). \tag{1.39}$$

$$i_c(t) = \sum_{k=1}^{\infty} I_k \cos\left(k\omega_0 t + \theta_k + k\frac{2\pi}{3}\right). \tag{1.40}$$

For $k = 1$, $i_a(t) = I_1 \cos(\omega_0 t + \theta_k)$, $i_b(t) = I_1 \cos(\omega_0 t + \theta_1 - \frac{2\pi}{3})$, and $i_c(t) = I_1 \cos(\omega_0 t + \theta_1 + \frac{2\pi}{3})$. Thus, the first harmonic is a positive sequence. For $k = 2$, $i_a(t) = I_2 \cos(2\omega_0 t + \theta_2)$, $i_b(t) = I_2 \cos(\omega_0 t + \theta_2 - \frac{4\pi}{3}) = I_2 \cos(\omega_0 t + \theta_2 + \frac{2\pi}{3})$, and $i_c(t) = I_2 \sin(\omega_0 t + \theta_2 + \frac{4\pi}{3}) = I_2 \cos(\omega_0 t + \theta_2 - \frac{2\pi}{3})$. Thus, the second harmonic is a negative sequence. For $k = 3$, $i_a(t) = I_3 \cos(3\omega_0 t + \theta_3 - 2\pi) = I_3 \cos(3\omega_0 t + \theta_3)$, $i_b(t) = I_3 \cos(\omega_0 t + \theta_3 - 2\pi) = I_3 \cos(3\omega_0 t + \theta_3)$, and $i_c(t) = I_3 \cos(\omega_0 t + \theta_3 + 2\pi) = I_3 \cos(3\omega_0 t + \theta_3)$, which makes the third harmonic a zero sequence. Table 1.1 lists the phase sequences for different order of harmonics.

1.6 Frequency Domain and Harmonic Domain

In this book, frequency and harmonic domains are used interchangeably. Nevertheless, there is a difference between the two. The frequency domain is generally the domain where the frequency can take on any real number. On the other hand, frequency can only take on an integer number in the harmonic domain. Thus, harmonic domain models can be expressed in matrix forms.

1.7 Power Definitions

1.7.1 Average Power

The average power is defined as

$$P_{avg} = \frac{1}{T}\int_{t}^{t+T} p(t)dt = \frac{1}{T}\int_{t}^{t+T} v(t)i(t)dt. \tag{1.41}$$

Substituting in the Fourier series of voltage and current yields

$$P_{avg} = \frac{1}{T}\int_{t}^{t+T} \left(\sum_{k=1}^{\infty} V_k \sin\left(k\omega t + \theta_{v,k}\right)\right)\left(\sum_{k=1}^{\infty} I_k \sin\left(k\omega t + \theta_{i,k}\right)\right) dt. \tag{1.42}$$

Since the average values of all the sinusoidal terms is zero, we will only have the time-invariant terms left in the summation, which is

$$P_{avg} = \sum_{k=1}^{\infty} V_k I_k \cos(\theta_{v,k} - \theta_{i,k}) = \sum_{k=1}^{\infty} V_k I_k \cos(\theta_k) = P_{1,avg} + P_{2,avg} + \cdots + P_{k,avg} + \cdots, \tag{1.43}$$

where $\cos(\theta_k)$ is called the displacement power factor for the k-th harmonic and $P_{k,avg}$ is the average power for the k-th harmonic.

Note that the fundamental active power $P_{1,avg}$ is sometimes used to represent the active power, ignoring powers associated with other harmonics.

1.7.2 Apparent and Reactive Power

The definition of active power is consensually agreed upon. However, the definitions of reactive power under unbalanced and nonsinusoidal conditions are still a debatable subject. Interested readers can refer to Prof. Emanuel's book [8] for a comprehensive review and discussion on this subject. Here are listed just a few well-known examples.

Budeanu's Method

C. I. Budeanu partitioned the squared harmonic current into two orthogonal terms:

$$I_k^2 = (I_k \cos\theta_k)^2 + (I_k \sin\theta_k)^2. \tag{1.44}$$

Thus,

$$S^2 = V^2 I^2 = \sum_{k=1}^{\infty} V_k^2 \left[\sum_{k=1}^{\infty} \left(I_k \cos\left(\theta_k\right)\right)^2\right] + \sum_{k=1}^{\infty} V_k^2 \left[\sum_{k=1}^{\infty} \left(I_k \sin\left(\theta_k\right)\right)^2\right]. \tag{1.45}$$

It can be shown that the apparent power can be written as three orthogonal components:

$$S^2 = P^2 + Q_B^2 + D_B^2, \tag{1.46}$$

where

$$Q_B = \sum_{k=1}^{\infty} V_k I_k \sin\theta_k \tag{1.47}$$

$$D_B = \sqrt{\sum_{m=1}^{\infty}\sum_{n=m+1}^{\infty} \left[\left(V_m I_n\right)^2 + \left(V_n I_m\right)^2 - 2V_m V_n I_m I_n \cos\left(\theta_m - \theta_n\right)\right]}. \tag{1.48}$$

Budeanu was the first one to recognize that a nonactive power is different from a reactive power. In fact, many books today still adopt Budeanu's definitions. Nevertheless, as proved by L. Czarnecki, Q_B and D_B do not possess the attributes that can be related to the power phenomena in circuits with nonsinusoidal waveforms [8].

Fryze's Method

S. Fryze defined an instantaneous active current, which mimics the voltage waveform, and can be expressed as

$$i_a = Gv = G\sum_{k=0}^{\infty} V_k \sin\left(k\omega t + \theta_{v,k}\right) = \sum_{k=0}^{\infty} I_k \sin\left(k\omega t + \theta_{v,k}\right). \tag{1.49}$$

The remaining component of the current is

$$i_b = i - i_a, \tag{1.50}$$

and is sometime called wattless current [8] and the following equation holds true:

$$I^2 = I_a^2 + I_b^2. \tag{1.51}$$

The apparent power, thus, is defined as

$$S^2 = P^2 + Q_F^2, \tag{1.52}$$

where $S = VI$, $P = VI_a$, and $Q_F = VI_b$. The power factor is

$$PF = \frac{P}{S} = \frac{I_a}{I}. \tag{1.53}$$

Czarnecki's Method

L. Czarnecki improved the work of Fryze and replaced i_b with the instantaneous reactive current, i_r and the instantaneous scattered current, i_s. The three currents are mutually orthogonal, and (1.54) resulted:

$$I^2 = I_a^2 + I_s^2 + I_r^2. \tag{1.54}$$

The apparent power is defined as

$$S^2 = P^2 + D_s^2 + Q_r^2, \tag{1.55}$$

where D_s is the scattered power and Q_r is the reactive power.

Emanuel's Method or IEEE Std. 1459-2010

Emanuel separated both P_1 and Q_1 from the rest of the powers. This can be achieved by partitioning the voltage and current into the following:

$$V^2 = V_1^2 + V_H^2 \tag{1.56}$$

$$I^2 = I_1^2 + I_H^2 \tag{1.57}$$

where V_1 and I_1 are the fundamental components of the voltage and current, respectively. Moreover, $V_H^2 = \sum_{k\neq 1}^{\infty} V_k^2$ and $I_H^2 = \sum_{k\neq 1}^{\infty} I_k^2$

$$S^2 = V^2I^2 = S_1^2 + D_I^2 + D_v^2 + S_H^2, \tag{1.58}$$

$$S_1 = V_1I_1 = \sqrt{P_1^2 + Q_1^2}, \tag{1.59}$$

$$D_I = V_1 I_H = V_1 \sqrt{\sum_{k \neq 1}^{\infty} I_k^2}, \tag{1.60}$$

$$D_V = V_H I_1 = I_1 \sqrt{\sum_{k \neq 1}^{\infty} V_k^2}, \tag{1.61}$$

and

$$S_H = V_H I_H = \sqrt{\sum_{k \neq 1}^{\infty} V_k^2 \sum_{k \neq 1}^{\infty} I_k^2}. \tag{1.62}$$

Note that this method was adopted in the IEEE Std. 1459-2010, Definitions for the Measurement of Electric Power Quantities [8]. For a three-phase system, in addition to the definitions mentioned above, one also needs to be aware of how the apparent power for each phase is formed to form a three-phase apparent power. Generally, there are three types, as listed below:

1. Arithmetic apparent power:

$$S_A = S_a + S_b + S_c. \tag{1.63}$$

   The associated power factor is $PF_A = \frac{P}{S_A}$, where $P = P_a + P_b + P_c$.
2. Vector apparent power:

$$S_v = \sqrt{P^2 + Q^2}, \tag{1.64}$$

   where $Q = Q_a + Q_b + Q_c$. The associated power factor, $PF_v = \frac{P}{S_v}$.
3. Effective apparent power:

$$S_e = 3V_e I_e, \tag{1.65}$$

   where $V_e = \sqrt{\frac{V_{ab}^2 + V_{bc}^2 + V_{ca}^2}{3}}$ and $I_e = \sqrt{\frac{I_a^2 + I_b^2 + I_c^2}{3}}$.

When the three-phase system is balanced, both definitions give identical results. However, for unbalanced conditions, the results differ: $S_A \geq S_V$ and $PF_A \leq PF_V$.

1.8 Harmonic Indices

Some of the commonly used indices for measuring the harmonic content of a waveform is discussed in this section. The total harmonic distortion (THD) and the total demand distortion (TDD) are the most commonly used ones. Both are measures of the effective value of a waveform and may be applied to either voltage or current [9].

1.8.1 Total Harmonic Distortion (THD)

THD is defined as the ratio of the sum of the powers of all harmonic components to the power of the fundamental frequency, and it is applied to both voltage and current. Note that the dc component is usually ignored for the THD calculation. THD for current, THD_I, can be stated as

$$THD_I = \frac{\sqrt{\sum_{k=2}^{\infty} I_k^2}}{I_1}, \tag{1.66}$$

where I_k is the k-th harmonic component of the current.

Similarly, THD for voltage, THD_V can be stated as

$$THD_V = \frac{\sqrt{\sum_{k=2}^{\infty} V_k^2}}{V_1}, \tag{1.67}$$

where V_k is the k-th harmonic component of the voltage.

THD and root-mean-square (RMS) values are directly linked. Since RMS current $I = \sqrt{\sum_{k=1}^{\infty} I_k^2}$, THD_I can be rewritten as

$$THD_I = \frac{\sqrt{\sum_{k=1}^{\infty} I_k^2 - I_1^2}}{I_1}, \tag{1.68}$$

and an equation linking THD_I and RMS values can be written as

$$I = I_1\sqrt{1 + THD_I^2}. \tag{1.69}$$

Equation (1.69) indicates that

1. The lower the THD_I values, the closer the true RMS value is to the fundamental current.
2. Line losses are proportional to the square of the RMS current. Thus, for a current distortion near 100%, the losses are almost twice those of a sinusoidal current.

THD can provide a good idea of how much losses or heat will be generated due to the voltage or current distortions. However, it is not a good indicator of the voltage stress within a capacitor as it is related to the peak value of the voltage waveform [9].

1.8.2 Total Demand Distortion (TDD)

THD value sometimes can be misleading. A small current may have a high THD value. However, it is not a significant threat to the system. For example, many adjustable speed drives will exhibit high THD values for the input current when they are operating at very light loads. This is not necessarily a significant concern because the magnitude of harmonic current is low, even though its relative current distortion is high [9]. Therefore, TDD[1] is defined as

$$TDD = \frac{\sqrt{\sum_{k=2}^{\infty} I_k^2}}{I_L}, \tag{1.70}$$

where I_L is the peak, or maximum, demand load current at the fundamental frequency component measured at the point of common coupling (PCC).

1.8.3 True Power Factor

The true power factor is defined as

$$PF_t = \frac{P_{avg}}{VI}. \tag{1.71}$$

1 The definition stated in (1.70) is based on IEEE Standard 519 1992. TDD defined in IEEE Standard 519 2014 is $TDD = \frac{\sqrt{\sum_{k=2}^{50} I_k^2}}{I_L}$.

In the absence of harmonics, (1.71) reduces to the displacement power factor as evident from (1.72)

$$\frac{P_{1,avg}}{V_1 I_1} = \cos(\theta_{v,1} - \theta_{i,1}), \tag{1.72}$$

where $\theta_{v,1}$ and $\theta_{i,1}$ are the phase angles of the fundamental voltage and current, respectively.

In the presence of harmonics, (1.71) can be expanded as

$$PF_t = \frac{P_{1,avg} + P_{2,avg} + P_{3,avg} + \cdots}{V_1\sqrt{1 + THD_V^2} I_1 \sqrt{1 + THD_I^2}}. \tag{1.73}$$

In most cases, the harmonic powers are small compared to the fundamental power, and the voltage distortion is also small [4].

$$PF_t \approx \frac{P_{1,avg}}{V_1 I_1 \sqrt{1 + THD_I^2}} = \frac{cos(\theta_{v,1} - \theta_{i,1})}{\sqrt{1 + THD_I^2}} \tag{1.74}$$

(1.74) indicates that the true power factor of a nonlinear load is limited by its THD_I. For example, the true power factor of a nonlinear load with $THD_I = 100\%$ can never exceed 0.707, no matter how good its displacement power factor is.

1.9 Detrimental Effects of Harmonics

In addition to causing the communication interference as mentioned in Section 1.1, harmonics are associated with many other detrimental effects.

1.9.1 Resonance

The inductive reactance of the ac power system (composed of generators, transmission lines, transformer, etc.) and the capacitive reactance of capacitor banks and underground cables produce resonance conditions. The resonant frequency of an LC circuit occurs when the inductive reactance equals the capacitive reactance. There are two types of resonance, which are parallel and series resonance.

Parallel Resonance

Parallel resonance occurs when the shunt capacitor appears in parallel with the equivalent system inductance, as depicted in Fig. 1.5. To clearly describe the concept of parallel resonance, the simplified diagram is illustrated in Fig. 1.6. At the resonant frequency, the apparent impedance of the parallel combination of the equivalent inductance and capacitance as seen from the harmonic current source becomes very large. The equivalent impedance Z_p is described in (1.75).

$$Z_p = \frac{X_c(X_{Leq} + R)}{X_c + X_{Leq} + R} = \frac{X_c(X_{Leq} + R)}{R} \approx \frac{X_{Leq}^2}{R} = QX_{Leq} = QX_C \tag{1.75}$$

where X_c, X_{Leq}, and R denote the reactances of the shunt capacitor, equivalent source and transformer inductors, and resistive load, respectively. The variable Q is defined as the quality factor of a resonant circuit determining the sharpness of the frequency response, specified as $Q = X_{Leq}/R = X_C/R$ and normally $R \ll X_{Leq}$. Note that the reactance in (1.75) is calculated at

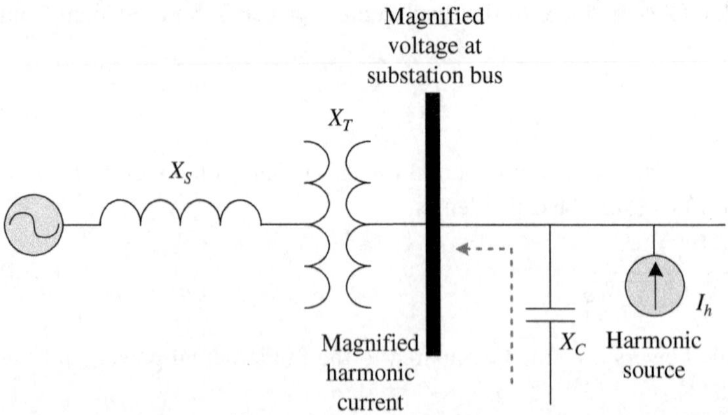

Figure 1.5 Shunt capacitor acts as parallel resonance in distribution system.

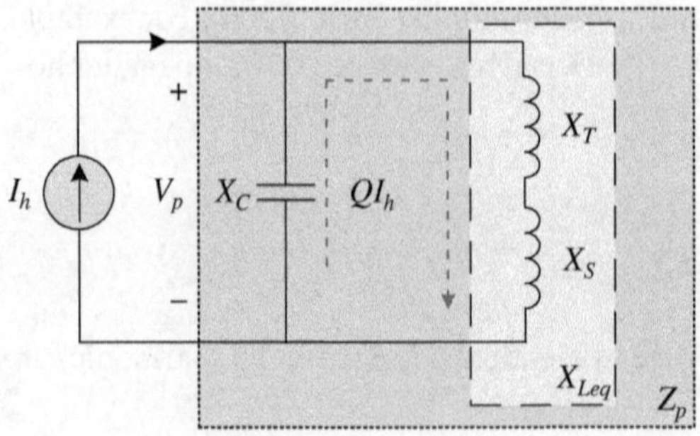

Figure 1.6 Simplified circuit of parallel resonance.

the resonant frequency. Determining Q is very crucial as it affects the voltage drop across the equivalent impedance. This can be observed through the relationship

$$V_p = Z_p I_h = QX_{Leq} I_h. \tag{1.76}$$

The voltage drop (V_p) is proportional to Q, meaning the larger Q causes the larger voltage drop at the resonant frequency. Furthermore, from (1.76) one may see that a small harmonic current can also cause a large voltage drop across the equivalent impedance. This behavior can be also found in the capacitor shunt. Assuming the current flowing to the capacitor shunt and power system as I_{res}, then

$$\begin{aligned} I_{res} &= \frac{V_p}{X_C} = \frac{QX_C I_h}{X_C} = QI_h \\ &= \frac{V_p}{X_{Leq}} = \frac{QX_{Leq} I_h}{X_{Leq}} = QI_h. \end{aligned} \tag{1.77}$$

From (1.77), one may observe that the current flowing through the capacitor shunt and power system is also enlarged Q times. Thus, it can cause capacitor failure and transformer overheating. To reduce the equivalent impedance for the parallel resonance, the resistive load should be added into the power system. This resistive load may be assumed as the lines or cables between the capacitor shunt and nearest transformer. The significant length of lines or cables can generate specified values of resistance. The frequency response of various values of

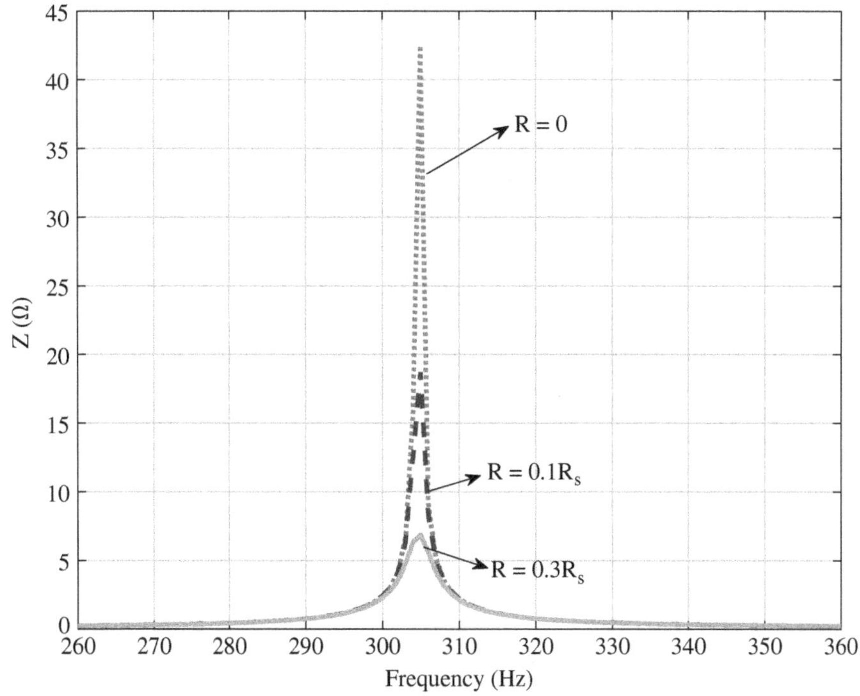

Figure 1.7 Resistive load variations on parallel resonance.

resistance to parallel resonance is shown in Fig. 1.7. It can be clearly seen that when there is no resistance between the capacitor shunt and transformer, the equivalent impedance is very sharp and high at the resonant frequency. When the resistive load is added, the magnitude can be greatly suppressed. If the resistance value is sufficient to provide the damping effect, then the catastrophic voltage and current failure can be avoided. Thus, one may conclude that the resistive load is able to damp the resonance, leading to a significant reduction in the harmonic distortion.

Example 1.1

For some known parameters: $L_{eq} = 3.949 \times 10^{-2}$ mH and $C = 6.9$ mF. When the grid resistances are varied as 0%, 10%, and 30% of the nominal grid resistance, $R_s = 2.747$ mΩ, simulate the impedance characteristics of parallel resonance circuit. Also, determine the parallel resonant frequency.

Solutions:

The transfer function of the equivalent impedance is obtained as

$$Z_p(\omega) = \frac{\left(R_s + j\omega L_{eq}\right)\left(\frac{1}{j\omega C}\right)}{R_s + j\omega L_{eq} + \frac{1}{j\omega C}}$$

Parallel resonance occurs when $X_{Leq} = X_C$. Thus the parallel resonant frequency can be obtained as

$$f_p = \frac{1}{2\pi\sqrt{L_{eq}C}} \tag{1.78}$$

As shown in Fig. 1.7, the parallel resonant frequency $f_p = 304.73$ Hz. □

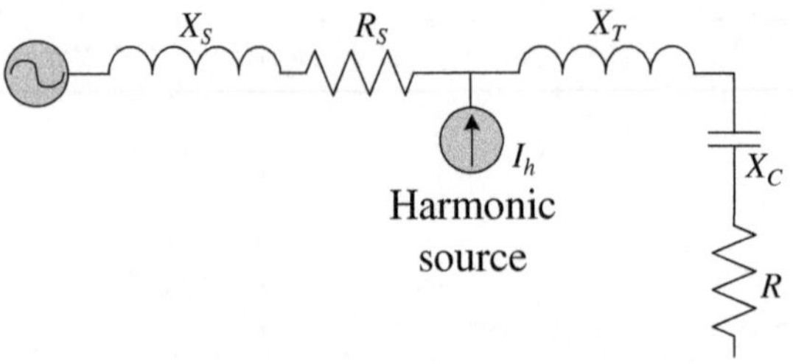

Figure 1.8 Series resonance in distribution system.

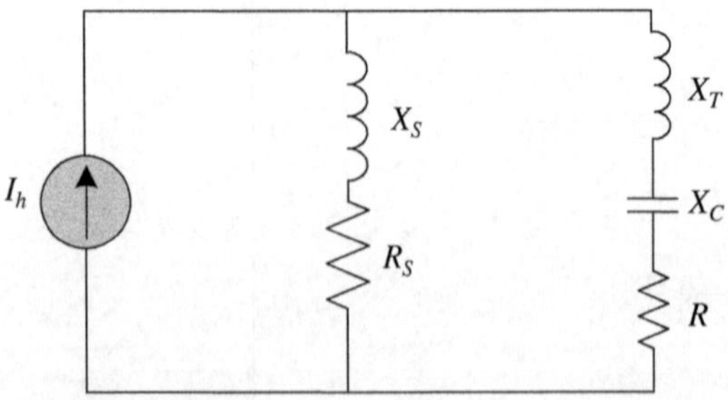

Figure 1.9 Simplified circuit of series resonance.

Series Resonance

A shunt capacitor and the inductance of a transformer or distribution line may appear as a series LC circuit, as depicted in Fig. 1.8. The simplified circuit is shown in Fig. 1.9. The overall impedance of the series resonance seen from the harmonic current source is very small (theoretically zero) and only limited by its resistance. Thus, the voltage at the capacitor is magnified and highly distorted, as can be seen from (1.79).

$$V_c = \frac{X_c}{X_T + X_c + R} V_h = \frac{X_c}{R} V_h \tag{1.79}$$

The equivalent impedance Z_p is described in (1.80).

$$Z_p = \frac{(R_s + X_s)\,(X_T + X_C + R)}{R_s + X_s + X_T + X_C + R} \tag{1.80}$$

Note that the resistances R and R_s are assumed to be very small. The negligible impedance of the series resonance is useful to absorb desired harmonic currents exhibiting the concept of a notch filter.

In a practical situation, the parallel resonance always appears in the power system due to the circuit topology. The parallel configuration is formed by the reactance of the source with the series connection of reactance of transformer and capacitor bank. Thus, the frequency response, depicted in Fig. 1.10, is composed by the parallel resonance and series resonance.

Example 1.2

For some known parameters: $R_s = 2.747$ mΩ, $L_s = 3.949 \times 10^{-2}$ mH, $L_T = 6.119 \times 10^{-2}$ mH, and $C = 6.9$ mF, simulate the impedance characteristics of parallel and series resonance circuits. Determine the parallel and series resonant frequencies.

Solutions:

The transfer function of the equivalent impedance is obtained as

$$Z_p(\omega) = \frac{(R_s + j\omega L_s)\left(j\omega L_T + \frac{1}{j\omega C} + R\right)}{R_s + j\omega L_s + j\omega L_T + \frac{1}{j\omega C} + R}.$$

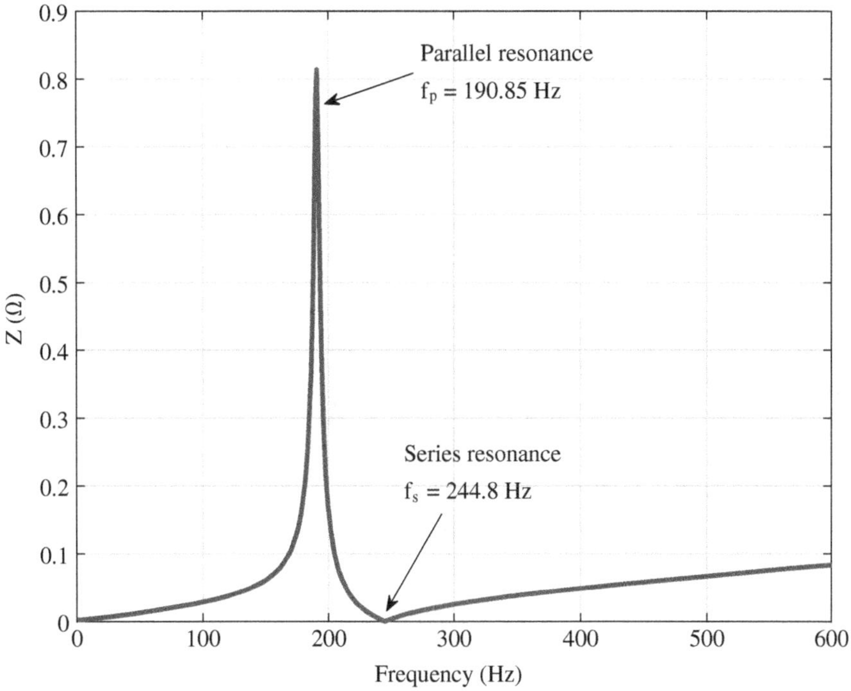

Figure 1.10 Series resonance frequency response.

As shown in Fig. 1.10, the parallel resonant frequency

$$f_p = \frac{1}{2\pi\sqrt{(L_T + L_S)\,C}}. \tag{1.81}$$

Meanwhile, the series resonant frequency can be obtained as

$$f_s = \frac{1}{2\pi\sqrt{L_T C}}. \tag{1.82}$$

Thus, $f_p = 190.85$ Hz and $f_s = 244.8$ Hz.

The problem of resonance is that small currents (voltages) generated by a harmonic source in the power system can be amplified significantly. Since switching the configuration of the capacitor banks (for a different power factor) changes the capacitive resonance of the system, more than one resonant frequency may exist. In addition, since motor load appears to be primarily inductive at harmonic frequencies, changes in motor loads on the system can result in a shift of the resonant frequency. □

1.9.2 Misoperations of Meters and Relays

Most of the watt-hour meters are designed to operate in the presence of purely sinusoidal voltage and current waveforms, and are calibrated in this way. This is because conventional watt-hour meters are based on the induction motor principles [10]. The rotor elements or the rotating disk inside the meter revolve at a speed proportional to the power flow. This disk then drives a series of gears that move dials on a register. Consequently, in the presence of harmonics, the conventional watt-hour meters will have an error in the power readings. Harmonics can also have detrimental

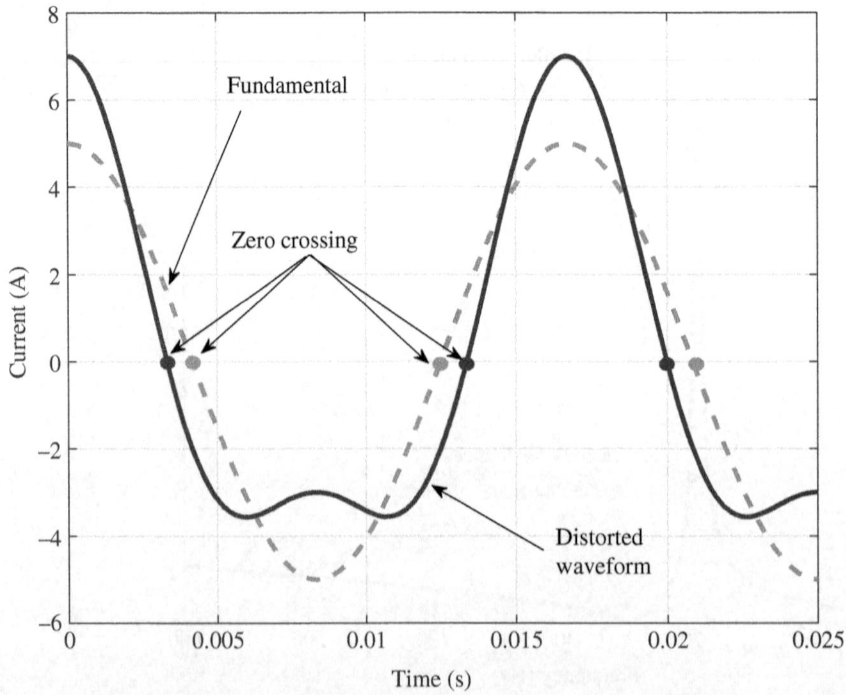

Figure 1.11 Second harmonics distortion effect for zero crossing detection.

effects on the operation of a protection relay. A relay is a device that will perform switching actions based on the information received from sensors, and is mainly used for power system protection. Harmonics can distort or degrade the operating characteristics of protective relays depending on the design features and principles of operation. Digital relays and algorithm that rely on sample data or zero crossings are particularly prone to error when harmonic distortion is present. Fig. 1.11 shows the zero crossings when a sinusoidal signal is contaminated with a second harmonic, as compared with a sinusoidal signal without any distortions. It can be clearly seen that the zero crossings are shifted when harmonic distortions are present, leading to the discrepancy with its fundamental signal.

1.9.3 Harmonics Impact on Motors

Harmonic voltage distortion at the motor terminals can cause pulsating torque, resulting in mechanical oscillations, which can lead to rotor shaft fatigue and accelerated aging of the shaft and connected mechanical parts. Moreover, the extra iron and copper losses at harmonic frequency also results in efficiency reduction.

1.9.4 Harmonics Impact on Transformers

The major effects of harmonics on transformers is that current harmonics cause an increase in copper losses and stray flux losses. In addition, voltage harmonics cause an increase in the iron losses. As a result, the effect of harmonics on the transformer is a reduction in both efficiency and life span.

1.10 Characteristic Harmonic and Non-Characteristic Harmonic

The characteristic harmonics in this book are referring to those produced by power converters under balanced conditions (or normal operation). For most of the situations, characteristic harmonics can be estimated by some analytical formulas. This can be seen by Examples 1.3 and 1.4.

Example 1.3 A square waveform, shown in Fig. 1.12, will only have odd harmonics due to its half-wave symmetry.

To simplify the analysis, we assume that $\omega = 1$. Then, a_k coefficient is found from

$$a_k = \frac{1}{\pi}\int_0^{2\pi} f(t)\cos(kt)\,dt = \frac{1}{\pi}\left[\int_0^{\pi} A\cos(kt)dt + \int_{\pi}^{2\pi} -A\cos(kt)dt\right] = 0. \tag{1.83}$$

Also, by inspection, the average (dc) value is zero. b_k coefficients are found from (1.84).

$$\begin{aligned} b_k &= \frac{1}{\pi}\int_0^{2\pi} f(t)\sin(kt)\,dt = \frac{1}{\pi}\left[\int_0^{\pi} A\sin(kt)dt + \int_{\pi}^{2\pi} -A\sin(kt)dt\right] \\ &= \frac{A}{k\pi}(1 - 2\cos(k\pi) + \cos(2k\pi)) \end{aligned} \tag{1.84}$$

for k = even, $b_k = 0$. For k = odd, (1.84) reduces to

$$b_k = \frac{4A}{k\pi}. \tag{1.85}$$

Thus, the k-th (characteristic) harmonic has an amplitude of $1/k$ in per unit of the fundamental amplitude, as listed in Table 1.2. □

Example 1.4 A three-phase, full-bridge rectifier, operated under balanced conditions, is shown in Fig. 1.13. The waveforms of the ac current are shown in Fig. 1.14. By means of Fourier analysis of the ac current, the fundamental frequency component of the phase A current, I_{a1}, has an RMS value [11]

$$I_{ac1} = \frac{1}{\pi}\sqrt{6}I_{dc}. \tag{1.86}$$

The harmonic components I_{ak} can be expressed in terms of the fundamental frequency component as

$$I_{ak} = \frac{I_{a1}}{k}, \tag{1.87}$$

where k is an odd integer number. □

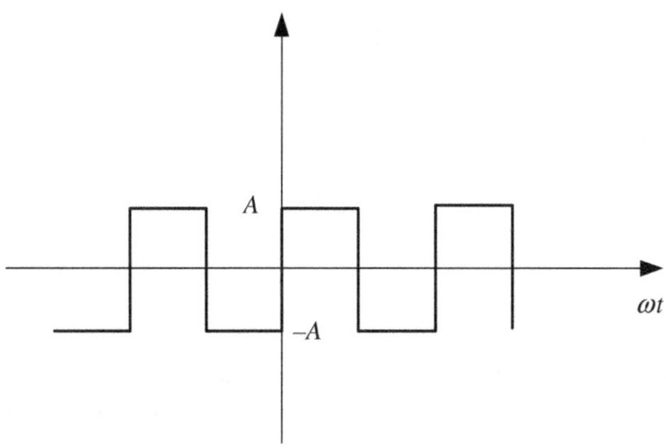

Figure 1.12 A square waveform.

Table 1.2 The harmonic amplitude of a square waveform.

Harmonic order	Amplitude in p.u.
1st	1
3rd	1/3
5th	1/5
7th	1/7
9th	1/9
⋮	
k-th	$1/k$
⋮	

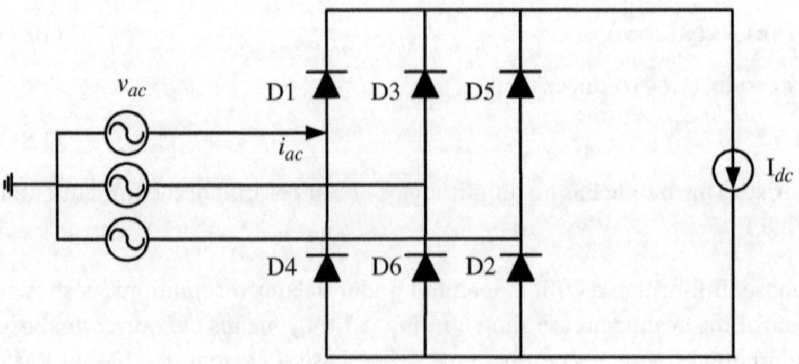

Figure 1.13 A three-phase, full-bridge rectifier.

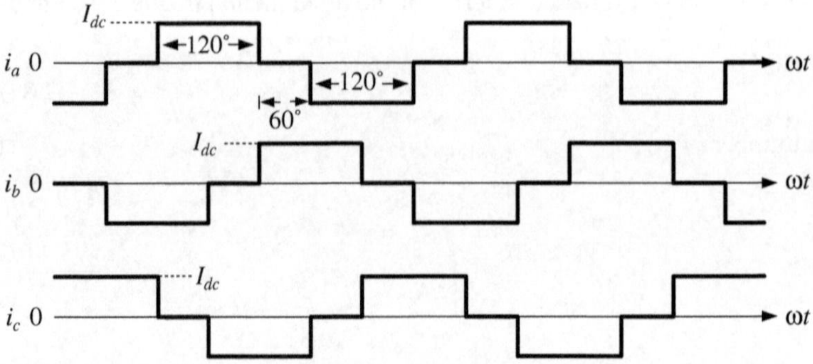

Figure 1.14 AC current waveform of a full bridge rectifier. Source: Modified from [11].

Non-characteristic harmonics are those produced by the power converter under unbalanced conditions, abnormal operation, or situations where other external harmonic sources are present. These non-characteristic harmonics usually require a harmonic power flow program or a full steady-state solver to obtain the value, and generally cannot be easily calculated by some analytical formula.

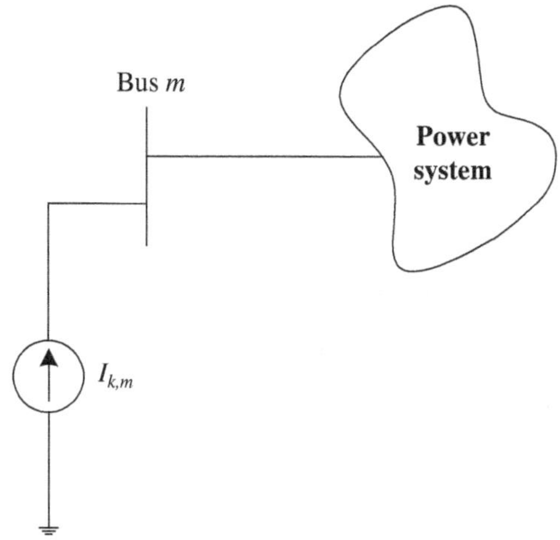

Figure 1.15 The harmonic current injection into bus m.

1.11 Harmonic Current Injection Method

One most common approach of modeling a power converter (or a nonlinear device in general) is to treat it as a harmonic current source, and its harmonic content can be obtained by taking a snapshot of the field measurement or simply determined by the characteristic harmonics. Thus, power system harmonic voltage is a direct (non-iterative) solution via (1.88).

$$|V_{k,m}| = |Z_{k,m}||I_{k,m}|, \tag{1.88}$$

where $I_{k,m}$ represents the injected k-th current harmonic to bus m, $V_{k,m}$ is the resulting k-th voltage harmonic at bus m, and $Z_{k,m}$ is the impedance matrix looking into bus m for the k-th harmonic, as shown in Fig. 1.15.

This method is often referred to as the harmonic current injection technique. The magnitude and the angle of each harmonic current injected into the power system by the harmonic device are assumed to be constant and known. Therefore, the following problems exist by using this approach in modeling a power converter:

1. It ignores completely the harmonic interactions between the power converters and the rest of the power network.
2. It ignores the harmonic interactions between the input side and the output side of the converter.
3. It neglects the fact that different controllers have different impacts on the system harmonics.

To be able to account for these three interactions, a full steady-state model of a converter is needed, and a simple harmonic current source is not sufficient.

1.12 Steady-State vs. Transient Response

There are two ways of decomposing the (total) response of a circuit. The first is to break it into a "natural response" and "forced response," and the second is to break it into a "transient response" and a "steady-state response." This can be illustrated by the step response of a RC circuit, as shown in Fig. 1.16.

By Kirchhoff's current law (KCL),

$$R_T C \frac{dv(t)}{dt} + v(t) = V_A, \text{for } t \geq 0. \tag{1.89}$$

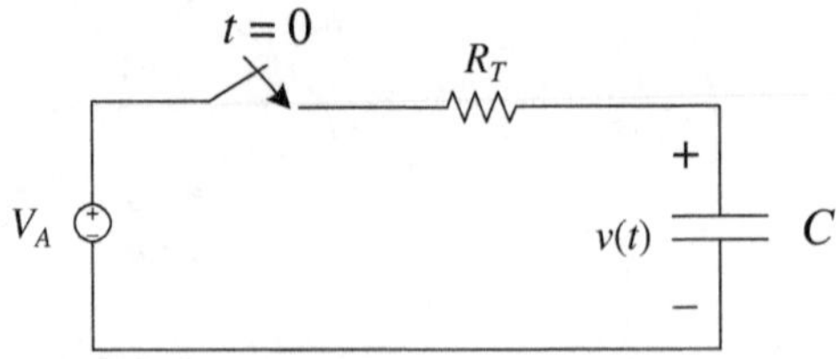

Figure 1.16 An RC circuit.

The solution to (1.89) is

$$v(t) = v_n + v_f, \tag{1.90}$$

where

$$v_n = v_o e^{-t/(R_T C)}, v_o = v(0), \tag{1.91}$$

and

$$v_f = V_A(1 - e^{-t/(R_T C)}), \tag{1.92}$$

where v_n is the natural response, and is the general solution of (1.89) when the input (i.e. V_A) is set to zero. v_f is the forced response because it is produced by the circuit when the forcing function (the input) is applied. It represents what the circuit is forced to do by the input excitation. The nature response eventually dies out along with the transient component of the forced response, leaving only the steady-state component of the forced response. Hence, the second way of looking at the complete response is to decompose it into transient and steady-state responses:

$$v = v_t + v_{ss}, \tag{1.93}$$

where

$$v_t = (v_o - V_A)e^{-t/R_T C}, \tag{1.94}$$

and $v_{ss} = V_A$.

The transient response v_t is temporary and is the portion of the complete response that decays to zero as time approaches infinity. The steady-state response v_{ss} is the portion of the complete response that remains after the transient response has died out. An electromagnetic transient program is a general purpose program, directly solving for the differential equations of a circuit. Thus, both transient and steady-state responses will be obtained in these programs.

1.13 Steady-State Modeling

In some situations, only the steady-state responses are of interest. For example, when one wants to perform harmonic analysis or find out the steady-state operating point for the design of a controller. It may be impractical to wait until the transient response of an electromagnetic program dies out. In particular, an under-damped circuit may experience an excessively long transient time. Consequently, there are research demands in finding techniques for quickly obtaining steady-state responses. The techniques generally are referred to as "accelerated" steady-state methods or simply steady-state methods. The steady-state methods are generally classified into frequency-domain, time-domain, and a hybrid of frequency- and time-domain methods. The following example (Example 1.5) is used to illustrate the time- and frequency-domain approaches in steady-state modeling.

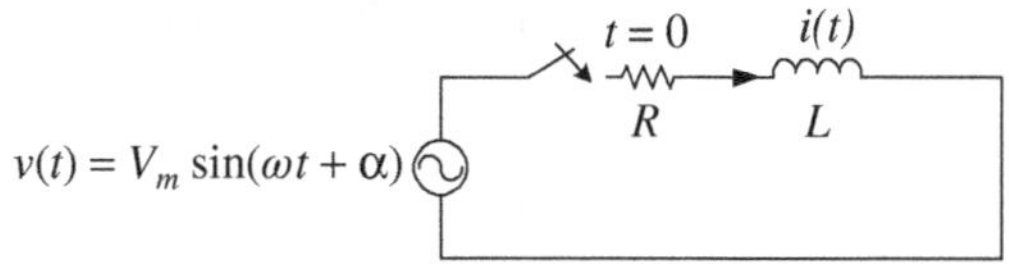

Figure 1.17 An *RL* circuit.

Example 1.5 An RL circuit is shown in Fig. 1.17. Assume that the breaker closes at time $t = 0$. By Kirchhoff's voltage law,

$$v(t) - Ri(t) - L\frac{di(t)}{dt} = 0 \tag{1.95}$$

$$\frac{L}{R}\frac{di(t)}{dt} + i(t) = \frac{V_m}{R}\sin(\omega t + \alpha). \tag{1.96}$$

The natural response is

$$\frac{L}{R}\frac{di(t)}{dt} + i(t) = 0. \tag{1.97}$$

The solution to (1.97) is

$$i_n = Ke^{-\frac{t}{\tau}}, \tag{1.98}$$

where $\tau = \frac{L}{R}$, and K is a constant.

The forced response is

$$\frac{L}{R}\frac{di(t)}{dt} + i(t) = \frac{V_m}{R}\sin(\omega t + \alpha). \tag{1.99}$$

If we assume

$$i_f(t) = a\cos(\omega t + \alpha) + b\sin(\omega t + \alpha) \tag{1.100}$$

and substitute (1.100) into (1.99), then (1.101) is resulted.

$$\left[\frac{L}{R}b\omega + a\right]\cos(\omega t + \alpha) + \left[b - \frac{L}{R}a\omega - \frac{V_m}{R}\right]\sin(\omega t + \alpha) = 0 \tag{1.101}$$

Consequently,

$$\frac{L}{R}b\omega + a = 0 \tag{1.102}$$

$$b - \frac{L}{R}a\omega - \frac{V_m}{R} = 0. \tag{1.103}$$

From (1.102) and (1.103), we obtain

$$a = -\frac{V_m}{\sqrt{R^2 + L^2\omega^2}}\frac{L\omega}{\sqrt{R^2 + L^2\omega^2}} \tag{1.104}$$

$$b = \frac{V_m}{\sqrt{R^2 + L^2\omega^2}}\frac{R}{\sqrt{R^2 + L^2\omega^2}}. \tag{1.105}$$

Thus,

$$i(t) = i_n(t) + i_f(t) \tag{1.106}$$

$$= Ke^{-\frac{t}{\tau}} + \frac{V_m}{R^2 + L^2\omega^2}\left[-L\omega\cos(\omega t + \alpha) + R\sin(\omega t + \alpha)\right]. \tag{1.107}$$

Let us define γ, as depicted in Fig. 1.18, and let $Z = \sqrt{R^2 + L^2\omega^2}$. Then, the following relations can be obtained:

$$\cos(\gamma) = \frac{R}{Z} \tag{1.108}$$

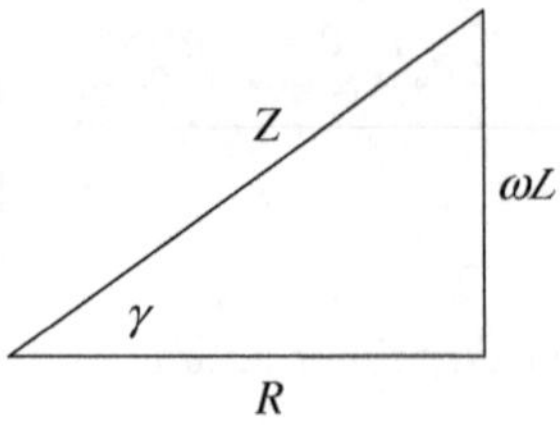

Figure 1.18 A triangle depicting the relationship between Z, R, and ωL.

$$\sin(\gamma) = \frac{\omega L}{Z}. \tag{1.109}$$

Let $I_m = \frac{V_m}{Z}$, we get

$$i(t) = Ke^{-\frac{t}{\tau}} + I_m \sin(\omega t + \alpha - \gamma). \tag{1.110}$$

If we set the initial condition $i(0)$ to be $I_m \sin(\alpha - \gamma)$, then K will be zero. This means that we can suppress or bypass the transient response if the initial conditions are properly set. Consequently, the art of the accelerated steady-state time-domain method is to find the proper initial conditions such that the steady-state is immediately obtained.

In terms of the frequency-domain method, the steady-state solution of Fig. 1.17 is simple because we can solve it by means of phasor analysis, which is

$$I_m = \frac{V_m \angle \alpha}{R + j\omega L} = \frac{V_m}{\sqrt{R^2 + \omega^2 L^2}} \angle \alpha - \tan^{-1}\left(\frac{\omega L}{R}\right). \tag{1.111}$$

Nevertheless, if the system contains a nonlinear device, the phasor analysis cannot be used, and a generalized frequency-domain method is needed, which is extensively reviewed in [12] and [13]. □

1.14 Large-Signal Modeling vs. Small-Signal Modeling

The concepts of large- and small-signal modeling can be illustrated with a tunnel-diode circuit shown in Fig. 1.19(a). A sinusoidal voltage source, $v_s(t) = V_m \cos(\omega t)$, is applied to the circuit, and we assume that the biasing circuit (i.e. the circuit without $v_s(t)$) has been designed properly so that there is only one operating point, N. The nonlinear $V - I$ characteristic of the tunnel diode is specified by

$$i = f(v), \tag{1.112}$$

and is plotted in Fig. 1.19(b).

Assume that the actual signal voltage, $v(t)$, and signal current, $i(t)$, are in the neighborhood of N, and can be denoted as

$$v(t) = V_N + \tilde{v}(t) \tag{1.113}$$

$$i(t) = I_N + \tilde{i}(t), \tag{1.114}$$

where V_N and I_N are obtained by setting $v_s(t) = 0$, and

$$I_N = f(V_N) \tag{1.115}$$

$$V_N = E_b - R_b I_N. \tag{1.116}$$

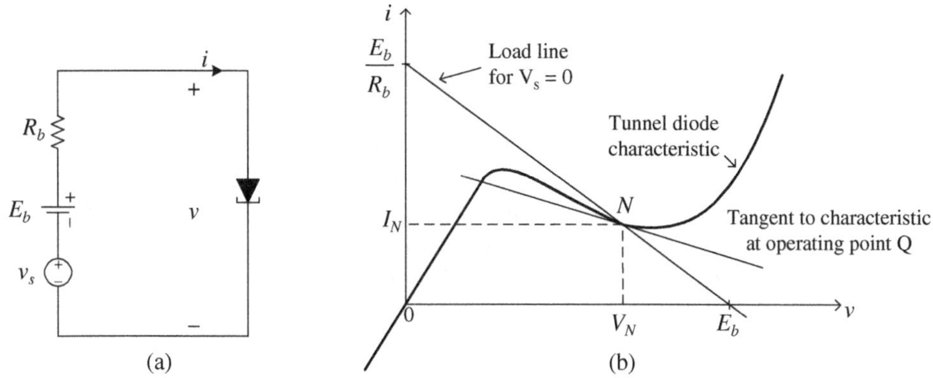

Figure 1.19 (a) Tunnel-diode circuit with signal source $v_s(t)$; (b) linear approximation to the diode characteristic at the operating point *N*.

Note that $\tilde{v}(t)$ and $\tilde{i}(t)$ bookkeep the displacement of instantaneous operating point away from (V_N, I_N) when the signal is applied. Substituting (1.113) and (1.114) into (1.112), one gets

$$i(t) = I_N + \tilde{i}(t) = f(V_N + \tilde{v}(t)). \tag{1.117}$$

Taking the first two terms of the Taylor series expansion of $f(V_N + \tilde{v}(t))$ about the point V_N, and neglecting the high-order terms, one obtains

$$i(t) \approx f(V_N) + \left.\frac{df}{dv}\right|_{V_N} \tilde{v}(t). \tag{1.118}$$

Geometrically, the nonlinear diode characteristic is replaced by the tangent line at the operating point. Comparing (1.113) to (1.118), one gets

$$\tilde{i} = \left.\frac{df}{dv}\right|_{V_N} \tilde{v}(t). \tag{1.119}$$

Let us denote

$$\left.\frac{df}{dv}\right|_{V_N} \tilde{v}(t) = G = \frac{1}{R}. \tag{1.120}$$

The quantity, $\left.\frac{df}{dv}\right|_{V_N}$ is called the small-signal model of the diode, which is equivalent to a resistor at the operating point of N. It is a small-signal model because this equivalent resistor model is only valid when the applied voltage v_s is small. Moreover, the small-signal model is heavily dependent on the operating point. The slopes of the tangent lines will be different and hence the equivalent resistors will be different at different operating points. This is in sharp contrast with the nonlinear characteristic of the diode, defined by (1.112). (1.112) is always valid, regardless of the value of the voltage v_s. Therefore, (1.112) is referred to as the large-signal model.

1.15 Discussion of IEEE Standard (STD) 519

In the power quality area, IEEE STD 519 is a prominent guideline to determine the harmonic control in an electrical power system. This standard defines the requirements that both parties (utilities and users) need to do to satisfy the predefined limits of current and voltage harmonics.

IEEE STD 519 1992 is sponsored by Transmission and Distribution Committee (TDC) of the IEEE Power Engineering Society (PES) and the Static Power Converter Committee (SPCC) of the IEEE Industrial Applications Society (IAS). This standard provides an in-depth discussion of harmonics and exhibits educational and informative knowledge. This standard is more focused on recommended practices for individual customers and utilities. It also recommends the methodology for evaluating new harmonic sources.

Unlike IEEE STD 519 1992, IEEE STD 519 2014 is only sponsored by TDC of the IEEE PES. Generally speaking, IEEE STD 519 2014 is the revised version of IEEE STD 519 1992. This standard is in compact form and does not discuss harmonic fundamentals. Some essential revisions are elucidated below.

Point of Common Coupling (PCC) Location

In IEEE STD 519 1992, the PCC is defined as the point between the nonlinear load and other loads as depicted in Fig. 1.20. In IEEE STD 519 2014, on the other hand, two types of PCC are distinguished, i.e. industrial users and commercial users. For industrial users, the PCC is located at the high-voltage (HV) side of the transformer as illustrated in Fig. 1.21(a). It is essential to observe that the PCC defined in IEEE STD 519 2014 is not applied for individual pieces of equipment or at locations within a user's facility. This clearly identifies that the PCC should not be at the device input. On the other hand, for commercial users the PCC is placed at the low-voltage (LV) side of the transformer as shown in Fig. 1.21(b).

The Responsibility of Both Parties (Utilities and Users)

In IEEE STD 519 1992, the responsibilities for utilities and users are not clearly defined. It states that the users must not degrade the voltage of the utility serving other users by requiring nonlinear currents from the utility. Moreover, it also recognizes the responsibility of the utilities to provide users with close to a sine wave of voltage. On the other hand, IEEE STD 519 2014 clearly describes the shared responsibility of harmonic control between utilities and users. It declares that all users must limit their harmonic current emissions to reasonable values determined in an equitable manner based on the inherent ownership stake each user has in the supply system. On the other hand, the utilities are responsible to decrease voltage distortion levels by modifying the supply system impedance characteristics as necessary.

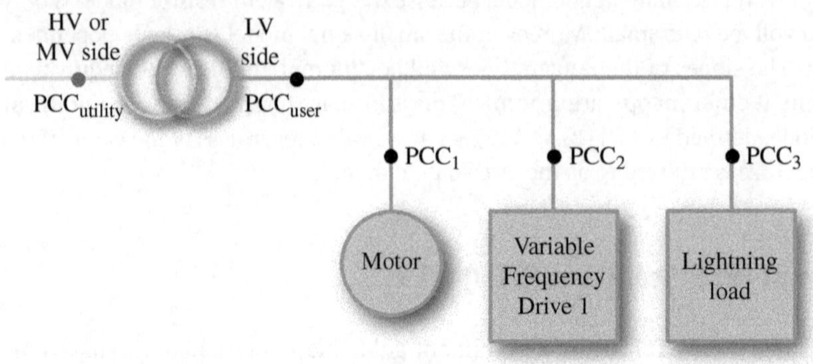

Figure 1.20 PCC based on IEEE STD 519 1992. Source: Based on "IEEE recommended practices and requirements for harmonic control in electrical power systems," IEEE Std 519–1992, pp. 1–112, 1993.

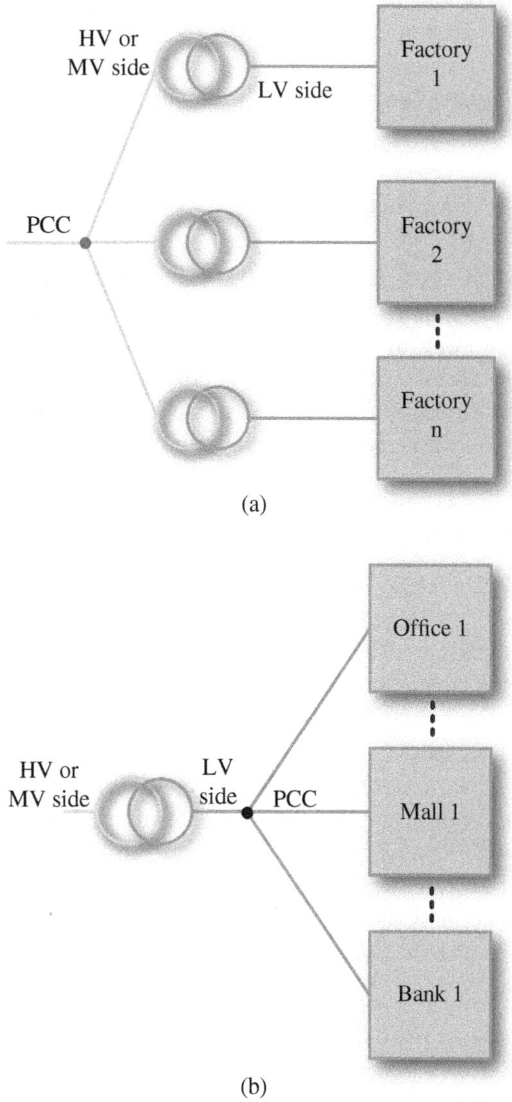

Figure 1.21 PCC for (a) industrial users; (b) commercial users based on IEEE STD 519 2014. Source: Based on IEEE Recommended Practice and Requirements for Harmonic Control in Electric Power Systems. (n.d.). IEEE. https://doi.org/10.1109/ieeestd.2014.6826459.

Harmonic Measurements

IEEE STD 519 2014 introduces the new term to analyze the harmonic characteristics based on statistical data. The individual harmonic measurements are aggregated based on RMS calculations and divided into two categories, i.e. very short-time and short-time harmonic measurements.

Very short-time harmonic values are assessed over a 3-second interval based on an aggregation of 15 consecutive 12 (10) cycle windows for a 60 (50) Hz power system. The individual harmonics in very short-term harmonic measurements are defined as

$$F_{n,vs} = \sqrt{\frac{1}{15}\sum_{i=1}^{15} F_{n,i}^2}, \tag{1.121}$$

where F denotes the voltage or current, n represents the harmonic order, and i is a simple counter. The subscript υs is the abbreviation of "very short."

Short-time harmonic values are assessed over a 10-minute interval based on an aggregation of 200 consecutive very-short time values for a specific frequency component. Thus, it can be formulated as

$$F_{n,sh} = \sqrt{\frac{1}{200}\sum_{j=1}^{200} F^2_{(n,\upsilon s)j}} \tag{1.122}$$

where j is a simple counter and subscript sh denotes "short." Note that F is in terms of the RMS value. Very short and short-time harmonic values will be accumulated over periods of one day and one week, respectively. For very short-time harmonic measurement, the 99th percentile value is calculated for each day for comparison with the recommended limits. For short-time harmonic measurement, the 95th and 99th percentile values are calculated each week for comparison with the recommended limits.

Recommended Harmonic Limits

IEEE STD 519 1992 defines the TDD as harmonic current distortion as a percentage of maximum demand load current (15 or 30 min demand). It states that there are no limits on harmonic orders. On the other hand, TDD in IEEE STD 519 2014 is limited up to 50th order. Other differences can be found in the voltage harmonic limitations. IEEE STD 519 2014 further defines the voltage distortion limit for the system with bus voltage $\leq$ 1 kV with the individual harmonic maximum 5.0% and THD 8.0% as listed in Table 1.3. For the current harmonic distortion limit, IEEE STD 519 2014 specifies the harmonic order observation up to 50th for each bus voltage level system, as shown in Tables 1.4, 1.5, and 1.6 and the additional current harmonic limit terms for the > 161 kV system for the ratio $I_{SC}/I_L < 25$ as presented in Table 1.6. Note that I_{SC} denotes maximum short-circuit current at PCC and I_L represents maximum demand load current (fundamental frequency component) at the PCC under normal load operating conditions.

At the PCC, users should limit their harmonic currents as follows:

- Daily 99th percentile very short-time (3 s) harmonic currents should be less than 2.0 times the values listed in Tables 1.4, 1.5, and 1.6.
- Weekly 99th percentile short-time (10 min) harmonic currents should be less than 1.5 times the values listed in Tables 1.4, 1.5, and 1.6.
- Weekly 95th percentile short-time (10 min) harmonic currents should be less than the values listed in Tables 1.4, 1.5, and 1.6.

Table 1.3 Voltage distortion limits.

Bus voltage V at PCC	Individual harmonic (%)	Total harmonic distortion THD (%)
$V \leq 1.0$ kV	5.0	8.0
1 kV $< V \leq 69$ kV	3.0	5.0
69 kV $< V \leq 161$ kV	1.5	2.5
161 kV $< V$	1.0	1.5

Table 1.4 Current distortion limits for systems rated 120 V through 69 kV.

	Maximum harmonic current distortion in percent of I_L					
	Individual harmonic order (odd harmonics)					
I_{SC}/I_L	$3 \leq h < 11$	$11 \leq h < 17$	$17 \leq h < 23$	$23 \leq h < 35$	$35 \leq h \leq 50$	TDD
< 20	4.0	2.0	1.5	0.6	0.3	5.0
20 ~ 50	7.0	3.5	2.5	1.0	0.5	8.0
50 ~ 100	10.0	4.5	4.0	1.5	0.7	12.0
100 ~ 1000	12.0	5.5	5.0	2.0	1.0	15.0
> 1000	15.0	7.0	6.0	2.5	1.4	20.0

Table 1.5 Current distortion limits for systems rated 69 kV through 161 kV.

	Maximum harmonic current distortion in percent of I_L					
	Individual harmonic order (odd harmonics)					
I_{SC}/I_L	$3 \leq h < 11$	$11 \leq h < 17$	$17 \leq h < 23$	$23 \leq h < 35$	$35 \leq h \leq 50$	TDD
< 20	2.0	1.0	0.75	0.3	0.15	2.5
20 ~ 50	3.5	1.75	1.25	0.5	0.25	4.0
50 ~ 100	5.0	2.25	2.0	0.75	0.35	6.0
100 ~ 1000	6.0	2.75	2.5	1.0	0.5	7.5
> 1000	7.5	3.5	3.0	1.25	0.7	10.0

Table 1.6 Current distortion limits for systems rated > 161 kV.

	Maximum harmonic current distortion in percent of I_L					
	Individual harmonic order (odd harmonics)					
I_{SC}/I_L	$3 \leq h < 11$	$11 \leq h < 17$	$17 \leq h < 23$	$23 \leq h < 35$	$35 \leq h \leq 50$	TDD
< 25	1.0	0.5	0.38	0.15	0.1	1.5
25 ~ 50	2.0	1.0	0.75	0.3	0.15	2.5
≥ 50	3.0	1.5	1.15	0.45	0.22	3.75

Some important notes related to the current distortion limits are as follows:

1. The even harmonics are limited to 25% of the odd harmonic limits.
2. Current distortions that result in a dc offset, e.g., half-wave converters, are not allowed.
3. All power generation equipment is limited to these values of current distortion, regardless of actual I_{SC}/I_L.

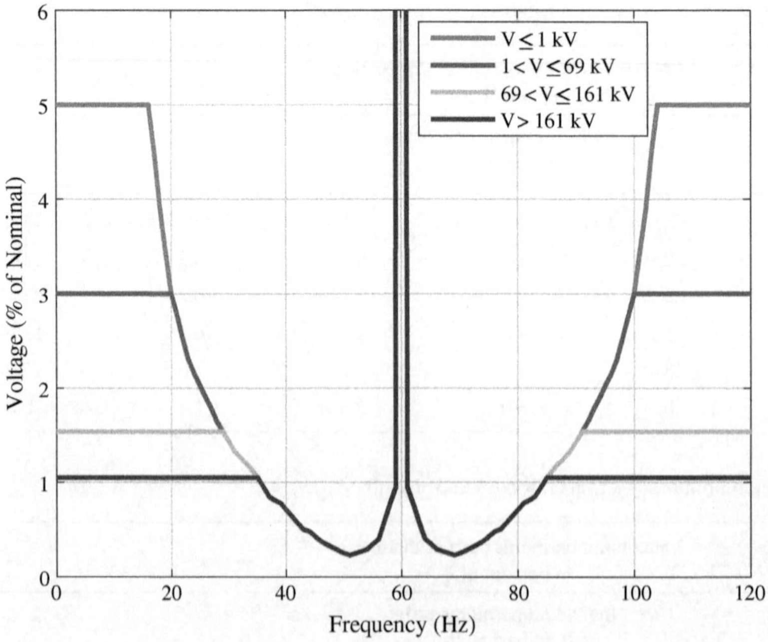

Figure 1.22 Interharmonic voltage limits based on flicker for frequencies up to 120 Hz for 60 Hz systems.

Interharmonic voltage

Interharmonic is a frequency component that is not an integer multiple of the frequency at which the supply system is operating (e.g., 50 Hz or 60 Hz). IEEE STD 519 2014 clearly describes the standard for interharmonic voltages. Fig. 1.22 presents the boundaries of various voltage level of interharmonic voltage for 60 Hz systems. These boundaries apply to weekly 95th percentile short-time harmonic voltages. For instance, if the PCC voltage is less than 1 kV, the allowable maximum voltage is 5% of its nominal value for the frequency less than 16 Hz or greater than 104 Hz. The limits becomes more restrictive for the frequency near to the fundamental frequency.

1.16 Supraharmonics

During recent decades, the power quality field has focused only on distorted frequencies below 2 kHz. This is due to the fact that consumer equipment does not emit the high-frequency component. However, as many pieces of equipment (e.g. electric vehicle charger, LED lamps, PV-inverters, storage-converters) are connected to the grid emitting the high-frequency range from 2 kHz to 150 kHz, the study of the high-frequency range has become prominent in order to avoid malfunctions in household devices and to preserve the lifetime of the devices.

Terminology

The word "supraharmonics" was first introduced during the IEEE Power and Energy Society General Meeting in 2013 [14]. Supraharmonics is defined as the harmonic distortion in the frequency range of 2 to 150 kHz. The lower limit (2 kHz) is taken from the upper limit of the frequency range where a rather comprehensive standard framework already exists. On the other hand, the upper limit is taken based on the lowest frequency of the radio broadcasting band, which also may produce the interference among the electronic devices [15].

Measuring Techniques

Some international organizations, such as International Electrotechnical Commission (IEC), European Committee for Electrotechnical Standardization (CENELEC), and IEEE are working to develop the standards, such as compatibility level, emission and immunity limits, and measurement techniques to cover supraharmonics. IEC 61000-4-7 presents the informative procedure to measure the frequency of 2-9 kHz, while IEC 61000-4-30 proposes a measurement methodology for frequency range of 9-150 kHz [16].

There are two ways to measure the supraharmonics: frequency domain and time domain methods. In the frequency domain method, two different principles are typically used: Discrete Fourier Transform (DFT) and heterodyne principle. Here, we only discuss the measuring techniques via FFT, which is illustrated in Fig. 1.23. For the first stage, the signal magnitude is altered to the specified level, so that it can be received by the computer or microprocessor. To isolate the measurement device to the real grid voltage, galvanic isolation is applied for protection purposes. Afterwards, the signal is filtered by an anti-aliasing filter (low pass filter) and signal filter (high pass filter) before it is converted to the digital signal via an A/D converter. The digital signal is filtered using a digital filter before the spectrum is derived from the DFT calculation. Commonly, only magnitude is considered, while the phase angle is discarded. Depending on the selected frequency resolution for DFT, the spectrum may be aggregated into a wider band. For example, in IEC 61000-4-7, the resulting spectra from 5 Hz bands are aggregated into 200 Hz bands. If more than one spectrum is recorded, these may be time-aggregated into longer intervals, e.g. from one spectrum per 200 ms to one spectrum per minute. The measurement data are then stored for later post-processing.

To capture the supraharmonics characteristics, the high pass filter (HPF) can be embedded within the current/voltage sensors for signal processing. The realization of HPF follows the standards of IEC 61000-4-7 and IEC 61000-4-30, which specifically recommend the application of HPF to reduce the effect of spectral leakage. Moreover, IEC 61000-4-7 recommends that the HPF should provide at least 55 dB damping at the fundamental frequency, while IEC 61000-4-30 suggests using HPF for 2 poles and 3 dB or lesser ringing at 1.5 kHz and higher frequencies.

The measuring techniques for the signal are based on these standards:

- **IEC 61000-4-7**: This standard proposes a measurement method of component in the frequency range of 2–9 kHz. Some extensions have been made by the researcher to cover the range of whole supraharmonics frequency. This standard suggests the sampled signal is performed in the interval of 200 ms, which is equal to 10 fundamental cycles of 50 Hz systems and 12 fundamental cycles of 60 Hz systems. Consequently, the frequency resolution is 5 Hz. Then the value is converted in the frequency domain via Fast Fourier Transform (FFT). The FFT results are

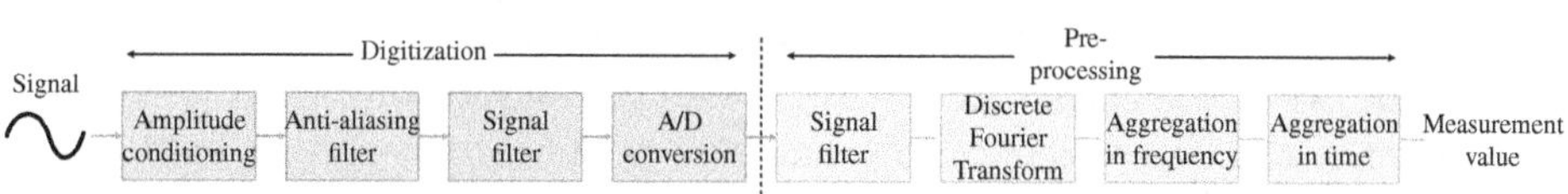

Figure 1.23 The process of DFT methods.

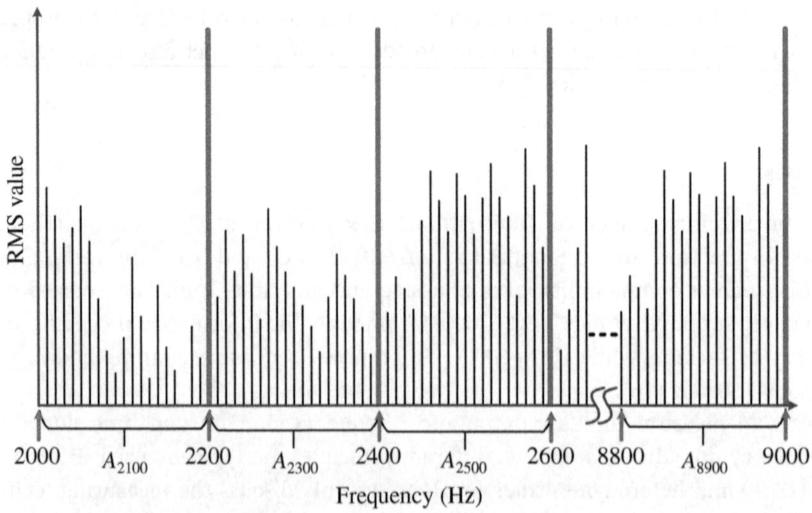

Figure 1.24 The grouping bands based on IEC 61000-4-7.

then grouped based on a predefined frequency band. This standard follows *Comité International Spécial des Pertubasi Radioeléctriques* (CISPR) 16-1, which specifies the bandwidth for the grouping signals should be fixed at 200 Hz bands. The illustration of the grouping signals based on this standard is depicted in Fig. 1.24.

If A_f is the RMS value of the component at frequency f, then the frequency separation between consecutive measured components A_f must be 5 Hz. The output A_b of each band is represented as

$$A_b = \sqrt{\sum_{f=b-95Hz}^{b+100Hz} A_f^2} \tag{1.123}$$

where b denotes the central frequency, whose values are 2100, 2300, 2500, …, 8900 Hz. In terms of filtering method, this standard suggests the utilization of a band pass filter shown in Fig. 1.25, which is able to attenuate the fundamental frequency and components above 9 kHz with an attenuation of the fundamental component of at least 55 dB.

In terms of sampling rate, this standard recommends the use of the Nyquist criterion, where the sampling rate must be at least greater than or equal to two times the observed frequency (9 kHz in this case). Thus, the sampling rate chosen is $f_s \geq 18$ kHz.

- **IEC 61000-4-30**: This standard presents some informative methods to locate the supraharmonics in the range of 9–150 kHz. The sampling rate of 1,024 kHz is recommended in this standard. The cascaded HPF and LPF are used to filter the signal as depicted in Fig. 1.26. The HPF is used to attenuate the fundamental and lower-order harmonics of the input signal. The HPF should be composed of 2 poles with 3 dB attenuation at 1.5 kHz or higher. On the other hand, LPF should be a 4th-order filter, with the cutoff frequency at 200 kHz.

  The filtered signal is then performed by 32 approximately equally-spaced measurements, taking each 10/12 cycle interval (10 cycles for 50 Hz system or 12 cycles for 60 Hz systems). Each measurement consists of 512 samples taken at 1,024 kHz sampling rate, which later is processed via FFT. Note that each measurement corresponds to 0.5 ms time window length in 2 kHz frequency resolution. Consequently, the length of the total gap is 184 ms. The magnitude of 32 measurement sets are calculated based on average, maximum, and minimum values.

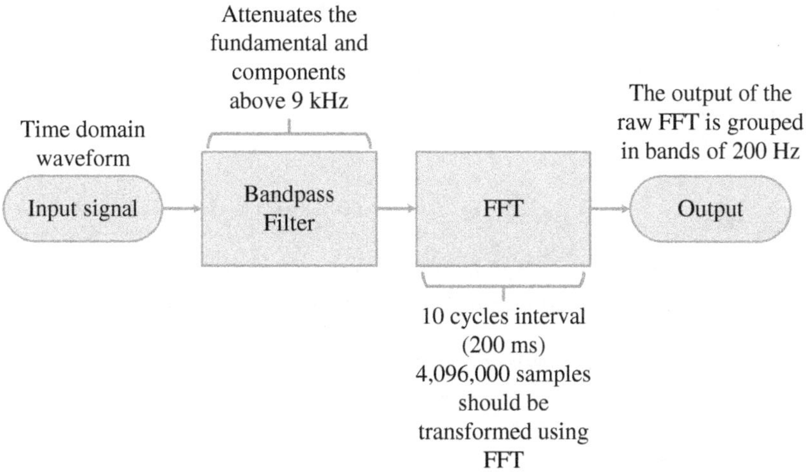

Figure 1.25 The measurement technique using IEC 61000-4-7 standard.

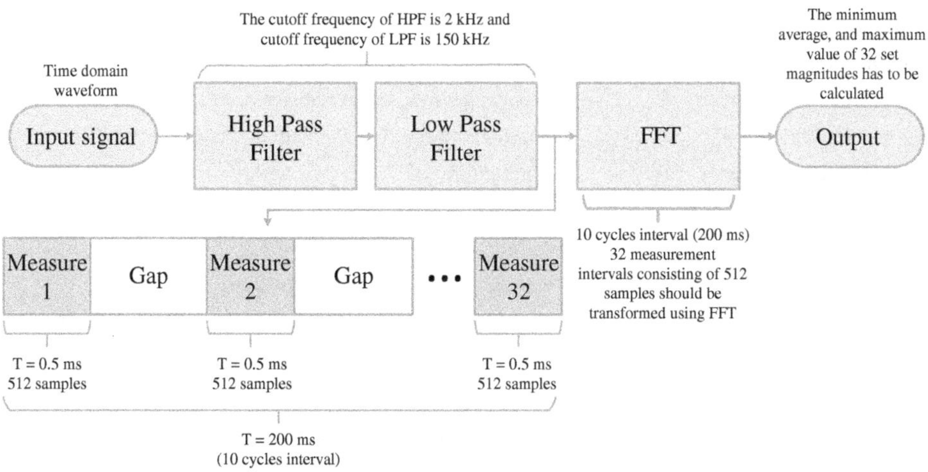

Figure 1.26 The measurement technique using IEC 61000-4-30 standard.

Sources of Emission

The supraharmonics emissions are originally generated from the smart electrical grids. The use of various technologies, such as electric vehicles (EV), smart metering, and distributed generators (DGs), encourages the grids to emit the supraharmonics. As the smart grids do not merely cover the energy transmission, data transmission over power line communication (PLC) is another source of supraharmonics. In addition, the effect of supraharmonics due to DGs originates from the interfacing of power converters to the grids [16].

The emission of PLC contributes to the short time period, while the power converters emit for the whole time period. As discussed in [17], [18], the passive power converters such as diode bridge rectifiers emit the harmonics in the lower frequency range, while the active power converters such as switch mode power supplies produce the harmonics in the higher frequency range. This is due to the pulse width modulation (PWM) switching pattern. Thus, power converters are considered to be a major source of supraharmonics emissions. Note that the method presented in Chapter 6 can accurately model any supraharmonic component that a power converter injects into the grid, without worrying about the leakage problem of DFT and violating the Nyquist criterion if the proper time step is not carefully chosen.

Effects

The effects of supraharmonics:

- **Additional heating:** Supraharmonics can lead to additional losses, rising the temperature of components, which results in lifetime shortage. In [19], the temperature of compact fluorescent

Table 1.7 Effect of supraharmonics in various pieces of equipment.

Equipment	Effect
Touch-dimmer lamps	Unintentional switching (between light steps, OFF, also ON)
Street lighting	Unintentional switch-on and switch-off
Traffic light	Malfunction
Traffic control system for public transportation buses	Malfunction
Solid state meters	Displaying wrong meter register values
Mains communicating system	Temporary or quasi-permanent loss of communication
Radio system for railway control	Malfunction
Contactless magnetic card reader	Malfunction of reading function
Home telephone	Malfunction of ringing
Heating systems	Incorrect alarms due to sensor faults
Routers	Loss of synchronization (40, 50, and 70 kHz) to the network
Notebooks	Disturbed cursor position (37 kHz)
Ceramic hobs	Incorrect relay switching
Coffee cooker	Incorrect control lamp function and malfunction
Professional hair-dryers	Switching ON/OFF spontaneously
Steam irons	Insufficient heat, water loss, incorrect control lamp blinking
Washing machines	Self-restart (some hours) after end of operation phase
Electronic ballasts	Audible noise
Induction cookers	Audible noise
PCs and lamps	Audible noise
PABX	Audible noise
Telephone (analogue)	Audible noise
TV and radio receiver	Audible noise (up to 20 kHz)
Video telecom circuits	Audible noise
Inductive train radio system	Audible noise
Keyless entry system	Malfunction
DC link capacitor in rectifier circuits	Thermal stress, ageing, lifetime reduction through larger HF currents

lamps is significantly rising about 1 K per 1% supraharmonic voltage. It becomes worse in the rectifier applications.

- **Audible noise:** The audible noise can be generated by the high switching frequency. Some evidence indicates that the presence of supraharmonics is able to create significant disturbing audible noise, such as the case where multiple devices operating at the same switching frequencies can cause the "chirping" noise. This has been reported in [20]: if multiple EVs are connected to the grid at the same time, significant *frequency beating* yielding an audible chirping noise can be produced. The frequency beating is a phenomenon caused by the interaction of two or multiple power converters due to the slight difference in switching frequency. Generally, a similar type of converter will never switch at exactly the same frequency, due to the impurities in the components and unavoidable flaws in the design process [21]. The illustration is given as follows: two power converters are electrically connected close to each other. The switching frequency of converter 1 is set at $f_1 = 10{,}000$ Hz, while converter 2 (same type as converter 1) has a switching frequency that is off by 0.01% of 10,000 Hz; $f_\Delta = 1$ Hz. Thus, the actual switching frequency of converter 2 becomes $f_2 = f_1 + f_\Delta = 10{,}001$ Hz. Adding two waveforms with slightly different frequencies will produce a beating frequency, causing the supraharmonic component to attenuate periodically with a beating frequency directly related to the difference in switching frequencies, expressed as

$$f_{beat} = |f_1 - f_2| = |f_\Delta| = \frac{1}{t_{beat}} \tag{1.124}$$

  where t_{beat} is the observed period of the beating oscillation.
- **Malfunction of equipment:** Some examples of equipment malfunction due to supraharmonics are: unintentional switching of touch dimmer lamps, charging interruption of EV, high errors in energy metering.
- **Malfunction of PLC:** Unacceptable damping of PLC signals has been observed in a low voltage grid with a very high penetration of single-phase and three-phase inverters.

More detailed examples of supraharmonics effects on various pieces of equipment are listed in Table 1.7.

Mitigation

As the harmonics are mostly generated from power converters and the most widely used power converter is the classical two-level voltage source converter (VSC), severe harmonics emissions may occur. Some attempts can be made to reduce the harmonics emissions as follows:

1. Replace the two-level VSC with multi-level VSC.
2. Employ a new modulation strategy. The spectrum of emission is originally concentrated in the switching frequency and its multiple integers. Thus, to distribute the energy, the modulation signal is slowly changed between ±10% of the fundamental switching frequency [22].
3. Add a passive filter between the converter and the grid [23].

2

Power Electronics Basics

"Engineering is the art of directing the great sources of power in nature for the use and convenience of man."

– Thomas Tredgold

Power electronics is the technology of converting electric power from one form to another using electronic power devices. Several types of solid state power semiconductor devices have been developed over the past 60 years, making it possible to build efficient power converters with excellent facility for the control of output parameters, such as voltage, current, or frequency. In a static power converter, the power semiconductor devices function as switches, which operate statically, that is, without moving contacts. The time durations of the turn-on and turn-off of these switches are controlled by the controllers to achieve power conversions, voltage and current transformations, power flow regulation, etc. The history of power electronics began with the introduction of the mercury arc rectifier, which was invented by Peter Cooper Hewitt in 1902. Then the metal tank rectifier, grid-controlled vacuum-tube rectifier, ignitron, phanotron, and thyratron were introduced gradually. These devices were applied for power control until the 1950s, when the first silicon-controlled rectifier (SCR) was developed by GE [24]. SCR was referred to by many names; the present-day name "thyristor" is a portmanteau word of *Thyr*-atron and trans-*istor*, which was adopted in 1962 [25]. Since the first SCR was developed, there have been tremendous advances in power semiconductor devices. This chapter will briefly discuss the types of semiconductor switches and their functionalities. Moreover, types of power electronic converters such as VSC and current source converter (CSC) will be briefly discussed in this chapter.

2.1 Some Basics

Most power electronic systems consist of two major modules: the power stage and the control stage. A generalized block diagram of a power electronic system is given in Fig. 2.1.

The function of the power stage is to perform the power conversion and processing of the energy from the input to the output by employing power switches. The control stage controls the amount of power transferred to the output. Generally, the converter is classified by the relationship between input and output, which are dc-to-dc, dc-to-ac, ac-to-dc, and ac-to-ac.[1] Regardless of the types of the

1 Note that some books refer to power electronic equipment doing dc-to-ac as inverters and ac-to-dc as converters. In this book, we do not make such a distinction, and refer to all of them as converters.

Harmonic Modeling of Voltage Source Converters using Basic Numerical Methods, First Edition.
Ryan Kuo-Lung Lian, Ramadhani Kurniawan Subroto, Victor Andrean, and Bing Hao Lin.

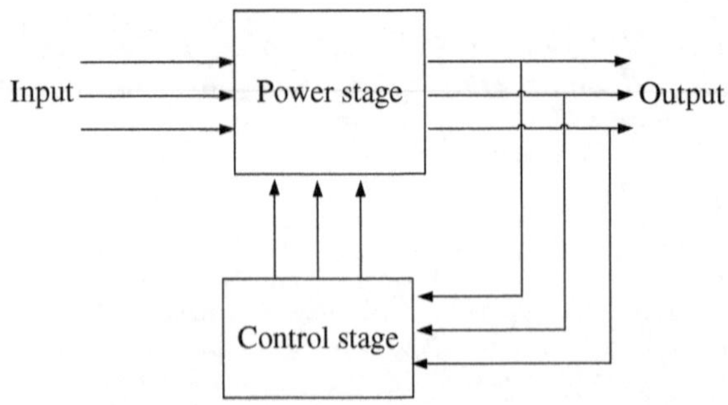

Figure 2.1 A generalized block diagram of a power electronic system.

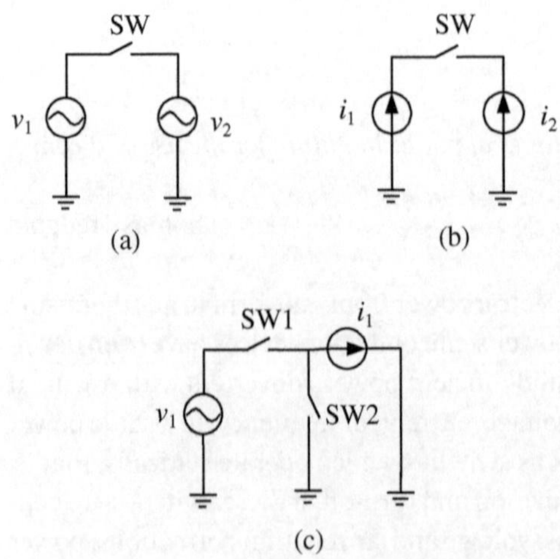

Figure 2.2 Various switched circuits.

conversion, any design of the power converter requires the fundamental laws of circuit theory to be followed — Kirchhoff's voltage law (KVL) and Kirchhoff's current law (KCL), which are mathematically stated in (2.1) and (2.2), respectively.

$$\sum v_{loop} = 0 \tag{2.1}$$

$$\sum i_{node} = 0, \tag{2.2}$$

where v_{loop} and i_{node} represent the loop voltages and nodal currents, respectively. Consequently, the switches, SW in Figs. 2.2(a) and (b), cannot be closed whereas the switches in Figs. 2.2(c) can be made to be closed and opened in a correct manner to obey the laws of the KVL and KCL. Note that in a very short time duration, a capacitor behaves as a voltage source and an inductor acts as a current source.

2.2 Semiconductors vs. Wide Bandgap Semiconductors

Power electronic devices generally consist of static switches, which are in general made of semiconductor material. A semiconductor material has an electrical conductivity value falling between that of a conductor and an insulator. Fig. 2.3 shows the energy bandgap for various materials. The electronic bandgap is the energy gap between the top of the valence band and the bottom of the conduction band in solid materials. Electrons can jump the gap to the conduction band by means of thermal or optical excitation. It is the bandgap that gives semiconductors the ability to switch currents on and off as desired in order to achieve a given electrical function.

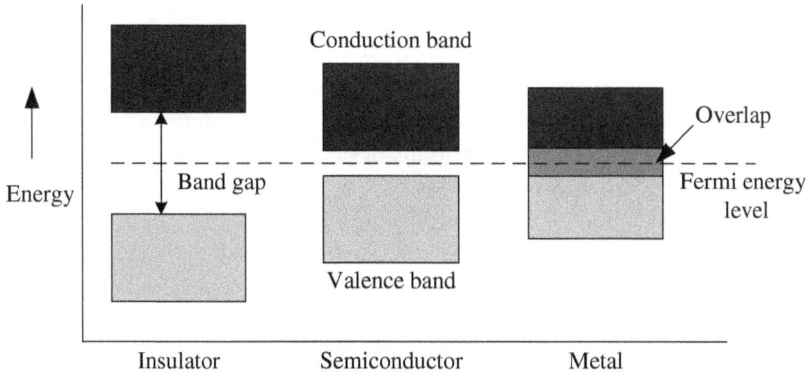

Figure 2.3 The bandgap energy for various materials.

Table 2.1 Physical characteristics of Si and main WGB semiconductors.

Property	Si	6H-SiC	4H-SiC	GaN
Bandgap (eV)	1.12	3.03	3.26	3.45
Electric breakdown field (kV/cm)	300	2500	2200	2000
Electron mobility (cm^2/V-s)	1500	500	1000	1250
Hole mobility (cm^2/V-s)	600	101	115	850
Thermal conductivity (W/cm-k)	1.5	4.9	4.9	1.3

The vast majority of power electronic switches are based on silicon (Si). Thus, the efficiency, cooling requirements, and power rating of the power converters are limited by the efficiency, thermal conductivity and voltage, and current limits imposed by Si. Consequently, the emergence of new power electronic devices based on wide bandgap (WBG) semiconductor materials will likely result in substantial improvements in the performance of power electronics converter systems in terms of higher blocking voltage, efficiency, and reliability as well as reduced thermal requirements [26]. Generally, Si has a bandgap in the range of 1–1.5 electronvolt (eV), whereas WBG materials have bandgaps in the range of 2–4 eV [27]. Table 2.1 lists some of the characteristics of Si and the main WBG semiconductors such as silicon carbides (4H-SiC and 6H-SiC) and gallium nitride (GaN) [26]. WBG semiconductors have electronic properties that fall in between those of conventional semiconductors and insulators. Thus, WBG has the capabilities to meet the higher performance demands of the evolving power equipment, operating at higher voltages and temperatures and enabling switching frequencies with greater efficiencies compared with Si devices. Along with performance improvements, WBG-based power electronics can be fabricated with a much smaller footprint compared with comparable Si devices due to decreased cooling requirements and smaller passive components. Nevertheless, the cost of WBG is still significantly higher than its Si counterpart. WBGs are expected to replace silicon for high voltage, high temperature, and high frequency application when their prices are in parity with those of Si.

2.3 Types of Static Switches

The characteristics of a power electronic converter essentially depend on the type of static switches. This section will briefly reviews some of the most commonly used switches in power converters. Fig. 2.4 lists some of the available Si and SiC switches on the market [28].

We only cover some of the most commonly used switches.

2.3.1 Uncontrolled Static Switch

The simplest uncontrolled static switch is the diode. A power diode is a two-terminal device whose circuit symbol is shown in Fig. 2.5(a). If a diode is present in an electrical circuit in such a way that

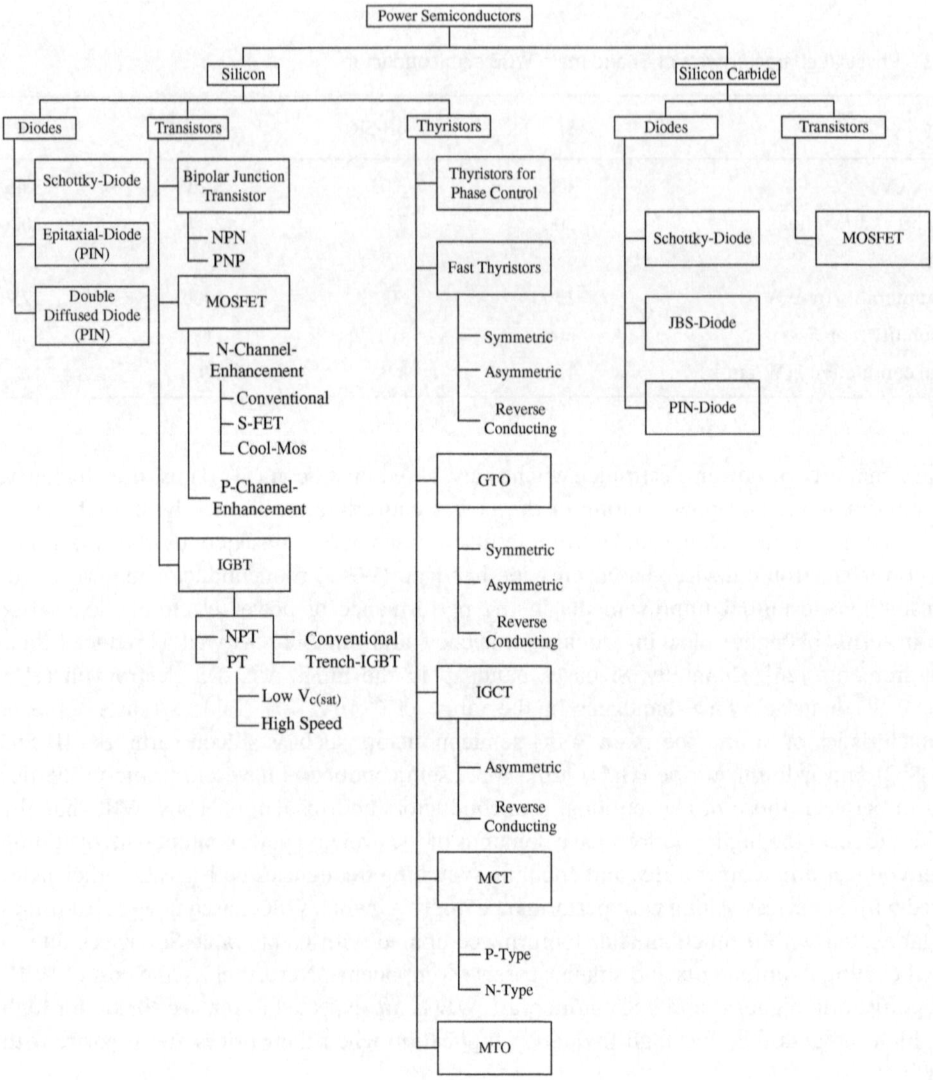

Figure 2.4 Available Si and SiC switches on the market. Source: Reprinted with permission from [28].

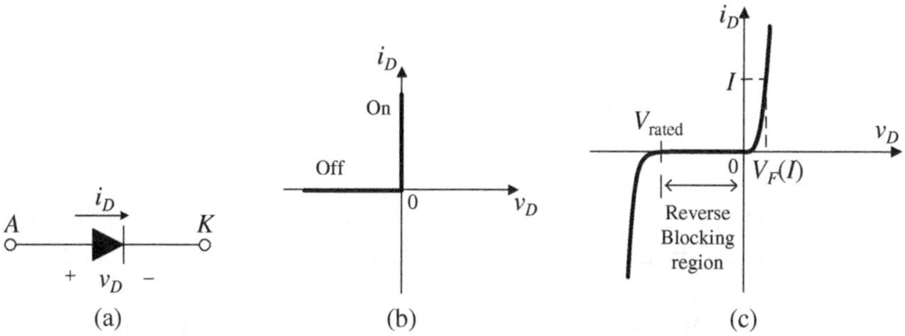

Figure 2.5 A diode: (a) symbol; (b) ideal $I - V$ characteristic; (c) practical $I - V$ characteristic.

its anode (terminal A) has a positive potential with respect to its cathode (terminal K), it is said to be forward-biased. An ideal diode conducts when forward-biased, with negligible voltage drop across it and forward current (I_D) flows through it. If, however, it is reverse-biased, ideally, it does not conduct. The current-voltage ($I - V$) characteristic for an ideal diode is shown in Fig. 2.5(b). A real diode will have a small forward voltage drop across it when it conducts and a small reverse leakage current when it is reverse-biased. The power terminals of the switch are A and K. It has no control terminal through which we can control its on and off switching operations. The switch blocks reverse voltage, but has no capability to block forward voltage. We can describe the ideal power diode as an uncontrolled static switch that turns on and turns off, depending on the operating point of the circuit.

2.3.2 Semi-Controllable Switches

Semi-controllable switches refer to switches that can be partially controlled. The SCR (or thyristor) is one such a device. It functions as a diode that does not conduct when forward biased until an appropriate control signal is applied to a third terminal, called the gate. The symbol and ideal $I - V$ characteristics for the SCR are shown in Figs. 2.6(a) and (b), respectively. The realistic switching characteristic curves are shown in Fig. 2.6(c), where v_{AK} and i_{SCR} are the voltage across the anode-cathode terminals and the current through the anode, respectively. It should be noted that once the SCR is turned on, the gate signal can be reminded. Thus, this device is also known as a latch device [29].

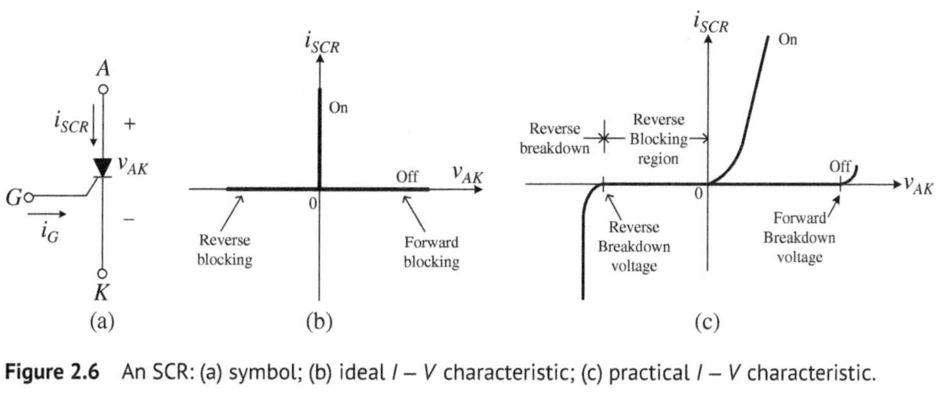

Figure 2.6 An SCR: (a) symbol; (b) ideal $I - V$ characteristic; (c) practical $I - V$ characteristic.

2.3.3 Controlled Switch

A controlled switch is one that can be turned on and off by activating and deactivating a control circuit. Some of the widely used fully controllable switches include:

Bipolar Junction Transistor (BJT)

The bipolar junction transistor (BJT) is a three layer (NPN or PNP) device. The symbol for the NPN BJT is shown in Fig. 2.7(a). The idealized $I - V$ characteristic for the BJT is shown in Fig. 2.7(b), and a typical $I - V$ characteristic for a BJT is shown in Fig. 2.7(b). The on state is achieved by providing sufficient base current to drive the BJT into saturation, and zero base current results in an off state. Generally, power BJTs have two distinct properties when compared to the low-power BJTs and logic transistors: large blocking voltages in the off-state and high forward current carrying capabilities in the on-state [29].

Metal Oxide Semiconductor Field-Effect Transistor (MOSFET)

Unlike the BJT, the metal oxide semiconductor field-effect transistor (MOSFET) belongs to the unipolar device family since it uses only the majority carriers in conduction. Thus, MOSFET switching speeds are faster than BJTs. The device symbol for the n-channel MOSFET is shown in Fig. 2.8(a). Figs. 2.8(b) and (c) show the idealized and practical $I - V$ characteristics. The MOSFET is a voltage-controlled device, which requires negligible current in its control terminal to maintain the on state. It requires the continuous application of a gate-to-source voltage of appropriate

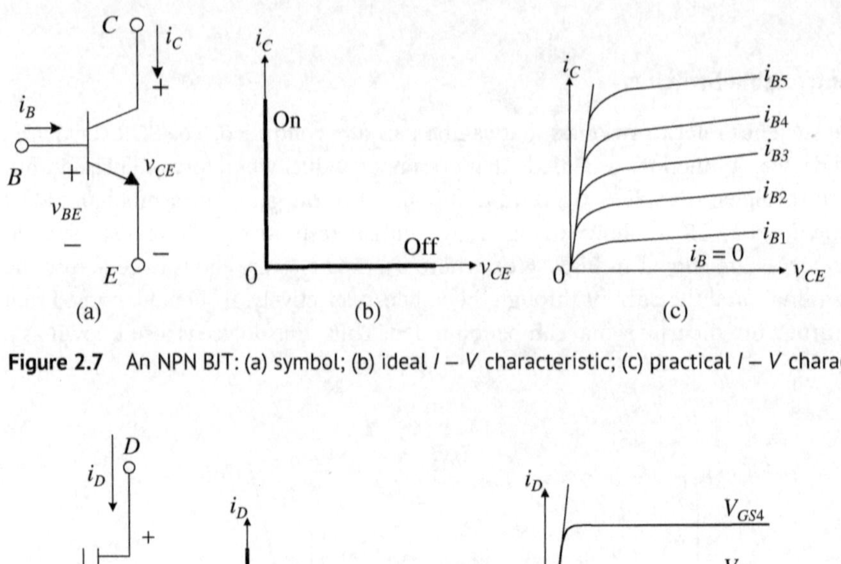

Figure 2.7 An NPN BJT: (a) symbol; (b) ideal $I - V$ characteristic; (c) practical $I - V$ characteristic.

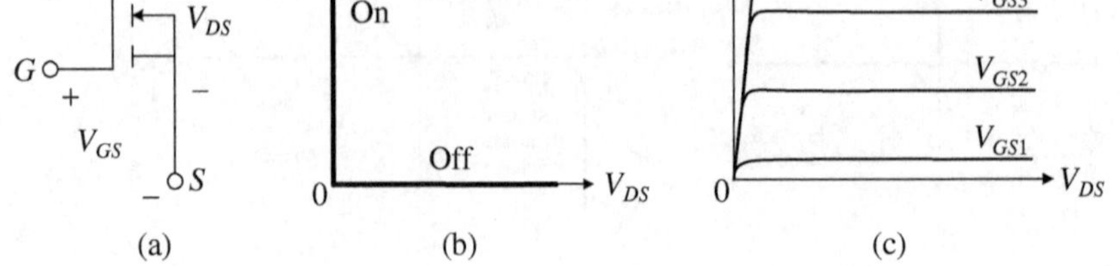

Figure 2.8 An N-channel MOSFET: (a) symbol; (b) ideal $I - V$ characteristic; (c) practical $I - V$ characteristic.

magnitude in order to be in the on state. Note that power MOSFETs inherently contain a built-in diode, often called the body diode. The switching speed of the body diode is much slower than that of the MOSFET. If the body diode is allowed to conduct, high peak currents can occur during the diode turn-off transition, and device failure can occur. Thus, the MOSFET is normally operated with $i \geq 0$ in the same manner as the BJT.

Gate Turn-off Thyristors (GTO)

In the conventional thyristor, the gate serves only to implement the turn-on switching and it has no role to play in the turn-off switching operation. On the other hand, the gate turn-off thyristor (GTO) can also be turned off by means of a reverse gate current pulse whose value is typically as large as one-third the anode current being turned off [11]. The circuit symbol for the GTO is shown in Fig. 2.9(a). Its ideal and practical $I - V$ characteristics are shown in Figs. 2.9(b) and (c), respectively.

Insulated Gate Bipolar Transistor (IGBT)

The circuit symbol for an insulated gate bipolar transistor (IGBT) is shown in Fig. 2.10(a), which can be seen as a device that combines BJT and MOSFET. Essentially, its collector-emitter characteristics are similar to those of the BJT but the control features are those of MOSFET. Similar to the MOSFET, IGBT is a voltage-controlled switch and it is a continuously controllable device. The on state voltage drop, on the other hand, is closer to that of a BJT. Unlike the MOSFET, the IGBT has no integral reverse diode, so it does not have reverse voltage blocking capability. The ideal and practical $I - V$ characteristics are shown in Figs. 2.10(b) and (c).

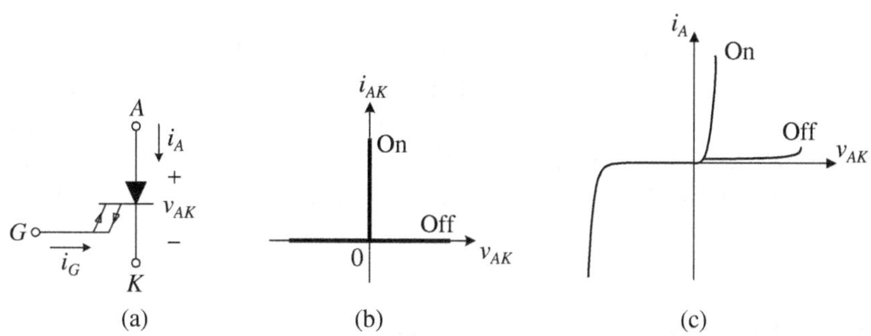

Figure 2.9 A GTO: (a) symbol; (b) ideal $I - V$ characteristic; (c) practical $I - V$ characteristic.

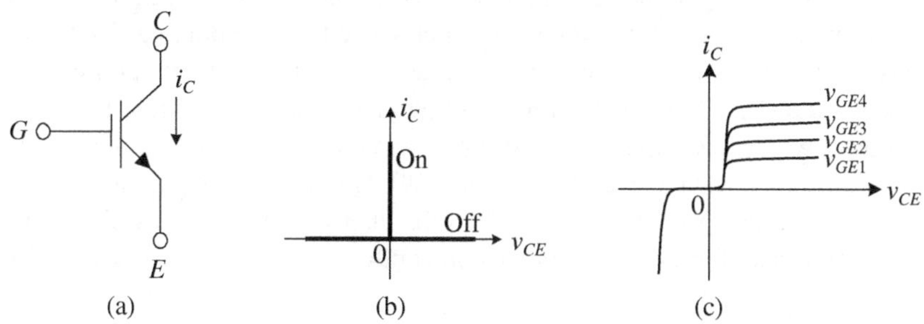

Figure 2.10 An IGBT: (a) symbol; (b) ideal $I - V$ characteristic; (c) practical $I - V$ characteristic.

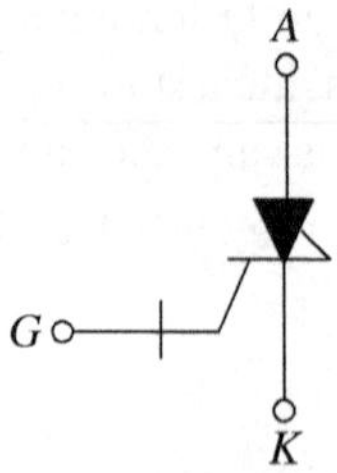

Figure 2.11 The IGCT symbol.

Insulated Gate Commutated Thyristor (IGCT)
The insulated gate commutated thyristor (IGCT), whose symbol is shown in Fig. 2.11, conceptually and structurally is a GTO switch with mitigated turn-off drive requirements. In addition, the IGCT has a lower on-state voltage drop and can also be switched faster compared to the GTO. In recent years, the IGCT has gained considerable attention for high-power converters due to its voltage/current handling capabilities [30]. The industry has seen the GTO thyristor being replaced by the IGCT [31]. The IGCT switches can be classified into asymmetrical, reverse-conducting, and symmetrical types [32]. The asymmetric IGCT is generally used when the reverse voltage-blocking capability is not required. The reverse-conducting IGCT integrates the freewheeling diode into one package, allowing reverse current conduction. The symmetrical IGCT is capable of reverse blocking.

Note that the emphasis of this book is on basic circuit operation rather on device performance. The behavior of a power converter circuit often is not affected significantly by the actual device used for switching, particularly if the voltage drop across a conducting switch is small compared to other circuit voltages. Therefore, the semiconductor switches are modeled as ideal switches so that circuit behavior can be emphasized. To account for the turn-on resistance of the switches, we simply model it as an ideal switch in series with a resistor. The switch transitions between turn-on and turn-off states are assumed to be instantaneous. Therefore, the ideal $I - V$ characteristics shown in Figs. 2.5(b)–2.10(b) are our primary concern when modeling a power converter for harmonic analysis. The dead-times inserted in the gating signals for preventing upper and lower switches of power converters, shorting out during the turn-on and turn-off transitions [11] are not considered in this book.

2.4 Combination of Switches

The static switches discussed so far can only conduct current in one direction, while the voltage blocking capability of some switches can be both directions. These basic switches can be used to construct more sophisticated switches. For example, a switch capable of conducting current in both forward and reverse directions while blocking the voltage in the forward direction can be constructed by means of a transistor in parallel with an antiparallel diode, as shown in Fig. 2.12. Another example is a triac, which is composed of two thyristors connected in antiparallel and is able to conduct currents and block voltage in both directions [29]. This is illustrated in Fig. 2.13(a). Similar functions can also be achieved by means of antiparalleled transistors, as shown in Fig. 2.13(b). Table 2.2 summarizes the directions of current conduction and voltage blocking for various switches [29].

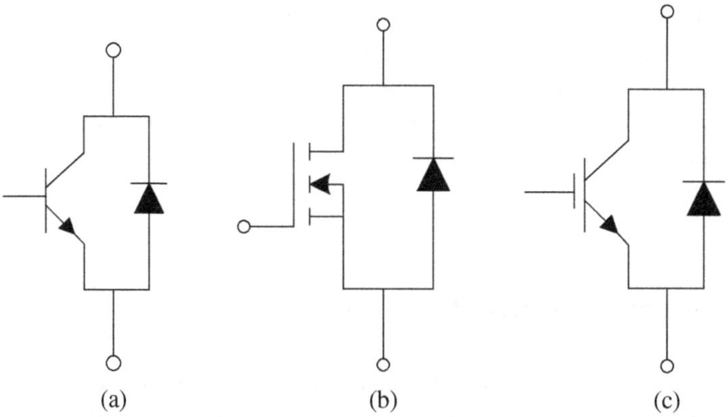

Figure 2.12 (a) a BJT and antiparallel diode; (b) a MOSFET and antiparallel diode; (c) an IGBT and antiparallel diode.

Figure 2.13 Some bidirectional switches: (a) a triac; (b) two transistors with antiparallel diodes.

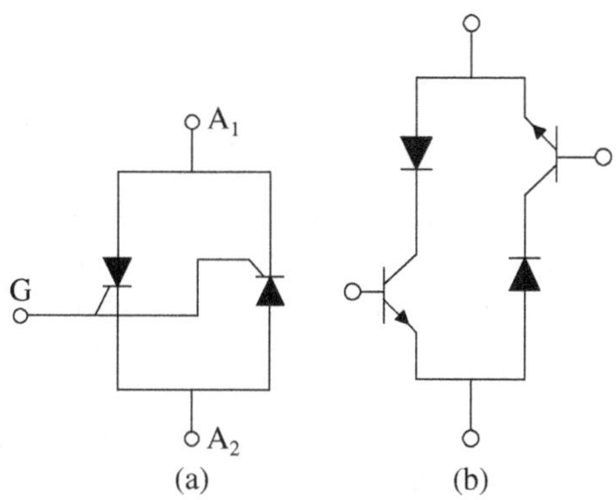

Table 2.2 Types of semiconductor static switches.

Current flow	Voltage blocking	Examples
Forward	Reverse	Diode
Forward	Forward	BJT, MOSFET, IGBT
Forward	Bidirectional	SCR, GTO
Bidirectional	Forward	Fig. 2.12
Bidirectional	Bidirectional	Fig. 2.13

2.5 Classification Based on Commutation Process

One widely used approach to the categorization of converters is based on the commutation process, defined as the transfer of current from branch i to branch j of a circuit, when the switch of branch i turns off while that of branch j turns on [30]. Based on this definition, the following two classes of converters are identified in the technical literature:

- *Line-Commutated Converter:* for a line-commutated converter, the electrical ac system dictates the commutation process. A diode-bridge rectifier is an example of a line-commutated converter. The line-commutated converter is also known as the naturally-commuted converter.
- *Forced-Commutated Converter:* for a forced-commutated converter, the transfer of current from one switch to another is done via a controlled process. Thus, in this type of converter, the switches can be controlled by external circuitry.

2.6 Voltage Source Converter vs. Current Source Converter

Among the conversion systems (i.e. dc-dc, dc-ac, ac-dc, and ac-ac), dc-ac and ac-dc converters can serve as a bridge between the ac and dc worlds, and they can be classified based on voltage and current waveforms at their dc sides. Thus, a CSC is a converter where the dc-side current retains the same polarity, and the direction of average power flow through the converter is determined by the polarity of the dc-side voltage. The dc side of a CSC is typically connected in series with a relatively large inductor that maintains the current continuity and is more representative of a current source, as shown in Fig. 2.14(a). For example, the conventional, six-pulse, thyristor-bridge rectifier is a CSC. Such a line-commutating converter must have an ac source connected to the converter. It consumes reactive power, and suffers from occasional commutation failures in the inverter mode of operation [6]. In a VSC, however, the dc-side voltage retains the same polarity, and the direction of the converter average power flow is determined by the polarity of the dc-side current. The dc-side terminals of a VSC are typically connected in parallel with a relatively large capacitor that resembles a voltage source, as shown in Fig. 2.14(b). Although CSCs have the advantages of inherent short-circuit protection and direct current controllability, they have limited dynamic performance because of the large dc inductor used in the dc-link [31].

Moreover, the forced-commutated CSC has not been as widely used for power system applications because it requires bipolar electronic switches. However, the power semiconductor industry has not yet fully established a widespread commercial supply of fast, fully controllable bipolar

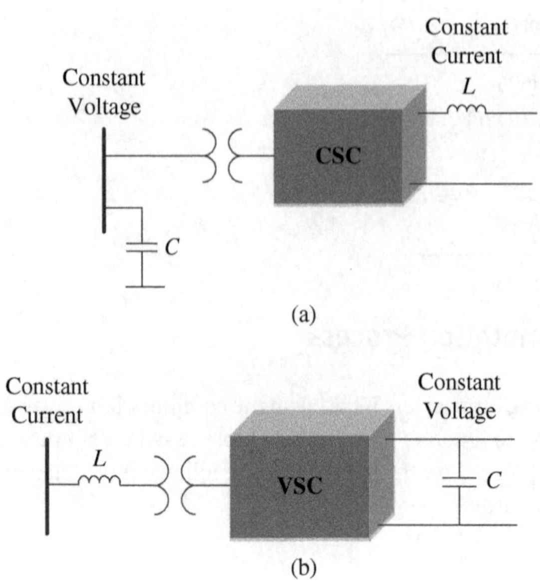

Figure 2.14 (a) Current source converter and (b) voltage source converter. Source: Modified from [33].

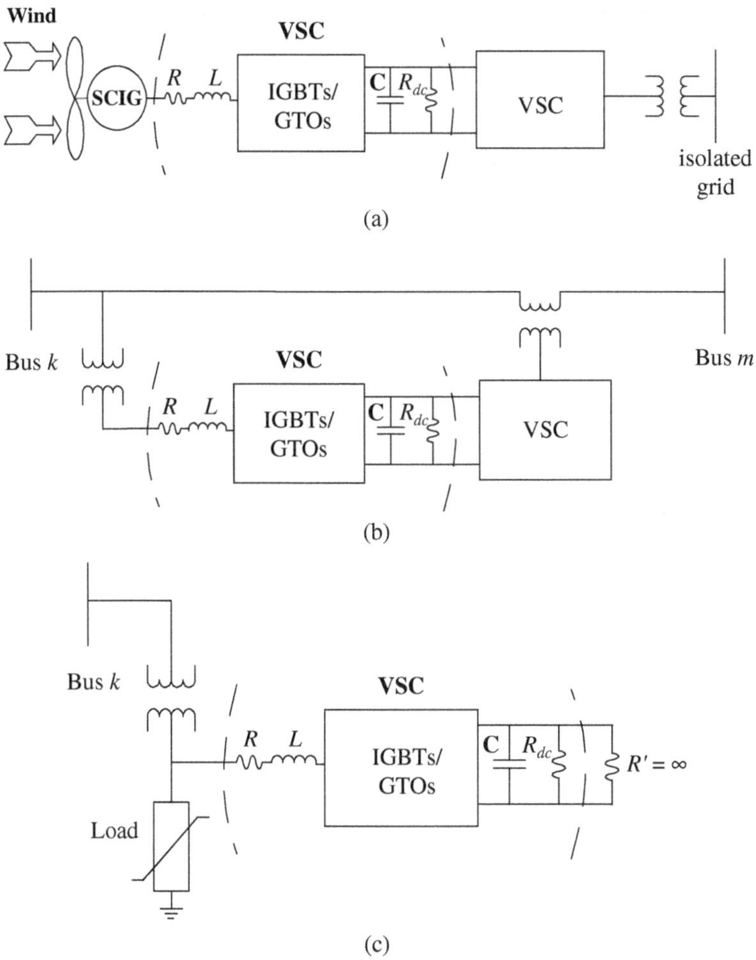

Figure 2.15 (a) Generation level: e.g. squirrel-cage induction generator (SCIG) wind energy conversion system; (b) transmission level: e.g. UPFC; (c) distribution level: e.g. shunt active filter.

switches. Although bipolar versions of the GTO and the IGCT are commercially available, they are limited in terms of switching speed and are mainly tailored for very high-power electronic converters [30]. On the other hand, a VSC requires reverse-conducting switches or switch cells, which can be realized through antiparallel connection of a GTO or IGBT. Moreover, a VSC has the advantage of rapid control of active and reactive power, and is able to connect to weak ac networks or even dead networks whereas the CSC requires the ac system to be strong [33]. Thus, for reasons of economics and performance, VSCs are often preferred over CSCs [6]. In fact, the applications of VSCs can be found in every aspect of power systems. For example, at the generation level, back-to-back VSCs are used as interfaces for distributed energy sources such as wind energy conversion systems (Fig. 2.15(a)). At the transmission level, VSC-based flexible ac transmission systems (FACTS) controllers are employed to improve system stability and control power flow. A good example is the unified power flow controller (UPFC) as shown in Fig. 2.15(b). At the distribution level, VSC-based custom power controllers, such as shunt active filters (Fig. 2.15(c)), are used to improve power quality.

However, like any other power converter, the VSC has a switching nature that produces harmonics. These harmonics can cause detrimental effects such as overheating of components, mechanical oscillations in motors, insulation or capacitor failure, and induced interference in communication circuits. Therefore, it is the main objective of this book to accurately model the harmonics produced by the VSCs.

3

Basic Numerical Iterative Methods

"Simplicity is the ultimate sophistication."

– Leonardo da Vinci

A numerical method is a mathematical tool designed to solve numerical problems. The implementation of a numerical method in a programming language is called a numerical algorithm. This chapter provides a basic review of some of the commonly used iterative numerical algorithms for solving nonlinear problems. Generally speaking, an iterative algorithm involves three steps, as shown in Fig. 3.1, and each step plays an important role. The first step is called initialization. This requires the unknown variables to be identified, and how to properly assign the estimated values for these unknown variables. In general, the closer the estimated values to the final solution, the faster the convergence and less likely to diverge. These estimated values in general can be obtained by applying the physical laws, existing knowledge about the system, and employing some simpler methods to get the first guess. The second step is the core of the algorithm, which is how the algorithm updates or improves the estimated values. For instance, Newton-type methods rely on gradients to improve the next estimated values. On the other hand, a meta-heuristic algorithm, called particle swarm optimization (PSO) updates its estimated values by using a linear combination of the global best estimated values and the current best estimated values. It is also this step that determines whether the algorithm is stable. The third step is the convergence checking, which is used to terminate the iteration. If the convergence criterion is set to be loose, the solution may converge prematurely. On the other hand, if it is set too tight, it may take a very long time to converge to the final values. The following subsections briefly go through the algorithms of Gauss–Seidel, Predictor-Corrector, Newton's method, and the PSO algorithm. In fact, being familiar with these methods allows one to handle many engineering problems requiring numerical iterations.

3.1 Definition of Error

The terminating condition in Fig. 3.1 requires the definition of error. When the error is smaller than a predefined tolerance value, then the convergence criterion is reached. An error can be intuitively defined by (3.1).

$$e_t = x_t - x_a, \tag{3.1}$$

where e_t represents the true error, x_t and x_a are the true and approximate values of the variable. Nevertheless, the true value is not known until the function is solved. Moreover, (3.1) takes no

Harmonic Modeling of Voltage Source Converters using Basic Numerical Methods, First Edition.
Ryan Kuo-Lung Lian, Ramadhani Kurniawan Subroto, Victor Andrean, and Bing Hao Lin.

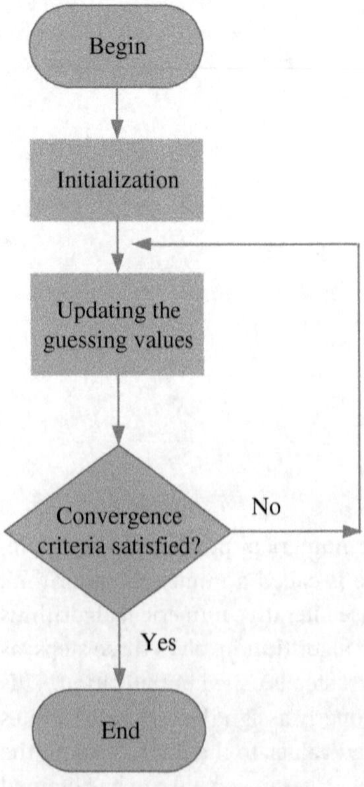

Figure 3.1 General concept of numerical iteration.

account of the order of magnitude of the value under examination. Hence, (3.2) is a more reasonable definition.

$$\epsilon_a = \frac{\text{approximate error}}{\text{approximation}} \times 100\%, \tag{3.2}$$

where ϵ_a represents the normalized approximate error.

Applying the concept of (3.2) to the iterative algorithm, we can estimate the error as the difference between previous and current approximations, and (3.3) is the result.

$$\epsilon_a = \frac{\text{current approximation} - \text{previous approximation}}{\text{current approximation}} \times 100\%, \tag{3.3}$$

3.2 The Gauss–Seidel Method

Because of its simplicity, the Gauss–Seidel method is the most commonly used iterative method. The following illustrates the concept of the Gauss–Seidel method. Assume that there are two unknown variables, x and y, that need to be solved. Moreover, x and y can be written as

$$x = f(x, y) \tag{3.4}$$

$$y = g(x, y). \tag{3.5}$$

The Gauss–Seidel method relies on the fact that the latest iterative values are used immediately in subsequent calculations. This is different from an alternative approach, called Gauss–Jacobi, where the new iterative values are retained for the next iteration. Fig. 3.2 illustrates the difference

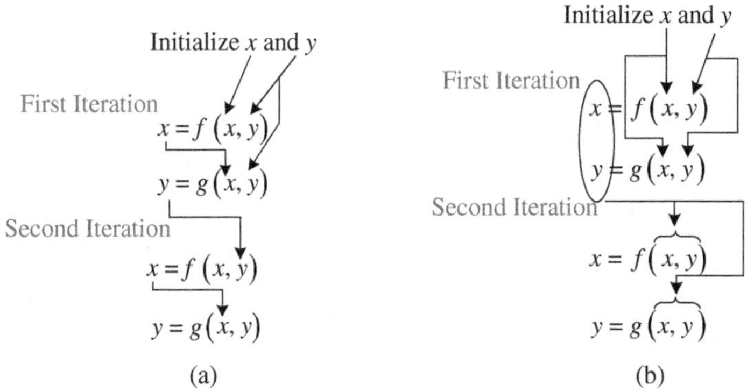

Figure 3.2 Gauss-type iteration – (a) Gauss–Seidel; (b) Gauss–Jacobi.

of the iteration process between Gauss–Seidel and Gauss–Jacobi methods. The iteration processes will stop when the convergence condition is reached, which can be defined in (3.6).

$$\epsilon_a = max\left(\left|\frac{x^{(i)} - x^{(i-1)}}{x^{(i)}}\right|, \left|\frac{y^{(i)} - y^{(i-1)}}{y^{(i)}}\right|\right) < \varepsilon, \tag{3.6}$$

where ε is a preset tolerance value. Alternatively, (3.7) can also be written by carefully selecting the value of ε_1.

$$\epsilon_a = Max\{|x^{(i)} - x^{(i-1)}|, |y^{(i)} - y^{(i-1)}|\} < \varepsilon_1. \tag{3.7}$$

Note that although there are certain cases where the Jacobi method is useful, Gauss–Seidel's utilization of the best available estimates usually makes it the method of preference. A simple analogy for the Gauss–Seidel algorithm describing Fig. 3.2 can be thought of as follows: A ball (i.e. (x, y)) covered with mud is passing between two players (i.e. $f(x, y)$ and $g(x, y)$). Each time a pass is made, the mud is patted away, slowly revealing the identity of the ball (i.e. the true solution found). Therefore, it can be expected that the convergence speed of the Gauss–Seidel method is low.

To generalize the Gauss–Seidel algorithm to a set of multi-variable nonlinear equations, (3.8) is considered.

$$f_k(x_1, x_2, \cdots, x_n) = 0, k = 1, 2, \ldots, n. \tag{3.8}$$

(3.8) can be expressed by

$$x_k = g_k(x_1, x_2, \cdots, x_n), k = 1, 2, \ldots, n. \tag{3.9}$$

The iteration is commenced as soon as a set of initial values are provided, and the estimated values of x_k at the ith iteration continues refining by substituting the most recent calculated solution, $(x_1^{(i)}, x_2^{(i)}, \ldots, x_{k-1}^{(i)})$ into (3.9), leading to (3.10).

$$x_k^{(i)} = g_k(x_1^{(i)}, x_2^{(i)}, \cdots, x_{k-1}^{(i)}, x_k^{(i-1)}, \cdots, x_n^{(i-1)}), k = 1, 2, \ldots, n. \tag{3.10}$$

The iteration continues until (3.11) is met.

$$\epsilon_a = max\left(\left|\frac{x_k^{(i)} - x_k^{(i-1)}}{x_k^{(i)}}\right|\right) < \varepsilon, k = 1, 2, \ldots, n. \tag{3.11}$$

Alternatively, (3.12) can also be written by carefully selecting the value of ε_1.

$$\epsilon_a = Max\{|x_k^{(i)} - x_k^{(i-1)}|\} < \varepsilon_1, k = 1, 2, \ldots, n. \tag{3.12}$$

3.3 Predictor-Corrector

The concept of the predictor-corrector method is somewhat similar to that of the Gauss–Seidel method. A predictor-corrector method essentially involves two sets of equations. The first set is used to predict values of the unknown variables. The second set, on the other hand, is to correct or improve the values predicted from the first set. Nevertheless, unlike the Gauss–Seidel method, the predictor-corrector method does not have a convergence condition to check, and the result is immediately outputted by the corrector. A predictor-corrector method has been much used for integrating ordinary differential equations, and typically uses an explicit method at the predictor step and an implicit method at the corrector step:

An ordinary differential equation, $\frac{dy}{dt} = f(x, t)$,[1] can be approximated by the following first-order difference equation:

1. Forward Euler method:

$$\dot{x}(t) = \frac{x(t+h) - x(t)}{h} \tag{3.13}$$

$$\Longrightarrow$$

$$x(t+h) = x(t) + h\dot{x}(t) \tag{3.14}$$

2. Backward Euler method:

$$\dot{x}(t+h) = \frac{x(t+h) - x(t)}{h} \tag{3.15}$$

$$\Longrightarrow$$

$$x(t+h) = x(t) + h\dot{x}(t+h) \tag{3.16}$$

3. Trapezoidal method:

$$\frac{\dot{x}(t) + \dot{x}(t+h)}{2} = \frac{x(t+h) - x(t)}{h} \tag{3.17}$$

$$\Longrightarrow$$

$$x(t+h) = x(t) + \frac{h}{2}[\dot{x}(t) + \dot{x}(t+h)] \tag{3.18}$$

Backward Euler and trapezoidal integration methods are called implicit integration because the unknowns, $x(t+h)$, appear at both sides of the equations. To solve this problem, one way is to employ the concept of the predictor-corrector. First, we use forward Euler to "predict" the value of $x(t+h)$ to obtain $\dot{x}(t+h)$. Then, it is substituted to (3.16) or (3.18) to "correct" the value of $x(t+h)$. This is illustrated by the following numerical example.

Example 3.1
Consider the differential equation $\dot{x} = x + t^2$ and apply the forward Euler formula with time step $h = 0.025$ and $x(0) = 1$. (a) Use forward Euler method. (b) Use predictor-corrector approach.

Solutions:

(a) Forward Euler method:

$x(t+h) = x(t) + h\dot{x}(t)$

$x(0) = 1$

$x(0.025) = 1 + 0.025 \times (1) = 1.025$

$x(0.05) = 1.025 + 0.025 \times (1.025 + (0.025)^2) = 1.050640625$

$\vdots$

1 This can be extended to a vector form.

(b) Predictor-Corrector approach:
First, we use forward Euler method to "predict" $x(t+h)$, and obtain $\dot{x}(t+h)$. Then, we substitute $\dot{x}(t+h)$ back to (3.16) to "correct" the value. The following delineates the steps.

$$
\begin{aligned}
&x(0) = 1\\
\text{predictor: } &x(0.025) = 1.025\\
&\dot{x}(0.025) = 1.025 + (0.025)^2 = 1.025625\\
\text{corrector: } &x(0.025) = 1 + (0.025)(1.025625) = 1.025640625\\
\text{predictor: } &x(0.05) = 1.025640625 + (0.025)(1.025640625 + (0.025)^2) = 1.051297265625\\
&\dot{x}(0.05) = (1.051297265625) + (0.05)^2 = 1.053797265625\\
\text{corrector: } &x(0.05) = 1.025640625 + (0.025)(1.053797265625) = 1.051985556640625\\
&\vdots
\end{aligned}
$$

The exact solution is

$$
\begin{aligned}
&x = 3e^t - t^2 - 2t - 2\\
&x(0) = 1\\
&x(0.025) = 1.0253203615732865\\
&x(0.05) = 1.0513132891280721\\
&\vdots
\end{aligned}
$$

It is evident that the Predictor-Corrector approach provides a better estimate than the forward Euler approach. □

The reader may intuitively think that the forward Euler method is sufficient for most of the problem as it is an explicit integration method, where $x(t+h)$ is directly a linear combination of $x(t)$ and $h\dot{x}(t)$. On the contrary, forward Euler is rarely used because it can become numerically unstable when the integration step is chosen inappropriately. On the other hand, back Euler and trapezoidal integration methods are numerically stable, and consequently, have been widely used in power system applications. The following numerical example illustrates the problem of the forward integration method.

Example 3.2 Consider the following differential equation,

$$\frac{dx}{dt} = -ax, x(0) = 1, a > 0. \tag{3.19}$$

The exact solution to (3.19) can be found analytically, and is stated in (3.20)

$$x(t) = e^{-at}. \tag{3.20}$$

Therefore, $x(t)$ is always stable.

If we use the forward Euler Method to integrate (3.19), which is restated here

$$x(t+h) = x(t) + h\dot{x}, \tag{3.21}$$

and can be expressed as

$$x_{n+1} = x_n - hax_n, \tag{3.22}$$

where x_{n+1} and x_n represent the values of x at $n+1$th and nth time step, respectively.

Finally, (3.23) results if subsequent terms are also evaluated recursively by (3.22).

$$x_{n+1} = (1 - ha)^{n+1} x_0, \tag{3.23}$$

The sequence of (3.23) is stable only if

$$|1 - ha| < 1, \tag{3.24}$$

which leads to (3.25)

$$0 < h < 2/a. \tag{3.25}$$

Hence, if h is chosen to be greater than $2/a$, an unstable solution will result, which contradicts the true solution. On the other hand, if the backward Euler method is employed, one obtains (3.26)

$$x(t+h) = x(t) - ahx(t+h). \tag{3.26}$$

Rearranging the equation leads to

$$x(t+h) = \frac{1}{1+ah}x(t), \tag{3.27}$$

which can be rewritten as

$$x_{n+1} = \frac{1}{1+ah}x_n. \tag{3.28}$$

Applying recurrence of (3.28) to the subsequent terms leads to

$$x_{n+1} = \left(\frac{1}{1+ah}\right)^{n+1} x_0. \tag{3.29}$$

(3.29) is always stable as long as $h > 0$. □

It is to be noted that the idea of the predictor and corrector is the "backbone" of the Kalman filter, which has been regarded as the optimal solution to many tracking and data prediction tasks. Fig. 3.3 briefly delineates the algorithm of the Kalman filter. As shown in the figure, in the predictor stage, the state and covariance from the previous stage, $\hat{\mathbf{x}}_{k-1}$ and $\mathbf{P}_{k-1}$ are used to predict the state and error covariance at the current time point ($\hat{\mathbf{x}}_k^-$ and $\mathbf{P}_k^-$) via (3.30) and (3.31).

$$\hat{\mathbf{x}}_k^- = \mathbf{A}\hat{\mathbf{x}}_{k-1} \tag{3.30}$$

$$\mathbf{P}_k^- = \mathbf{A}\mathbf{P}_{k-1}\mathbf{A}^T + \mathbf{Q}, \tag{3.31}$$

where $\mathbf{A}$ is defined as the system matrix and $\mathbf{Q}$ is the covariance matrix of state noise. Then, the Kalman gain, $\mathbf{K}_k$, which can be regarded as the factor that determines how much the prediction

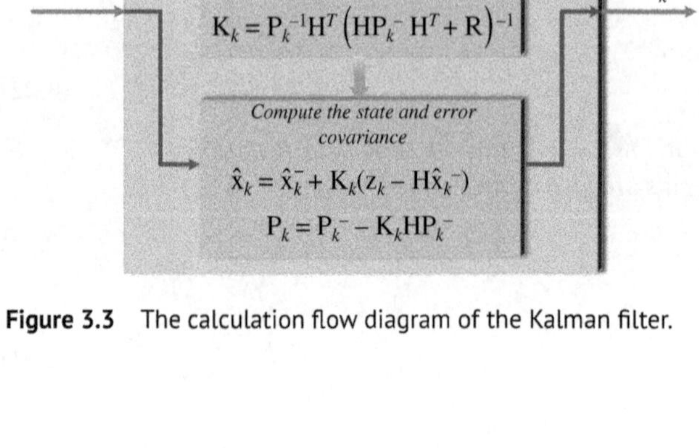

Figure 3.3 The calculation flow diagram of the Kalman filter.

should be calibrated, and is stated in (3.32)

$$\mathbf{K}_k = \mathbf{P}_k^- \mathbf{H}^T \left(\mathbf{H}\mathbf{P}_k^- \mathbf{H}^T + \mathbf{R}\right)^{-1} \tag{3.32}$$

where **H** represents state-to-measurement matrix and **R** denotes covariance matrix of measurement noise. Finally, in the correction stage, the estimated state to be outputted is computed via (3.33), and the covariance is updated by (3.34).

$$\hat{\mathbf{x}}_k = \hat{\mathbf{x}}_k^- + \mathbf{K}_k \left(\mathbf{z}_k - \mathbf{H}\mathbf{x}_k^-\right) \tag{3.33}$$

$$\mathbf{P}_k = \mathbf{P}_k^- - \mathbf{K}_k \mathbf{H}\mathbf{P}_k^- \tag{3.34}$$

Note that we did not go through the details of the derivation of (3.30)–(3.34) since it is beyond the scope of this book. Readers who are interested in this topic can refer to the well-written introductory book on Kalman filter by P. Kim [34].

3.4 Newton's Method

Newton's method, also known as the Newton–Raphson method, named after Isaac Newton and Joseph Raphson, is a method for finding successively better approximations to the roots (or zeroes) of a real-valued function. It is a very powerful method as the convergence of the solution is quadratic, which means that the difference between the root and the approximation is squared at each iteration step. The following describes some of the common applications of Newton's method.

3.4.1 Root Finding

To find the root of a function, $f(x)$, one starts with an initial guess, x_0, which is assumed to be reasonably close to the root, as shown in Fig. 3.4. Then, the next better approximation, x_1, can be

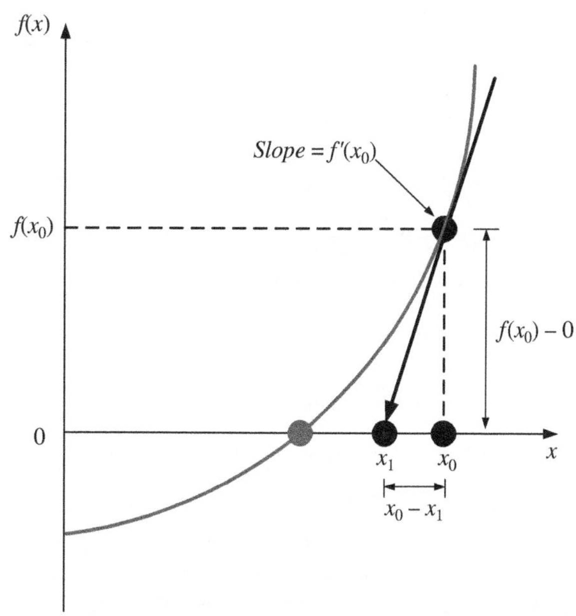

Figure 3.4 Root-finding by Newton's method.

found by calculating the x-intercept of the tangent line at $f(x_0)$.

$$x_1 = x_0 - \frac{1}{f'(x_0)} f(x_0) \tag{3.35}$$

Extending (3.35) to the n-th iteration,

$$x_{n+1} = x_n - \frac{1}{f'(x_n)} f(x_n). \tag{3.36}$$

Moreover, extending the scalar-form of (3.36) to vector-form, (3.36) is changed to (3.37).

$$\zeta^{(n+1)} = \zeta^{(n)} - \mathbf{J}^{-1}\mathbf{M}(\zeta^{(n)}), \tag{3.37}$$

where ζ is the unknown vectors, $\mathbf{J}$ is the Jacobian matrix, which is defined as $\frac{\partial \mathbf{M}}{\partial \zeta}$, and $\mathbf{M}$, called the mismatch equation, meaning that as long as $\mathbf{M}$ is different from zero or greater than a preset tolerance value, the iteration continues. It is worth mentioning that Fig. 3.4 is a graph that represents the concept of Newton's method well. Thus, to determine whether a problem can be solved by Newton's method, one should see whether it can be recast as a root-finding problem. This point is illustrated in Sections 3.4.2, 3.4.3, 3.4.4, and 3.4.5.

3.4.2 Numerical Integration

As mentioned in Section 3.3, backward Euler and trapezoidal methods are implicit integration, and it is shown that predictor-corrector can be one viable solution. Another approach is to use Newton's formula, and the trick is to turn (3.16) and (3.18) into a root-finding problem. (3.16) and (3.18) can be rewritten as (3.38) and (3.39), respectively.

$$x(t+h) - x(t) - h\dot{x}(t+h) = 0 \tag{3.38}$$

$$x(t+h) - x(t) - \frac{h}{2}[\dot{x}(t) + \dot{x}(t+h)] = 0 \tag{3.39}$$

The unknown variables in both (3.38) and (3.39) are $x(t+h)$. The mismatch equations are simply $x(t+h) - x(t) - h\dot{x}(t+h)$ and $x(t+h) - x(t) - \frac{h}{2}[\dot{x}(t) + \dot{x}(t+h)]$ for backward Euler and trapezoidal methods, respectively. Therefore, Newton's formulations to solve $x(t+h)$ of (3.38) and (3.39) can be formulated in (3.40).

$$\zeta^{(n+1)} = \zeta^{(n)} - J^{-1}M(\zeta^{(n)}), \tag{3.40}$$

where $\zeta = x(t+h)$,

$$J = 1 - h\frac{\partial \dot{x}(t+h)}{\partial x(t+h)}, \tag{3.41}$$

$$M = x(t+h) - x(t) - h\dot{x}(t+h) \tag{3.42}$$

for backward Euler method.

$$J = 1 - \frac{h}{2}\frac{\partial \dot{x}(t+h)}{\partial x(t+h)}, \tag{3.43}$$

$$M = x(t+h) - x(t) - \frac{h}{2}[\dot{x}(t) + \dot{x}(t+h)] \tag{3.44}$$

for the trapezoidal rule method.

Fig. 3.5 depicts the implicit integration employing Newton's method at each time step. At $t = 0$, the initial values x_0 are given. Then, the value at the next time step, t_1, x_1, is computed by means of (3.40). This process is repeated for the following time steps until the final time is reached. Note that

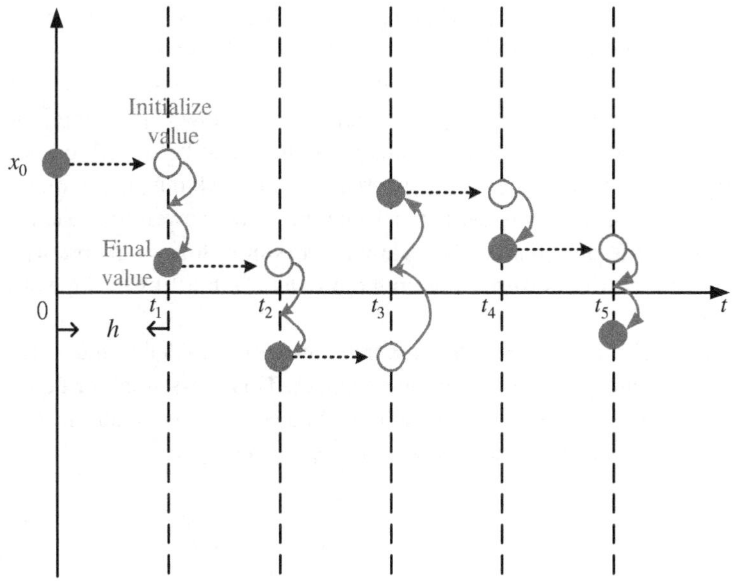

Figure 3.5 Numerical integration using Newton's method (Solid circles: outputted final values. Empty circles: Initialized values Arrows: iterations).

since continuity should preserve for a real physical system, the initial conditions for each time step can simply be the final values of the previous time step, as shown in Fig. 3.5. Moreover, due to the quadratic convergence of Newton's method, it usually takes fewer than 5 (typically 2–3) iterations to attain the solution. Hence, this approach of numerical integration is quite frequently used. In fact, most of the electromagnetic transient programs employ such a concept for solving differential equations.

3.4.3 Power Flow

Transmission-level operations under normal balanced three-phase steady-state conditions require [35]: (1) generation equals the demand plus losses; (2) bus voltage magnitudes remain close to rated values; (3) generators operate within specified real and reactive power limits; (4) transmission lines and transformers are not overloaded [35]. A power-flow (also called load flow) program is the basic tool for investigating these requirements. The principal information obtained from the power-flow study is the magnitude and phase angle of the voltage at each bus, and the real and reactive power flowing in each line.

Conventional nodal or loop analysis is not suitable for power-flow studies because the input data for loads are normally given in terms of power, rather than impedance. Moreover, generators are considered as power sources, not voltage or current sources. The power-flow problem is therefore formulated as a set of nonlinear algebraic equations, which require numerical iterative methods such as Gauss–Seidel or Newton's methods to obtain the solution.

The power-flow problem begins with identifying the known and unknown variables in the system. The known and unknown variables are dependent on the type of bus. A bus with predetermined active and reactive powers is called a load bus. A bus with at least one generator connected to it is called a generator bus. The exception is one arbitrarily-selected bus that has a generator. This bus is referred to as the slack bus.

In the power-flow problem, it is assumed that the real power (P) and reactive power (Q) at each load bus are known. For this reason, load buses are also known as PQ buses. For generator buses, it is assumed that the real power generated and the voltage magnitude, $|V|$, is known. For the slack bus, it is assumed that the voltage magnitude and its phase angle (δ) are known. Therefore, for each load bus, both the voltage magnitude and angle are unknown and must be solved for. For each generator bus, the voltage angle must be solved for, whereas for the slack bus, there are no variables that need to be solved. The slack bus is used to balance the active power and reactive power during load flow calculation. The slack bus can be regarded as to supply for system losses by emitting or absorbing active and/or reactive power to and from the system. Table 3.1 lists the types of buses used in a power-flow study.

To employ Newton's method to solve for a power-flow problem, one needs to derive the active and reactive powers in terms of voltage magnitudes and phase angles. This is essential for deriving the Jacobian matrix. The following describes the derivations of power-flow formulation: The relationship between the injected node current and the node voltage is given by

$$I_i = \sum_{j=1}^{n} Y_{ij} V_j \tag{3.45}$$

where I_i is the net current injected into the network at bus i, Y_{ij} is the bus admittance matrix, which can be expressed as $Y_{ij} = |Y_{ij}|\angle\theta_{ij}$, and V_j is the voltage at node j, which can be expressed as $V_j = |V_j|\angle\delta_j$. Hence,

$$I_i = \sum_{j=1}^{n} |Y_{ij}||V_j|\angle(\theta_{ij} + \delta_j) \tag{3.46}$$

The complex power for the i-th bus can be stated in (3.47).

$$S_i = V_i I_i^* \tag{3.47}$$

$$S_i^* = V_i^* I_i \tag{3.48}$$

$$\Rightarrow P_i - jQ_i = (|V_i|\angle - \delta_i)(\sum_{j=1}^{n} |Y_{ij}|V_j|\angle\theta_{ij} + \delta_j) \tag{3.49}$$

$$\begin{cases} P_i = \sum_{j=1}^{n} |V_i||V_j||Y_{ij}|\cos(\theta_{ij} - \delta_i + \delta_j) \\ Q_i = -\sum_{j=1}^{n} |V_i||V_j||Y_{ij}|\sin(\theta_{ij} - \delta_i + \delta_j) \end{cases} \tag{3.50}$$

Now, it is ready to apply Newton's method to solve for power-flow problem. The Newton's formulation in vector form is stated in

$$\zeta^{(i+1)} = \zeta^{(i)} - \mathbf{J}^{-1}[\mathbf{M}]. \tag{3.51}$$

Table 3.1 Various load buses.

Bus type	**Generator bus**		**Load bus**
Bus name	PV bus	Slack bus	PQ bus
Given	P and $|V|$	$|V|$ and δ	P and Q
Solve for	δ	None	$|V|$ and δ

Here, ζ is the unknown vector, and

$$\zeta = \begin{bmatrix} |\mathbf{V}| \\ \delta \end{bmatrix}, \tag{3.52}$$

where $|\mathbf{V}|$ is a vector of voltage magnitudes and δ is a vector of voltage angles. The mismatch vector, $\mathbf{M}$ is expressed as

$$\mathbf{M} = \begin{bmatrix} \mathbf{P} - \mathbf{P_{ref}} \\ \mathbf{Q} - \mathbf{Q_{ref}} \end{bmatrix}, \tag{3.53}$$

where $\mathbf{P_{ref}}$ and $\mathbf{Q_{ref}}$ are the vectors of scheduled active and reactive powers, respectively.

The slack bus is usually designated as node 1. Hence, there is no need to solve for $|V_1|$ and δ_1. Similarly, for the PV buses, there is no need to solve for their voltage magnitudes. For simplicity, we assume that all the buses, except bus 1, are PQ buses.

Therefore,

$$\zeta = \begin{bmatrix} |V_2| \\ |V_3| \\ \vdots \\ |V_n| \\ \hline \delta_1 \\ \delta_2 \\ \vdots \\ \delta_n \end{bmatrix}_{2(n-1)\times 1} \tag{3.54}$$

The mismatch equations,

$$\mathbf{M} = \begin{bmatrix} P_2 - P_{2ref} \\ P_3 - P_{3ref} \\ \vdots \\ P_n - P_{nref} \\ \hline Q_2 - Q_{2ref} \\ Q_3 - Q_{3ref} \\ \vdots \\ Q_n - Q_{nref} \end{bmatrix}_{2(n-1)\times 1} \tag{3.55}$$

The Jacobian Matrix is

$$\mathbf{J} = \begin{bmatrix} \frac{\partial M_1}{\partial |V_2|} \cdots \frac{\partial M_1}{\partial |V_n|} & \frac{\partial M_1}{\partial \delta_2} \cdots \frac{\partial M_1}{\partial \delta_n} \\ \frac{\partial M_2}{\partial |V_2|} \cdots \frac{\partial M_2}{\partial |V_n|} & \frac{\partial M_2}{\partial \delta_2} \cdots \frac{\partial M_2}{\partial \delta_n} \\ \vdots & \vdots \\ \frac{\partial M_{2n-1}}{\partial |V_2|} \cdots \frac{\partial M_{2n-1}}{\partial |V_n|} & \frac{\partial M_{2n-1}}{\partial \delta_2} \cdots \frac{\partial M_{2n-1}}{\partial \delta_n} \end{bmatrix}_{2(n-1)\times 2(n-1)} \tag{3.56}$$

Referring back to (3.50), one can obtain each element of the Jacobian matrix. Let us first derive the expressions for the terms, $\frac{\partial M_1}{\partial |V_2|}, \frac{\partial M_1}{\partial |V_3|}, \frac{\partial M_1}{\partial \delta_2}, \frac{\partial M_1}{\partial \delta_3}, \frac{\partial M_n}{\partial |V_2|}, \frac{\partial M_n}{\partial |V_3|}, \frac{\partial M_n}{\partial \delta_2}$, and $\frac{\partial M_n}{\partial \delta_3}$. From these expressions, we can obtain the general expression for all the elements of the Jacobian matrix.

$$\frac{\partial M_1}{\partial |V_2|} = \frac{\partial (P_2 - P_{2ref})}{\partial |V_2|} = \frac{\partial P_2}{\partial |V_2|} \rightarrow \text{P and V have the same indices} \tag{3.57}$$

$$\frac{\partial P_2}{\partial |V_2|} = |V_1||Y_{21}|\cos(\theta_{21} - \delta_2 + \delta_1) + 2|V_2||Y_{22}|\cos\theta_{22} + |V_3||Y_{23}|\cos(\theta_{23} - \delta_2 + \delta_3) + \cdots \tag{3.58}$$

Thus, $\frac{\partial P_i}{\partial |V_i|} = 2|V_i||Y_{ii}|\cos\theta_{ii} + \sum_{j\neq i}|V_j||Y_{ij}|\cos(\theta_{ij} - \delta_i + \delta_j)$

$$\frac{\partial M_1}{\partial |V_3|} = \frac{\partial (P_2 - P_{2ref})}{\partial |V_3|} = \frac{\partial P_3}{\partial |V_3|} \rightarrow \text{P and V have different indices} \tag{3.59}$$

$$\frac{\partial P_2}{\partial |V_3|} = |V_2||Y_{23}|\cos(\theta_{23} - \delta_2 + \delta_3) \tag{3.60}$$

Thus, $\frac{\partial P_i}{\partial |V_j|} = |V_i||Y_{ij}|\cos(\theta_{ij} - \delta_i + \delta_j)$ for $j \neq i$

$$\frac{\partial M_1}{\partial \delta_2} = \frac{\partial (P_2 - P_{2ref})}{\partial \delta_2} = \frac{\partial P_2}{\partial \delta_2} \rightarrow \text{P and } \delta \text{ have the same indices} \tag{3.61}$$

$$\frac{\partial P_2}{\partial \delta_2} = |V_2||V_1||Y_{21}|\sin(\theta_{21} - \delta_2 + \delta_1) + |V_2||V_3||Y_{23}|\sin(\theta_{23} - \delta_2 + \delta_3) + \cdots \tag{3.62}$$

Thus, $\frac{\partial P_i}{\partial \delta_i} = \sum_{j\neq i}|V_i||V_j||Y_{ij}|\sin(\theta_{ij} - \delta_i + \delta_j)$

$$\frac{\partial M_1}{\partial \delta_3} = \frac{\partial (P_2 - P_{2ref})}{\partial \delta_3} = \frac{\partial P_2}{\partial \delta_3} \rightarrow \text{P and V have different indices} \tag{3.63}$$

$$\frac{\partial P_2}{\partial \delta_3} = -|V_2||V_3||Y_{23}|\sin(\theta_{23} - \delta_2 + \delta_3) \tag{3.64}$$

Thus, $\frac{\partial P_i}{\partial \delta_j} = -|V_i||V_j||Y_{ij}|\sin(\theta_{ij} - \delta_i + \delta_j)$ for $j \neq i$

$$\frac{\partial M_n}{\partial |V_2|} = \frac{\partial (Q_2 - Q_{2ref})}{\partial |V_2|} = \frac{\partial Q_2}{\partial |V_2|} \rightarrow \text{Q and } |V| \text{ have the same indices} \tag{3.65}$$

$$\frac{\partial Q_2}{\partial |V_2|} = -|V_1||Y_{21}|\sin(\theta_{21} - \delta_2 + \delta_1) - 2|V_2||Y_{22}|\sin(\theta_{22} - |V_3||Y_{23}|\sin(\theta_{23} - \delta_2 + \delta_3) + \cdots \tag{3.66}$$

Thus, $\frac{\partial Q_i}{\partial |V_i|} = -2|V_i||Y_{ii}|\sin\theta_{ii} - \sum_{j\neq i}|V_j||Y_{ij}|\sin(\theta_{ij} - \delta_i + \delta_j)$

$$\frac{\partial M_n}{\partial |V_3|} = \frac{\partial (Q_2 - Q_{2ref})}{\partial |V_3|} = \frac{\partial Q_2}{\partial |V_3|} \rightarrow \text{Q and } |V| \text{ have different indices} \tag{3.67}$$

$$\frac{\partial Q_2}{\partial |V_3|} = -|V_2||Y_{23}|\sin(\theta_{23} - \delta_2 + \delta_3) \tag{3.68}$$

Thus, $\frac{\partial Q_i}{\partial |V_j|} = -|V_i||Y_{ij}|\sin(\theta_{ij} - \delta_i + \delta_j)$ for $j \neq i$

$$\frac{\partial M_n}{\partial \delta_2} = \frac{\partial (Q_2 - Q_{2ref})}{\partial \delta_2} = \frac{\partial Q_2}{\partial \delta_2} \rightarrow \text{Q and } \delta \text{ have the same indices} \tag{3.69}$$

$$\frac{\partial Q_2}{\partial \delta_2} = |V_2||V_1||Y_{21}|\cos(\theta_{21} - \delta_2 + \delta_1) + |V_2||V_3||Y_{23}|\cos(\theta_{23} - \delta_2 + \delta_3) + \cdots \tag{3.70}$$

Thus, $\frac{\partial Q_i}{\partial \delta_i} = \sum_{j \neq i} |V_i||V_j||Y_{ij}| \cos(\theta_{ij} - \delta_i + \delta_j) \rightarrow$ Q and δ have different indices

$$\frac{\partial M_n}{\partial \delta_3} = \frac{\partial (Q_2 - Q_{2ref})}{\partial \delta_3} = \frac{\partial Q_2}{\partial \delta_3} \tag{3.71}$$

$$\frac{\partial Q_2}{\partial \delta_3} = -|V_2||V_3||Y_{23}| \cos(\theta_{23} - \delta_2 + \delta_3) \tag{3.72}$$

Thus, $\frac{\partial Q_i}{\partial \delta_j} = -|V_i||V_j||Y_{ij}| \cos(\theta_{ij} - \delta_i + \delta_j)$ for $j \neq i$

The readers are encouraged to use the above derivations to solve for power-flow problem examples from a power-system textbook such as [35] and [36].

3.4.4 Harmonic Power Flow

In 1982, Profs. Xia and Heydt published a classic paper, extending the conventional power flow to permit the inclusion of nonlinear loads. This is illustrated with the following example. A four bus system shown in Fig. 3.6, has a nonlinear device, installed at bus 3. The $V - I$ characteristic of the nonlinear device is assumed to be known, and is specified as $i = g(v)$. Moreover, the active and reactive power for the nonlinear device is specified as shown in Fig. 3.6.

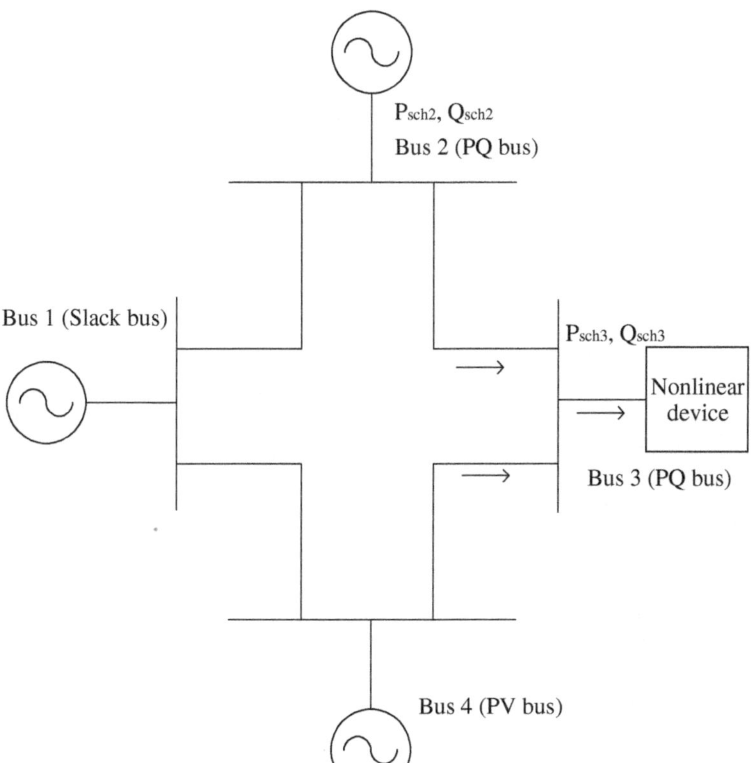

Figure 3.6 A four-bus system containing a nonlinear device.

The mismatch equations associated with Fig. 3.6 can be divided into two parts. The first part is the power flow constraints, which are

$$P_2 - P_{2sch} = 0 \tag{3.73}$$

$$Q_2 - Q_{2sch} = 0 \tag{3.74}$$

$$P_3 - P_{3sch} = 0 \tag{3.75}$$

$$Q_3 - Q_{3sch} = 0 \tag{3.76}$$

$$P_4 - P_{4sch} = 0. \tag{3.77}$$

The second part is pertaining to the KCL for the nonlinear device at each harmonic order, k, as shown in Fig. 3.7.

$$I_{1,k} + I_{2,k} - I_{3,k} = 0, \text{for } k = 2, 3, 4, \ldots, n \tag{3.78}$$

Assume that the nonlinear device can be controlled via voltage control signals, whose amplitudes (V_c) and phase angles (β) can be adjusted to change the injections of the real and reactive powers. Hence, we can write that

$$I_{3,k} = F(\mathbf{V_c}, \beta) \tag{3.79}$$

$$V_{3,k} = G(\mathbf{V_c}, \beta), \tag{3.80}$$

where $\mathbf{V_c} = \left[|V_{c,1}| \;\; |V_{c,2}| \;\; \cdots \;\; |V_{c,n}|\right]^T$, and $\beta = \left[\beta_1 \;\; \beta_2 \;\; \cdots \;\; \beta_n\right]^T$.

Consequently, Newton's formulation can be written as

$$\zeta^{(i+1)} = \zeta^{(i)} - \mathbf{J}^{-1}\mathbf{M}, \tag{3.81}$$

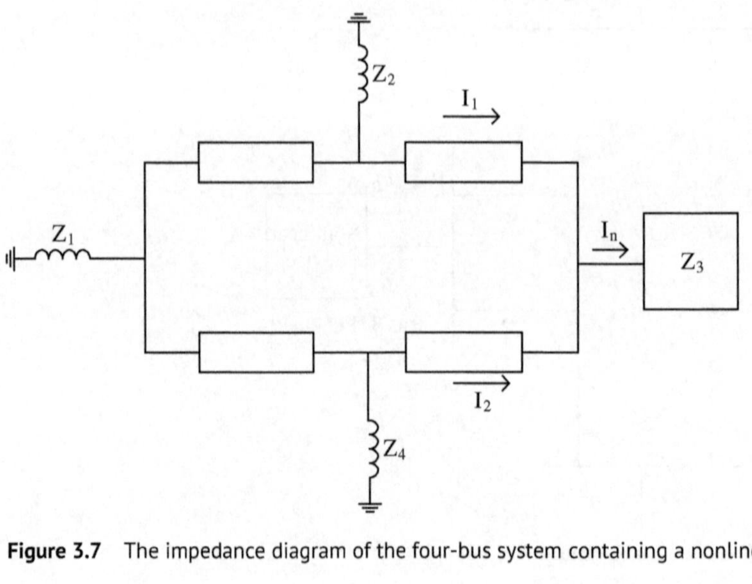

Figure 3.7 The impedance diagram of the four-bus system containing a nonlinear device.

where $\zeta = \begin{bmatrix} |V_{2,1}| \\ \delta_{2,1} \\ |V_{c,1}| \\ \beta_1 \\ \delta_{4,1} \\ |V_{c,2}| \\ \beta_2 \\ \vdots \\ |V_{c,3}| \\ \beta_n \end{bmatrix}$, $\mathbf{M} = \begin{bmatrix} P_2 - P_{2sch} \\ P_3 - P_{3sch} \\ P_4 - P_{4sch} \\ Q_2 - Q_{2sch} \\ Q_3 - Q_{3sch} \\ Re\{I_{1,2} + I_{2,2} - I_{3,2}\} \\ Im\{I_{1,2} + I_{2,2} - I_{3,2}\} \\ \vdots \\ Re\{I_{1,n} + I_{2,n} - I_{3,n}\} \\ Im\{I_{1,n} + I_{2,n} - I_{3,n}\} \end{bmatrix}$, and $\mathbf{J} = \frac{\partial M}{\partial \zeta}$.

Note that depending on the definitions of the power, different Jacobian matrices will result. For example, P and Q may be defined by only the fundamental components of the voltage and current or defined by other advanced power theory discussed in Section 1.7.

3.4.5 Shooting Method

In numerical analysis, the shooting method is a method for solving a boundary value problem by converting it to the solution of an initial value problem. Essentially, we "shoot" out trajectories in different directions until we find a trajectory that has the desired boundary value. The following example illustrates the concept of the shooting method.

The temperature (T) of a 20 m thin rod can be described by (3.82).

$$\frac{d^2T}{dx^2} + 0.05(40 - T) = 0, \tag{3.82}$$

where x denotes the position and $T(x)$ represents the temperature of position x. If T at 0 m and T at 20 m are known to be $T(0) = 30°C$ and $T(20) = 400°C$, respectively, a second-order differential equation with two boundary conditions is imposed.

Alternatively, we can solve the problem by converting (3.82) into an initial-value problem. First, (3.82) can be expressed as two first-order differential equations.

$$y = \frac{dT}{dx} \tag{3.83}$$

$$\frac{dy}{dx} = 0.05(T - 40) \tag{3.84}$$

To solve for (3.83) and (3.84), we need to know the initial conditions for $T(0)$ and $y(0)$. We know $T(0) = 30°C$, but do not know $y(0)$. What should we do? The numerical approach is to simply guess it, and use some iterative technique to update this value until convergence is reached.

The following steps illustrate this:

1. Set $y(0) = 2.5$, we can then use backward Euler's method, trapezoidal rule method, or other higher-order integration methods such as Runge–Kutta method [37] to integrate (3.83) and (3.84), and obtain $T(20) = 91.54 \neq 400$.
2. Set $y(0) = 5$, and obtain $T(20) = 580.9 \neq 400$.
3. Set $y(0) = 4.0759$, and obtain $T(20) = 400 = 400$. The solution is found!

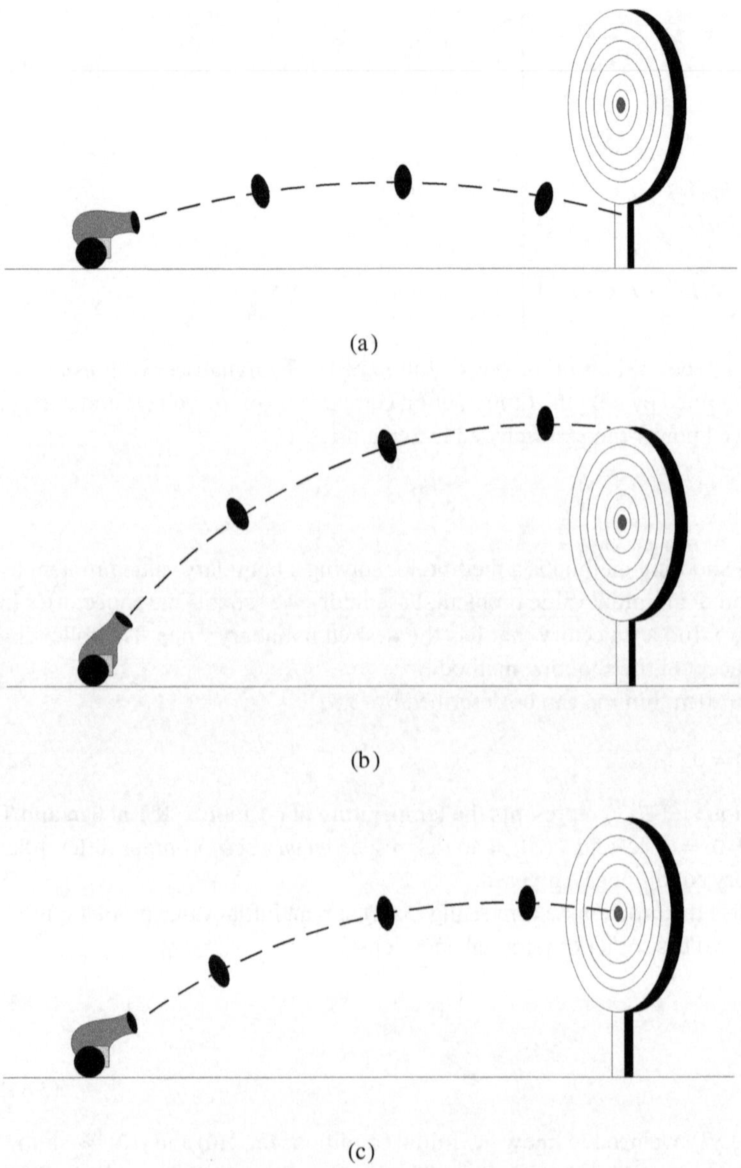

Figure 3.8 (a) undershoot; (b) overshoot; (c) bullseyes.

The above steps are analogues to a cannon shooting at a target. Fig. 3.8 shows how the cannon adjusts its firing angle until it finally hits the target.

In 1972, two USA researchers, Thomas Aprille and Timothy Trick [38], recognized that finding the periodic steady-state solution of a circuit can be formulated as a two-point boundary-value problem in which the solution over one period, T, is required to satisfy the periodic constraint:

$$\mathbf{x}(t + T) = \mathbf{x}(t), \tag{3.85}$$

where t is an arbitrary time point, which is usually set to 0.

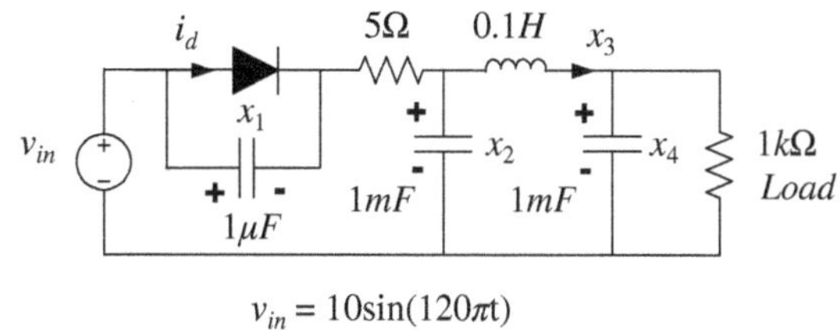

Figure 3.9 The dc power supply circuit extracted. Source: Reprinted with permission from [38].

$\mathbf{x}(0)$ is the initial value and $\mathbf{x}(T)$ is the final value. Consequently, the shooting method can be utilized to obtain the steady-state solution. The problem can be recast as to find the "correct" initial condition sought for the circuit's differential equations such that the solution at the end of one period matches the initial condition.

To illustrate this point, the nonlinear dc power supply example presented in the classic paper of Aprille and Trick [38] is simulated. The schematic diagram of the nonlinear dc power supply is plotted in Fig. 3.9.

Note that the diode is characterized by a nonlinear function, which is stated in (3.86)

$$i_d = 10^{-6}(e^{40x_1} - 1) \tag{3.86}$$

The state-space model for Fig. 3.9 is given as

$$\begin{aligned}
\dot{x}_1 &= \frac{1}{10^{-6}}\left\{\frac{1}{5}[-x_1 - x_2 + 10\sin 60(2\pi)t] - i_d\right\}\\
\dot{x}_2 &= \frac{1}{10^{-3}}\left\{\frac{1}{5}[-x_1 - x_2 + 10\sin 60(2\pi)t] - x_3\right\}\\
\dot{x}_3 &= \frac{1}{0.1}[x_2 - x_4]\\
\dot{x}_4 &= \frac{1}{10^{-3}}[x_3 - \frac{x_4}{1000}]\\
i_d &= 10^{-6}[e^{40x_1} - 1].
\end{aligned} \tag{3.87}$$

To use the shooting method to find out the initial condition, which can immediately lead to steady-state, one can formulate the following Newton's formula:

$$\zeta^{(n+1)} = \zeta^{(n)} - \mathbf{J}^{-1}\mathbf{M}(\zeta^{(n)}), \tag{3.88}$$

where $\zeta = \mathbf{x}(0)$ is the unknown vector, $\mathbf{x}(0) = \begin{bmatrix}x_1(0) & x_2(0) & x_3(0) & x_4(0)\end{bmatrix}^T$, $\mathbf{M} = \mathbf{x}(0) - \mathbf{x}(T)$, and $\mathbf{J} = \frac{\partial \mathbf{M}}{\partial \zeta} = \mathbb{I} - \frac{\partial \mathbf{x}(T)}{\partial \mathbf{x}(0)}$.

$\mathbf{x}(T)$ itself is easy to obtain by using some numerical integration methods. The obtained initial conditions, which can immediately lead to steady-state, are

$$\begin{bmatrix}x_1(0) & x_2(0) & x_3(0) & x_4(0)\end{bmatrix}^T = \begin{bmatrix}-9.0756 & 9.567 & 0.009 & 9.1028\end{bmatrix}^T. \tag{3.89}$$

Fig. 3.10 shows the steady-state waveforms of one period when all these initial conditions are set as (3.89). It can be noted that $\mathbf{x}(0) = \mathbf{x}(T)$ after one period, as shown in the figure.

On the other hand, if the initial conditions are arbitrarily set to zero, i.e.

$$\begin{bmatrix}x_1(0) & x_2(0) & x_3(0) & x_4(0)\end{bmatrix}^T = \begin{bmatrix}0 & 0 & 0 & 0\end{bmatrix}^T, \tag{3.90}$$

then $\mathbf{x}(0) \neq \mathbf{x}(T)$, as shown in Fig. 3.11.

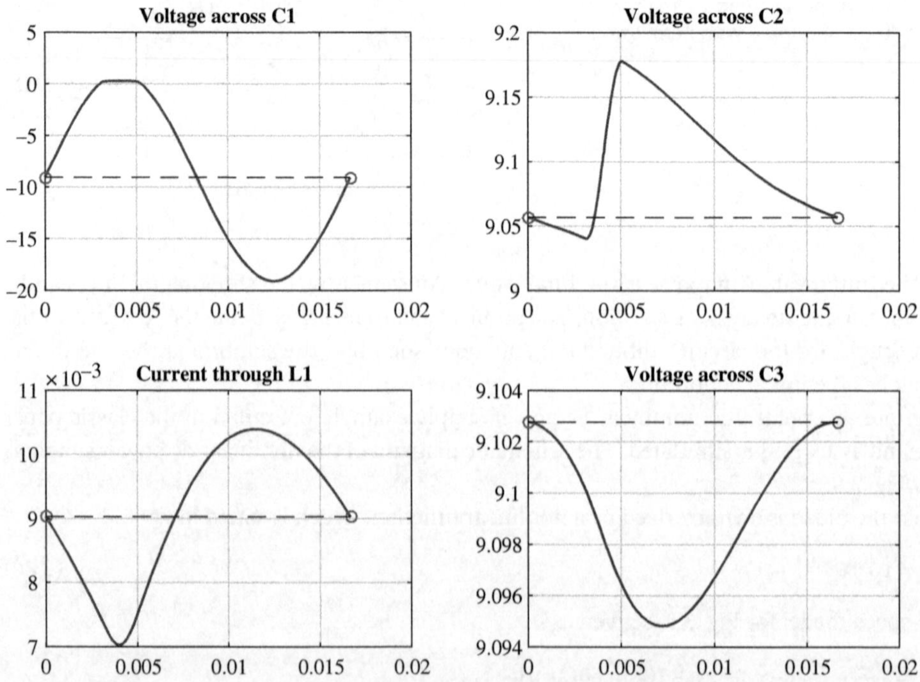

Figure 3.10 Steady-state waveforms of a nonlinear dc-dc power supply.

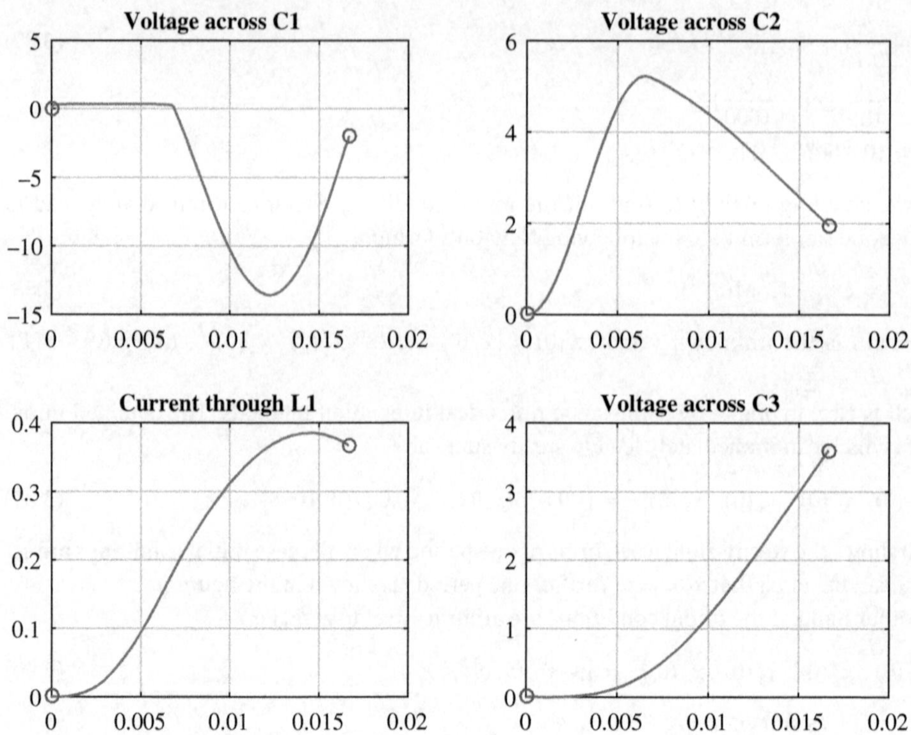

Figure 3.11 Nonperiodic waveforms of a nonlinear dc-dc power supply.

The MATLAB codes for generating waveforms of Fig. 3.10 are listed as follows:

```
x00=[0 0 0 0];
counter=0;
x0=x00;
x_prev=x0;
h1=linspace(0,1/60,500);
h=h1(2);
error=1;
while error >(1e-5)
[t,x]=ode23s('pow_supp_eqn',h1,x0);
G=[-40*exp(40*x(1,1))-2e5 -2e5 0 0;-200 -200 -1000 0;0 10 0 -10;...
    0 0 1000 -1];
phi=inv(eye(4)-h*G);
G_old=G;
stm_old=phi;
for k=2:1:500
    G=[-40*exp(40*x(k,1))-2e5 -2e5 0 0;-200 -200 -1000 0;0 10 0 -10;...
        0 0 1000 -1];
         phi=inv(eye(4)-h*G/2)*(eye(4)+h*G_old/2)*phi;
        G_old=G;
        end

         x_new=x_prev'-inv(eye(4)-phi)*(x_prev'-x(500,:)');

    error=max(abs((x_new-x_prev')));
    x0=x_new';
    x_prev=x_new';
    counter=counter+1;
 end
subplot(221); plot(t,x(:,1),t(1),x(1,1),'ro',t(500),x(500,1),'ro');...
    title('Voltage across C1');hold on;
plot(linspace(t(1),t(500),10),ones(1,10)*x(1,1),'k--');hold off;
grid on;
subplot(222); plot(t,x(:,2),t(1),x(1,2),'ro',t(500),x(500,2),'ro');...
    title('Voltage across C2');grid on;hold on;
plot(linspace(t(1),t(500),10),ones(1,10)*x(1,2),'k--');hold off;
subplot(223); plot(t,x(:,3),t(1),x(1,3),'ro',t(500),x(500,3),'ro');...
    title('Current through L1');grid on;hold on;
plot(linspace(t(1),t(500),10),ones(1,10)*x(1,3),'k--');hold off;
subplot(224); plot(t,x(:,4),t(1),x(1,4),'ro',t(500),x(500,4),'ro');...
    title('Voltage across C3');grid on;hold on;
plot(linspace(t(1),t(500),10),ones(1,10)*x(1,4),'k--');hold off;
```

Note that `pow_supp_eqn` needs to contain the state space equations of (3.87).

3.4.6 Advantages of Newton's Method

The main advantage of Newton's method is its rate of convergence. When the method converges, it does so quadratically. It is better to illustrate the concept of quadratic convergence with a numerical example.

Example 3.3 Find the root of

$$f(x) = x - x^{1/3} - 2 = 0 \tag{3.91}$$

by means of Gauss–Seidel method and Newton's method, and compare their rate of convergence.

To use Gauss–Seidel method, we can rewrite (3.91) as

$$x = x^{1/3} + 2 \tag{3.92}$$

If the initial condition $x^{(0)}$ is set to 3, then the following shows the value at iteration i, $x^{(i)}$.

$x^{(0)} = 3$
$x^{(1)} = 3^{1/3} + 2 = 3.4422495703$
$x^{(2)} = 3.4422495703^{1/3} + 2 = 3.5098974493$
$x^{(3)} = 3.5197243050$
$\vdots$

Thus, the approximate relative error at ith iteration is as follows:

$\varepsilon_a^{(1)} = 12.8477\%$
$\varepsilon_a^{(2)} = 1.9273\%$
$\varepsilon_a^{(3)} = 0.2792\%$
$\vdots$

On the other hand, for Newton's formulation,

$$x^{(i+1)} = x^{(i)} - J^{-1}M \tag{3.93}$$

where $M = x - x^{1/3} - 2$ and $J = \frac{\partial M}{\partial x} = 1 - 1/3x^{-2/3}$.

If the initial condition $x(0)$ is also set to 3, then the following shows the value at iteration i, $x^{(i)}$.

$x^{(0)} = 3$
$x^{(1)} = x^{(0)} = 3 - \frac{3-3^{1/3}-2}{1-\frac{1}{3}(3)^{-2/3}} = 3.52664429313903$
$x^{(2)} = 3.52138014739733$
$x^{(3)} = 3.52137970680457$
$\vdots$

The approximate error at ith iteration for Newton's method is as follows:

$\varepsilon_a^{(1)} = 14.9333\%$
$\varepsilon_a^{(2)} = 0.1495\%$
$\varepsilon_a^{(3)} = 1.2513 \times 10^{-5}\%$
$\vdots$

Table 3.2 summarizes the approximate errors for Gauss–Seidel and Newton's methods. It clearly shows that the error of Newton's method decreases quadratically (square of the error) whereas that of the Gauss–Seidel method decreases linearly. □

Table 3.2 The convergence rate of Gauss–Seidel and Newton's methods.

Gauss–Seidel method		**Newton's method**	
Error at each iteration	Error difference	Error at each iteration	Error difference
$\varepsilon_a^{(1)} = 12.8477\%$	N/A	$\varepsilon_a^{(1)} = 14.9333\%$	N/A
$\varepsilon_a^{(2)} = 1.9273\%$	$\frac{\varepsilon_a^{(2)}}{\varepsilon_a^{(1)}} = 0.15$	$\varepsilon_a^{(1)} = 0.1495\%$	$\frac{\varepsilon_a^{(2)}}{\varepsilon_a^{(1)}} \approx 0.01$
$\varepsilon_a^{(3)} = 0.2792\%$	$\frac{\varepsilon_a^{(3)}}{\varepsilon_a^{(2)}} = 0.0145 \approx 0.015$	$\varepsilon_a^{(1)} = 1.2512 \times 10^{-5}\%$	$\frac{\varepsilon_a^{(3)}}{\varepsilon_a^{(2)}} = 0.84 \times 10^{-4} \approx (0.01)^2$
Linear convergence		Quadratic convergence	

3.4.7 Quasi-Newton Method

The amount of computation required to carry out a Newton–Raphson solution for certain systems is extensive because it requires evaluation and inversion of a very large Jacobian matrix as well as evaluation of the mismatch equations in each iteration.

A Quasi-Newton method was introduced so that the full Jacobian matrix is only calculated once to start the iteration process. Refinement of the Jacobian matrix is achieved by introducing correction terms in the iteration loop. This avoids re-evaluating the partial derivatives and yields the inverse of the Jacobian matrix by simple matrix-vector multiplications. The total arithmetic calculations required is $O(n^2)$, rather than $O(n^3)$ as required by the original Newton's method [37]. Nevertheless, the downside of the Quasi-Newton method is that it is no longer quadratic convergent but rather super-linear convergent, which is better than linear convergent but worse than quadratic convergent. A simple derivation of this method is as follows. The secant approximation for the Jacobian matrix can be stated as:

$$\mathbf{J}^{(i)}(\mathbf{x}^{(i)} - \mathbf{x}^{(i-1)}) = \mathbf{M}^{(i)} - \mathbf{M}^{(i-1)}, \tag{3.94}$$

where $\mathbf{J}^{(i)} = \mathbf{J}(\mathbf{x}^{(i)})$, $\mathbf{M}^{(i)} = \mathbf{M}(\mathbf{x}^{(i)})$, and $\mathbf{M}^{(i-1)} = \mathbf{M}(\mathbf{x}^{(i-1)})$

Newton's method can be written as

$$\mathbf{J}^{(i-1)}(\mathbf{x}^{(i)} - \mathbf{x}^{(i-1)}) = -\mathbf{M}^{(i-1)} \tag{3.95}$$

If we take the difference between (3.94) and (3.95), we will get

$$(\mathbf{J}^{(i)} - \mathbf{J}^{(i-1)})(\mathbf{x}^{(i)} - \mathbf{x}^{(i-1)}) = \mathbf{M}^{(i)} \tag{3.96}$$

Equation (3.96) is undetermined, and one possible solution is

$$\mathbf{J}^{(i)} - \mathbf{J}^{(i-1)} = \frac{\mathbf{M}^{(i)}(\mathbf{x}^{(i)} - \mathbf{x}^{(i-1)})^T}{||\mathbf{x}^{(i)} - \mathbf{x}^{(i-1)}||_2^2}. \tag{3.97}$$

Hence, an iterative form for approximating the Jacobian matrix can be found as

$$\mathbf{J}^{(i)} = \mathbf{J}^{(i-1)} + \frac{\mathbf{M}^{(i)}(\mathbf{x}^{(i)} - \mathbf{x}^{(i-1)})^T}{||\mathbf{x}^{(i)} - \mathbf{x}^{(i-1)}||_2^2}. \tag{3.98}$$

Then, we can apply the Sherman–Morrison formula as stated in (3.99) to find out the inverse of the Jacobian matrix, denoted as $\mathbf{J_{inv}}$ via simple matrix multiplication as stated in (3.100).

$$(\mathbf{A} + \mathbf{u}\mathbf{v}^T)^{-1} = \mathbf{A}^{-1} - \frac{\mathbf{A}^{-1}\mathbf{u}\mathbf{v}^T\mathbf{A}^{-1}}{1 + \mathbf{v}^T\mathbf{A}^{-1}\mathbf{u}}, \tag{3.99}$$

provided that $\mathbf{A} \in R^{n\times n}$ is non-singular and $\mathbf{u}, \mathbf{v} \in R^n$ are nonzero vectors with $\mathbf{v}^T\mathbf{A}^{-1}\mathbf{u} \neq -1$.

$$\mathbf{J_{inv}}^{(i)} = \mathbf{J_{inv}}^{(i-1)} - \frac{\mathbf{J_{inv}}^{(i-1)}\mathbf{M}^i(\mathbf{x}^{(i)} - \mathbf{x}^{(i-1)})^T\mathbf{J_{inv}}^{(i-1)}}{||\mathbf{x}^{(i)} - \mathbf{x}^{(i-1)}||_2^2 + (\mathbf{x}^{(i)} - \mathbf{x}^{(i-1)})^T\mathbf{J_{inv}}^{(i-1)}\mathbf{M}^{(i)}} \tag{3.100}$$

Another possible solution for approximating the Jacobian matrix is stated in (3.101) [39].

$$\mathbf{J}^{(i)} - \mathbf{J}^{(i-1)} = \frac{[\mathbf{M}^{(i)} - \mathbf{M}^{(i-1)} - \mathbf{J}^{(i-1)}(\mathbf{x}^{(i)} - \mathbf{x}^{(i-1)})](\mathbf{x}^{(i)} - \mathbf{x}^{(i-1)})^T}{||\mathbf{x}^{(i)} - \mathbf{x}^{(i-1)}||_2^2} \tag{3.101}$$

Its inverse can also be obtained by (3.99). A pseudo-code is listed below for implementing such a Quasi-Newton method [37].

Step 1: Initialize (i.e. $\zeta^{(0)}$) to obtain the approximate Jacobian and mismatch equations $\mathbf{J}^{(0)}$ and $\mathbf{M}^{(0)}$

Step 2: Set $\mathbf{P}^{(0)} = (\mathbf{J}^{(0)})^{-1}$

Step 3: while *error* > ε

i) Begin
ii) Evaluate $\mathbf{M}^{(k)}$

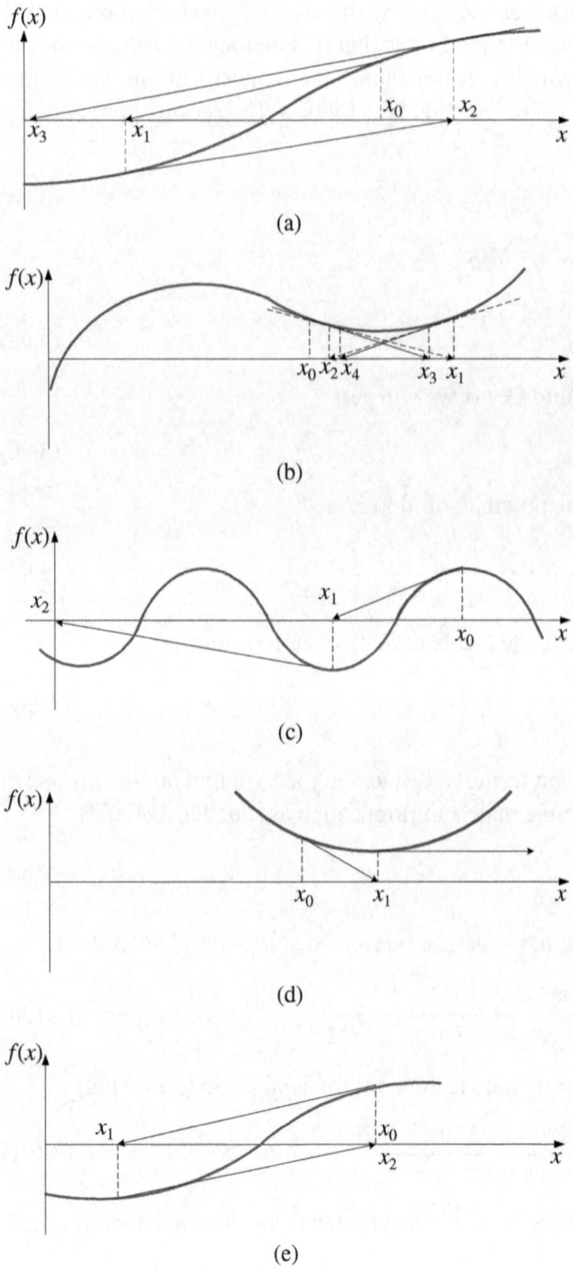

Figure 3.12 Cases where Newton's method exhibits poor convergence.

iii) $\mathbf{P}^{(k)} = \mathbf{P}^{(k-1)} + \frac{(\mathbf{z}_k - \mathbf{P}^{(k-1)}\mathbf{y}_k)\mathbf{z}_k^T\mathbf{P}^{(k-1)}}{\mathbf{z}_k^T\mathbf{P}^{(k-1)}\mathbf{y}_k}$
where $\mathbf{z}_k = \zeta^{(k)} - \zeta^{(k-1)}$, and $\mathbf{y}_k = \mathbf{M}^{(k)} - \mathbf{M}^{(k-1)}$
iv) $\zeta^{(k+1)} = \zeta^{(k)} - \mathbf{P}^{(k)}\mathbf{M}^{(k)}$
v) Set $error = max||\frac{\zeta^{(k+1)} - \zeta^{(k)}}{\zeta^{(\mathbf{k})}}||$
vi) End

3.4.8 Limitation of Newton's Method

Sections 3.4.1 to 3.4.5 are just a portion of numerous applications of Newton's method. Despite its popularity, one needs to be aware of its inherent limitations and problems.

Fig. 3.12(a) depicts the case where an inflection point (that is, $f''(x) = 0$) occurs in the vicinity of a root. Notice that iterations beginning at x_0 progressively diverge from the root. Fig. 3.12(b) shows that a near-zero slope is reached, and the true solution is far away from it. Fig. 3.12(c) shows how an initial guess that is close to one root can jump to a location several roots away. This tendency to move away from the area of interest is because near-zero slopes are encountered. Obviously, a zero slop ($f'(x) = 0$) is truly a disaster as shown in Fig. 3.12(d) because it causes division by zero in Newton's formula. Graphically, it means that the solution shoots off horizontally and never hits the x-axis [40]. Fig. 3.12 (e) illustrates that Newton's method may also suffer from the problem of cycling where it oscillates around a local maximum or minimum.

Generally, there is no general convergence criterion for Newton's method. Its convergence depends on the nature of the function and on the accuracy of the initial guess. The only remedy is to have an initial guess that is sufficiently close to the root. And for some functions, no guess will work. Good guesses are usually predicated on knowledge of the physical problem setting.

3.5 PSO

PSO is a very popular meta-heuristic method, and was developed by Kennedy and Eberhart [41] in 1995, based on the swarm behavior such as fish and birds schooling in nature. It is quite similar to another meta-heuristic method, called genetic algorithm (GA), which was developed in the 1960s by John Holland. Although PSO has many similarities with GA, it is much simpler as it does not use mutation or crossover operations. Instead, it uses real-number randomness and the global communication among the swarm particles [42]. Essentially, the algorithm searches the space of an objective function by adjusting the trajectories of individual agents, called particles.

The PSO algorithm follows the concept presented in Fig. 3.1. (3.102) and (3.103) constitute the PSO algorithm. The term, $v^{(k+1)}$ can be considered as the correction term to the value of $x^{(i)}$ in order to produce $x^{(i+1)}$. Thus, PSO and other iterative algorithms such as Newton's method only differ in the way the correction term is updated. For PSO, the correction term is updated by (3.103).

$$x^{(i+1)} = x^{(i)} + v^{(i+1)}, \tag{3.102}$$

and

$$v^{(i+1)} = wv^{(k)} + c_1r_1(p_{best} - x) + c_2r_2(g_{best} - x), \tag{3.103}$$

where w is the weighting coefficient, g_{best} represents the current global best, and p_{best} represents individual's own best local position in history. Moreover, c_1 and c_2 are the acceleration constants,

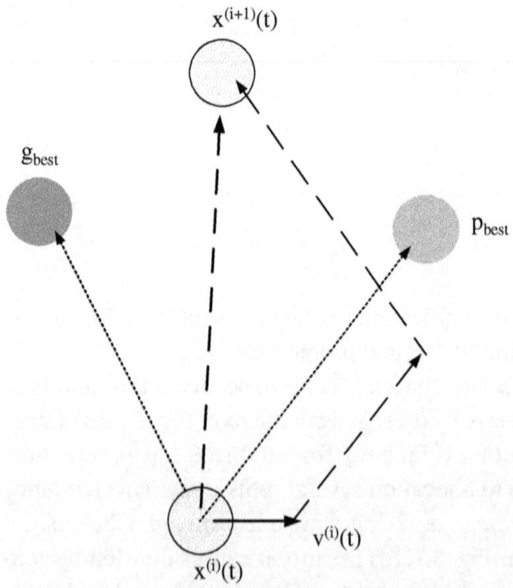

Figure 3.13 Pictorial diagram of the PSO algorithm describing (3.102) and (3.103).

which are typically set to 2. r_1 and r_2 are the random values ranging between 0 and 1. As Fig. 3.13 shows, the particle movement has two major components: stochastic and deterministic. The particle is attracted toward g_{best} and p_{best}, while at the same time it has a tendency to move randomly. The randomness is to ensure that local optimization can be avoided.

A sample of PSO code for carrying out parameter estimation is provided in Chapter 13.

4

Matrix Exponential

"No problem can be solved from the same level of consciousness that created it."

– Albert Einstein

If one chooses to model a power converter in time-domain, one needs to be familiar with basic numerical methods for evaluating integration and matrix exponentials. This chapter is mainly devoted to matrix exponentials. In fact, it has been shown by Van Loan in [43] that the exponential of a certain block triangular matrix can replace the need for computing the integrals.

It is proved in [43] that if a block triangular matrix, $\mathbf{A}$ is defined as

$$\mathbf{A} = \begin{bmatrix} \mathbf{A}_1 & \mathbf{B}_1 & \mathbf{C}_1 & \mathbf{D}_1 \\ \mathbf{0} & \mathbf{A}_2 & \mathbf{B}_2 & \mathbf{C}_2 \\ \mathbf{0} & \mathbf{0} & \mathbf{A}_3 & \mathbf{B}_3 \\ \mathbf{0} & \mathbf{0} & \mathbf{0} & \mathbf{0} \end{bmatrix}, \tag{4.1}$$

then for $t \geq 0$,

$$e^{\mathbf{A}} = \begin{bmatrix} \mathbf{F}_1 & \mathbf{G}_1 & \mathbf{H}_1 & \mathbf{J}_1 \\ \mathbf{0} & \mathbf{F}_2 & \mathbf{G}_2 & \mathbf{H}_2 \\ \mathbf{0} & \mathbf{0} & \mathbf{F}_3 & \mathbf{G}_3 \\ \mathbf{0} & \mathbf{0} & \mathbf{0} & \mathbf{0} \end{bmatrix}. \tag{4.2}$$

where

$$\mathbf{F}_i(t) = e^{\mathbf{A}_i t}, i = 1, 2, 3, 4. \tag{4.3}$$

$$\mathbf{G}_i(t) = \int_0^t e^{\mathbf{A}_i(t-\tau)}\mathbf{B}_i e^{\mathbf{A}_{i+1}\tau} d\tau, i = 1, 2, 3. \tag{4.4}$$

$$\mathbf{H}_i(t) = \int_0^t e^{\mathbf{A}_i(t-\tau)}\mathbf{C}_i e^{\mathbf{A}_{i+2}\tau} d\tau + \int_0^t \int_0^\tau e^{\mathbf{A}_i(t-\tau)}\mathbf{B}_i e^{\mathbf{A}_{i+1}(\tau-r)}\mathbf{B}_{i+1} e^{\mathbf{A}_{i+2}r} dr d\tau, i = 1, 2. \tag{4.5}$$

$$\begin{aligned} \mathbf{J}_1(t) = & \int_0^t e^{\mathbf{A}_1(t-\tau)}\mathbf{D}_1 e^{\mathbf{A}_4\tau} d\tau + \int_0^t \int_0^\tau e^{\mathbf{A}_1(t-\tau)}[\mathbf{C}_1 e^{\mathbf{A}_3(\tau-r)}\mathbf{B}_3 + \mathbf{B}_1 e^{\mathbf{A}_2(\tau-r)}\mathbf{C}_2] e^{\mathbf{A}_4 r} dr d\tau \\ & + \int_0^t \int_0^\tau \int_0^r e^{\mathbf{A}_1(t-\tau)}\mathbf{B}_1 e^{\mathbf{A}_2(\tau-r)}\mathbf{B}_2 e^{\mathbf{A}_3(r-w)}\mathbf{B}_3 e^{\mathbf{A}_4(w)} dw dr d\tau \end{aligned} \tag{4.6}$$

Harmonic Modeling of Voltage Source Converters using Basic Numerical Methods, First Edition.
Ryan Kuo-Lung Lian, Ramadhani Kurniawan Subroto, Victor Andrean, and Bing Hao Lin.

Hence, one can avoid evaluating the convolution integral when obtaining the solution to the state-space equation, $\dot{\mathbf{x}} = \mathbf{Ax} + \mathbf{Bu}$, which is

$$\mathbf{x}(t) = e^{\mathbf{A}t}\mathbf{x}(0) + \int_0^t e^{\mathbf{A}(t-\tau)}\mathbf{Bu}(\tau)d\tau, \tag{4.7}$$

by formulating the state-space equation into an augmented form as stated in (4.8). This can be done by treating the input as one of the system state and evaluating the matrix exponential of the state matrix.

$$\dot{\bar{\mathbf{x}}} = \bar{\mathbf{A}}\bar{\mathbf{x}}, \tag{4.8}$$

where

$$\bar{\mathbf{x}}(t) = \begin{bmatrix} \mathbf{x} \\ \mathbf{u} \end{bmatrix} \tag{4.9}$$

$$\bar{\mathbf{A}} = \begin{bmatrix} \mathbf{A} & \mathbf{B} \\ \mathbf{0} & \mathbf{\Omega} \end{bmatrix}, \tag{4.10}$$

and $\mathbf{\Omega}$ is the state matrix, describing the dynamics of the input. Hence, (4.7) is equivalent to

$$x(t) = \begin{bmatrix} \mathbb{I} & \mathbf{0} \end{bmatrix} e^{\bar{\mathbf{A}}t}\bar{x} \tag{4.11}$$

Sections 4.1–4.2 describes the properties of an exponential matrix and methods on how to evaluate it.

For some differential equations, which cannot be obtained by matrix exponential, numerical integration is the only way to obtain the solution. Some aspects of numerical integration has already been addressed in Chapter 3, and their properties have been briefly discussed. In Section 4.6, another commonly used integration method, called the Runge–Kutta method, will be discussed. This integration method does not involve iteration, yet is stable and accurate, which is suitable for evaluating differential equations whose system matrix is time-varying.

4.1 Definition of Matrix Exponential

For a differential equation,

$$\dot{x} = ax, \tag{4.12}$$

where a is a constant, its solution is $e^{at}x(0)$.

In general, e^a can be defined by the Taylor series and stated as

$$e^a = 1 + a + \frac{a^2}{2!} + \cdots + \frac{a^q}{q!} + \cdots \tag{4.13}$$

Thus, e^{at} is

$$e^{at} = 1 + at + \frac{a^2t^2}{2!} + \cdots + \frac{a^qt^q}{q!} + \cdots \tag{4.14}$$

Following the same reason, the matrix exponential, $e^{\mathbf{A}}$, where $\mathbf{A}$ is an n by n matrix, is defined as

$$e^{\mathbf{A}} = \mathbb{I} + \mathbf{A} + \frac{\mathbf{A}^2}{2!} + \cdots + \frac{\mathbf{A}^q}{q!} + \cdots \tag{4.15}$$

Based on the definitions of (4.15), the following properties of the matrix exponential can be obtained.

- $e^0 = \mathbb{I}$
- $\frac{e^{\mathbf{A}t}}{dt} = \mathbf{A}e^{\mathbf{A}t} = e^{\mathbf{A}t}\mathbf{A}$
- $e^{\mathbf{A}t}e^{\mathbf{A}\tau} = e^{\mathbf{A}(t+\tau)} = e^{\mathbf{A}\tau}e^{\mathbf{A}t}$
- $(e^{\mathbf{A}t})^{-1} = e^{-\mathbf{A}t}$
- $det(e^{\mathbf{A}t}) = e^{tr(\mathbf{A})t}$
- $\lambda(e^{\mathbf{A}t}) = e^{\lambda(\mathbf{A})t}$, where λ is a scalar

Note that one may use (4.15) to evaluate matrix exponential. However, the results are greatly affected by the truncation of the arithmetic and truncations of the series [44]. Thus, there have been several better alternatives in evaluating matrix exponential, which will be discussed in the next section.

4.2 Evaluation of Matrix Exponential

In this section, we will go through some of the numerical methods that, we felt, are important to know or we have used in this book for evaluating matrix exponential. These include inverse Laplace transform, Cauchy–Hamilton method, Padé approximation, scaling and squaring, Krylov subspace methods, and restarted Krylov subspace methods. In fact, matrix exponential is an interesting topic of its own. Readers who are interested in this topic, can refer to the papers [45] and [44] and the book [46] written by Van Loan.

4.2.1 Inverse Laplace Transform

A system of autonomous linear differential equations is represented as

$$\dot{x} = \mathbf{A}\mathbf{x}. \tag{4.16}$$

The Laplace transform of (4.16) is

$$sX(s) - \mathbf{x}(0) = \mathbf{A}X(s). \tag{4.17}$$

Hence,

$$X(s) = (s\mathbb{I} - \mathbf{A})^{-1}\mathbf{x}(0). \tag{4.18}$$

Then, the inverse transformation leads to

$$\mathbf{x}(t) = \mathcal{L}^{-1}\left((s\mathbb{I} - \mathbf{A})^{-1}\right)\mathbf{x}(0), \tag{4.19}$$

where $\mathcal{L}^{-1}(\cdot)$ represents inverse Laplace transformation.

Example 4.1 Find the solution of the following homogeneous differential equation

$$\dot{x} = \mathbf{A}\mathbf{x}, \tag{4.20}$$

where $\mathbf{A} = \begin{bmatrix} 0 & 1 \\ -1 & 0 \end{bmatrix}$.

$$s\mathbb{I} - \mathbf{A} = \begin{bmatrix} s & -1 \\ 1 & s \end{bmatrix}. \tag{4.21}$$

Hence,

$$(s\mathbb{I} - \mathbf{A})^{-1} = \begin{bmatrix} \frac{s}{s^2+1} & \frac{1}{s^2+1} \\ \frac{-1}{s^2+1} & \frac{s}{s^2+1} \end{bmatrix} \tag{4.22}$$

The state transition matrix, $\mathbf{\Phi}(t) = e^{\mathbf{A}t}$ is obtained as follows

$$\mathbf{\Phi}(t) = \mathcal{L}^{-1}\left((s\mathbb{I} - \mathbf{A})^{-1}\right) = \begin{bmatrix} \cos t & \sin t \\ -\sin t & \cos t \end{bmatrix}. \tag{4.23}$$

Consequently,

$$\mathbf{x}(t) = \begin{bmatrix} \cos t & \sin t \\ -\sin t & \cos t \end{bmatrix} \mathbf{x}(0). \tag{4.24}$$

Note that (4.24) is a general form of a simple harmonic oscillator with radian frequency of 1 rad/s. □

4.2.2 Cayley–Hamilton Method

Consider a square matrix **A** with dimension n by n and a characteristic polynomial

$$\Delta(s) = |s\mathbb{I} - \mathbf{A}| = s^n + c_{n-1}s^{n-1} + \cdots + c_0, \tag{4.25}$$

where $c_{n-1}, \ldots, c_0$ are constants, corresponding to coefficients of $s^{n-1}, \ldots, s^0$, respectively, and define a corresponding matrix polynomial, formed by substituting **A** for s above

$$\Delta(\mathbf{A}) = \mathbf{A}^n + c_{n-1}\mathbf{A}^{n-1} + \cdots + c_0\mathbb{I}. \tag{4.26}$$

The Cayley–Hamilton theorem states that every matrix satisfies its own characteristic equation, that is

$$\Delta(\mathbf{A}) = 0 \tag{4.27}$$

This can be illustrated by the following numerical example.

Example 4.2 The matrix $\mathbf{A} = \begin{bmatrix} 1 & 1 \\ 0 & 1 \end{bmatrix}$ has characteristic equation

$$\lambda^2 - 2\lambda + 1 = 0. \tag{4.28}$$

The Cayley–Hamilton theorem implies that

$$\mathbf{A}^2 - 2\mathbf{A} + \mathbb{I} = 0. \tag{4.29}$$

To verify this, we have

$$\mathbf{A}^2 - 2\mathbf{A} + \mathbb{I} = \begin{bmatrix} 1 & 2 \\ 0 & 1 \end{bmatrix} - 2\begin{bmatrix} 1 & 1 \\ 0 & 1 \end{bmatrix} + \begin{bmatrix} 1 & 0 \\ 0 & 1 \end{bmatrix} = \begin{bmatrix} 0 & 0 \\ 0 & 0 \end{bmatrix}. \tag{4.30}$$

To see how it can help to solve the matrix exponential, the following can be observed. For a 2×2 matrix **A**, the characteristic equation has the form of (4.31)

$$\lambda^2 + b\lambda + c = 0, \tag{4.31}$$

By the Cayley–Hamilton theorem,

$$\mathbf{A}^2 + b\mathbf{A} + c\mathbb{I} = 0. \tag{4.32}$$

Hence,

$$\mathbf{A}^2 = -b\mathbf{A} - c\mathbb{I}. \tag{4.33}$$

This allows us to replace the $\mathbf{A}^2$ term in the series for $e^{\mathbf{A}}$ with a linear combination of $\mathbf{A}$ and $\mathbb{I}$.

Similarly, we have that

$$\mathbf{A}^3 = (\mathbf{A}^2)\mathbf{A} = (-b\mathbf{A} - c\mathbb{I})\mathbf{A} = -b\mathbf{A}^2 - c\mathbf{A} = -b(-b\mathbf{A} - c\mathbb{I}) - c\mathbf{A} = (b^2 - c)\mathbf{A} - bc\mathbb{I}, \tag{4.34}$$

$\mathbf{A}^3$ can also be written as a linear combination of $\mathbf{A}$ and $\mathbb{I}$. This argument continues for higher powers of $\mathbf{A}$, so that eventually we find that we may write $e^{\mathbf{A}}$ as a linear combination of $\mathbf{A}$ and $\mathbb{I}$, rather than as an infinite linear combination of $\mathbb{I}$ and all positive powers of $\mathbf{A}$. In other words, the expression simplifies to

$$e^{\mathbf{A}} = \mathbb{I} + \mathbf{A} + \frac{1}{2}\mathbf{A}^2 + \frac{1}{6}\mathbf{A}^3 + \cdots = a_1\mathbf{A} + a_0\mathbb{I}, \tag{4.35}$$

for some constants a_1 and a_0. Thus, to find the matrix $e^{\mathbf{A}}$, we just need to find these two constants. Note that for an $n \times n$ matrix, there will be n constants. To find these constants, we can use eigenvalues. □

Suppose λ is an eigenvalue for $\mathbf{A}$, and λ also satisfies the characteristic equation for $\mathbf{A}$. Thus, we can write e^{λ} as a linear combination of λ and 1. Moreover, it will be exactly the same linear combination. Hence, we also have

$$e^{\lambda} = a_1\lambda + a_0. \tag{4.36}$$

This is the sort of equation we can solve. Thus, there are three cases to be considered (for a 2×2 matrix).

- **Case 1:** $\mathbf{A}$ has two distinct real eigenvalues, λ_1 and λ_2. Then, we have two equations

$$\begin{aligned} e^{\lambda_1} &= a_1\lambda_1 + a_0 \\ e^{\lambda_2} &= a_1\lambda_2 + a_0 \end{aligned} \tag{4.37}$$

This is a linear system of two equations in two unknowns, so we can easily solve for a_1 and a_0, and then set $e^{\mathbf{A}} = a_1\mathbf{A} + a_0\mathbb{I}$.

Example 4.3 The matrix $\mathbf{A} = \begin{bmatrix} -5 & 2 \\ -6 & 2 \end{bmatrix}$ has eigenvalues $\lambda_1 = -1$ and $\lambda_2 = -2$. Hence, we have

$$\begin{cases} -a_1 + a_0 = e^{-1} \\ -2a_1 + a_0 = e^{-2} \end{cases} \tag{4.38}$$

The solution is

$$\begin{aligned} a_1 &= e^{-1} - e^{-2}, \\ a_0 &= 2e^{-1} - e^{-2} \end{aligned} \tag{4.39}$$

It follows that

$$\begin{aligned} e^{\mathbf{A}} &= a_1\mathbf{A} + a_0\mathbb{I} \\ &= (e^{-1} - e^{-2})\begin{bmatrix} -5 & 2 \\ -6 & 2 \end{bmatrix} + (2e^{-1} - e^{-2})\begin{bmatrix} 1 & 0 \\ 0 & 1 \end{bmatrix} \\ &= \begin{bmatrix} -3e^{-1} + 4e^{-2} & 2e^{-1} - 2e^{-2} \\ 6e^{-2} - 6e^{-1} & 4e^{-1} - 3e^{-2} \end{bmatrix} \end{aligned} \tag{4.40}$$

□

- **Case 2: A** has a repeated eigenvalue λ. We have two unknowns, but we only have one equation. To get a second equation, we differentiate (4.36) with respect to λ. Thus, our two equations are

$$\begin{aligned} e^{\lambda} &= a_1\lambda + a_0 \\ e^{\lambda} &= a_1 \end{aligned} \tag{4.41}$$

Example 4.4 Continued from **Example** 4.2, we get

$$\begin{aligned} e^{1} &= a_1 + a_0, \\ e^{1} &= a_1 \end{aligned} \tag{4.42}$$

Thus,

$$e^{\mathbf{A}} = e\begin{bmatrix} 1 & 1 \\ 0 & 1 \end{bmatrix}. \tag{4.43}$$

□

- **Case 3: A** has complex eigenvalues. These are necessarily conjugates of one another, and therefore not equal. This gives two equations in two (complex) unknowns. Setting real and imaginary parts equal gives four equations in four unknowns, which can easily be solved.

4.2.3 Padé Approximation

Padé approximation is an extension of the concept of Taylor approximation with rational functions rather than polynomials.

If the rational function is given as follows:

$$R_{M,N}(x) = \frac{\sum_{k=0}^{M} a_k x^k}{1 + \sum_{k=0}^{N} b_k x^k}, \tag{4.44}$$

it is called the Padé of type (M, N). We can then try to match the Padé approximate with truncated Taylor series,

$$f(x) = \sum_{k=0}^{M+N} c_k x^k \tag{4.45}$$

If we consider only the special case of a diagonal rational approximation, $M = N = n$ [47], then

$$c_0 + c_1 x + \cdots + c_{2n} x^{2n} = \frac{a_0 + a_1 x + \cdots + a_n x^n}{1 + b_1 x + \cdots + b_n x^n}. \tag{4.46}$$

$$(c_0 + c_1 x + \cdots + c_{2n} x^{2n})(1 + b_1 x + \cdots + b_n x^n) = a_0 + a_1 x + \cdots + a_n x^n \tag{4.47}$$

Matching coefficients, we have

$$\begin{bmatrix} b_1 \\ \vdots \\ b_n \end{bmatrix} = \begin{bmatrix} c_n & \cdots & c_1 \\ \vdots & \ddots & \vdots \\ c_{2n-1} & \cdots & c_n \end{bmatrix}^{-1} \begin{bmatrix} -c_{n+1} \\ \vdots \\ -c_{2n} \end{bmatrix}, \tag{4.48}$$

and

$$\begin{bmatrix} a_0 \\ a_1 \\ \vdots \\ a_n \end{bmatrix} = \begin{bmatrix} c_0 & & & 0 \\ c_1 & c_0 & & \\ \vdots & \vdots & \ddots & \\ c_n & c_{n-1} & \cdots & c_0 \end{bmatrix} \begin{bmatrix} 1 \\ b_1 \\ \vdots \\ b_n \end{bmatrix} \tag{4.49}$$

Example 4.4 An exponential function, $f(x) = e^x$. Find the Padé approximation for

(a) $n = 1$

Let us consider the Taylor series defined in (4.13) as

$$e^x = 1 + x + \frac{x^2}{2} + \ldots + \frac{x^q}{q!}. \tag{4.50}$$

From (4.50), one may see that $c_1 = 1$ and $c_2 = \frac{1}{2}$. Hence, using (4.48), $b_1 = \frac{-c_2}{c_1} = -\frac{1}{2}$. Consequently,

$$\begin{bmatrix} a_0 \\ a_1 \end{bmatrix} = \begin{bmatrix} 1 & 0 \\ 1 & 1 \end{bmatrix} \begin{bmatrix} 1 \\ -\frac{1}{2} \end{bmatrix} = \begin{bmatrix} 1 \\ \frac{1}{2} \end{bmatrix}. \tag{4.51}$$

Thus,

$$e^x \approx \frac{1 + \frac{x}{2}}{1 - \frac{x}{2}} \tag{4.52}$$

(b) $n = 2$

From (4.50), one may obtain $c_3 = \frac{1}{6}$ and $c_4 = \frac{1}{24}$. Hence, b_1 and b_2 can be yielded as

$$\begin{bmatrix} b_1 \\ b_2 \end{bmatrix} = \begin{bmatrix} \frac{1}{2} & 1 \\ \frac{1}{6} & \frac{1}{2} \end{bmatrix}^{-1} \begin{bmatrix} -\frac{1}{6} \\ -\frac{1}{24} \end{bmatrix} = \begin{bmatrix} -\frac{1}{2} \\ \frac{1}{12} \end{bmatrix} \tag{4.53}$$

Consequently, a_1, a_2, and a_3 can be obtained as

$$\begin{bmatrix} a_0 \\ a_1 \\ a_2 \end{bmatrix} = \begin{bmatrix} 1 & 0 & 0 \\ 1 & 1 & 0 \\ \frac{1}{2} & 1 & 1 \end{bmatrix} \begin{bmatrix} 1 \\ -\frac{1}{2} \\ \frac{1}{12} \end{bmatrix} = \begin{bmatrix} 1 \\ \frac{1}{2} \\ \frac{1}{12} \end{bmatrix}. \tag{4.54}$$

Thus,

$$e^x \approx \frac{1 + \frac{x}{2} + \frac{x^2}{12}}{1 - \frac{x}{2} + \frac{x^2}{12}}. \tag{4.55}$$

Similar to the Taylor series, we can also extend the scalar exponential to matrix exponential. The (p, q) Padé approximation to $e^{\mathbf{A}}$ is defined by (4.56) [45].

$$R_{pq}(\mathbf{A}) = [\mathbf{D}_{pq}(\mathbf{A})]^{-1}\mathbf{N}_{pq}(\mathbf{A}), \tag{4.56}$$

where

$$\mathbf{N}_{pq}(\mathbf{A}) = \sum_{j=0}^{p} \frac{(p+q-j)!p!}{(p+q)!j!(p-j)!}\mathbf{A}^j \tag{4.57}$$

and

$$\mathbf{D}_{pq}(\mathbf{A}) = \sum_{j=0}^{q} \frac{(p+q-j)!p!}{(p+q)!j!(q-j)!}(-\mathbf{A})^j. \tag{4.58}$$

Thus, the (2,2) Padé approximation to $e^{\mathbf{A}}$ is

$$e^{\mathbf{A}} \approx \left[\mathbb{I} - \frac{\mathbf{A}}{2} + \frac{\mathbf{A}^2}{12}\right]^{-1} \left[\mathbb{I} + \frac{\mathbf{A}}{2} + \frac{\mathbf{A}^2}{12}\right] \tag{4.59}$$

Non-singularity of $\mathbf{D}_{pq}(\mathbf{A})$ is assured if p and q are large enough or if the eigenvalues of $\mathbf{A}$ are negative [45]. Similar to the Taylor series method, roundoff error also makes Padé approximation unreliable. Generally, Padé approximation is used when $||\mathbf{A}||$ is not too large, and according to [45], the diagonal approximants ($p = q$) are preferred over the off-diagonal approximants ($p \neq q$) because the former takes less computation efforts and is more accurate than the latter.

4.2.4 Scaling and Squaring

The round-off error difficulties and the computing costs of the Taylor and Padé approximation increase when $||\mathbf{A}||$ is large. The scaling and squaring method is a widely used algorithm to compute the exponential matrix. In fact, MATLAB's built-in function `expm` uses a scaling and squaring method together with the Padé approximation. It is employed to alleviate the round-off error difficulties and high computation costs of evaluating Taylor and Padé approximation for matrix with large norm [44].

The key of this method is based on the following property:

$$e^{\mathbf{A}} = \left(e^{\mathbf{A}/m}\right)^m \tag{4.60}$$

where m is a positive integer value, whose value is normally chosen to be a power of two, for which $e^{\mathbf{A}/m}$ can be reliably and efficiently computed. Normally, we will keep the value of $||\mathbf{A}|| \,/m \leq 1$ so that it can be satisfactorily computed by the Taylor or Padé approximations. For instance, $e^{\mathbf{A}}$ can be written as $e^{(\mathbf{A}/2^j)^{2^j}}$, where j is a positive integer value. $e^{(\mathbf{A}/2^j)}$ itself can be evaluated by diagonal Padé approximation with q order (i.e. $\mathbf{R}_{qq}$). A method called inverse error analysis [44] can be used to determine the value of q and j.

4.3 Krylov Subspace Method

In many applications, one may not need to solve the exponential of the entire size of a matrix $\mathbf{A}$ $\left(e^{\mathbf{A}}\right)$, yet only the product of $e^{\mathbf{A}}\mathbf{v}$ is desired. This holds true when obtaining the solution of a homogeneous differential equation like (4.16). Nevertheless, $\mathbf{A}$ may be a large and sparse matrix, but e^{A} is typically dense. Thus, it is desired to avoid computing this whole size of $\mathbf{A}$. Krylov subspace method (KSM) is known as a powerful method to deal with multiplication of exponential matrix with a vector for large size and sparse matrix. The KSM has been developed in the early 1980s for iteratively solving the linear problem $\mathbf{Av} = \mathbf{b}$ for a large and sparse $\mathbf{A}$ matrix [48].

The approximated solutions obtained from KSM are spanned by the vectors $\{\mathbf{v}, \mathbf{Av}, \mathbf{A}^2\mathbf{v}, \dots, \mathbf{A}^{m-1}\mathbf{v}\}$, for m is typically smaller than the size of $\mathbf{A}$. To construct the spanned vectors, two algorithms can be used, which typically depends on the symmetry of the matrix. The Lanczos method is used if $\mathbf{A}$ is symmetric. On the other hand, if $\mathbf{A}$ is non-symmetric, Arnoldi iteration is employed. As the power converter model is generally composed by non-symmetric matrix (which will be further discussed in Chapter 5), only Arnoldi iteration is described in this book.

The products of Arnoldi iteration are a Hessenberg matrix $\left(\mathbf{H}_m\right)$ and an orthonormal basis matrix $\left(\mathbf{V}_m\right)$. The Hessenberg matrix is an upper triangular matrix with size $m \times m$, which typically represents the projection of linear transformation $\mathbf{A}$ onto the subspace $\mathbf{K}_m$, with respect to the basis $\mathbf{V}_m$. The subspace $\mathbf{K}_m$ refers to the Krylov subspace as stated in (4.61).

$$\mathbf{K}_m = span\left\{\mathbf{v}, \mathbf{Av}, \mathbf{A}^2\mathbf{v}, \cdots, \mathbf{A}^{m-1}\mathbf{v}\right\} \tag{4.61}$$

On the other hand, $\mathbf{V}_m$ defines an $n \times m$ orthonormal basis matrix expressed in (4.62)

$$\mathbf{V}_m = \begin{bmatrix}\mathbf{v}_1 & \mathbf{v}_2 & \cdots & \mathbf{v}_m\end{bmatrix}_{n\times m}. \tag{4.62}$$

Figure 4.1 Arnoldi algorithm.

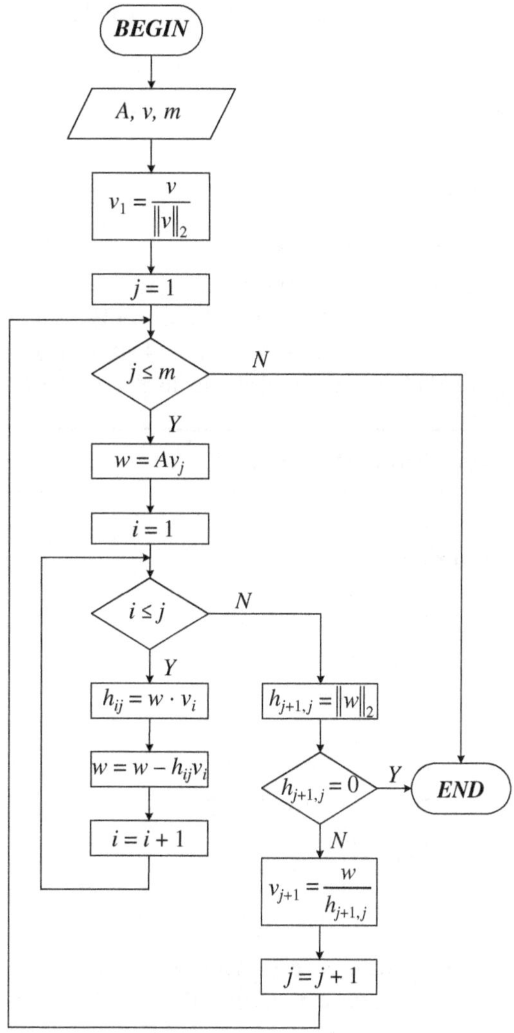

Note that the first basis of $\mathbf{V}_m$ is obtained from $\mathbf{v}_1 = \frac{\mathbf{v}}{\|\mathbf{v}\|_2}$, while $\mathbf{v}_2, \dots, \mathbf{v}_m$ are yielded from Arnoldi iteration shown in Algorithm 1, and the flow chart of Fig. 4.1.

Consequently, due to the orthonormal merit of $\mathbf{V}_m$, the Hessenberg matrix is equivalent with

$$\mathbf{H}_m = \mathbf{V}_m^T \mathbf{A} \mathbf{V}_m \tag{4.63}$$

Since $\mathbf{V}_m^T \mathbf{V}_m = \mathbb{I}$, the multiplication of exponential matrix with a vector can be extended into

$$e^{\mathbf{A}}\mathbf{v} = \mathbf{V}_m^T \mathbf{V}_m e^{\mathbf{A}}\mathbf{v}. \tag{4.64}$$

Moreover, since $\mathbf{v}_1 = \frac{\mathbf{v}}{\|\mathbf{v}\|_2}$ and $\mathbf{v}_j^T \mathbf{v} = 0$ for any $j > 1$,

$$\mathbf{V}_m^T \mathbf{v} = \| \mathbf{v}\|_2 \mathbf{e}_1, \tag{4.65}$$

where $\mathbf{e}_1$ is the first column of $\mathbb{I}$. Defining the optimal Krylov approximation to $y(t) = e^{\mathbf{A}t}\mathbf{v}$ as

$$\mathbf{y}_{opt} = \mathbf{V}_m \mathbf{w}_{opt}, \tag{4.66}$$

there exists a value, τ such that $\mathbf{w}_{opt}$ can be approximated as stated in (4.67)

$$\mathbf{w}_{opt} \approx (\mathbf{V}_m^T\mathbf{V}_m)^{-1}\mathbf{V}_m^T y(\tau) = \mathbf{V}_m^T e^{\mathbf{A}\tau}\mathbf{v}. \tag{4.67}$$

Substituting (4.65) into (4.67), one gets

$$\mathbf{w}_{opt} \approx \| \mathbf{v}\|_2 \mathbf{V}_m^T e^{\mathbf{A}\tau}\mathbf{V}_m e_1. \tag{4.68}$$

Consequently,

$$e^{\mathbf{A}t}\mathbf{v} \approx \| \mathbf{v}\|_2 \mathbf{V}_m e^{\mathbf{H}_m t}\mathbf{e}_1, \tag{4.69}$$

where $\mathbf{H}_m$ is a Hessenberg matrix, which is almost a triangular matrix with zero entries below the first subdiagonal. Moreover, the size of $\mathbf{H}_m$ is $m \times m$, where m is much smaller than n, which is the size of $\mathbf{A}$. Thus, the advantage of (4.69) is that the large sparse $e^{\mathbf{A}t}$ problem is replaced by a small dense $e^{\mathbf{H}_m t}$ problem, which effectively reduces the computation time.

Algorithm 1 Arnoldi algorithm

1: $v_1 = {}^{v}/_{\|v\|_2}$
2: **for** $j = 1, 2, \dots, m$ **do**
3: $\quad w = Av_j$
4: $\quad$ **for** $i = 1, 2, \dots, j$ **do**
5: $\quad\quad h_{ij} = wv_i$
6: $\quad\quad w = w - h_{ij}v_i$
7: $\quad$ **end for**
8: $\quad h_{j+1,j} = \|w\|_2$
9: $\quad$ **if** $h_{j+1,j} = 0$ **then stop**
10: $\quad v_{j+1} = {}^{w}/_{h_{j+1,j}}$
11: **end for**

The following code is a sample code for KSM. The inputs of this function are `A`, `v`, and `t`, which represent the system matrix, initial condition vector, and time, respectively. `y` is the output, which is the product of KSM. The size of Hessenberg matrix is user-defined, so in the sample code below, we define it as `m`, whose value is 30 or `n` if it is less than 30.

```
function y = expv(t,A,v)
%% Initialization
[n,n] = size(A);
m = min(n,30);
t = abs(t);
beta = norm(v);
y = v;
Vm = zeros(n,m+1);
H = zeros(m+2,m+2);

Vm(:,1) = (1/beta)*y;
%% Arnoldi process
for j = 1:m
    p = A*Vm(:,j);
    for i = 1:j
        H(i,j) = Vm(:,i)'*p;        % Hessenberg matrix
        p = p-H(i,j)*Vm(:,i);
    end
    s = norm(p);
```

```
        H(j+1,j) = s;
        Vm(:,j+1) = (1/s)*p;
end

H(m+2,m+1)=1;
F=expm(t*H(1:m,1:m));
y=Vm(:,1:m)*(beta*F(1:m,1));
```

4.4 Krylov Space Method with Restarting

Based on the work of [49], the speed of the convergence depends on the norm of $\mathbf{A}\tau$. Since $\mathbf{V}_m e^{\mathbf{H}_m \tau} \mathbf{e}_1$ is a polynomial approximation, it is more accurate when the norm of $\mathbf{H}_m \tau$ is not too large. Nevertheless, if τ is chosen too small, the computation time may be long. Thus, it is important to find the appropriate value of τ to speed up the convergence.

The obvious way to monitor the error resulted by KSM is by calculating its residual. Botchev, *et. al.* [50] has proposed the method to check the KSM residual by calculating the difference of the output from KSM and the exact solution of ordinary differential equation, as stated in

$$\mathbf{r}_m(t) = \dot{\mathbf{y}}_{opt} - \mathbf{A}\mathbf{y} \tag{4.70}$$

Consequently, the residual can also be expressed as

$$\mathbf{r}_m(t) = h_{m+1,m} \mathbf{e}_m^{\mathbf{T}} e^{t\mathbf{H}_m} \left(\beta \mathbf{e}_1\right) \tag{4.71}$$

where $\mathbf{e}_m = \begin{bmatrix} 0 & \dots & 0 & 1 \end{bmatrix}_{1\times m}^T$.

The procedure of solving the exponential matrix via KSM with restarting technique based on [50] can be summarized as follows:

1. The size of Hessenberg matrix is determined by the residual value. The user only needs to specify the maximum size of Hessenberg matrix.
2. The mechanism of computing the residual is performed at every iteration in the Arnoldi iteration, described in the previous section.
3. The calculation of residual is done by dividing the time interval into six regions in particular and sorting the maximum value of each residual. If the maximum value of the residual is below

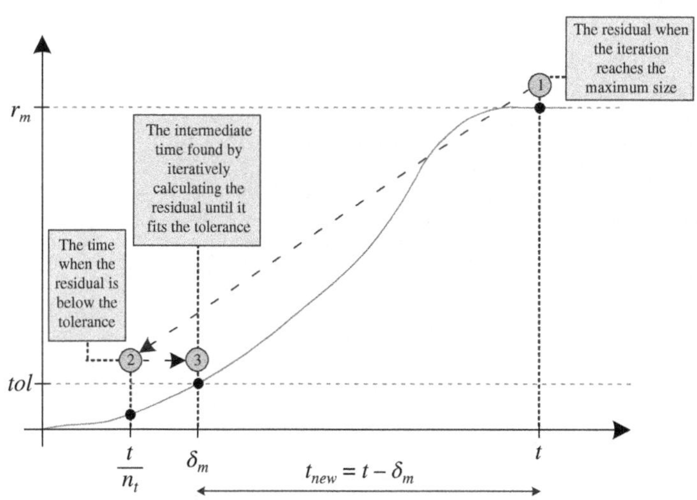

Figure 4.2 The mechanism of finding intermediate time for Krylov subspace method with restarting technique.

the specified tolerance, the Arnoldi iteration stops and the Hessenberg matrix $(\mathbf{H}_m)$ and the orthonormal basis matrix $(\mathbf{V}_m)$ result.

4. If the maximum size of the Hessenberg matrix is reached, but the maximum residual value is still greater than the specified tolerance, then the restarting technique is performed.
5. Restarting technique is taken by reducing the time interval, such that with the maximum size of Hessenberg matrix, the residual can be kept below the specified tolerance. Hence, the output of the restarting technique here is restarting time interval δ_m required to yield the sufficient residual, which is less than the specified tolerance. Fig. 4.2 illustrates the principle of the KSM with restarting technique.

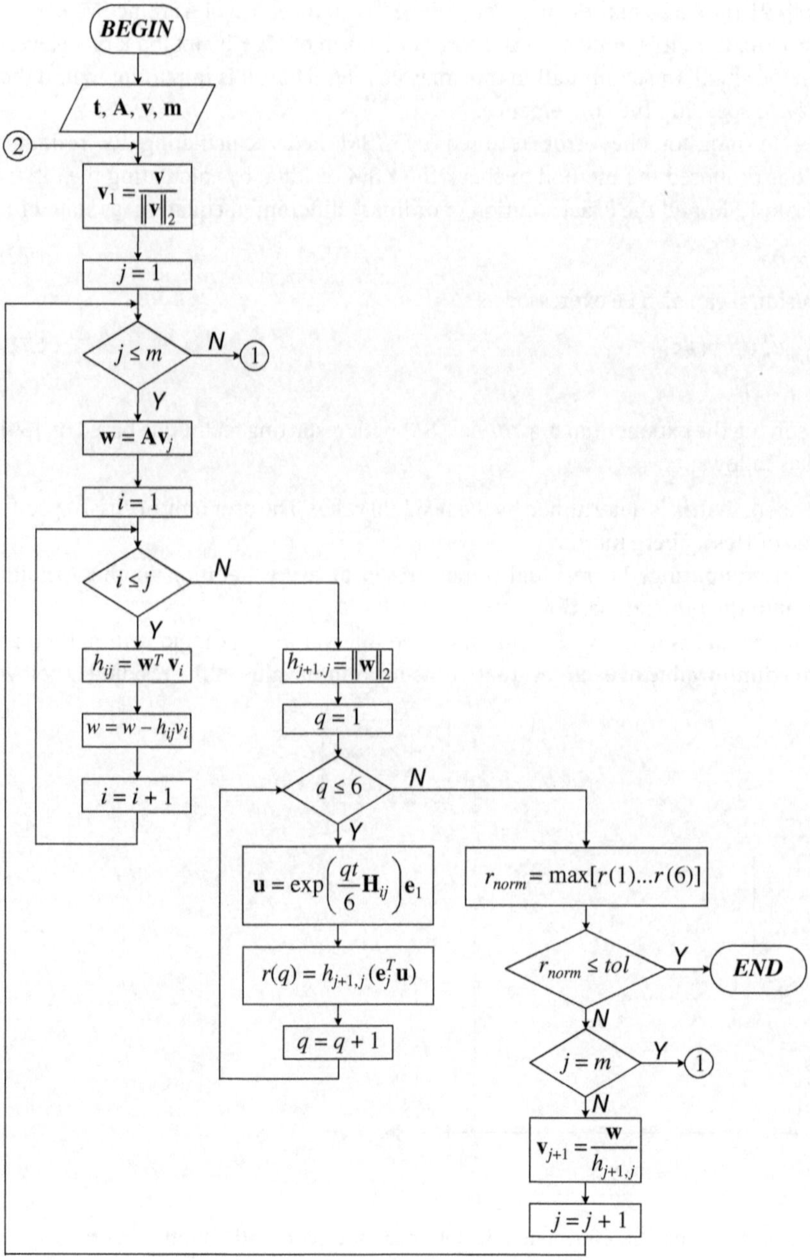

Figure 4.3 The flowchart of Krylov subspace method with restarting technique.

6. Once δ_m is found by the restarting technique, the iteration for computing the exponential matrix will be continued and the parameters are updated to $t_{new} = t - \delta_m$ and $\mathbf{v}_{new}$, whose value is yielded from $\mathbf{v}_{new} = \|\mathbf{v}\|\,\mathbf{V}_m e^{\delta_m \mathbf{H}_m}\mathbf{e}_1$.
7. Re-compute the exponential matrix with the input t_{new}, $\mathbf{v}_{new}$, and $\mathbf{A}$ via KSM with restarting until δ_m is not produced.

The flowcharts of KSM with restarting technique are depicted in Figs. 4.3 and 4.4.

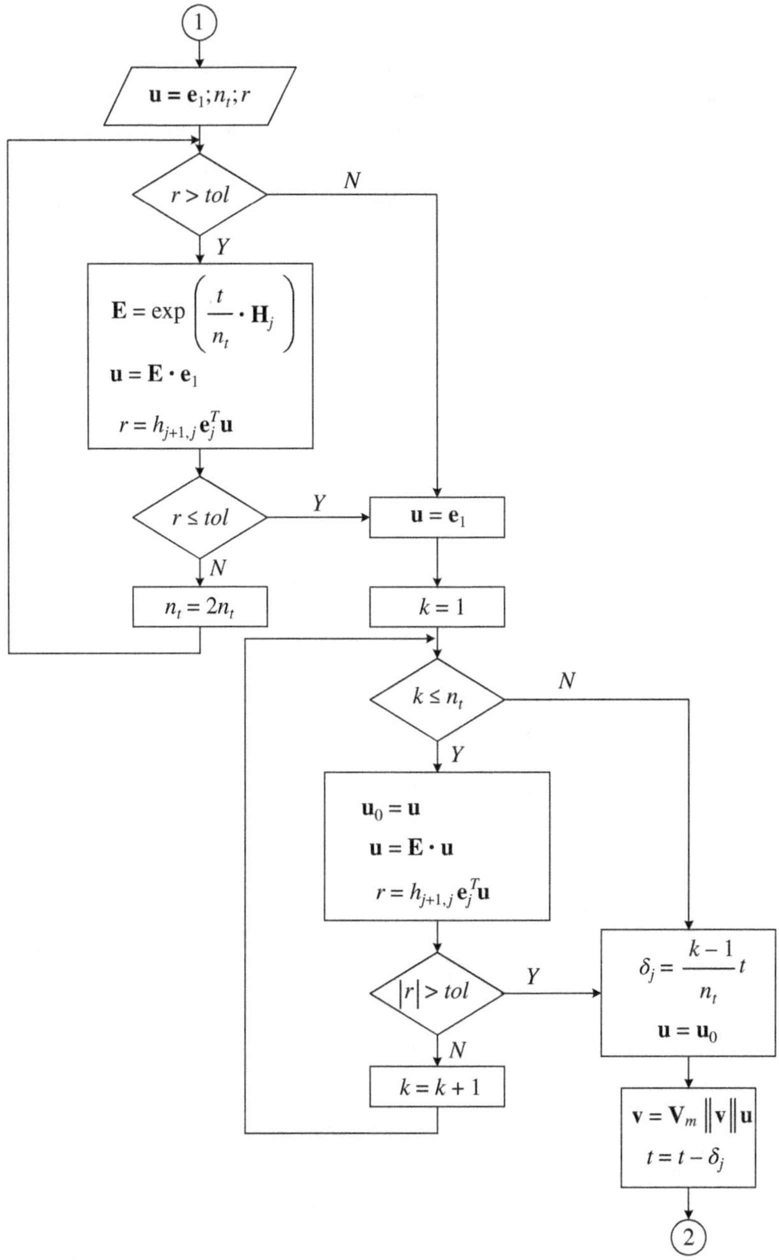

Figure 4.4 The flowchart of Krylov subspace method with restarting technique (*cont'*).

4.5 Application of Augmented Matrix on DC-DC Converters

In this section, we will show how we can use the homogeneous state-space equation of (4.8) to obtain the dynamic model of a dc-dc converter. Large-signal modeling will be discussed in later chapters. The small-signal model (SSM) is the essential tool for controller design. The most well-known SSMs are based on time-average (TA) and sampled-data approaches. Since the SSM based on TA has been thoroughly discussed in [51], we will show how a sampled-data model can be easily obtained by the following example.

Example 4.5 A dc-dc boost converter, operating at continuous conduction mode (CCM) is used as an illustrating example. Fig. 4.5 shows the schematic diagram of a boost converter. When *SW* is turned on, the resulting circuit is shown in Fig. 4.6(a). On the other hand, when *SW* is off, the resulting circuit is shown in Fig. 4.6(b). The boost parameters are listed in Table 4.1.

Assume that the turn-on time is from 0 to DT_{sw}, where D is the duty cycle, and T_{sw} is the switching period. Hence, the state-space equation during this interval is

$$\frac{d}{dt}\begin{bmatrix} i_L \\ v_C \end{bmatrix} = \begin{bmatrix} -\frac{r}{L} & 0 \\ 0 & -\frac{1}{RC} \end{bmatrix}\begin{bmatrix} i_L \\ v_C \end{bmatrix} + \begin{bmatrix} \frac{1}{L} \\ 0 \end{bmatrix} v_s \tag{4.72}$$

The differential equation for the turn-off period ($DT_{sw} < t \le T_{sw}$)

$$\frac{d}{dt}\begin{bmatrix} i_L \\ v_C \end{bmatrix} = \begin{bmatrix} -\frac{r}{L} & -\frac{1}{L} \\ \frac{1}{C} & -\frac{1}{RC} \end{bmatrix}\begin{bmatrix} i_L \\ v_C \end{bmatrix} + \begin{bmatrix} \frac{1}{L} \\ 0 \end{bmatrix} v_s. \tag{4.73}$$

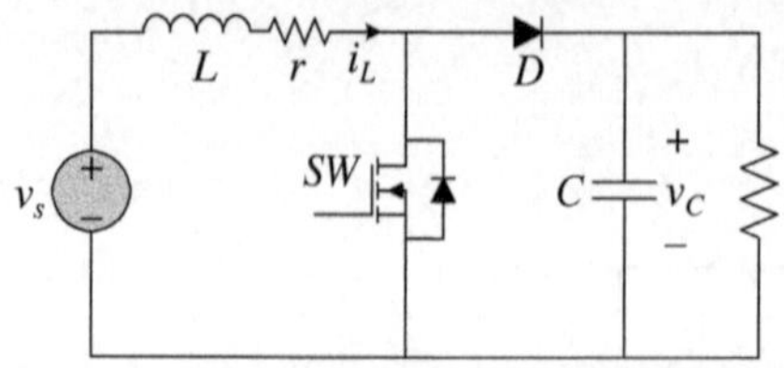

Figure 4.5 Schematic diagram of dc-dc boost converter.

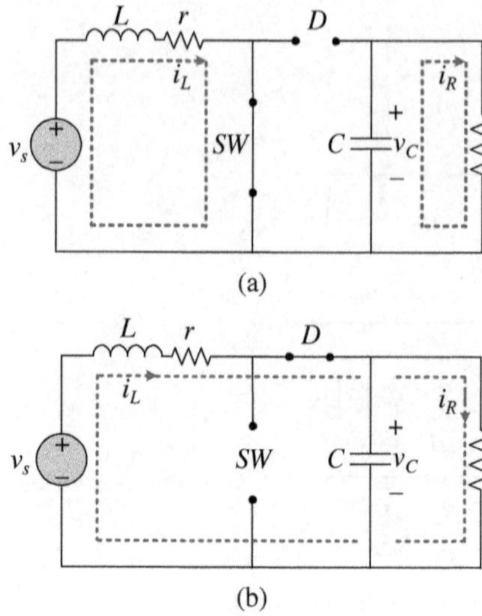

Figure 4.6 The operation modes of dc-dc boost converter: (a) ON-mode; (b) OFF-mode.

Table 4.1 DC-DC boost converter parameters.

Components	Symbol	Value	Unit
Input voltage	V_{in}	15	V
Inductor	L	58	μH
Capacitor	C	5.5	μF
Resistor	R	18.6	Ω
Internal resistor in inductor	r	0.025	Ω
Duty ratio	D	0.4	

Recognizing that the voltage source can be written as $\frac{dv_s}{dt} = 0$, with $v_s(0)$ set to v_s, one can rewrite (4.72) and (4.73) to (4.74) and (4.75), respectively.

$$\frac{d\bar{\mathbf{x}}}{dt} = \underbrace{\begin{bmatrix} -\frac{r}{L} & 0 & \frac{1}{L} \\ 0 & -\frac{1}{RC} & 0 \\ 0 & 0 & 0 \end{bmatrix}}_{\bar{\mathbf{A}}_{\mathbf{on}}} \bar{\mathbf{x}} \tag{4.74}$$

$$\frac{d\bar{\mathbf{x}}}{dt} = \underbrace{\begin{bmatrix} -\frac{r}{L} & -\frac{1}{L} & \frac{1}{L} \\ \frac{1}{C} & -\frac{1}{RC} & 0 \\ 0 & 0 & 0 \end{bmatrix}}_{\bar{\mathbf{A}}_{\mathbf{off}}} \bar{\mathbf{x}} \tag{4.75}$$

where $\bar{\mathbf{x}} = \begin{bmatrix} \mathbf{x} & v_s \end{bmatrix}^T$ and $\mathbf{x} = \begin{bmatrix} i_L & v_C \end{bmatrix}$ Hence, a sample point at any instant can be obtained by computing the matrix exponential of $e^{\bar{\mathbf{A}}_{\mathbf{on}}t}$ and $e^{\bar{\mathbf{A}}_{\mathbf{off}}t}$. For example, to obtain $\bar{\mathbf{x}}(T)$,

$$\bar{\mathbf{x}}(T) = \bar{\mathbf{\Phi}}(T)\bar{\mathbf{x}}(0), \tag{4.76}$$

where $\bar{\Phi}(T) = e^{\bar{\mathbf{A}}_{\mathbf{off}}(T-t_{on})}e^{\bar{\mathbf{A}}_{\mathbf{on}}(t_{on})}$, where T is the period and is equal to T_{sw} in this case, and t_{on} is the turn-on time, and $\bar{\mathbf{x}}(0)$ is the value of $\bar{\mathbf{x}}$ at time $t = 0$. Generally, the form of $\bar{\mathbf{\Phi}}$ is described as

$$\bar{\mathbf{\Phi}} = \begin{bmatrix} A_p & B_p \\ 0 & 1 \end{bmatrix} \tag{4.77}$$

(4.76) can be partitioned into

$$\begin{bmatrix} \mathbf{x}(T) \\ v_s(T) \end{bmatrix} = \begin{bmatrix} \mathbf{A_p} & \mathbf{B_p} \\ 0 & 1 \end{bmatrix} \begin{bmatrix} \mathbf{x}(0) \\ v_s(0) \end{bmatrix}. \tag{4.78}$$

At steady-state condition $\mathbf{x}_{ss} = \mathbf{x}(T) = \mathbf{x}(0)$; thus, one may easily obtain the steady-state variables as (4.79)

$$\mathbf{x}_{ss} = \left(\mathbf{I} - \mathbf{A_p}\right)^{-1}\mathbf{B_p}v_s(0). \tag{4.79}$$

Substituting the parameters in Table 4.1, one may obtain the steady-state values as

$$\mathbf{x}_{ss} = \begin{bmatrix} 0.8706 \\ 25.4727 \end{bmatrix}.$$

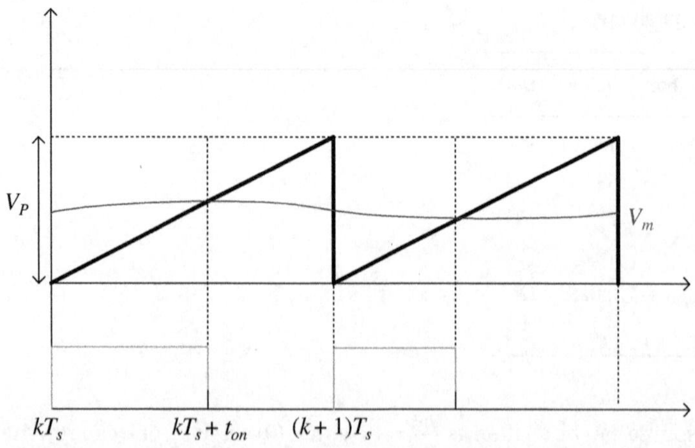

Figure 4.7 Subinterval diagram of a PWM converter.

The augmented steady-state vector is represented as

$$\bar{\mathbf{x}}_{ss} = \begin{bmatrix}\mathbf{x}_{ss} & v_s\end{bmatrix}^T = \begin{bmatrix}0.8706 & 25.4727 & 15\end{bmatrix}^T$$

The SSM can be derived by taking the partial derivatives of (4.76) with respect to $\mathbf{x}(0)$, v_s, and v_m, respectively. Note that v_m is the modulating signal as shown in Fig. 4.7. Hence, the following SSM for the boost converter can be written as follows:

$$\Delta\bar{\mathbf{x}}[(k+1)T_s] = \frac{\partial\bar{\mathbf{x}}[(k+1)T_s]}{\partial\bar{\mathbf{x}}[kT_s]}\Delta\bar{\mathbf{x}}(kT_s) + \frac{\partial\bar{\mathbf{x}}[(k+1)T_s]}{\partial v_s[kT_s]}\Delta v_s[kT_s] + \frac{\partial\bar{\mathbf{x}}[(k+1)T_s]}{\partial v_m[kT_s]}\Delta v_m[kT_s], \tag{4.80}$$

where

$$\frac{\partial\bar{\mathbf{x}}[(k+1)T_s]}{\partial\bar{\mathbf{x}}[kT_s]} = \bar{\mathbf{\Phi}}, \tag{4.81}$$

$$\frac{\partial\bar{\mathbf{x}}[(k+1)T_s]}{\partial v_s[kT_s]} = \bar{\mathbf{\Phi}}\begin{bmatrix}0\\0\\1\end{bmatrix}, \tag{4.82}$$

$$\frac{\partial\bar{\mathbf{x}}[(k+1)T_s]}{\partial v_m[kT_s]} = \frac{\partial\bar{\mathbf{x}}[(k+1)T_s]}{\partial t_{on}}\frac{\partial t_{on}}{\partial v_m[kT_s]}, \tag{4.83}$$

and
$$\frac{\partial\bar{\mathbf{x}}[(k+1)T_s]}{\partial t_{on}} = \left[-\bar{\mathbf{A}}_{\mathbf{off}}\cdot\bar{\mathbf{\Phi}} + e^{\bar{\mathbf{A}}_{\mathbf{off}}(T-t_{on})}\cdot\bar{\mathbf{A}}_{\mathbf{on}}\cdot e^{\bar{\mathbf{A}}_{\mathbf{on}}t_{on}}\right]\cdot\bar{\mathbf{x}}_{ss} \tag{4.84}$$

To obtain $\frac{\partial t_{on}}{\partial v_m[kT_s]}$, the switching constraint equation stated in (4.85) is first obtained by observing Fig. 4.7.

$$\frac{V_P}{T_{sw}}t_{on} - v_m[t_{on}] = 0 \tag{4.85}$$

Hence, $\frac{\partial t_{on}}{\partial v_m[t_{on}]}$ can be obtained by taking the derivative of (4.85) with respect to t_{on}, which yields

$$\frac{\partial t_{on}}{\partial v_m[t_{on}]} = \frac{T_{sw}}{V_P}. \tag{4.86}$$

Finally, $\frac{\partial t_{on}}{\partial v_m[kT_s]} = \frac{T_{sw}}{V_P} e^{j\omega t_{on}}$.

From (12.41), the input-to-output transfer function in the z-domain can be stated as

$$G_{ux} = \frac{\tilde{\bar{\mathbf{x}}}(z)}{\tilde{v}_s(z)} = \left(z\mathbb{I} - \frac{\partial \bar{\mathbf{x}}[(k+1)T_s]}{\partial \bar{\mathbf{x}}[kT_s]} \right)^{-1} \frac{\partial \bar{\mathbf{x}}[(k+1)T_s]}{\partial v_s[kT_s]}. \tag{4.87}$$

Similarly, the control-to-output transfer function in the z-domain can be stated as

$$G_{dx} = \frac{\tilde{\bar{\mathbf{x}}}(z)}{\tilde{v}_m(z)} = \left(z\mathbb{I} - \frac{\partial \bar{\mathbf{x}}[(k+1)T_s]}{\partial \bar{\mathbf{x}}[kT_s]} \right)^{-1} \frac{\partial \bar{\mathbf{x}}[(k+1)T_s]}{\partial v_m[kT_s]}. \tag{4.88}$$

Hence, the (4.87) and (4.88) can be expressed by this following code:

```
%% Boost Converter Small Signal Modeling
clear all
clc
%% Initialization parameter
Vs=15;                                          % Input voltage
L=58e-6;                                        % Inductor
C=5.5e-6;                                       % Capacitor
r=0.025;                                        % Internal resistance in inductor
R=18.6;                                         % Load
fsw=40e3;                                       % Switching frequency
T=1/fsw;                                        % Switching period
D=0.4;                                          % Duty ratio
ton=T*D;                                        % turn-on time
s=tf('s');
z=tf('z',1/fsw);
w=2*pi*fsw;
%% On and Off state matrix
Aon=[-r/L, 0,1/L;0,-1/(R*C),0;0,0,0];           % ON-mode
Aoff=[-r/L, -1/L,1/L;1/C,-1/(R*C),0;0,0,0];     % OFF-mode
%% Solution
Phi=expm(Aoff*(T-ton))*expm(Aon*ton);           % Phi_check
Ap=Phi(1:2,1:2);                                % Ap
Bp=Phi(1:2,3);                                  % Bp
xss=inv(eye(2)-Ap)*Bp*Vs;                       % Xss
xss_check=[xss;Vs];                             % Xss_check
%% Partial derivatives of x_check respect to modulating signal
dQQ=-Aoff*Phi+Aon*expm(Aoff*(T-ton))*expm(Aon*ton);
%% Small signal model transfer function
% Transfer function duty-to-state variables
Gdx=inv(z*eye(3)-Phi)*dQQ*xss_check*T*z;
% Transfer function input-to-state variables
Gux=inv(z*eye(3)-Phi)*Phi*[0;0;1];
```

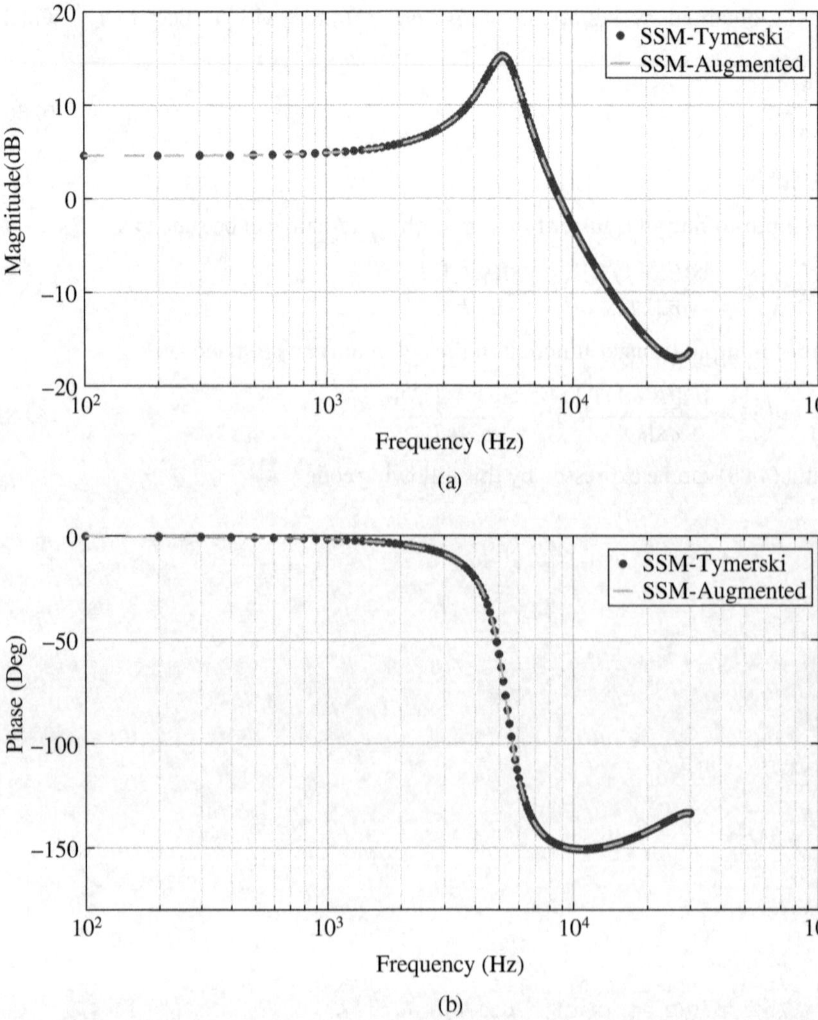

Figure 4.8 Frequency response of input-to-output voltage: (a) magnitude; (b) phase.

Figs. 4.8 and 4.9 show the frequency response of the input-to-output voltage and control-to-output voltage, respectively, of a boost converter, whose switching frequency is 40 kHz. On the same figures, we also plot the sampled-data model proposed by Tymerski [52], in which the model is conducted by directly evaluating the convolution integrals. Obviously, the sampled-data model obtained by treating the input as the state variables leads to much more compact expressions.

4.6 Runge–Kutta Methods

The matrices, which we have discussed so far, to be evaluated exponentially are all constant matrices. According to [53] and [54], a linear time-varying (LTV) system

$$\frac{d\mathbf{x}}{dt} = \mathbf{A}(t)\mathbf{x}(t) \tag{4.89}$$

will have a closed-form solution, if

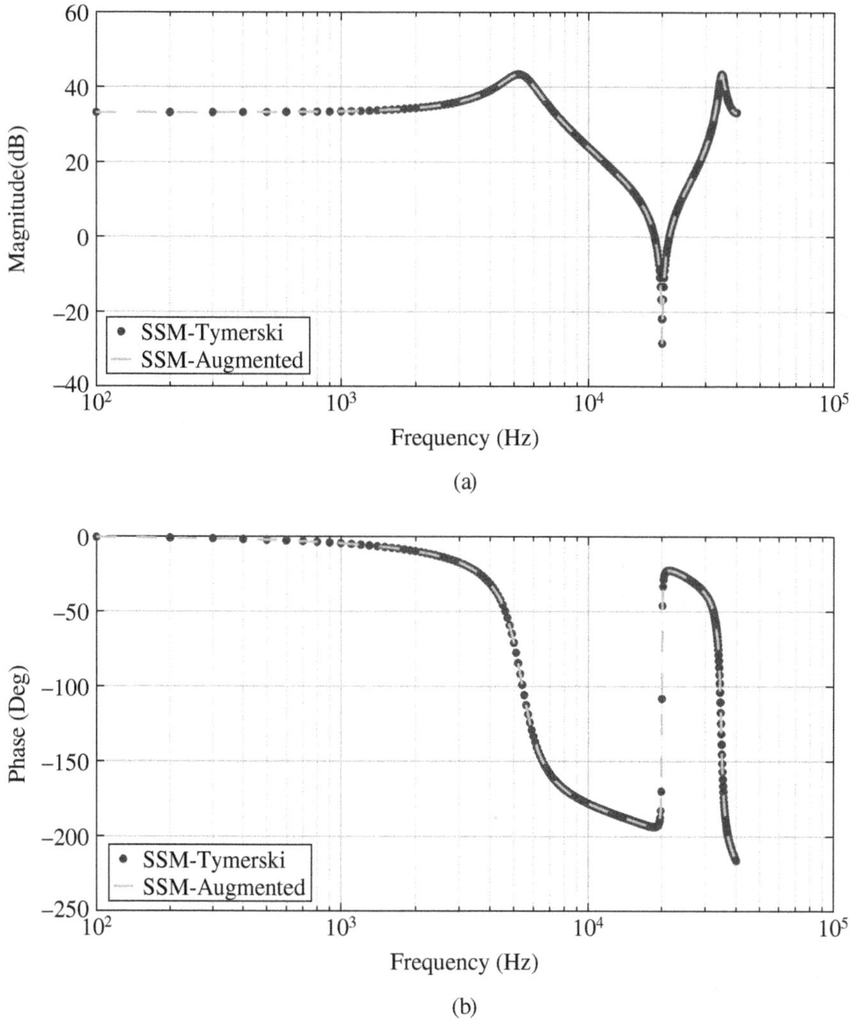

Figure 4.9 Frequency response of control-to-output voltage: (a) magnitude; (b) phase.

1) $\mathbf{A}(t) = \mathbf{A}$, a constant matrix.
2) $\mathbf{A}(t) = \mathbf{\Lambda}(t)$, a diagonal matrix.
3) $\mathbf{A}(t) = \mathbf{\Delta}(t)$, a triangular matrix.
4) $\mathbf{A}(t)$ is a member of the commutative class, i.e.

$$\mathbf{A}(t)\left\{\int_{t_0}^{t}\mathbf{A}(\tau)d\tau\right\} = \left\{\int_{t_0}^{t}\mathbf{A}(\tau)d\tau\right\}\mathbf{A}(t). \tag{4.90}$$

5) $\mathbf{A}(t)$ is a member of $\mathbb{A}_h$ class, i.e.

$$\mathbb{A}_h \triangleq \left\{ \mathbf{A}(t)|\ \mathbf{A}_1\mathbf{A}(t) - \mathbf{A}(t)\mathbf{A}_1 = \frac{d}{dt}\frac{\mathbf{A}(t)}{h(t)}, \forall t \right\} \tag{4.91}$$

where $\mathbf{A}_1$ is a constant matrix and $h(t)$ is a non-zero time function.

6) If $\mathbf{A}(t)$ can be decomposed into

$$\mathbf{A}(t) = \mathbf{A}_1(t) + \mathbf{A}_2(t). \tag{4.92}$$

where $\mathbf{A}_1(t)$ belongs to 1), 2), 3), 4), and 5). Let $\mathbf{\Psi}_1(t)$ be the solution of

$$\frac{d}{dt}\mathbf{\Psi}_1(t) = \mathbf{A}_1(t)\mathbf{\Psi}_1(t),\ \mathbf{\Psi}_1(0) = \mathbb{I} \tag{4.93}$$

If $\mathbf{F}_2(t)$, defined by

$$\mathbf{F}_2(t) = \mathbf{\Psi}_1^{-1}(t)\mathbf{A}_2(t)\mathbf{\Psi}_1(t) \tag{4.94}$$

also belongs to 1), 2), 3), 4), and 5), then the state transition matrix, $\mathbf{\Psi}(t)$ is

$$\mathbf{\Psi}(t) = \mathbf{\Psi}_1(t)\mathbf{\Psi}_2(t) \tag{4.95}$$

where $\mathbf{\Psi}_2(t)$ is the solution of

$$\frac{d}{dt}\mathbf{\Psi}_2(t) = \mathbf{F}_2(t)\mathbf{\Psi}_2(t),\ \mathbf{\Psi}_2(0) = \mathbb{I}. \tag{4.96}$$

7) If

$$\mathbf{A}(t) = \left[\mathbf{M}(t)\mathbf{D}(t) + \frac{d}{dt}\mathbf{M}(t)\right]\mathbf{M}^{-1}(t) \tag{4.97}$$

where $\mathbf{D}(t)$ is an arbitrary $n \times n$ diagonal matrix and $\mathbf{M}(t)$ is an $n \times n$ nonsingular matrix whose column $\mathbf{M}_i(t)$ must satisfy the equation

$$\frac{d}{dt}\mathbf{M}_i(t) = \mathbf{B}_i(t)\mathbf{M}_i(t) \tag{4.98}$$

where $\mathbf{B}_i(t)$ is an arbitrary $n \times n$ matrix, $i = 1, 2, \dots, n$.

Therefore, if $\mathbf{A}(t)$ does not satisfy any of the above categories, we can only evaluate (4.89) by means of numerical integration methods. We have discussed some of the one-step integration methods in Chapter 3. In this section, we will discuss more general and accurate methods, called Runge–Kutta methods. A generalized form of

$$y_{i+1} = y_i + \phi(x_i, y_i, h)h, \tag{4.99}$$

where $\phi(x_i, y_i, h)$ is called an increment function. The increment function [40] can be written in a general form as

$$\phi = a_1k_1 + a_2k_2 + \cdots + a_nk_n, \tag{4.100}$$

where the a's are constants and the k's are

$$\begin{aligned}
k_1 &= f(x_i, y_i)\\
k_2 &= f(x_i + p_1h, y_i + q_{11}k_1h)\\
k_3 &= f(x_i + p_2h, y_i + q_{21}k_1h + q_{22}k_2h)\\
&\vdots\\
k_n &= f(x_i + p_{n-1}h, y_i + q_{n-1,1}k_1h + q_{n-1,2}k_2h + \cdots + q_{n-1,n-1}k_{n-1}h)
\end{aligned}$$

To find the values of a's, p's, and q's, we will illustrate it using the second-order of Runge–Kutta methods.

The second order version of (4.99) is

$$y_{i+1} = y_i + (a_1k_1 + a_2k_2)h, \tag{4.101}$$

where

$$k_1 = f(x_i, y_i) \tag{4.102}$$

and

$$k_2 = f(x_i + p_1 h, y_i + q_{11} k_1 h). \tag{4.103}$$

The Taylor series of k_2 yields

$$f(x_i + p_1 h, y_i + q_{11} k_1 h) = f(x_i, y_i) + p_1 h \frac{\partial f}{\partial x} + q_{11} k_1 h \frac{\partial f}{\partial y} + \cdots . \tag{4.104}$$

This result along with (4.102) can be substituted into (4.101) to yield

$$y_{i+1} = y_i + \left[a_1 f(x_i, y_i) + a_2 f(x_i, y_i) \right] h + \left[a_2 p_1 \frac{\partial f}{\partial x} + a_2 q_{11} f(x_i, y_i) \frac{\partial f}{\partial y} \right] h^2 + \cdots . \tag{4.105}$$

On the other hand, the second-order Taylor series for y_{i+1} is

$$y_{i+1} = y_i + f(x_i, y_i) h + \frac{f'(x_i, y_i)}{2!} h^2, \tag{4.106}$$

where

$$f'(x_i, y_i) = \frac{\partial f(x, y)}{\partial x} + \frac{\partial f(x, y)}{\partial y} \frac{dy}{dx}. \tag{4.107}$$

Comparing (4.105) and (4.106), the following holds true:

$$\begin{aligned} a_1 + a_2 &= 1 \\ a_2 p_1 &= 0.5 \\ a_2 q_{11} &= 0.5 \end{aligned}$$

There are four unknowns but we only have three equations. Thus, one unknown needs to be assumed to determine the other three and there is a family of second-order methods rather than a single one. Similar derivations can be applied for the other order of the Runge–Kutta methods. Among them, the most popular one is the fourth order. To solve differential equation (4.108)

$$\dot{\mathbf{x}} = \mathbf{A}(t)\mathbf{x} = f(t, \mathbf{x}), \tag{4.108}$$

we use the fourth order Runge–Kutta method with Gill coefficients:

$$\mathbf{x}_{i+1} = \mathbf{x}_i + \frac{h}{6} \left\{ \mathbf{k}_1 + 2\left(1 - \frac{1}{\sqrt{2}}\right) \mathbf{k}_2 + 2\left(1 + \frac{1}{\sqrt{2}}\right) \mathbf{k}_3 + \mathbf{k}_4 \right\}, \tag{4.109}$$

where $\mathbf{x}_i = \mathbf{x}(t_i)$, i is the ith interval, and $h = t_{i+1} - t_i$ is step size. The vectors, $\mathbf{k}_1$, $\mathbf{k}_2$, $\mathbf{k}_3$, and $\mathbf{k}_4$ are given by

$$\mathbf{k}_1 = f(t_i, \mathbf{x}_i) \tag{4.110}$$

$$\mathbf{k}_2 = f\left(t_i + \frac{1}{2}h, \mathbf{x}_i + \frac{1}{2}h\mathbf{k}_1\right) \tag{4.111}$$

$$\mathbf{k}_3 = f\left(t_i + \frac{1}{2}h, \mathbf{x}_i + \left(-\frac{1}{2} + \frac{1}{\sqrt{2}}\right) h\mathbf{k}_1 + \left(1 - \frac{1}{2}\right) h\mathbf{k}_2\right) \tag{4.112}$$

$$\mathbf{k}_4 = f\left(t_i + h, \mathbf{x}_i - \frac{1}{\sqrt{2}} h\mathbf{k}_2 + \left(1 + \frac{1}{2}\right) h\mathbf{k}_3\right). \tag{4.113}$$

Combining equations, one can have

$$\mathbf{k}_1 = \mathbf{A}(t)\mathbf{x}_i \tag{4.114}$$

$$\mathbf{k}_2 = \mathbf{E}_{\mathbf{rk}}(t)\mathbf{x}_i \tag{4.115}$$

$$\mathbf{k}_3 = \mathbf{F}_{\mathbf{rk}}(t)\mathbf{x}_i \tag{4.116}$$

$$\mathbf{k}_4 = \mathbf{G}_{\mathbf{rk}}(t)\mathbf{x}_i \tag{4.117}$$

where

$$\mathbf{E}_{\mathbf{rk}}(t) = \mathbf{A}\left(t_i + \frac{h}{2}\right)\left\{\mathbb{I} + \frac{1}{2}h\mathbf{A}(t_i)\right\}$$

$$\mathbf{F}_{\mathbf{rk}}(t) = \mathbf{A}\left(t_i + \frac{h}{2}\right)\left\{\mathbb{I} + \left(-\frac{1}{2} + \frac{1}{\sqrt{2}}\right)h\mathbf{A}(t_i) + \left(1 - \frac{1}{\sqrt{2}}\right)h\mathbf{E}_{\mathbf{rk}}(t_i)\right\}$$

$$\mathbf{G}_{\mathbf{rk}}(t) = \mathbf{A}(t_i + h)\left\{\mathbb{I} - \frac{h}{\sqrt{2}}\mathbf{E}_{\mathbf{rk}}(t_i) + \left(1 + \frac{1}{\sqrt{2}}\right)h\mathbf{F}_{\mathbf{rk}}(t_i)\right\}.$$

Combining equations, one has

$$\mathbf{x}_{i+1} = \mathbf{K}_{\mathbf{rk}}(t_i)\mathbf{x}_i, \tag{4.118}$$

where

$$\mathbf{K}_{\mathbf{rk}}(t_i) = \mathbb{I} + \frac{h}{6}\left\{\mathbf{A}(t_i) + 2\left(1 - \frac{1}{\sqrt{2}}\right)\mathbf{E}_{\mathbf{rk}}(t_i) + 2\left(1 + \frac{1}{\sqrt{2}}\right)\mathbf{F}_{\mathbf{rk}}(t_i) + \mathbf{G}_{\mathbf{rk}}(t_i).\right\}$$

Using (4.118), the following expressions can be immediately written out:

$$\mathbf{x}(\tau) = \mathbf{K}_{\mathbf{rk}}(\tau - h)\mathbf{K}_{\mathbf{rk}}(\tau - 2h)\dots\mathbf{K}_{\mathbf{rk}}(0)\mathbf{x}(0) = \left\{\prod_{i=1}^{N}\mathbf{K}_{\mathbf{rk}}(\tau - ih)\right\}\mathbf{x}(0). \tag{4.119}$$

Hence,

$$\boldsymbol{\Psi}(\tau, 0) = \left\{\prod_{i=1}^{N}\mathbf{K}_{\mathbf{rk}}(\tau - ih)\right\}. \tag{4.120}$$

5

Modeling of Voltage Source Converters

"While the appearances may vary, the essence remains unchanged"

– Zhuangzi (369 BC — 286 BC)

As mentioned in Chapter 2 and reported in [31], VSCs have the highest market share and are widely used in industries. This chapter is to derive the state space equations of various types of VSCs. The state space equations will be used in later chapters for harmonic model derivations.

5.1 Single-Phase Two-Level VSCs

The schematic circuit of a single-phase VSC is depicted in Fig. 5.1. Assuming the diodes remain appropriately biased, an equivalent simplified circuit may be analyzed which contains only four fully controlled bi-directional switches.

5.1.1 Switching Functions

In order to specify the states of the bi-directional switches, it is convenient to associate a switching function with each switch, SW. Therefore, a switching function S_i corresponding to SW_i, is defined to be 1 when SW_i is closed, and 0 when it is open. Thus, four switching functions exist for a single-phase VSC. To ensure the VSC terminal voltage, $v_t(t)$, is always uniquely defined as a function of the dc voltage, it is necessary that the two switches in each leg of the VSC are switched in a complementary fashion. As a result, the switching functions S_3 and S_4 in the lower half legs may be inferred from S_1 and S_2 in the upper half legs. Typically, two PWM techniques are applied to generate switching functions of a single-phase two-level VSC, namely bipolar and unipolar PWMs.

Bipolar PWM

Bipolar PWM employs a modulating signal to drive the switches in each leg. This modulation technique allows the diagonally opposite switches, i.e. S_1, S_4 and S_2, S_3, to work as pairs according to the modulating and carrier waveforms. It is worth mentioning that S_2, S_3 are the complementary of S_1, S_4. Table 5.1 lists the VSC terminal voltage, v_t, generated by bipolar PWM. As a consequence, the output voltage v_t is illustrated in Fig.5.2. It is called "bipolar PWM," since the output voltage varies between v_{dc} and $-v_{dc}$.

Harmonic Modeling of Voltage Source Converters using Basic Numerical Methods, First Edition.
Ryan Kuo-Lung Lian, Ramadhani Kurniawan Subroto, Victor Andrean, and Bing Hao Lin.

Table 5.1 Bipolar Switching Functions.

S_1	S_2	v_t	Comments
0	1	$-v_{dc}$	S_2 and S_3 on
1	0	v_{dc}	S_1 and S_4 on

Unipolar PWM

Unipolar PWM applies two modulating signals to drive switches in each leg. The modulating signal of leg B is shifted 180° to the modulating signal of leg A. Thus, the modulating signal of leg B is $-v_{mod}$. Consequently, the following rules apply to unipolar PWM:

- If $v_{\text{mod}} \geq v_{carrier}$, then $v_{AN} = v_{dc}$
- If $v_{\text{mod}} < v_{carrier}$, then $v_{AN} = 0$
- If $-v_{\text{mod}} \geq v_{carrier}$, then $v_{BN} = v_{dc}$
- If $-v_{\text{mod}} < v_{carrier}$, then $v_{BN} = 0$

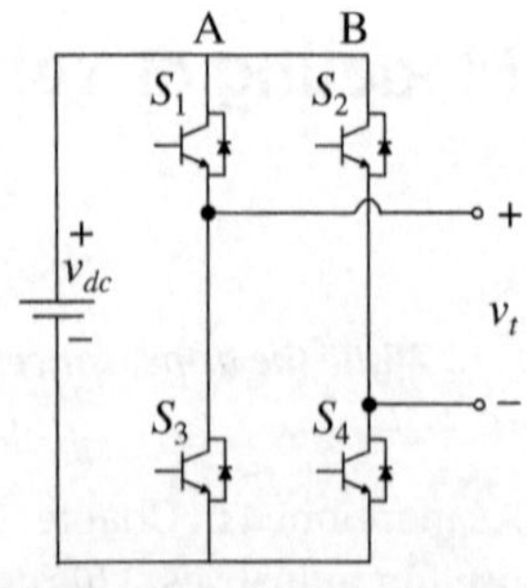

Figure 5.1 Single-phase, two-level VSC.

Fig. 5.3 illustrates the modulating technique of unipolar PWM and the corresponding output voltage yielded by this technique.

The VSC terminal voltage, v_t, generated by unipolar PWM for various modes is listed in Table 5.2. Fig. 5.4 depicts the harmonic characteristics comparison of v_t between bipolar and unipolar PWM for the 50 Hz fundamental frequency and frequency modulation index $m_f = 101$. It can be clearly seen that the significant harmonics of bipolar PWM occur at m_f, $2m_f$, $3m_f$, and so on. On the other hand, the significant harmonics of unipolar PWM happened at $2m_f$, $4m_f$, and so on. This can conclude that unipolar PWM shows better THD compared to bipolar PWM.

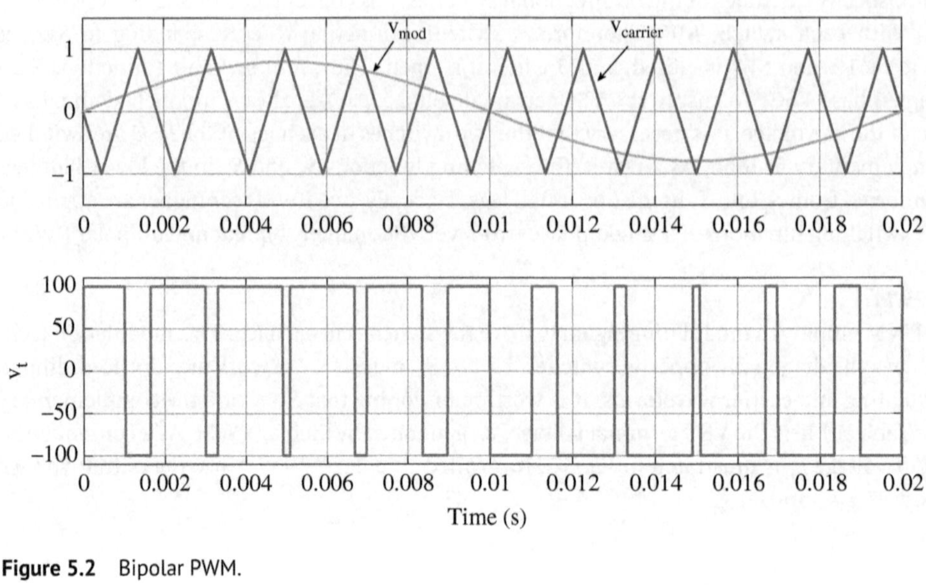

Figure 5.2 Bipolar PWM.

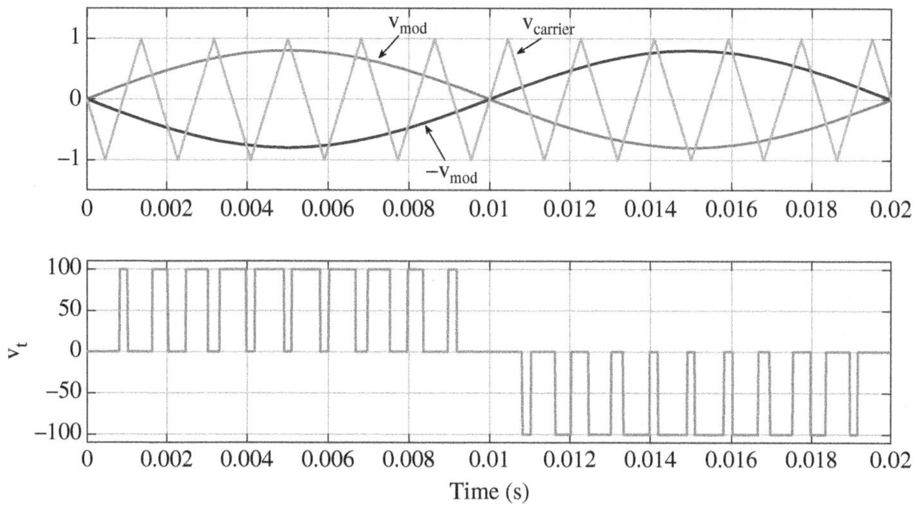

Figure 5.3 Unipolar PWM.

Table 5.2 Unipolar Switching Functions.

S_1	S_2	v_t	**Comments**
0	0	0	S_3 and S_4 on
0	1	$-v_{dc}$	S_3 on and S_4 off
1	0	v_{dc}	S_3 off and S_4 on
1	1	0	S_3 and S_4 off

5.1.2 Switched Circuits

As listed in Table 5.2, there are 4 modes affecting the behavior of the VSC for unipolar PWM. However, only two modes provides the terminal voltage, while the others cause zero terminal voltage. Fig. 5.5 depicts the interfacing of a single-phase, two-level VSC with AC and DC systems. In the ac side, the grid voltage (v_s) is connected to a line filter, composed of inductor (L_s) and resistor (R_s). In the dc side, VSC output voltage is filtered by a dc link capacitor (C) in parallel with a dc resistor (R_{dc}).

When the switch condition $S_1 = 1$ and $S_2 = 0$, we get (5.1) and (5.2)

$$L_s \frac{di_s}{dt} = -R_s i_s - v_{dc} + v_s \tag{5.1}$$

$$C\frac{dv_{dc}}{dt} = -\frac{v_{dc}}{R_{dc}} + i_s + i_{dc} \tag{5.2}$$

On the other hand, when the switch condition $S_1 = 0$ and $S_2 = 1$, the dynamics of the grid current and dc voltage can be obtained as

$$L_s \frac{di_s}{dt} = -R_s i_s + v_{dc} + v_s \tag{5.3}$$

$$C\frac{dv_{dc}}{dt} = -\frac{v_{dc}}{R_{dc}} - i_s + i_{dc} \tag{5.4}$$

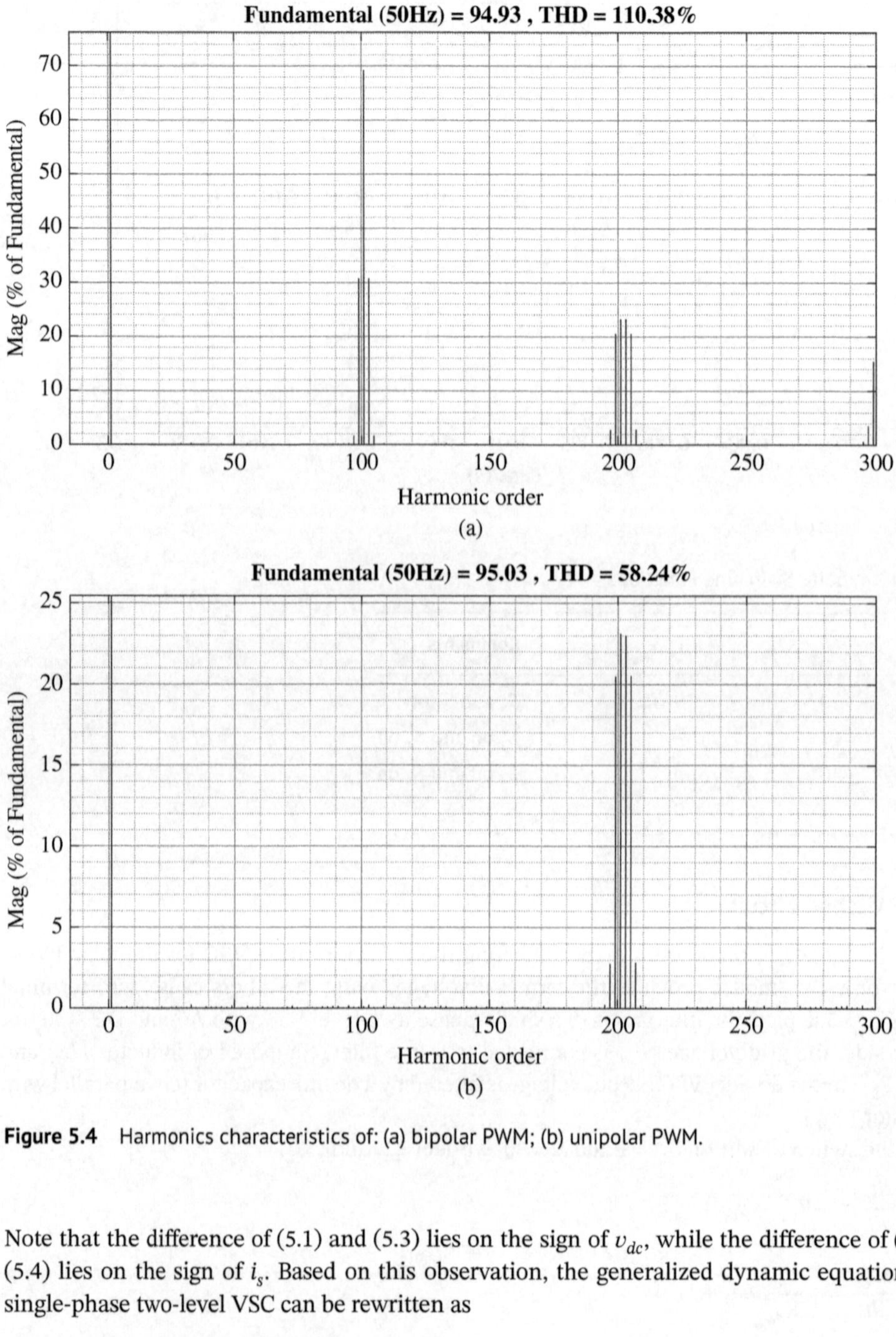

Figure 5.4 Harmonics characteristics of: (a) bipolar PWM; (b) unipolar PWM.

Note that the difference of (5.1) and (5.3) lies on the sign of v_{dc}, while the difference of (5.2) and (5.4) lies on the sign of i_s. Based on this observation, the generalized dynamic equations of the single-phase two-level VSC can be rewritten as

$$\frac{di_s}{dt} = -\frac{R_s}{L_s}i_s - \frac{(S_1 - S_2)}{L_s}v_{dc} + \frac{1}{L_s}v_s$$
$$\frac{dv_{dc}}{dt} = \frac{(S_1 - S_2)}{C}i_s - \frac{1}{R_{dc}C}v_{dc} + \frac{1}{C}i_{dc} \tag{5.5}$$

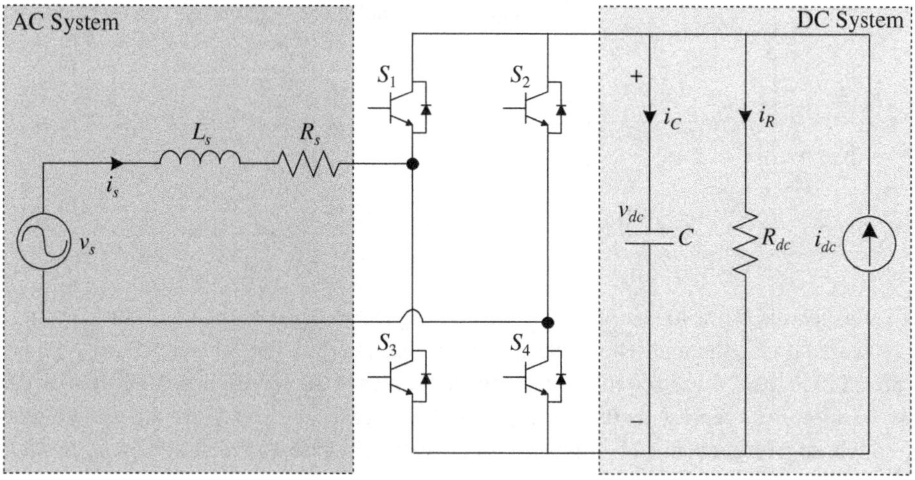

Figure 5.5 Interfacing of a single-phase, two-level VSC with AC and DC systems.

Rearranging (5.5) into a general matrix form yields

$$\dot{\mathbf{x}} = \mathbf{A}\mathbf{x} + \mathbf{B}\mathbf{u} \tag{5.6}$$

where $\mathbf{x} = \begin{bmatrix} i_s \\ v_{dc} \end{bmatrix}$; $\mathbf{A} = \begin{bmatrix} -\frac{R_s}{L_s} & -\frac{(S_1-S_2)}{L_s} \\ \frac{(S_1-S_2)}{C} & -\frac{1}{R_{dc}C} \end{bmatrix}$; $\mathbf{B} = \begin{bmatrix} \frac{1}{L_s} & 0 \\ 0 & \frac{1}{C} \end{bmatrix}$; and $\mathbf{u} = \begin{bmatrix} v_s \\ i_{dc} \end{bmatrix}$.

Note that here $\mathbf{u}$ represents the source input, rather than the control input.

5.2 Three-Phase Two-Level VSCs

Fig. 5.6 shows the schematic diagram of a three-phase two-level VSC. As can be seen, it is equipped with two switches for each leg/phase. Similar to the single-phase VSC, the upper and lower switches work in a complementary fashion. Hence, we denote the upper switches of phases a, b, and c as

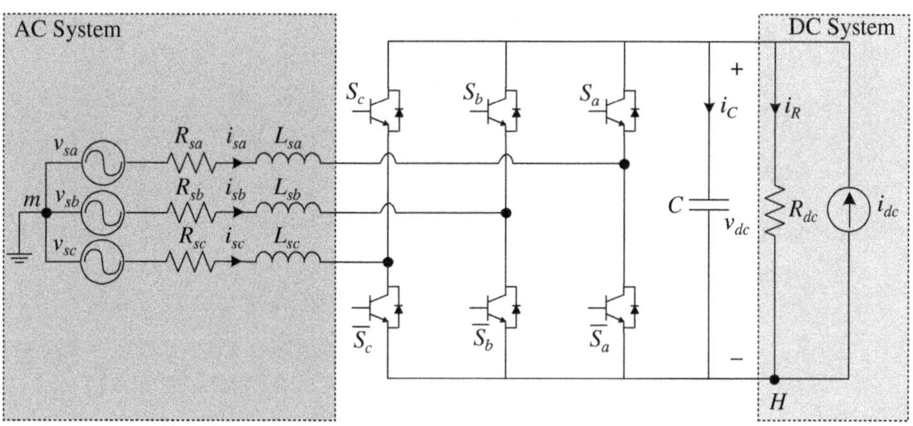

Figure 5.6 Three-phase, two-level VSC.

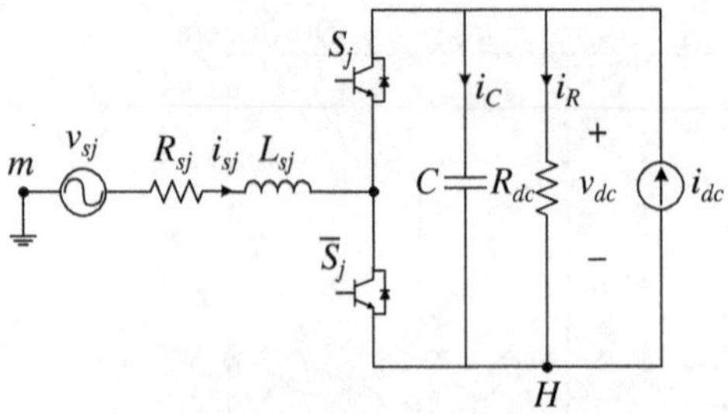

Figure 5.7 Equivalent single-phase two-level VSC.

S_a, S_b, and S_c, respectively. It is to be noted that the floating DC node is denoted by H, and ground is denoted by m. Following the mathematical derivations done in [55], the equivalent single-phase VSC, as depicted in Fig. 5.7, is employed to obtain the dynamics of the three-phase VSC. It is worth to note that the subscript j represents the phase $j \in \{a, b, c\}$. Hence, v_{sj} and i_{sj} are the grid voltage and current of phase-j, respectively. The inductor and resistor of phase-j are denoted by L_{sj} and R_{sj}, respectively. $\overline{S}_j$ is the complementary switch of S_j. C and R_{dc} are the capacitor and resistor at dc side whose currents are represented by i_C and i_R, respectively. Two combinations can be made for circuit analysis, i.e $S_j = 1$ and $S_j = 0$, as illustrated in Fig. 5.8.

1. When $S_j = 1$:

$$v_{sj} = R_{sj}i_{sj} + L_{sj}\frac{di_{sj}}{dt} + v_{dc} + V_{Hm} \tag{5.7}$$

2. When $S_j = 0$:

$$v_{sj} = R_{sj}i_{sj} + L_{sj}\frac{di_{sj}}{dt} + V_{Hm} \tag{5.8}$$

Combining (5.7) and (5.8), results in

$$L_{sj}\frac{di_{sj}}{dt} = -R_{sj}i_{sj} - S_jv_{dc} - V_{Hm} + v_{sj} \tag{5.9}$$

Extending (5.9) in the abc-frame gives

$$\begin{cases} L_{sa}\dfrac{di_{sa}}{dt} = -R_{sa}i_{sa} - S_av_{dc} - V_{Hm} + v_{sa} \\ L_{sb}\dfrac{di_{sb}}{dt} = -R_{sb}i_{sb} - S_bv_{dc} - V_{Hm} + v_{sb} \\ L_{sc}\dfrac{di_{sc}}{dt} = -R_{sc}i_{sc} - S_cv_{dc} - V_{Hm} + v_{sc} \end{cases} \tag{5.10}$$

Adding the equations in (5.10) obtains

$$\begin{aligned} V_{Hm} = &-\frac{1}{3}\left(L_{sa}\frac{di_{sa}}{dt} + L_{sb}\frac{di_{sb}}{dt} + L_{sc}\frac{di_{sc}}{dt}\right) - \frac{1}{3}\left(R_{sa}i_{sa} + R_{sb}i_{sb} + R_{sc}i_{sc}\right) \\ &-\frac{1}{3}\left(S_a + S_b + S_c\right)v_{dc} + \frac{1}{3}\left(v_{sa} + v_{sb} + v_{sc}\right) \end{aligned} \tag{5.11}$$

In the dc side, we have

$$C\frac{dv_{dc}}{dt} = \sum_{j=a}^{c} S_ji_{sj} - \frac{1}{R_{dc}}v_{dc} + i_{dc} \tag{5.12}$$

Accordingly, substituting (5.11) into each equation in (5.10) and incorporating (5.12) yields the state-space form (5.13).

$$\mathbf{L}\dot{\mathbf{x}} = \mathbf{A}\mathbf{x} + \mathbf{B}\mathbf{u} \tag{5.13}$$

where

$$\mathbf{L} = \begin{bmatrix} 2L_{sa} & -L_{sb} & -L_{sc} & 0 \\ -L_{sa} & 2L_{sb} & -L_{sc} & 0 \\ -L_{sa} & -L_{sb} & 2L_{sc} & 0 \\ 0 & 0 & 0 & C \end{bmatrix}_{4\times4}$$

$$\mathbf{x} = \begin{bmatrix} i_{sa} & i_{sb} & i_{sc} & v_{dc} \end{bmatrix}^T_{4\times1}$$

$$\mathbf{A} = \begin{bmatrix} -2R_{sa} & R_{sb} & R_{sc} & -2S_a + S_b + S_c \\ R_{sa} & -2R_{sb} & R_{sc} & S_a - 2S_b + S_c \\ R_{sa} & R_{sb} & -2R_{sc} & S_a + S_b - 2S_c \\ S_a & S_b & S_c & -\frac{1}{R_{dc}} \end{bmatrix}_{4\times4}$$

$$\mathbf{B} = \begin{bmatrix} 2 & -1 & -1 & 0 \\ -1 & 2 & -1 & 0 \\ -1 & -1 & 2 & 0 \\ 0 & 0 & 0 & 1 \end{bmatrix}_{4\times4}$$

$$\mathbf{u} = \begin{bmatrix} v_{sa} & v_{sb} & v_{sc} & i_{dc} \end{bmatrix}^T_{4\times1}$$

Note that the singularity problem will occur since **L** matrix is not full-rank. Thus, to avoid this issue, we consider the phase current relationship

$$i_{sa} + i_{sb} + i_{sc} = 0 \tag{5.14}$$

Substituting $i_{sc} = -i_{sa} - i_{sb}$ into (5.13) gives

$$\mathbf{L}_r\dot{\mathbf{x}}_r = \mathbf{A}_r\mathbf{x}_r + \mathbf{B}_r\mathbf{u} \tag{5.15}$$

where

$$\mathbf{L}_r = \begin{bmatrix} 2L_{sa} + L_{sc} & -L_{sb} + L_{sc} & 0 \\ -L_{sa} + L_{sc} & 2L_{sb} + L_{sc} & 0 \\ 0 & 0 & C \end{bmatrix}_{3\times3}$$

$$\mathbf{x}_r = \begin{bmatrix} i_{sa} & i_{sb} & v_{dc} \end{bmatrix}^T_{3\times1}$$

$$\mathbf{A}_r = \begin{bmatrix} -2R_{sa} - R_{sc} & R_{sb} - R_{sc} & -2S_a + S_b + S_c \\ R_{sa} - R_{sc} & -2R_{sb} - R_{sc} & S_a - 2S_b + S_c \\ (S_a - S_c) & (S_b - S_c) & -\frac{1}{R_{dc}} \end{bmatrix}_{3\times3}$$

$$\mathbf{B}_r = \begin{bmatrix} 2 & -1 & -1 & 0 \\ -1 & 2 & -1 & 0 \\ 0 & 0 & 0 & 1 \end{bmatrix}_{3\times4}$$

Although the analysis done above can greatly simplify the model derivations, some readers may find it difficult to comprehend, as Fig. 5.7 is not the circuit that is in real operation. To illustrate that the above derivations are indeed valid, we will demonstrate how one would derive the VSC model by directly analyzing Fig. 5.6 and show that the two approaches yield the same VSC model.

To validate the dynamics of the three-phase two-level VSC, mesh analysis is conducted. Considering all the switching combinations as listed in Table 5.3.

Table 5.3 Switching Combinations of Two-Level VSC.

Mode	S_a	S_b	S_c
1	0	0	0
2	0	0	1
3	0	1	0
4	0	1	1
5	1	0	0
6	1	0	1
7	1	1	0
8	1	1	1

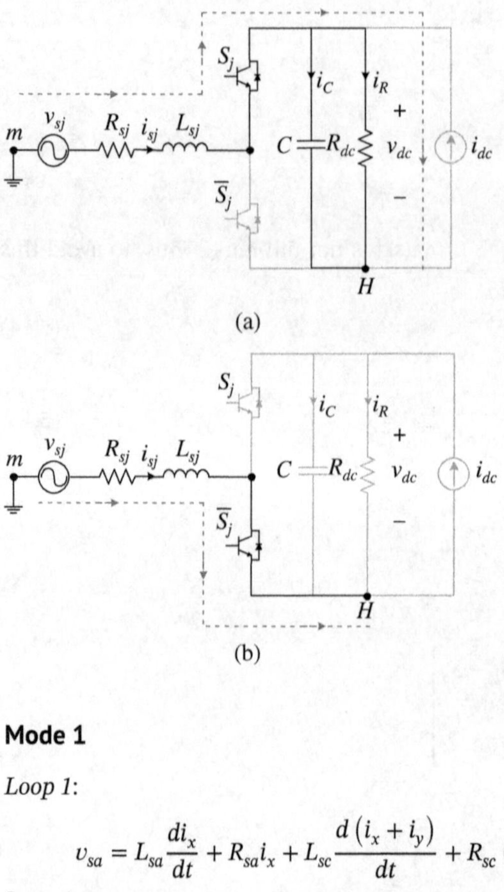

Figure 5.8 The operation modes of two-level VSC when (a) S_j is on; (b) S_j is off.

Mode 1

Loop 1:

$$v_{sa} = L_{sa}\frac{di_x}{dt} + R_{sa}i_x + L_{sc}\frac{d\left(i_x + i_y\right)}{dt} + R_{sc}\left(i_x + i_y\right) + v_{sc} \tag{5.16}$$

Loop 2:

$$v_{sb} = L_{sb}\frac{di_y}{dt} + R_{sb}i_y + L_{sc}\frac{d\left(i_x + i_y\right)}{dt} + R_{sc}\left(i_x + i_y\right) + v_{sc} \tag{5.17}$$

Moreover, as shown in Fig. 5.9, the current in each phase can be formulated as

$$i_x = i_{sa} \tag{5.18}$$

$$i_y = i_{sb} \tag{5.19}$$

$$i_{sc} = -\left(i_x + i_y\right) \tag{5.20}$$

Substituting (5.18) and (5.19) into (5.16) and (5.17) yields

$$v_{sa} = \left(L_{sa} + L_{sc}\right)\frac{di_{sa}}{dt} + \left(R_{sa} + R_{sc}\right)i_{sa} + L_{sc}\frac{di_{sb}}{dt} + R_{sc}i_{sb} + v_{sc} \tag{5.21}$$

$$v_{sb} = L_{sc}\frac{di_{sa}}{dt} + R_{sc}i_{sa} + \left(L_{sb} + L_{sc}\right)\frac{di_{sb}}{dt} + \left(R_{sb} + R_{sc}\right)i_{sb} + v_{sc} \tag{5.22}$$

Introducing *Loop 3* by subtracting (5.22) from (5.21) results in

$$v_{sa} - v_{sb} = L_{sa}\frac{di_{sa}}{dt} + R_{sa}i_{sa} - L_{sb}\frac{di_{sb}}{dt} - R_{sb}i_{sb} \tag{5.23}$$

The sum of (5.21) and (5.23) produces

$$\left(2L_{sa} + L_{sc}\right)\frac{di_{sa}}{dt} + \left(-L_{sb} + L_{sc}\right)\frac{di_{sb}}{dt} = \left(-2R_{sa} - R_{sc}\right)i_{sa} + \left(R_{sb} - R_{sc}\right)i_{sb} + 2v_{sa} - v_{sb} - v_{sc} \tag{5.24}$$

Subtracting (5.23) from (5.22) yields

$$\left(-L_{sa} + L_{sc}\right)\frac{di_{sa}}{dt} + \left(2L_{sb} + L_{sc}\right)\frac{di_{sb}}{dt} = \left(R_{sa} - R_{sc}\right)i_{sa} + \left(-2R_{sb} - R_{sc}\right)i_{sb} - v_{sa} + 2v_{sb} - v_{sc} \tag{5.25}$$

The current in dc side can be formulated as

$$i_C = -i_R + i_{dc} \tag{5.26}$$

Thus, the dc voltage dynamics can be obtained as

$$C\frac{dv_{dc}}{dt} = -\frac{v_{dc}}{R_{dc}} + i_{dc} \tag{5.27}$$

Accordingly, (5.24), (5.25), and (5.27) justify (5.15) when $S_a = 0$, $S_b = 0$, and $S_c = 0$.

Mode 2

Loop 1:

$$v_{sa} = L_{sa}\frac{di_x}{dt} + R_{sa}i_x - v_{dc} + L_{sc}\frac{d\left(i_x + i_y\right)}{dt} + R_{sc}\left(i_x + i_y\right) + v_{sc} \tag{5.28}$$

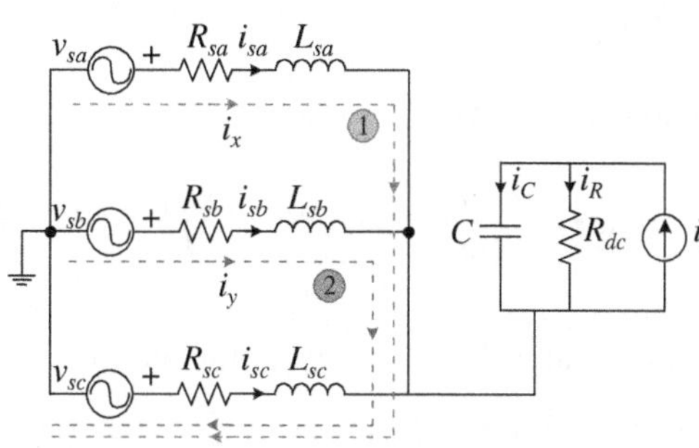

Figure 5.9 Schematic circuit of VSC at Mode 1.

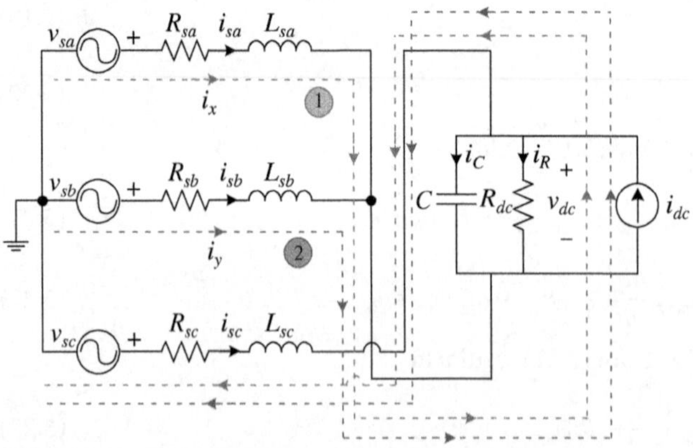

Figure 5.10 Schematic circuit of VSC at Mode 2.

Loop 2:

$$v_{sb} = L_{sb}\frac{di_y}{dt} + R_{sb}i_y - v_{dc} + L_{sc}\frac{d\left(i_x + i_y\right)}{dt} + R_{sc}\left(i_x + i_y\right) + v_{sc} \tag{5.29}$$

Moreover, as shown in Fig. 5.10, the current in each phase can be formulated as

$$i_x = i_{sa} \tag{5.30}$$

$$i_y = i_{sb} \tag{5.31}$$

$$i_{sc} = -\left(i_x + i_y\right) \tag{5.32}$$

Substituting (5.30) and (5.31) into (5.28) and (5.29) yields

$$v_{sa} = \left(L_{sa} + L_{sc}\right)\frac{di_{sa}}{dt} + \left(R_{sa} + R_{sc}\right)i_{sa} + L_{sc}\frac{di_{sb}}{dt} + R_{sc}i_{sb} - v_{dc} + v_{sc} \tag{5.33}$$

$$v_{sb} = L_{sc}\frac{di_{sa}}{dt} + R_{sc}i_{sa} + \left(L_{sb} + L_{sc}\right)\frac{di_{sb}}{dt} + \left(R_{sb} + R_{sc}\right)i_{sb} - v_{dc} + v_{sc} \tag{5.34}$$

Introducing *Loop 3* by subtracting (5.34) from (5.33) results in

$$v_{sa} - v_{sb} = L_{sa}\frac{di_{sa}}{dt} + R_{sa}i_{sa} - L_{sb}\frac{di_{sb}}{dt} - R_{sb}i_{sb} \tag{5.35}$$

The sum of (5.33) and (5.35) produces

$$\left(2L_{sa} + L_{sc}\right)\frac{di_{sa}}{dt} + \left(-L_{sb} + L_{sc}\right)\frac{di_{sb}}{dt} = \left(-2R_{sa} - R_{sc}\right)i_{sa} + \left(R_{sb} - R_{sc}\right)i_{sb} + v_{dc} + 2v_{sa} - v_{sb} - v_{sc} \tag{5.36}$$

Subtracting (5.35) from (5.34) yields

$$\left(-L_{sa} + L_{sc}\right)\frac{di_{sa}}{dt} + \left(2L_{sb} + L_{sc}\right)\frac{di_{sb}}{dt} = \left(R_{sa} - R_{sc}\right)i_{sa} + \left(-2R_{sb} - R_{sc}\right)i_{sb} + v_{dc} - v_{sa} + 2v_{sb} - v_{sc} \tag{5.37}$$

The current in dc side can be formulated as

$$i_x + i_y + i_C + i_R - i_{dc} = 0 \tag{5.38}$$

Thus, the dc voltage dynamics can be obtained as

$$C\frac{dv_{dc}}{dt} = -i_{sa} - i_{sb} - \frac{v_{dc}}{R_{dc}} + i_{dc} \tag{5.39}$$

Accordingly, (5.36), (5.37), and (5.39) justify (5.15) when $S_a = 0$, $S_b = 0$, and $S_c = 1$.

Mode 3

Loop 1:

$$v_{sa} = L_{sa}\frac{di_x}{dt} + R_{sa}i_x + L_{sc}\frac{d\left(i_x + i_y\right)}{dt} + R_{sc}\left(i_x + i_y\right) + v_{sc} \tag{5.40}$$

Loop 2:

$$v_{sb} = L_{sb}\frac{di_y}{dt} + R_{sb}i_y + v_{dc} + L_{sc}\frac{d\left(i_x + i_y\right)}{dt} + R_{sc}\left(i_x + i_y\right) + v_{sc} \tag{5.41}$$

Moreover, as shown in Fig. 5.11, the current in each phase can be formulated as

$$i_x = i_{sa} \tag{5.42}$$

$$i_y = i_{sb} \tag{5.43}$$

$$i_{sc} = -\left(i_x + i_y\right) \tag{5.44}$$

Substituting (5.42) and (5.43) into (5.40) and (5.41) yields

$$v_{sa} = \left(L_{sa} + L_{sc}\right)\frac{di_{sa}}{dt} + \left(R_{sa} + R_{sc}\right)i_{sa} + L_{sc}\frac{di_{sb}}{dt} + R_{sc}i_{sb} + v_{sc} \tag{5.45}$$

$$v_{sb} = L_{sc}\frac{di_{sa}}{dt} + R_{sc}i_{sa} + \left(L_{sb} + L_{sc}\right)\frac{di_{sb}}{dt} + \left(R_{sb} + R_{sc}\right)i_{sb} + v_{dc} + v_{sc} \tag{5.46}$$

Introducing *Loop 3* by subtracting (5.46) from (5.45) results in

$$v_{sa} - v_{sb} = L_{sa}\frac{di_{sa}}{dt} + R_{sa}i_{sa} - L_{sb}\frac{di_{sb}}{dt} - R_{sb}i_{sb} - v_{dc} \tag{5.47}$$

The sum of (5.45) and (5.47) produces

$$\left(2L_{sa} + L_{sc}\right)\frac{di_{sa}}{dt} + \left(-L_{sb} + L_{sc}\right)\frac{di_{sb}}{dt} = \left(-2R_{sa} - R_{sc}\right)i_{sa} + \left(R_{sb} - R_{sc}\right)i_{sb} + v_{dc} + 2v_{sa} - v_{sb} - v_{sc} \tag{5.48}$$

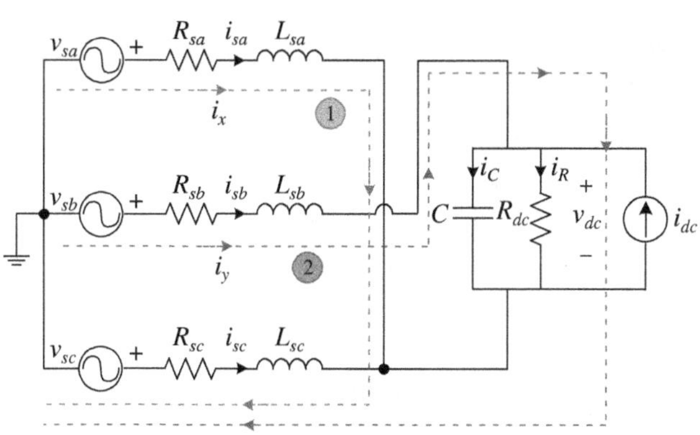

Figure 5.11 Schematic circuit of VSC at Mode 3.

Subtracting (5.47) from (5.46) yields

$$\left(-L_{sa}+L_{sc}\right)\frac{di_{sa}}{dt}+\left(2L_{sb}+L_{sc}\right)\frac{di_{sb}}{dt}=\left(R_{sa}-R_{sc}\right)i_{sa}+\left(-2R_{sb}-R_{sc}\right)i_{sb}-2v_{dc}-v_{sa}+2v_{sb}-v_{sc} \tag{5.49}$$

The current in dc side can be formulated as

$$i_y = i_C + i_R - i_{dc} \tag{5.50}$$

Thus, the dc voltage dynamics can be obtained as

$$C\frac{dv_{dc}}{dt} = -i_{sb} - \frac{v_{dc}}{R_{dc}} + i_{dc} \tag{5.51}$$

Accordingly, (5.48), (5.49), and (5.51) justify (5.15) when $S_a = 0$, $S_b = 1$, and $S_c = 0$.

Mode 4

Loop 1:

$$v_{sa} = L_{sa}\frac{di_x}{dt} + R_{sa}i_x - v_{dc} + L_{sc}\frac{d\left(i_x+i_y\right)}{dt} + R_{sc}\left(i_x+i_y\right) + v_{sc} \tag{5.52}$$

Loop 2:

$$v_{sb} = L_{sb}\frac{di_y}{dt} + R_{sb}i_y + L_{sc}\frac{d\left(i_x+i_y\right)}{dt} + R_{sc}\left(i_x+i_y\right) + v_{sc} \tag{5.53}$$

Moreover, as shown in Fig. 5.12, the current in each phase can be formulated as

$$i_x = i_{sa} \tag{5.54}$$

$$i_y = i_{sb} \tag{5.55}$$

$$i_{sc} = -\left(i_x + i_y\right) \tag{5.56}$$

Substituting (5.54) and (5.55) into (5.52) and (5.53) yields

$$v_{sa} = \left(L_{sa}+L_{sc}\right)\frac{di_{sa}}{dt} + \left(R_{sa}+R_{sc}\right)i_{sa} + L_{sc}\frac{di_{sb}}{dt} + R_{sc}i_{sb} - v_{dc} + v_{sc} \tag{5.57}$$

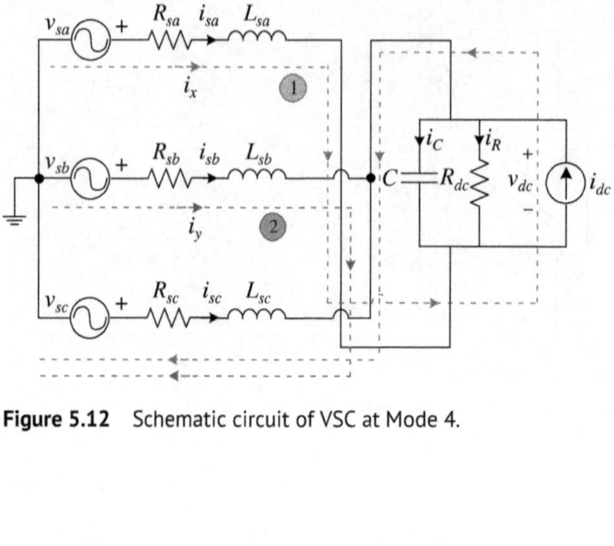

Figure 5.12 Schematic circuit of VSC at Mode 4.

$$v_{sb} = L_{sc}\frac{di_{sa}}{dt} + R_{sc}i_{sa} + \left(L_{sb} + L_{sc}\right)\frac{di_{sb}}{dt} + \left(R_{sb} + R_{sc}\right)i_{sb} + v_{sc} \tag{5.58}$$

Introducing *Loop 3* by subtracting (5.58) from (5.57) results in

$$v_{sa} - v_{sb} = L_{sa}\frac{di_{sa}}{dt} + R_{sa}i_{sa} - L_{sb}\frac{di_{sb}}{dt} - R_{sb}i_{sb} - v_{dc} \tag{5.59}$$

The sum of (5.57) and (5.59) produces

$$\left(2L_{sa} + L_{sc}\right)\frac{di_{sa}}{dt} + \left(-L_{sb} + L_{sc}\right)\frac{di_{sb}}{dt} = \left(-2R_{sa} - R_{sc}\right)i_{sa} + \left(R_{sb} - R_{sc}\right)i_{sb} + 2v_{dc} + 2v_{sa} - v_{sb} - v_{sc} \tag{5.60}$$

Subtracting (5.59) from (5.58) yields

$$\left(-L_{sa} + L_{sc}\right)\frac{di_{sa}}{dt} + \left(2L_{sb} + L_{sc}\right)\frac{di_{sb}}{dt} = \left(R_{sa} - R_{sc}\right)i_{sa} + \left(-2R_{sb} - R_{sc}\right)i_{sb} - v_{dc} - v_{sa} + 2v_{sb} - v_{sc} \tag{5.61}$$

The current in dc side can be formulated as

$$i_x + i_C + i_R - i_{dc} = 0 \tag{5.62}$$

Thus, the dc voltage dynamics can be obtained as

$$C\frac{dv_{dc}}{dt} = -i_{sa} - \frac{v_{dc}}{R_{dc}} + i_{dc} \tag{5.63}$$

Accordingly, (5.60), (5.61), and (5.63) justify (5.15) when $S_a = 0$, $S_b = 1$, and $S_c = 1$.

Mode 5

Loop 1:

$$v_{sa} = L_{sa}\frac{di_x}{dt} + R_{sa}i_x + v_{dc} + L_{sc}\frac{d\left(i_x + i_y\right)}{dt} + R_{sc}\left(i_x + i_y\right) + v_{sc} \tag{5.64}$$

Loop 2:

$$v_{sb} = L_{sb}\frac{di_y}{dt} + R_{sb}i_y + L_{sc}\frac{d\left(i_x + i_y\right)}{dt} + R_{sc}\left(i_x + i_y\right) + v_{sc} \tag{5.65}$$

Moreover, as shown in Fig. 5.13, the current in each phase can be formulated as

$$i_x = i_{sa} \tag{5.66}$$

$$i_y = i_{sb} \tag{5.67}$$

$$i_{sc} = -\left(i_x + i_y\right) \tag{5.68}$$

Substituting (5.66) and (5.67) into (5.64) and (5.65) yields

$$v_{sa} = \left(L_{sa} + L_{sc}\right)\frac{di_{sa}}{dt} + \left(R_{sa} + R_{sc}\right)i_{sa} + L_{sc}\frac{di_{sb}}{dt} + R_{sc}i_{sb} + v_{dc} + v_{sc} \tag{5.69}$$

$$v_{sb} = L_{sc}\frac{di_{sa}}{dt} + R_{sc}i_{sa} + \left(L_{sb} + L_{sc}\right)\frac{di_{sb}}{dt} + \left(R_{sb} + R_{sc}\right)i_{sb} + v_{sc} \tag{5.70}$$

Introducing *Loop 3* by subtracting (5.70) from (5.69) results in

$$v_{sa} - v_{sb} = L_{sa}\frac{di_{sa}}{dt} + R_{sa}i_{sa} - L_{sb}\frac{di_{sb}}{dt} - R_{sb}i_{sb} + v_{dc} \tag{5.71}$$

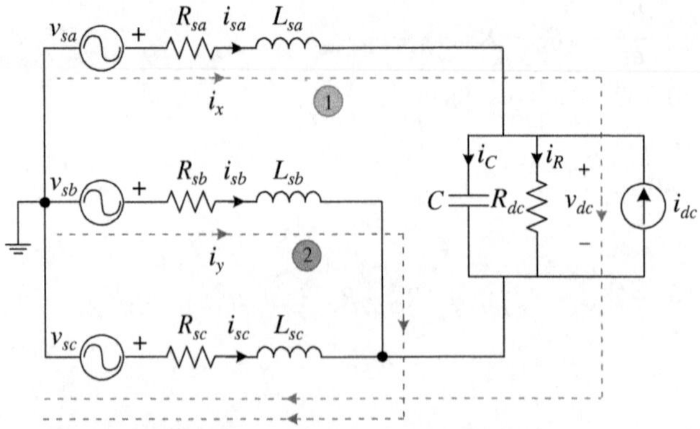

Figure 5.13 Schematic circuit of VSC at Mode 5.

The sum of (5.69) and (5.71) produces

$$(2L_{sa}+L_{sc})\frac{di_{sa}}{dt}+(-L_{sb}+L_{sc})\frac{di_{sb}}{dt}=(-2R_{sa}-R_{sc})\,i_{sa}+(R_{sb}-R_{sc})\,i_{sb}-2v_{dc}+2v_{sa}-v_{sb}-v_{sc} \tag{5.72}$$

Subtracting (5.71) from (5.70) yields

$$(-L_{sa}+L_{sc})\frac{di_{sa}}{dt}+(2L_{sb}+L_{sc})\frac{di_{sb}}{dt}=(R_{sa}-R_{sc})\,i_{sa}+(-2R_{sb}-R_{sc})\,i_{sb}+v_{dc}-v_{sa}+2v_{sb}-v_{sc} \tag{5.73}$$

The current in dc side can be formulated as

$$i_x = i_C + i_R - i_{dc} \tag{5.74}$$

Thus, the dc voltage dynamics can be obtained as

$$C\frac{dv_{dc}}{dt}=i_{sa}-\frac{v_{dc}}{R_{dc}}+i_{dc} \tag{5.75}$$

Accordingly, (5.72), (5.73), and (5.75) justify (5.15) when $S_a = 1$, $S_b = 0$, and $S_c = 0$.

Mode 6

Loop 1:

$$v_{sa}=L_{sa}\frac{di_x}{dt}+R_{sa}i_x+L_{sc}\frac{d\left(i_x+i_y\right)}{dt}+R_{sc}\left(i_x+i_y\right)+v_{sc} \tag{5.76}$$

Loop 2:

$$v_{sb}=L_{sb}\frac{di_y}{dt}+R_{sb}i_y-v_{dc}+L_{sc}\frac{d\left(i_x+i_y\right)}{dt}+R_{sc}\left(i_x+i_y\right)+v_{sc} \tag{5.77}$$

Moreover, as shown in Fig. 5.14, the current in each phase can be formulated as

$$i_x = i_{sa} \tag{5.78}$$

$$i_y = i_{sb} \tag{5.79}$$

$$i_{sc} = -\left(i_x+i_y\right) \tag{5.80}$$

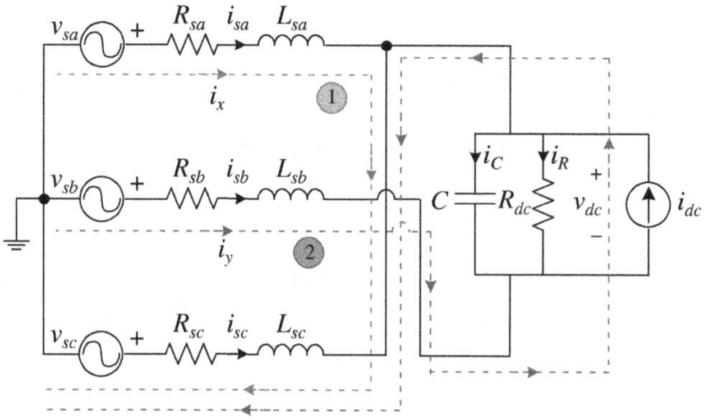

Figure 5.14 Schematic circuit of VSC at Mode 6.

Substituting (5.78) and (5.79) into (5.76) and (5.77) yields

$$v_{sa} = \left(L_{sa} + L_{sc}\right)\frac{di_{sa}}{dt} + \left(R_{sa} + R_{sc}\right)i_{sa} + L_{sc}\frac{di_{sb}}{dt} + R_{sc}i_{sb} + v_{sc} \tag{5.81}$$

$$v_{sb} = L_{sc}\frac{di_{sa}}{dt} + R_{sc}i_{sa} + \left(L_{sb} + L_{sc}\right)\frac{di_{sb}}{dt} + \left(R_{sb} + R_{sc}\right)i_{sb} - v_{dc} + v_{sc} \tag{5.82}$$

Introducing *Loop 3* by subtracting (5.82) from (5.81) results in

$$v_{sa} - v_{sb} = L_{sa}\frac{di_{sa}}{dt} + R_{sa}i_{sa} - L_{sb}\frac{di_{sb}}{dt} - R_{sb}i_{sb} + v_{dc} \tag{5.83}$$

The sum of (5.81) and (5.83) produces

$$\left(2L_{sa} + L_{sc}\right)\frac{di_{sa}}{dt} + \left(-L_{sb} + L_{sc}\right)\frac{di_{sb}}{dt} = \left(-2R_{sa} - R_{sc}\right)i_{sa} + \left(R_{sb} - R_{sc}\right)i_{sb} - v_{dc} + 2v_{sa} - v_{sb} - v_{sc} \tag{5.84}$$

Subtracting (5.83) from (5.82) yields

$$\left(-L_{sa} + L_{sc}\right)\frac{di_{sa}}{dt} + \left(2L_{sb} + L_{sc}\right)\frac{di_{sb}}{dt} = \left(R_{sa} - R_{sc}\right)i_{sa} + \left(-2R_{sb} - R_{sc}\right)i_{sb} + 2v_{dc} - v_{sa} + 2v_{sb} - v_{sc} \tag{5.85}$$

The current in dc side can be formulated as

$$i_y + i_C + i_R - i_{dc} = 0 \tag{5.86}$$

Thus, the dc voltage dynamics can be obtained as

$$C\frac{dv_{dc}}{dt} = -i_{sb} - \frac{v_{dc}}{R_{dc}} + i_{dc} \tag{5.87}$$

Accordingly, (5.84), (5.85), and (5.87) justify (5.15) when $S_a = 1$, $S_b = 0$, and $S_c = 1$.

Mode 7

Loop 1:

$$v_{sa} = L_{sa}\frac{di_x}{dt} + R_{sa}i_x + v_{dc} + L_{sc}\frac{d\left(i_x + i_y\right)}{dt} + R_{sc}\left(i_x + i_y\right) + v_{sc} \tag{5.88}$$

Loop 2:

$$v_{sb} = L_{sb}\frac{di_y}{dt} + R_{sb}i_y + v_{dc} + L_{sc}\frac{d\left(i_x + i_y\right)}{dt} + R_{sc}\left(i_x + i_y\right) + v_{sc} \tag{5.89}$$

Moreover, as shown in Fig. 5.15, the current in each phase can be formulated as

$$i_x = i_{sa} \tag{5.90}$$

$$i_y = i_{sb} \tag{5.91}$$

$$i_{sc} = -\left(i_x + i_y\right) \tag{5.92}$$

Substituting (5.90) and (5.91) into (5.88) and (5.89) yields

$$v_{sa} = \left(L_{sa} + L_{sc}\right)\frac{di_{sa}}{dt} + \left(R_{sa} + R_{sc}\right)i_{sa} + L_{sc}\frac{di_{sb}}{dt} + R_{sc}i_{sb} + v_{dc} + v_{sc} \tag{5.93}$$

$$v_{sb} = L_{sc}\frac{di_{sa}}{dt} + R_{sc}i_{sa} + \left(L_{sb} + L_{sc}\right)\frac{di_{sb}}{dt} + \left(R_{sb} + R_{sc}\right)i_{sb} + v_{dc} + v_{sc} \tag{5.94}$$

Introducing *Loop 3* by subtracting (5.94) from (5.93) results in

$$v_{sa} - v_{sb} = L_{sa}\frac{di_{sa}}{dt} + R_{sa}i_{sa} - L_{sb}\frac{di_{sb}}{dt} - R_{sb}i_{sb} \tag{5.95}$$

The sum of (5.93) and (5.95) produces

$$\left(2L_{sa} + L_{sc}\right)\frac{di_{sa}}{dt} + \left(-L_{sb} + L_{sc}\right)\frac{di_{sb}}{dt} = \left(-2R_{sa} - R_{sc}\right)i_{sa} + \left(R_{sb} - R_{sc}\right)i_{sb} - v_{dc} + 2v_{sa} - v_{sb} - v_{sc} \tag{5.96}$$

Subtracting (5.95) from (5.94) yields

$$\left(-L_{sa} + L_{sc}\right)\frac{di_{sa}}{dt} + \left(2L_{sb} + L_{sc}\right)\frac{di_{sb}}{dt} = \left(R_{sa} - R_{sc}\right)i_{sa} + \left(-2R_{sb} - R_{sc}\right)i_{sb} - v_{dc} - v_{sa} + 2v_{sb} - v_{sc} \tag{5.97}$$

The current in dc side can be formulated as

$$i_x + i_y + i_{dc} = i_C + i_R \tag{5.98}$$

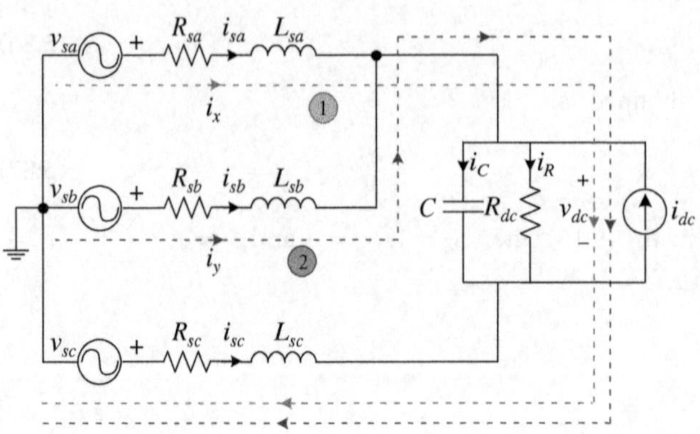

Figure 5.15 Schematic circuit of VSC at Mode 7.

Thus, the dc voltage dynamics can be obtained as

$$C\frac{dv_{dc}}{dt} = i_{sa} + i_{sb} - \frac{v_{dc}}{R_{dc}} + i_{dc} \tag{5.99}$$

Accordingly, (5.96), (5.97), and (5.99) justify (5.15) when $S_a = 1$, $S_b = 1$, and $S_c = 0$.

Mode 8

Loop 1:

$$v_{sa} = L_{sa}\frac{di_x}{dt} + R_{sa}i_x + L_{sc}\frac{d\left(i_x + i_y\right)}{dt} + R_{sc}\left(i_x + i_y\right) + v_{sc} \tag{5.100}$$

Loop 2:

$$v_{sb} = L_{sb}\frac{di_y}{dt} + R_{sb}i_y + L_{sc}\frac{d\left(i_x + i_y\right)}{dt} + R_{sc}\left(i_x + i_y\right) + v_{sc} \tag{5.101}$$

Moreover, as shown in Fig. 5.16, the current in each phase can be formulated as

$$i_x = i_{sa} \tag{5.102}$$

$$i_y = i_{sb} \tag{5.103}$$

$$i_{sc} = -\left(i_x + i_y\right) \tag{5.104}$$

Substituting (5.102) and (5.103) into (5.100) and (5.101) yields

$$v_{sa} = \left(L_{sa} + L_{sc}\right)\frac{di_{sa}}{dt} + \left(R_{sa} + R_{sc}\right)i_{sa} + L_{sc}\frac{di_{sb}}{dt} + R_{sc}i_{sb} + v_{sc} \tag{5.105}$$

$$v_{sb} = L_{sc}\frac{di_{sa}}{dt} + R_{sc}i_{sa} + \left(L_{sb} + L_{sc}\right)\frac{di_{sb}}{dt} + \left(R_{sb} + R_{sc}\right)i_{sb} + v_{sc} \tag{5.106}$$

Introducing *Loop 3* by subtracting (5.106) from (5.105) results in

$$v_{sa} - v_{sb} = L_{sa}\frac{di_{sa}}{dt} + R_{sa}i_{sa} - L_{sb}\frac{di_{sb}}{dt} - R_{sb}i_{sb} \tag{5.107}$$

The sum of (5.105) and (5.107) produces

$$\left(2L_{sa} + L_{sc}\right)\frac{di_{sa}}{dt} + \left(-L_{sb} + L_{sc}\right)\frac{di_{sb}}{dt} = \left(-2R_{sa} - R_{sc}\right)i_{sa} + \left(R_{sb} - R_{sc}\right)i_{sb} + 2v_{sa} - v_{sb} - v_{sc} \tag{5.108}$$

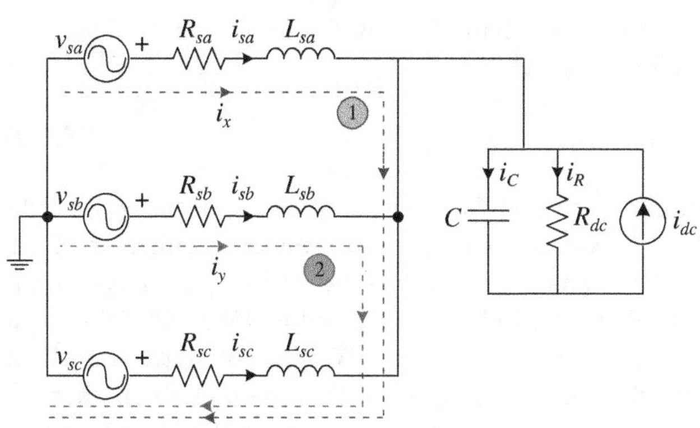

Figure 5.16 Schematic circuit of VSC at Mode 8.

Subtracting (5.107) from (5.106) yields

$$\left(-L_{sa}+L_{sc}\right)\frac{di_{sa}}{dt}+\left(2L_{sb}+L_{sc}\right)\frac{di_{sb}}{dt}=\left(R_{sa}-R_{sc}\right)i_{sa}+\left(-2R_{sb}-R_{sc}\right)i_{sb}-v_{sa}+2v_{sb}-v_{sc} \tag{5.109}$$

The current in dc side can be formulated as

$$i_C+i_R=i_{dc} \tag{5.110}$$

Thus, the dc voltage dynamics can be obtained as

$$C\frac{dv_{dc}}{dt}=-\frac{v_{dc}}{R_{dc}}+i_{dc} \tag{5.111}$$

Accordingly, (5.108), (5.109), and (5.111) justify (5.15) when $S_a = 1$, $S_b = 1$, and $S_c = 1$.

5.3 Three-Phase Multilevel Voltage Source Converter

The AC output voltage of a two-level VSC is limited by the maximum voltage blocking capability of its switching devices. Hence, multilevel VSCs (MVSCs) have been developed to overcome this limitation. Compared with the two-level converters, MVSCs have lower common mode voltage, lower voltage stress on power switches, lower dv/dt ratio, and better harmonic contents [56]. Some of the most widely used MVSCs include diode clamped multilevel converters, flying capacitor multilevel converters, cascaded multilevel converters, and modular multilevel converters. This section is devoted to the mathematical derivations of these four MVSCs.

5.3.1 Multilevel PWM

The modulation strategy for multilevel converters can be classified into three types, specifically, carrier-based pulse width modulation (PWM), space vector modulation, and selective harmonic elimination. Among them, the carrier-based modulation strategy (or multicarrier PWM) is widely used in industry [57]. Multicarrier PWM strategy is similar to that of the sinusoidal PWM strategy except for the fact that several carriers are used. Multicarrier PWM is one in which several triangular carrier signals are compared with one sinusoidal modulating signal. The number of carriers required to produce l-level output is $l - 1$. All carriers have the same frequency and amplitude.

The amplitude modulation index is defined as [32]

$$m_a=\frac{V_m}{V_{cr}(l-1)} \tag{5.112}$$

where $0 \leq m_a \leq 1$, V_m is the peak amplitude of the modulating signal v_m and V_{cr} is the peak amplitude of each carrier signal. The modulating signal is continuously compared with each of the carrier signals and whenever the modulating signal is greater than the carrier signal, pulse is generated. The carrier-based PWM can be further classified into phase-shifted PWM (PS-PWM) and level-shifted PWM (LS-PWM). There are three techniques for LS-PWM: (a) in-phase disposition (IPD), where all carriers are in phase; (b) alternative phase opposite disposition (APOD), where all carriers are alternatively in opposite disposition; and (c) phase opposite disposition (POD), where all carriers above the zero reference are in phase but in opposition with those below the zero reference [58], [59]. In the PS-PWM, all the triangular carriers have the same frequency and the same peak-to-peak amplitude, but there is a phase shift between any two adjacent carrier waves, given by

$$\phi_{cr} = \frac{360^\circ}{l-1} \tag{5.113}$$

The MATLAB code of generating PS-PWM is listed as follows:

```
function[SST,SSseq] = else_sw_time_CMC(level,Mag,delta,freq,mf,Ts)
%% The function to obtain the switching time and switching pattern using...
%% Phase-shifted modulation

N=(level-1)/2;
th2=360/(level-1);
for i=1:N
    th(1,i)=th2*(i-1);
end
th=[th(1,:); th(1,:); th(1,:); th(1,:); th(1,:); th(1,:)];
Fs=1e5;
SN=Ts*Fs;
tt=linspace(0,Ts,SN);

%% Find switching time and switching pattern
ST=zeros(length(tt),1+6*N);
k=1;
for m=1:length(Mag)
    for n=1:N
        mag=Mag(m);
        delta1=delta(m);
        trith=th(m,n);
        for ii=1:length(tt)
            ST(ii,1+N*(m-1)+n) = fST(tt(ii),mag,freq,delta1,mf,trith);
            ST(ii,1)=tt(ii);
        end
        for iii=1:length(tt)-1
            if (ST(iii,1+N*(m-1)+n)>=0 && ST(iii+1,1+N<(m-1)+n)<=0) || ...
                    (ST(iii,1+N*(m-1)+n)<=0 && ST(iii+1,1+N*(m-1)+n)>=0)
                SSX(k,N*(m-1)+n)=fzero(@fST,[tt(iii) tt(iii+1)], ...
                    optimset('Display','off'),mag,freq,delta1,mf,trith);
                k=k+1;
            end
        end
        k=1;
    end
end

SSS=[];
for i=1:size(SSX,2)
    SSS=[SSS;SSX(1:size(SSX,1),i)];
end

SSS=[SSS;1/60];

SSS=sort(SSS);
A=find(SSS==0);
SSS=SSS(length(A)+1:length(SSS),1);
%%
for m=1:length(Mag)
        for n=1:N
            mag=Mag(m);
            delta1=delta(m);
            trith=th(m,n);
```

```
                for ii=1:length(SSS)
                if fST(SSS(ii,1),mag,freq,delta1,mf,trith)>0
                    SSS(ii,1+N*(m-1)+n) = 1;
                else
                     SSS(ii,1+N*(m-1)+n) = 0;
                end
                end
            end
end
%%
delta_t=1e-15;
delta_T=[-delta_t delta_t];
for t=1:2
    dt=delta_T(t);
    for m=1:length(Mag)
        for n=1:N
            mag=Mag(m);
            delta1=delta(m);
            trith=th(m,n);
            for ii=1:length(SSS)
                if t==1
                    SSSold(ii,1) = SSS(ii)+dt;
                    if fST(SSSold(ii,1),mag,freq,delta1,mf,trith)>0
                        SSSold(ii,1+N*(m-1)+n) = 1;
                    else
                        SSSold(ii,1+N*(m-1)+n) = 0;
                    end
                else
                    SSSnew(ii,1) = SSS(ii)+dt;
                    if fST(SSSnew(ii,1),mag,freq,delta1,mf,trith)>0
                        SSSnew(ii,1+N*(m-1)+n) = 1;
                    else
                        SSSnew(ii,1+N*(m-1)+n) = 0;
                    end
                end
            end
        end
    end
end

SSSEND=[SSSold;SSSnew];
SSSEND=sortrows(SSSEND,1);
%%
SST=SSS(:,1);
SSseq=SSSold(:,2:size(SSS,2));
```

Fig. 5.17 illustrates multicarrier modulation for $l = 3$, $m_a = 5/6$, and frequency modulation index, $m_f = 15$ with IPD, APOD, and POD techniques.

5.3.2 Diode Clamped Multilevel VSCs

The simplest diode clamped multi-level voltage source converter (DCM-VSC) is a three-level one, which is also called neutral point clamped (NPC) multilevel converter. NPC converter was invented by Nabae in 1981 [60]. It was later extended to more than three-level for other high power applications. Different from other multilevel converters such as flying capacitors or cascaded multilevel converters, DCM-VSCs have forbidden states [61], in which the output voltage is different if the current direction is different under the same switching function state. Fig. 5.18 shows a three-level DCM-VSC where the upper switches are assumed to be complementary with the lower switches in each leg.

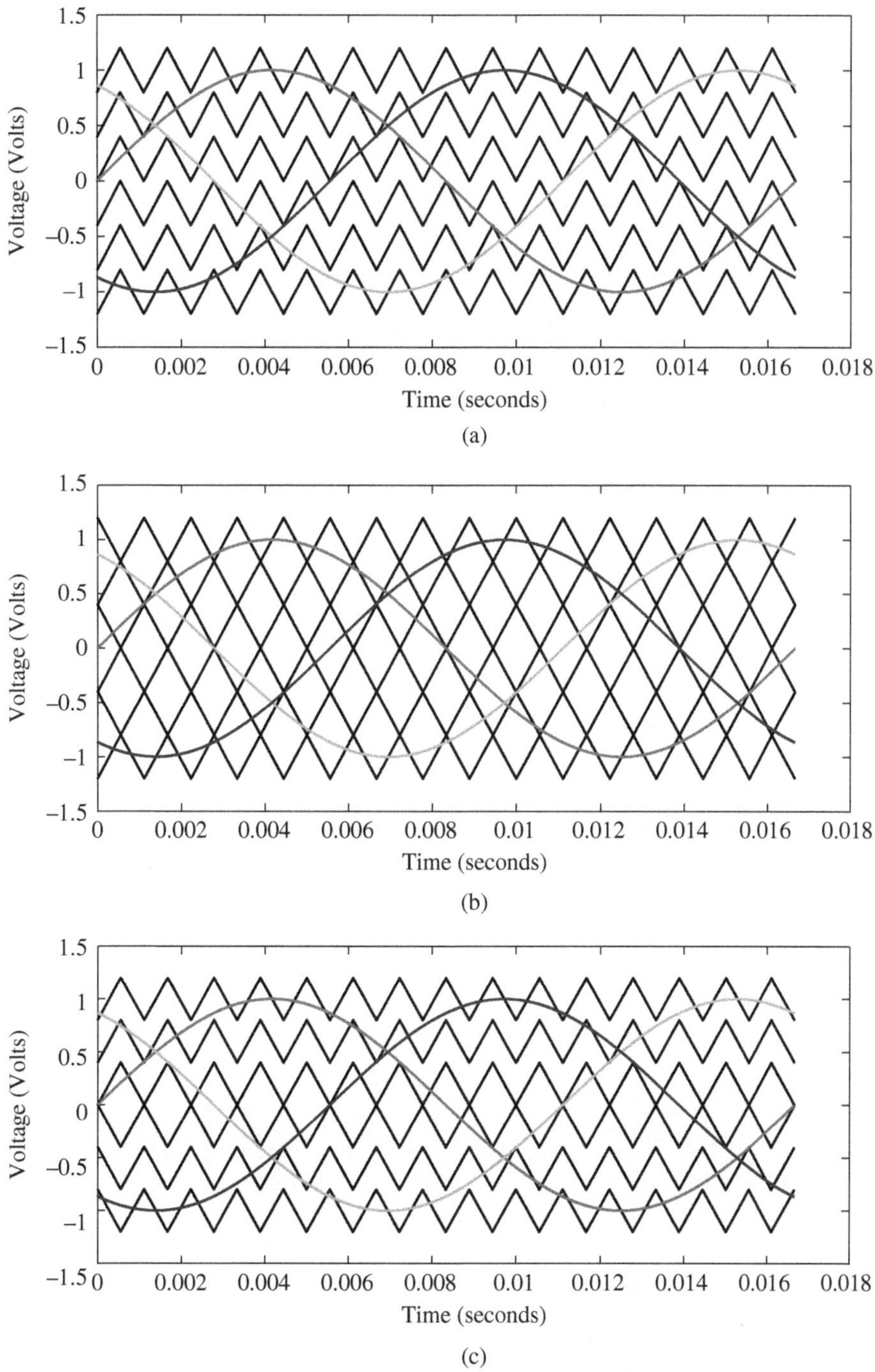

Figure 5.17 Multicarrier Modulation Techniques: (a) In-Phase Disposition; (b) Alternative Phase Opposite Disposition; (c) Phase Opposite Disposition.

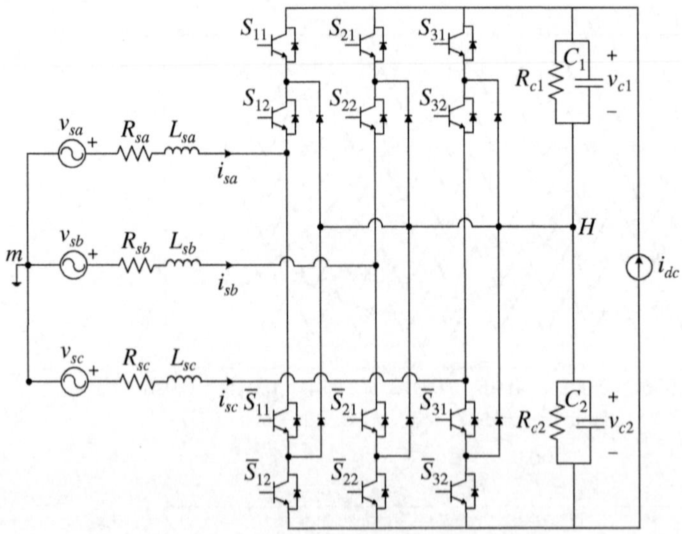

Figure 5.18 Three-phase, three-level DCM-VSC.

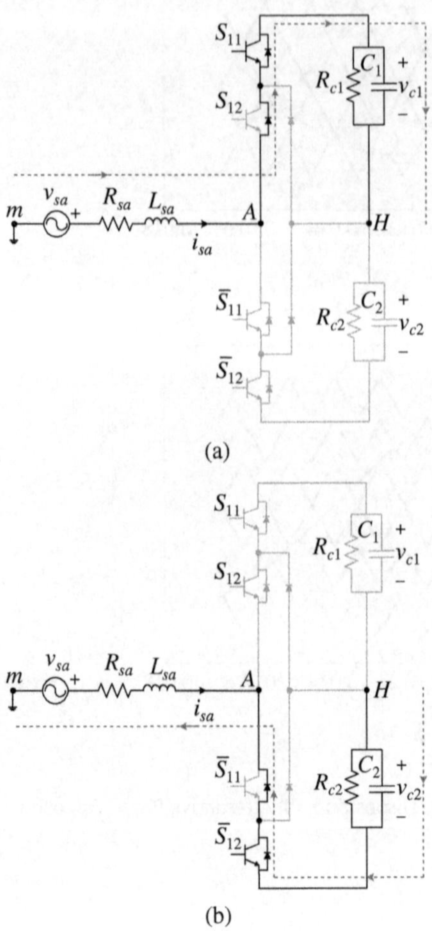

Figure 5.19 Forbidden states of DCM-VSC when (a) current flows into the converter; (b) current flows out from the converter.

To illustrate the forbidden states, Fig. 5.19 is referred, and without loss of generality, only phase a leg of the VSC is shown. When S_{11} turns on, and S_{12} turns off and if the current flows as shown in Fig. 5.19(a), the voltage across node A to its floating node (V_{AH}) becomes $V_{AH} = v_{c1}$. However, when the current flows in the reverse direction, the circuit becomes that depicted in Fig. 5.19(b) and $V_{AH} = -v_{c2}$. Consequently, $S_{11} = 1$ and $S_{12} = 0$ constitute a forbidden state because the output voltage is different when the current direction is different. In contrast, when $S_{11} = 0$, $S_{12} = 1$ the circuit remains that shown in Fig. 5.20(a), regardless of the direction of current flow. Such a state is a legitimate state. Table 5.4 lists all the possible legitimate switching states for phase a and Fig. 5.20 illustrates all the possible legitimate switching states for phase a.

Table 5.4 Legitimate switching states of phase a.

S_{11}	S_{12}	V_{AH}
1	1	v_{C1}
0	1	0
0	0	$-v_{C2}$

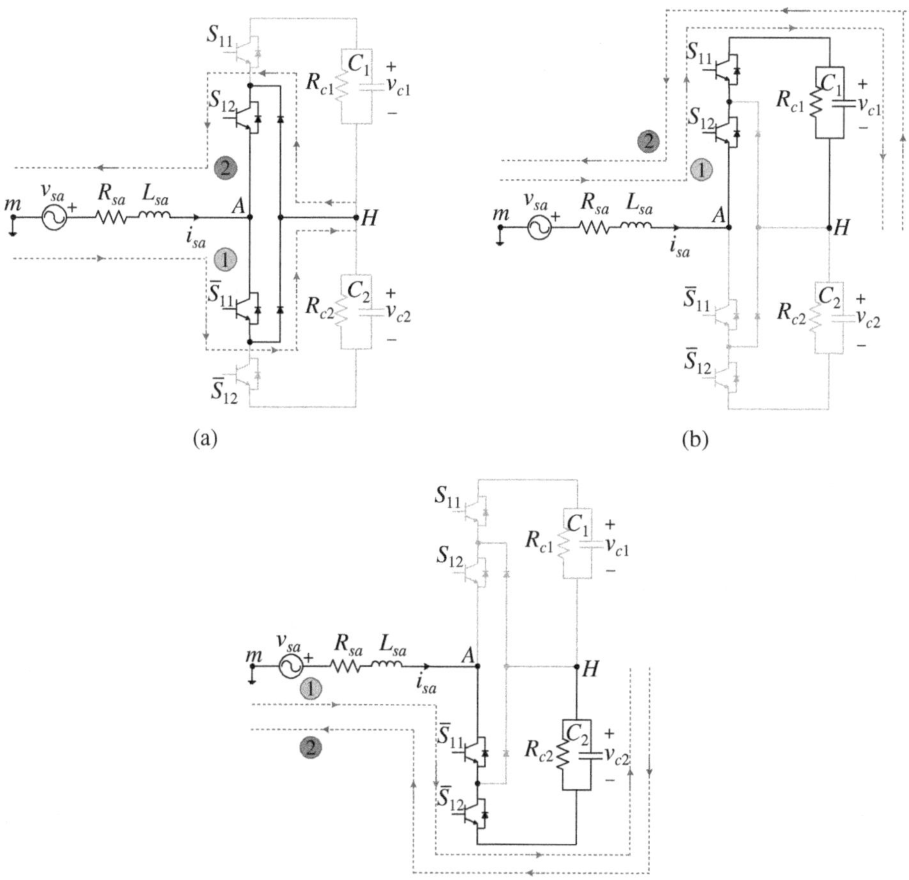

Figure 5.20 The operation modes of DCM-VSC when (a) S_{11} is off and S_{12} is on; (b) S_{11} and S_{12} are on; (c) S_{11} and S_{12} are off.

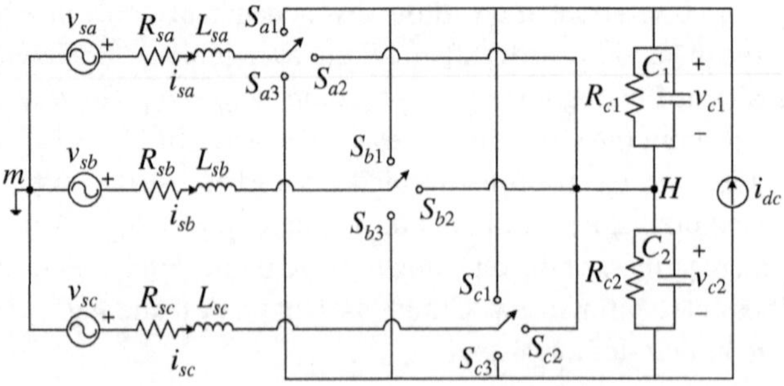

Figure 5.21 The equivalent single-pole, triple-throw model for a three-level DCM-VSC.

Table 5.5 Switching status of S_a.

	S_{a1}	S_{a2}	S_{a3}
S_a connected to S_{a1}	1	0	0
S_a connected to S_{a2}	0	1	0
S_a connected to S_{a3}	0	0	1

To have a better representation of all these states, Fig. 5.18 is replaced with Fig. 5.21 where all the bidirectional switches are replaced with single-pole triple-throw switches [62]. S_a in Fig. 5.21 can be represented by three switching states, S_{a1}, S_{a2}, S_{a3}. When S_a is connected to S_{a1} in Fig. 5.21, S_{a1}=1, S_{a2}=0, S_{a3}=0. Table 5.5 summarizes the switching status for S_a. Similar rules can be applied to S_b and S_c.

Based on Table 5.5, one can derive a three-level, three-phase, three-wire model of a DCM-VSC. The following equations can be obtained (see Figure 5.22).

When S_j is connected to S_{j1},

$$L_{sj}\frac{di_{sj}}{dt} = -R_{sj}i_{sj} - v_{c1} + v_{sj} - V_{Hm} \tag{5.114}$$

When S_j is connected to S_{j2},

$$L_{sj}\frac{di_{sj}}{dt} = -R_{sj}i_{sj} + v_{sj} - V_{Hm} \tag{5.115}$$

When S_j is connected to S_{j3},

$$L_{sj}\frac{di_{sj}}{dt} = -R_{sj}i_{sj} + v_{c2} + v_{sj} - V_{Hm} \tag{5.116}$$

The general KVL can be obtained as

$$L_{sj}\frac{di_{sj}}{dt} = -R_{sj}i_{sj} - S_{j1}v_{c1} + S_{j3}v_{c2} + v_{sj} - V_{Hm} \tag{5.117}$$

In the *abc*-frame, the dynamics can be extended into

$$\begin{cases} L_{sa}\dfrac{di_{sa}}{dt} = -R_{sa}i_{sa} - S_{a1}v_{c1} + S_{a3}v_{c2} + v_{sa} - V_{Hm} \\ L_{sb}\dfrac{di_{sb}}{dt} = -R_{sb}i_{sb} - S_{b1}v_{c1} + S_{b3}v_{c2} + v_{sb} - V_{Hm} \\ L_{sc}\dfrac{di_{sc}}{dt} = -R_{sc}i_{sc} - S_{c1}v_{c1} + S_{c3}v_{c2} + v_{sc} - V_{Hm} \end{cases} \tag{5.118}$$

Adding each equation in (5.118) yields

$$V_{Hm} = -\frac{1}{3}\left(L_{sa}\frac{di_{sa}}{dt} + L_{sb}\frac{di_{sb}}{dt} + L_{sc}\frac{di_{sc}}{dt}\right) - \frac{1}{3}\left(R_{sa}i_{sa} + R_{sb}i_{sb} + R_{sc}i_{sc}\right) - \frac{1}{3}\left(S_{a1} + S_{b1} + S_{c1}\right)v_{c1} + \frac{1}{3}\left(S_{a3} + S_{b3} + S_{c3}\right)v_{c2} + \frac{1}{3}\left(v_{sa} + v_{sb} + v_{sc}\right) \tag{5.119}$$

Substituting (5.119) into (5.118) and forming into matrix form gives

$$\mathbf{L}\frac{d\mathbf{i}}{dt} = \mathbf{A}\mathbf{x} + \mathbf{B}\mathbf{u} \tag{5.120}$$

where

$$\mathbf{L} = \begin{bmatrix} 2L_{sa} & -L_{sb} & -L_{sc} \\ -L_{sa} & 2L_{sb} & -L_{sc} \\ -L_{sa} & -L_{sb} & 2L_{sc} \end{bmatrix}_{3\times3}$$

$$\mathbf{i} = \begin{bmatrix} i_{sa} & i_{sb} & i_{sc} \end{bmatrix}^T_{3\times1}$$

$$\mathbf{A} = \begin{bmatrix} -2R_{sa} & R_{sb} & R_{sc} & -2S_{a1} + S_{b1} + S_{c1} & 2S_{a3} - S_{b3} - S_{c3} \\ R_{sa} & -2R_{sb} & R_{sc} & S_{a1} - 2S_{b1} + S_{c1} & -S_{a3} + 2S_{b3} - S_{c3} \\ R_{sa} & R_{sb} & -2R_{sc} & S_{a1} + S_{b1} - 2S_{c1} & -S_{a3} - S_{b3} + 2S_{c3} \end{bmatrix}_{3\times5}$$

$$\mathbf{x} = \begin{bmatrix} i_{sa} & i_{sb} & i_{sc} & v_{c1} & v_{c2} \end{bmatrix}^T_{5\times1}$$

$$\mathbf{B} = \begin{bmatrix} 2 & -1 & -1 & 0 \\ -1 & 2 & -1 & 0 \\ -1 & -1 & 2 & 0 \end{bmatrix}_{3\times4}$$

$$\mathbf{u} = \begin{bmatrix} v_{sa} & v_{sb} & v_{sc} & i_{dc} \end{bmatrix}^T_{4\times1}$$

On the other hand, applying KCL on DC side of converter, one get the current dynamics as

$$C_1\frac{dv_{c1}}{dt} = S_{a1}i_{sa} + S_{b1}i_{sb} + S_{c1}i_{sc} - \frac{v_{c1}}{R_{c1}} + i_{dc} \tag{5.121}$$

$$C_2\frac{dv_{c2}}{dt} = -S_{a3}i_{sa} - S_{b3}i_{sb} - S_{c3}i_{sc} - \frac{v_{c2}}{R_{c2}} + i_{dc} \tag{5.122}$$

To overcome the singularity issue, one can employ the current relationship as

$$i_{sa} + i_{sb} + i_{sc} = 0 \tag{5.123}$$

From (5.123), one can substitute phase *c* current $i_{sc} = -i_{sa} - i_{sb}$ into (5.120) resulting in

$$\mathbf{L}_r\frac{d\mathbf{i}_r}{dt} = \mathbf{A}_r\mathbf{x}_r + \mathbf{B}_r\mathbf{u} \tag{5.124}$$

where

$$\mathbf{L}_r = \begin{bmatrix} 2L_{sa} + L_{sc} & -L_{sb} + L_{sc} \\ -L_{sa} + L_{sc} & 2L_{sb} + L_{sc} \end{bmatrix}_{2\times2}$$

$$\mathbf{i}_r = \begin{bmatrix} i_{sa} & i_{sb} \end{bmatrix}^T_{2\times1}$$

$$\mathbf{A}_r = \begin{bmatrix} -2R_{sa} - R_{sc} & R_{sb} - R_{sc} & -2S_{a1} + S_{b1} + S_{c1} & 2S_{a3} - S_{b3} - S_{c3} \\ R_{sa} - R_{sc} & -2R_{sb} - R_{sc} & S_{a1} - 2S_{b1} + S_{c1} & -S_{a3} + 2S_{b3} - S_{c3} \end{bmatrix}_{2\times4}$$

$$\mathbf{x}_r = \begin{bmatrix} i_{sa} & i_{sb} & v_{c1} & v_{c2} \end{bmatrix}^T_{4\times1}$$

$$\mathbf{B}_r = \begin{bmatrix} 2 & -1 & -1 & 0 \\ -1 & 2 & -1 & 0 \end{bmatrix}_{2\times4}$$

The dc side dynamics can be modified into

$$C_1 \frac{dv_{c1}}{dt} = \left(S_{a1} - S_{c1}\right) i_{sa} + \left(S_{b1} - S_{c1}\right) i_{sb} - \frac{v_{c1}}{R_{c1}} + i_{dc} \tag{5.125}$$

$$C_2 \frac{dv_{c2}}{dt} = -\left(S_{a3} - S_{c3}\right) i_{sa} - \left(S_{b3} - S_{c3}\right) i_{sb} - \frac{v_{c2}}{R_{c2}} + i_{dc} \tag{5.126}$$

Thus, combining (5.125), (5.126), and (5.120) into matrix form yields

$$\dot{\mathbf{x}}_r = \mathbf{F}\mathbf{x}_r + \mathbf{G}\mathbf{u} \tag{5.127}$$

where

$$\mathbf{F} = \begin{bmatrix} \multicolumn{4}{c}{\mathbf{L}_r^{-1}\mathbf{A}_r} \\ \frac{(S_{a1}-S_{c1})}{C_1} & \frac{(S_{b1}-S_{c1})}{C_1} & -\frac{1}{R_{c1}C_1} & 0 \\ \frac{-(S_{a3}-S_{c3})}{C_2} & \frac{-(S_{b3}-S_{c3})}{C_2} & 0 & -\frac{1}{R_{c2}C_2} \end{bmatrix}_{4\times4}$$

$$\mathbf{G} = \begin{bmatrix} \multicolumn{4}{c}{\mathbf{L}_r^{-1}\mathbf{B}_r} \\ 0 & 0 & 0 & \frac{1}{C_1} \\ 0 & 0 & 0 & \frac{1}{C_2} \end{bmatrix}_{4\times4}$$

5.3.3 Flying Capacitor Multilevel VSCs

The topology of a three-phase l-level flying capacitor VSC (FC-VSC) is depicted in Fig. 5.23. Each phase of a FC-VSC is composed of N series cells, in which each cell consists of two IGBTs (S_j and $\overline{S}_j, j \in a, b, c$). The cells are connected to each other through an auxiliary capacitor C_j. The number of level (l) in FC-VSC determines the number of N-cells through (5.128)

$$l = N + 1 \tag{5.128}$$

To derive the FCM-VSC, let us first look at an equivalent three-level single-phase flying capacitor as shown in Fig. 5.24. Since the number of switches is four, the dynamics of four operation modes, listed in Table 5.6, are derived.

When S_{j2} and S_{j1} are ON, as shown in Fig. 5.25(a), one may derive the KVL as

$$L_{sj} \frac{di_{sj}}{dt} = v_{sj} - R_{sj} i_{sj} - v_{dc} - V_{Hm} \tag{5.129}$$

However, when S_{j1} is OFF, as depicted in Fig. 5.25(b), the dynamics can be obtained as

$$L_{sj} \frac{di_{sj}}{dt} = v_{sj} - R_{sj} i_{sj} + v_{cj1} - v_{dc} - V_{Hm} \tag{5.130}$$

Figure 5.22 The operation modes of DCM-VSC when the switch is connected to (a) S_{j1}; (b) S_{j2}; (c) S_{j3}.

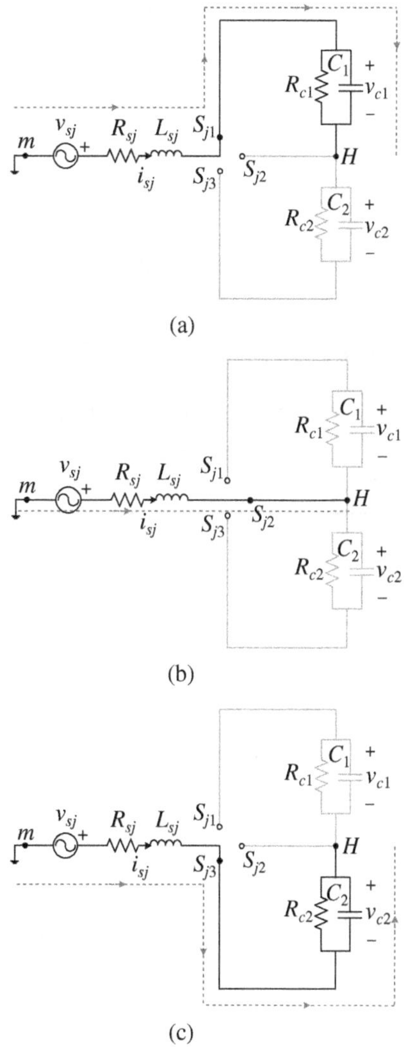

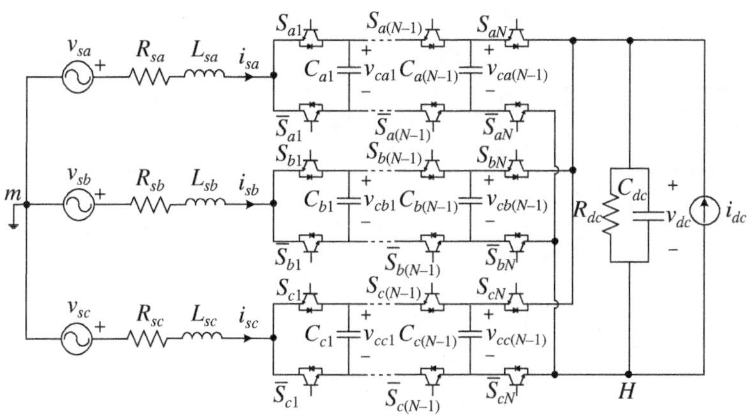

Figure 5.23 Schematic diagram of an l-level FC-VSC.

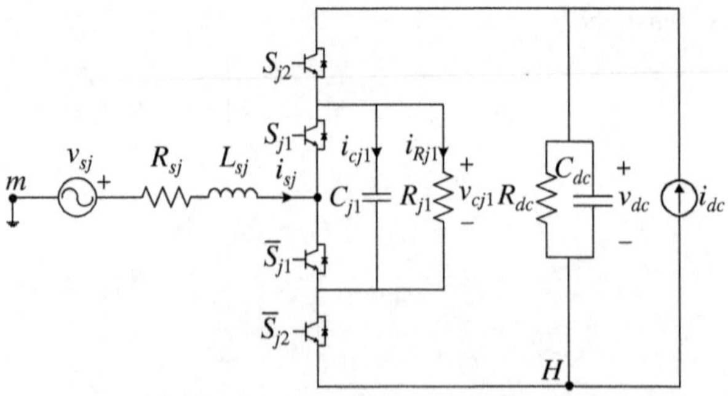

Figure 5.24 Equivalent circuit of single-phase FC-VSC.

Table 5.6 Operation modes in three-level single-phase FC-VSC.

Mode	S_{j1}	S_{j2}
1	1	1
2	0	1
3	1	0
4	0	0

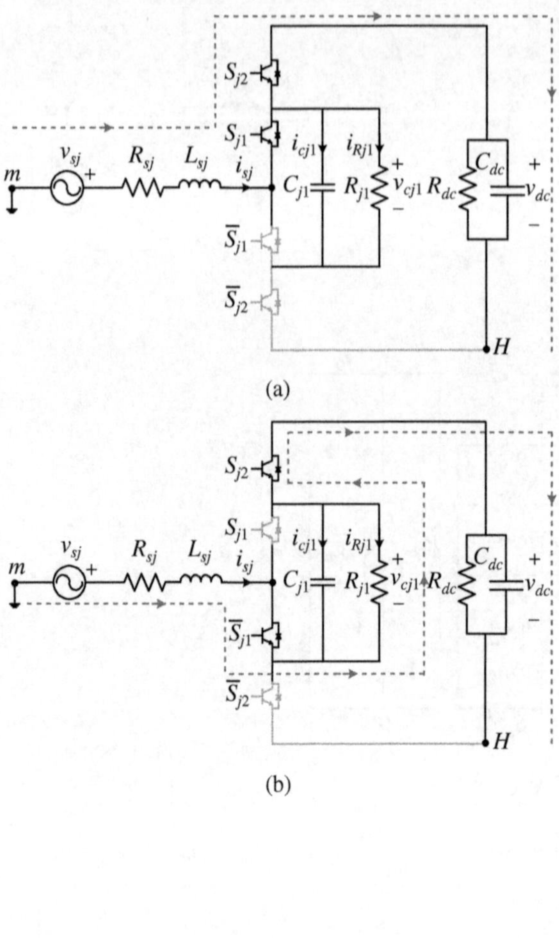

Figure 5.25 The operation modes of FC-VSC when S_{j2} is ON and (a) S_{j1} is ON; (b) S_{j1} is OFF.

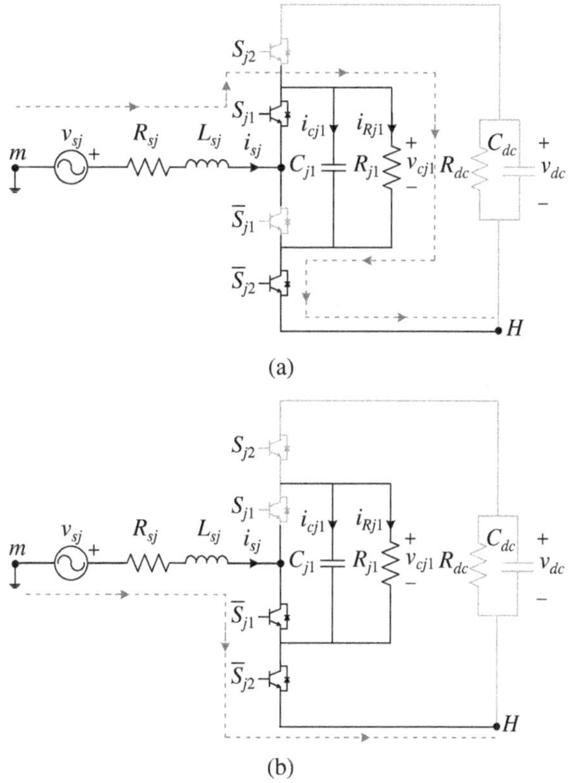

Figure 5.26 The operation modes of FC-VSC when S_{j2} is OFF and (a) S_{j1} is ON; (b) S_{j1} is OFF.

When S_{j2} is OFF and S_{j1} is ON, as shown in Fig. 5.26(a), the KVL can be derived as

$$L_{sj}\frac{di_{sj}}{dt} = v_{sj} - R_{sj}i_{sj} - v_{cj1} - V_{Hm} \tag{5.131}$$

On the other hand, when S_{j2} and S_{j1} are OFF, as illustrated in Fig. 5.26(b), the dynamics can be obtained as

$$L_{sj}\frac{di_{sj}}{dt} = v_{sj} - R_{sj}i_{sj} - V_{Hm} \tag{5.132}$$

From (5.129) to (5.132), it can be noticed that the existences of the capacitor voltage and dc voltage are affected by the switching patterns. Thus, by combining (5.129) to (5.132), the general KVL can be obtained as

$$L_{sj}\frac{di_{sj}}{dt} = -R_{sj}i_{sj} - \left(S_{j1} - S_{j2}\right)v_{cj1} - S_{j2}v_{dc} - V_{Hm} + v_{sj} \tag{5.133}$$

In the *abc*-frame, the dynamics can be extended into

$$\begin{cases} L_{sa}\dfrac{di_{sa}}{dt} = -R_{sa}i_{sa} - \left(S_{a1} - S_{a2}\right)v_{ca1} - S_{a2}v_{dc} - V_{Hm} + v_{sa} \\ L_{sb}\dfrac{di_{sb}}{dt} = -R_{sb}i_{sb} - \left(S_{b1} - S_{b2}\right)v_{cb1} - S_{b2}v_{dc} - V_{Hm} + v_{sb} \\ L_{sc}\dfrac{di_{sc}}{dt} = -R_{sc}i_{sc} - \left(S_{c1} - S_{c2}\right)v_{cc1} - S_{c2}v_{dc} - V_{Hm} + v_{sc} \end{cases} \tag{5.134}$$

Adding the equations in (5.134) yields

$$\begin{aligned}V_{Hm} = &-\frac{1}{3}\left(L_{sa}\frac{di_{sa}}{dt} + L_{sb}\frac{di_{sb}}{dt} + L_{sc}\frac{di_{sc}}{dt}\right) - \frac{1}{3}\left(R_{sa}i_{sa} + R_{sb}i_{sb} + R_{sc}i_{sc}\right)\\ &-\frac{1}{3}\left[\left(S_{a1} - S_{a2}\right)v_{ca1} + \left(S_{b1} - S_{b2}\right)v_{cb1} + \left(S_{c1} - S_{c2}\right)v_{cc1}\right] - \frac{1}{3}\left(S_{a2} + S_{b2} + S_{c2}\right)v_{dc}\\ &+\frac{1}{3}\left(v_{sa} + v_{sb} + v_{sc}\right)\end{aligned} \tag{5.135}$$

Substituting (5.135) into (5.134) and forming into a matrix form, produces

$$\mathbf{L}\frac{d\mathbf{i}}{dt} = \mathbf{A}\mathbf{x} + \mathbf{B}\mathbf{u} \tag{5.136}$$

where

$$\mathbf{L} = \begin{bmatrix} 2L_{sa} & -L_{sb} & -L_{sc} \\ -L_{sa} & 2L_{sb} & -L_{sc} \\ -L_{sa} & -L_{sb} & 2L_{sc} \end{bmatrix}_{3\times3}$$

$$\mathbf{i} = \begin{bmatrix} i_{sa} & i_{sb} & i_{sc} \end{bmatrix}_{3\times1}^{T}$$

$$\mathbf{A} = \begin{bmatrix} \mathbf{A_1} & \mathbf{A_2} & \mathbf{A_3} \end{bmatrix}_{3\times7}$$

$$\mathbf{A_1} = \begin{bmatrix} -2R_{sa} & R_{sb} & R_{sc} \\ R_{sa} & -2R_{sb} & R_{sc} \\ R_{sa} & R_{sb} & -2R_{sc} \end{bmatrix}_{3\times3}$$

$$\mathbf{A_2} = \begin{bmatrix} -2\left(S_{a1} - S_{a2}\right) & S_{b1} - S_{b2} & S_{c1} - S_{c2} \\ S_{a1} - S_{a2} & -2\left(S_{b1} - S_{b2}\right) & S_{c1} - S_{c2} \\ S_{a1} - S_{a2} & S_{b1} - S_{b2} & -2\left(S_{c1} - S_{c2}\right) \end{bmatrix}_{3\times3}$$

$$\mathbf{A_3} = \begin{bmatrix} -2S_{a2} + S_{b2} + S_{c2} \\ S_{a2} - 2S_{b2} + S_{c2} \\ S_{a2} + S_{b2} - 2S_{c2} \end{bmatrix}_{3\times1}$$

$$\mathbf{x} = \begin{bmatrix} i_{sa} & i_{sb} & i_{sc} & v_{ca1} & v_{cb1} & v_{cc1} & v_{dc} \end{bmatrix}_{7\times1}^{T}$$

$$\mathbf{B} = \begin{bmatrix} 2 & -1 & -1 & 0 \\ -1 & 2 & -1 & 0 \\ -1 & -1 & 2 & 0 \end{bmatrix}_{3\times4}$$

$$\mathbf{u} = \begin{bmatrix} v_{sa} & v_{sb} & v_{sc} & i_{dc} \end{bmatrix}_{4\times1}^{T}$$

On the other hand, by applying KCL in each cell and dc side of the converter, one may obtain the submodule (SM) and dc-link capacitor voltage dynamics as

$$C_{j1}\frac{dv_{cj1}}{dt} = \left(S_{j1} - S_{j2}\right)i_{sj} - \frac{v_{cj1}}{R_{j1}} \tag{5.137}$$

$$C_{dc}\frac{dv_{dc}}{dt} = S_{a2}i_{sa} + S_{b2}i_{sb} + S_{c2}i_{sc} - \frac{v_{dc}}{R_{dc}} + i_{dc} \tag{5.138}$$

Generalizing (5.137) into the *abc*-frame produces

$$\begin{cases} C_{a1}\dfrac{dv_{ca1}}{dt} = \left(S_{a1} - S_{a2}\right) i_{sa} - \dfrac{v_{ca1}}{R_{a1}} \\ C_{b1}\dfrac{dv_{cb1}}{dt} = \left(S_{b1} - S_{b2}\right) i_{sb} - \dfrac{v_{cb1}}{R_{b1}} \\ C_{c1}\dfrac{dv_{cc1}}{dt} = \left(S_{c1} - S_{c2}\right) i_{sc} - \dfrac{v_{cc1}}{R_{c1}} \end{cases} \tag{5.139}$$

As **L** is prone to have the singularity issue, the modification is made by using the relationship of phase current (5.140)

$$i_{sa} + i_{sb} + i_{sc} = 0 \tag{5.140}$$

Consequently, the reduced state-space model can be obtained by substituting phase *c* current as $i_{sc} = -i_{sa} - i_{sb}$ into the first and second rows of (5.136) as

$$\mathbf{L}_r\frac{d\mathbf{i}_r}{dt} = \mathbf{A}_r\mathbf{x}_r + \mathbf{B}_r\mathbf{u} \tag{5.141}$$

where

$$\mathbf{L}_r = \begin{bmatrix} 2L_{sa} + L_{sc} & -L_{sb} + L_{sc} \\ -L_{sa} + L_{sc} & 2L_{sb} + L_{sc} \end{bmatrix}_{2\times2}$$

$$\mathbf{i}_r = \begin{bmatrix} i_{sa} & i_{sb} \end{bmatrix}^T_{2\times1}$$

$$\mathbf{A}_r = \begin{bmatrix} \mathbf{A}_1 & \mathbf{A}_2 & \mathbf{A}_3 \end{bmatrix}_{2\times6}$$

$$\mathbf{A}_1 = \begin{bmatrix} -2R_{sa} - R_{sc} & R_{sb} - R_{sc} \\ R_{sa} - R_{sc} & -2R_{sb} - R_{sc} \end{bmatrix}_{2\times2}$$

$$\mathbf{A}_2 = \begin{bmatrix} -2\left(S_{a1} - S_{a2}\right) & S_{b1} - S_{b2} & S_{c1} - S_{c2} \\ S_{a1} - S_{a2} & -2\left(S_{b1} - S_{b2}\right) & S_{c1} - S_{c2} \end{bmatrix}_{2\times3}$$

$$\mathbf{A}_3 = \begin{bmatrix} -2S_{a2} + S_{b2} + S_{c2} \\ S_{a2} - 2S_{b2} + S_{c2} \end{bmatrix}_{2\times1}$$

$$\mathbf{x}_r = \begin{bmatrix} i_{sa} & i_{sb} & v_{ca1} & v_{cb1} & v_{cc1} & v_{dc} \end{bmatrix}^T_{6\times1}$$

$$\mathbf{B}_r = \begin{bmatrix} 2 & -1 & -1 & 0 \\ -1 & 2 & -1 & 0 \end{bmatrix}_{2\times4}$$

(5.139) and (5.138) can be modified into

$$C_{c1}\frac{dv_{cc1}}{dt} = -\left(S_{c1} - S_{c2}\right) i_{sa} - \left(S_{c1} - S_{c2}\right) i_{sb} - \frac{v_{cc1}}{R_{c1}} \tag{5.142}$$

$$C_{dc}\frac{dv_{dc}}{dt} = \left(S_{a2} - S_{c2}\right) i_{sa} + \left(S_{b2} - S_{c2}\right) i_{sb} - \frac{v_{dc}}{R_{dc}} + i_{dc} \tag{5.143}$$

Thus, incorporating (5.141), (5.142), and (5.143) into a matrix form yields

$$\dot{\mathbf{x}}_r = \mathbf{F}\mathbf{x}_r + \mathbf{G}\mathbf{u} \tag{5.144}$$

where

$$\mathbf{F} = \begin{bmatrix} & & & \mathbf{L}_r^{-1}\mathbf{A}_r & & \\ \frac{S_{a1}-S_{a2}}{C_{a1}} & 0 & -\frac{1}{R_{a1}C_{a1}} & 0 & 0 & 0 \\ 0 & \frac{S_{b1}-S_{b2}}{C_{b1}} & 0 & -\frac{1}{R_{b1}C_{b1}} & 0 & 0 \\ -\frac{(S_{c1}-S_{c2})}{C_{c1}} & -\frac{(S_{c1}-S_{c2})}{C_{c1}} & 0 & 0 & -\frac{1}{R_{c1}C_{c1}} & 0 \\ \frac{S_{a2}-S_{c2}}{C_{dc}} & \frac{S_{b2}-S_{c2}}{C_{dc}} & 0 & 0 & 0 & -\frac{1}{R_{dc}C_{dc}} \end{bmatrix}_{6\times 6}$$

$$\mathbf{G} = \begin{bmatrix} & \mathbf{L}_r^{-1}\mathbf{B}_r & & \\ 0 & 0 & 0 & 0 \\ 0 & 0 & 0 & 0 \\ 0 & 0 & 0 & 0 \\ 0 & 0 & 0 & \frac{1}{C_{dc}} \end{bmatrix}_{6\times 4}$$

Now let us extend the three-level FC-VSC into l-level FC-VSC by generalizing (5.134) as

$$\begin{cases} L_{sa}\dfrac{di_{sa}}{dt} = -R_{sa}i_{sa} - \mathbf{S_a}\mathbf{v_{ca}} - S_{aN}v_{dc} - V_{Hm} + v_{sa} \\ L_{sb}\dfrac{di_{sb}}{dt} = -R_{sb}i_{sb} - \mathbf{S_b}\mathbf{v_{cb}} - S_{bN}v_{dc} - V_{Hm} + v_{sb} \\ L_{sc}\dfrac{di_{sc}}{dt} = -R_{sc}i_{sc} - \mathbf{S_c}\mathbf{v_{cc}} - S_{cN}v_{dc} - V_{Hm} + v_{sc} \end{cases} \tag{5.145}$$

where

$$\begin{aligned} \mathbf{S_a} &= \left[\left(S_{a1} - S_{a2}\right) \ \ \ldots \ \ \left(S_{aN-1} - S_{aN}\right)\right]_{1\times N-1} \\ \mathbf{S_b} &= \left[\left(S_{b1} - S_{b2}\right) \ \ \ldots \ \ \left(S_{bN-1} - S_{bN}\right)\right]_{1\times N-1} \\ \mathbf{S_c} &= \left[\left(S_{c1} - S_{c2}\right) \ \ \ldots \ \ \left(S_{cN-1} - S_{cN}\right)\right]_{1\times N-1} \\ \mathbf{v_{ca}} &= \left[v_{ca1} \ \ \cdots \ \ v_{caN-1}\right]^T_{N-1\times 1} \\ \mathbf{v_{cb}} &= \left[v_{cb1} \ \ \cdots \ \ v_{cbN-1}\right]^T_{N-1\times 1} \\ \mathbf{v_{cc}} &= \left[v_{cc1} \ \ \cdots \ \ v_{ccN-1}\right]^T_{N-1\times 1} \end{aligned}$$

The dynamics of the capacitor module at each SM can be generalized as

$$\begin{cases} C_{ak}\dfrac{dv_{cak}}{dt} = \left(S_{ak} - S_{ak+1}\right) i_{sa} - \dfrac{v_{cak}}{R_{ak}} \\ C_{bk}\dfrac{dv_{cbk}}{dt} = \left(S_{bk} - S_{bk+1}\right) i_{sb} - \dfrac{v_{cbk}}{R_{bk}} \\ C_{ck}\dfrac{dv_{cck}}{dt} = \left(S_{ck} - S_{ck+1}\right) i_{sc} - \dfrac{v_{cck}}{R_{c1}} \end{cases} \tag{5.146}$$

where $k \in [1, N-1]$. Furthermore, the dynamics of the dc capacitor is represented as

$$C_{dc}\frac{dv_{dc}}{dt} = S_{aN}i_{sa} + S_{bN}i_{sb} + S_{cN}i_{sc} - \frac{v_{dc}}{R_{dc}} + i_{dc} \tag{5.147}$$

Extending the three-level FC-VSC for l-level FC-VSC, (5.141) can be modified into

$$\mathbf{L}_r\frac{d\mathbf{i}_r}{dt} = \mathbf{A}_r\mathbf{x}_r + \mathbf{B}_r\mathbf{u} \tag{5.148}$$

where

$$\mathbf{L}_r = \begin{bmatrix} 2L_{sa} + L_{sc} & -L_{sb} + L_{sc} \\ -L_{sa} + L_{sc} & 2L_{sb} + L_{sc} \end{bmatrix}_{2\times2}$$

$$\mathbf{A}_r = \begin{bmatrix} \mathbf{A_1} & \mathbf{A_2} & \mathbf{A_3} \end{bmatrix}_{2\times3N}$$

$$\mathbf{A_1} = \begin{bmatrix} -2R_{sa} - R_{sc} & R_{sb} - R_{sc} \\ R_{sa} - R_{sc} & -2R_{sb} - R_{sc} \end{bmatrix}_{2\times2}$$

$$\mathbf{A_2} = \begin{bmatrix} -2\mathbf{S_a} & \mathbf{S_b} & \mathbf{S_c} \\ \mathbf{S_a} & -2\mathbf{S_b} & \mathbf{S_c} \end{bmatrix}_{2\times3N-3}$$

$$\mathbf{A_3} = \begin{bmatrix} -2S_{aN} + S_{bN} + S_{cN} \\ S_{aN} - 2S_{bN} + S_{cN} \end{bmatrix}_{2\times1}$$

$$\mathbf{x}_r = \begin{bmatrix} i_{sa} & i_{sb} & \mathbf{v_{ca}} & \mathbf{v_{cb}} & \mathbf{v_{cc}} & v_{dc} \end{bmatrix}^T_{3N\times1}$$

$$\mathbf{B}_r = \begin{bmatrix} 2 & -1 & -1 & 0 \\ -1 & 2 & -1 & 0 \end{bmatrix}_{2\times4}$$

Combining all of the equations, the state-space representation of l-level FC-VSC can be denoted as

$$\dot{\mathbf{x}}_r = \mathbf{F}\mathbf{x}_r + \mathbf{G}\mathbf{u} \tag{5.149}$$

where

$$\mathbf{F} = \begin{bmatrix} & & \mathbf{L}_r^{-1}\mathbf{A}_r & & & \\ \mathbf{F_a} & 0 & -\mathbf{F_{va}} & 0 & 0 & 0 \\ 0 & \mathbf{F_b} & 0 & -\mathbf{F_{vb}} & 0 & 0 \\ -\mathbf{F_c} & -\mathbf{F_c} & 0 & 0 & -\mathbf{F_{vc}} & 0 \\ \frac{S_{aN}-S_{cN}}{C_{dc}} & \frac{S_{bN}-S_{cN}}{C_{dc}} & 0 & 0 & 0 & -\frac{1}{R_{dc}C_{dc}} \end{bmatrix}_{3N\times3N}$$

$$\mathbf{F_a} = \begin{bmatrix} \frac{S_{a1}-S_{a2}}{C_{a1}} & \cdots & \frac{S_{aN-1}-S_{aN}}{C_{aN-1}} \end{bmatrix}^T_{N-1\times1}$$

$$\mathbf{F_b} = \begin{bmatrix} \frac{S_{b1}-S_{b2}}{C_{b1}} & \cdots & \frac{S_{bN-1}-S_{bN}}{C_{bN-1}} \end{bmatrix}^T_{N-1\times1}$$

$$\mathbf{F_c} = \begin{bmatrix} \frac{S_{c1}-S_{c2}}{C_{c1}} & \cdots & \frac{S_{cN-1}-S_{cN}}{C_{cN-1}} \end{bmatrix}^T_{N-1\times1}$$

$$\mathbf{F_{va}} = diag\left(\begin{bmatrix} \frac{1}{R_{a1}C_{a1}} & \cdots & \frac{1}{R_{aN-1}C_{aN-1}} \end{bmatrix}\right)_{N-1\times N-1}$$

$$\mathbf{F_{vb}} = diag\left(\begin{bmatrix} \frac{1}{R_{b1}C_{b1}} & \cdots & \frac{1}{R_{bN-1}C_{bN-1}} \end{bmatrix}\right)_{N-1\times N-1}$$

$$\mathbf{F_{vc}} = diag\left(\begin{bmatrix} \frac{1}{R_{c1}C_{c1}} & \cdots & \frac{1}{R_{cN-1}C_{cN-1}} \end{bmatrix}\right)_{N-1\times N-1}$$

$$\mathbf{G} = \begin{bmatrix} & \mathbf{L}_r^{-1}\mathbf{B}_r & & \\ 0 & 0 & 0 & 0 \\ \vdots & \vdots & \vdots & \vdots \\ 0 & 0 & 0 & 0 \\ 0 & 0 & 0 & \frac{1}{C_{dc}} \end{bmatrix}_{3N\times4}$$

5.3.4 Cascaded Multi-Level VSCs

Cascaded multi-level voltage source converter (CML-VSC) is one of the most well-known converter topologies for high-power medium-voltage applications [63]. This converter consists of multiple modules of single-phase H-bridge converter. As illustrated in Fig 5.27, each of H-bridge converter has one capacitor installed in each module. Thus, this configuration allows the converter to be used for higher voltage level applications. Each module is equipped with four active switches, in which the upper and lower switches work reciprocally. Note that, $\overline{S}$ indicates the complementary of switch S. For example, when the switch S_{a1} is on, then the switch $\overline{S}_{a1}$ is off. The relation between the number of modules with level is described as

$$l = 2N + 1 \tag{5.150}$$

where N is the number of H-bridge modules of each phase. Thus, the voltage level l is an odd number. The operating voltage, harmonic requirements, and manufacturing cost are the factors to determine the number of modules in CML-VSC. This converter offers numerous features compared to the conventional two-level VSC. The main advantages are modular structure, lower voltage THD, and lower dv/dt [32].

Let us consider that the converter is connected to the three-phase voltage source v_{sa}, v_{sb}, and v_{sc} with the line parameters R_{sj} and L_{sj} for $j \in a, b, c$. Assume the line parameters

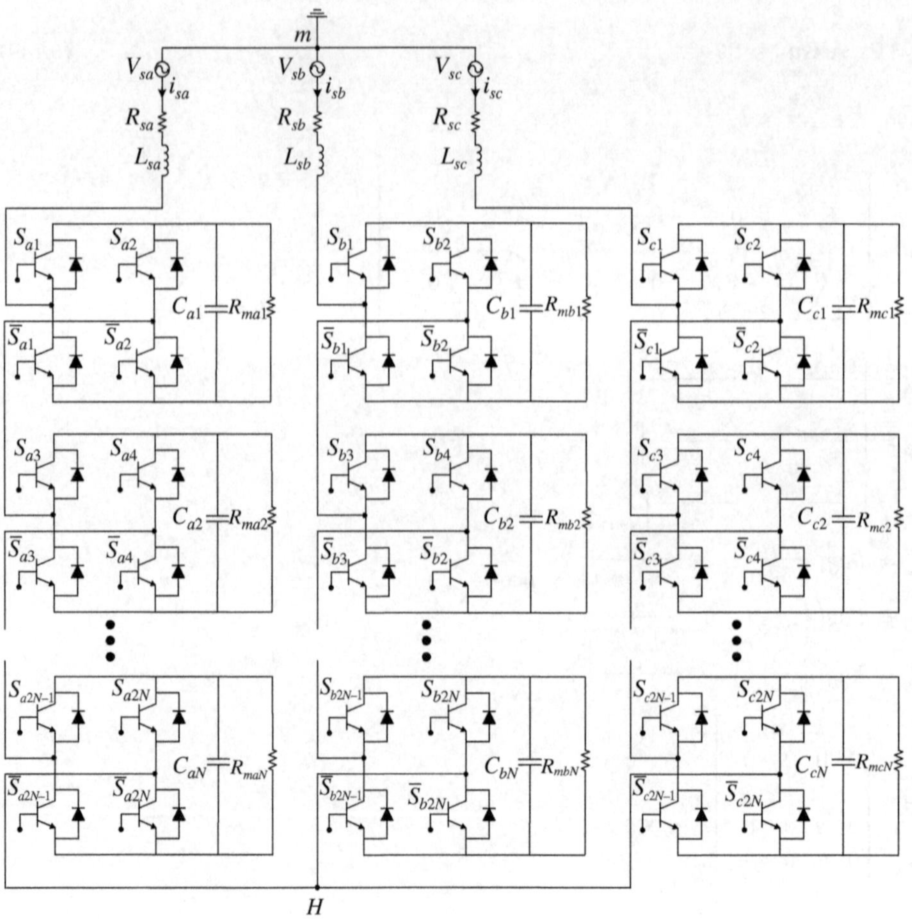

Figure 5.27 Schematic circuit of three-phase l-level CML-VSC.

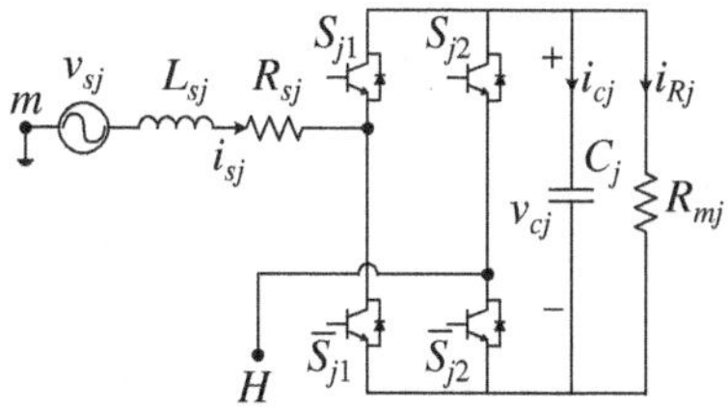

Figure 5.28 Equivalent circuit of single-phase CML-VSC.

Table 5.7 Operation modes in three-level single-phase CML-VSC.

Mode	S_{j1}	S_{j2}
1	1	1
2	0	1
3	1	0
4	0	0

of each phase and the capacitor of each module in CML-VSC are not identical as shown in Fig 5.27.

To simplify the analysis, let us consider the equivalent circuit of single-phase CML-VSC as depicted in Fig. 5.28. Note that C_j and R_{mj} are submodule capacitor and resistor. $\overline{S}_{j1}$ and $\overline{S}_{j2}$ are the complementary of switches S_{j1} and S_{j2}, respectively.

Since the number of switches is four, four operation modes, listed in Table 5.7, are investigated. When S_{j2} and S_{j1} are ON, as depicted in Fig. 5.29(a), one may derive the KVL as

$$L_{sj}\frac{di_{sj}}{dt} = -R_{sj}i_{sj} + v_{sj} - V_{Hm} \tag{5.151}$$

Nevertheless, when S_{j1} is OFF, as illustrated in Fig. 5.29(b), the dynamics can be obtained as

$$L_{sj}\frac{di_{sj}}{dt} = -R_{sj}i_{sj} + v_{cj} + v_{sj} - V_{Hm} \tag{5.152}$$

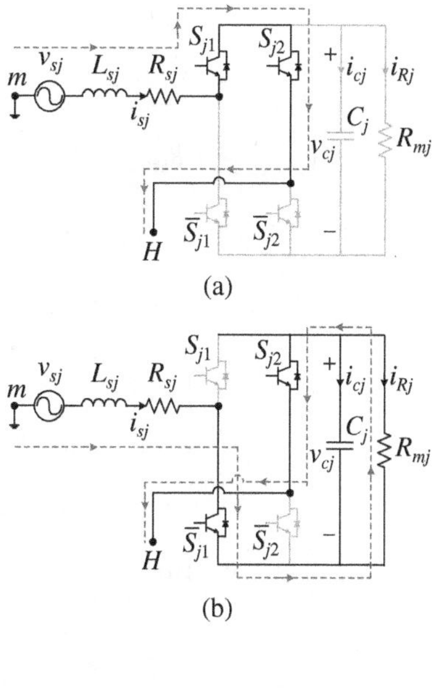

Figure 5.29 The operation modes of CML-VSC when S_{j2} is ON and (a) S_{j1} is ON; (b) S_{j1} is OFF.

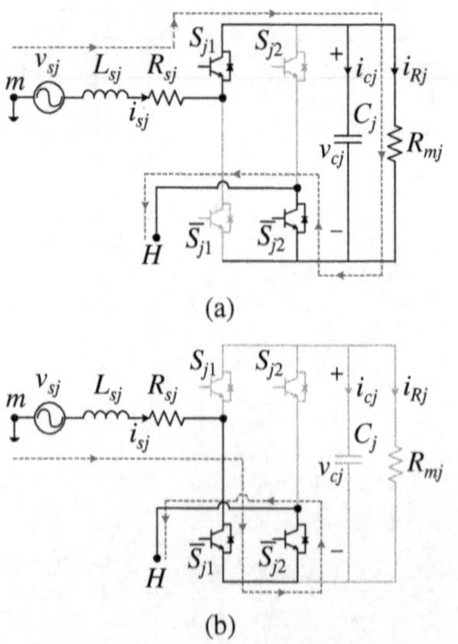

Figure 5.30 The operation modes of CML-VSC when S_{j2} is OFF and (a) S_{j1} is ON; (b) S_{j1} is OFF.

When S_{j2} is OFF and S_{j1} is ON, as depicted in Fig. 5.30(a), the KVL can be derived as

$$L_{sj}\frac{di_{sj}}{dt} = -R_{sj}i_{sj} - v_{cj} + v_{sj} - V_{Hm} \tag{5.153}$$

On the other hand, when S_{j1} is OFF, as shown in Fig. 5.30(b), the dynamics be obtained as

$$L_{sj}\frac{di_{sj}}{dt} = -R_{sj}i_{sj} + v_{sj} - V_{Hm} \tag{5.154}$$

According to (5.151) – (5.154), it can be observed that the capacitor voltage is influenced by the switching patterns. Hence, by combining (5.151) – (5.154), the general KVL can be formulated as

$$L_{sj}\frac{di_{sj}}{dt} = -R_{sj}i_{sj} - \left(S_{j1} - S_{j2}\right) v_{cj} + v_{sj} - V_{Hm} \tag{5.155}$$

In the *abc*-frame, (5.155) can be extended into

$$\begin{cases} L_{sa}\dfrac{di_{sa}}{dt} = -R_{sa}i_{sa} - \left(S_{a1} - S_{a2}\right) v_{ca} + v_{sa} - V_{Hm} \\ L_{sb}\dfrac{di_{sb}}{dt} = -R_{sb}i_{sb} - \left(S_{b1} - S_{b2}\right) v_{cb} + v_{sb} - V_{Hm} \\ L_{sc}\dfrac{di_{sc}}{dt} = -R_{sc}i_{sc} - \left(S_{c1} - S_{c2}\right) v_{cc} + v_{sc} - V_{Hm} \end{cases} \tag{5.156}$$

Adding each equation in (5.156) gives

$$\begin{aligned} V_{Hm} = &-\frac{1}{3}\left(L_{sa}\frac{di_{sa}}{dt} + L_{sb}\frac{di_{sb}}{dt} + L_{sc}\frac{di_{sc}}{dt}\right) - \frac{1}{3}\left(R_{sa}i_{sa} + R_{sb}i_{sb} + R_{sc}i_{sc}\right) \\ &- \frac{1}{3}\left[\left(S_{a1} - S_{a2}\right) v_{ca} + \left(S_{b1} - S_{b2}\right) v_{cb} + \left(S_{c1} - S_{c2}\right) v_{cc}\right] \\ &+ \frac{1}{3}\left(v_{sa} + v_{sb} + v_{sc}\right) \end{aligned} \tag{5.157}$$

Substituting (5.157) into (5.156) and forming into a matrix form results in

$$\mathbf{L}\frac{d\mathbf{i}}{dt} = \mathbf{A}\mathbf{x} + \mathbf{B}\mathbf{u} \tag{5.158}$$

where

$$\mathbf{L} = \begin{bmatrix} 2L_{sa} & -L_{sb} & -L_{sc} \\ -L_{sa} & 2L_{sb} & -L_{sc} \\ -L_{sa} & -L_{sb} & 2L_{sc} \end{bmatrix}_{3\times 3}$$

$$\mathbf{i} = \begin{bmatrix} i_{sa} & i_{sb} & i_{sc} \end{bmatrix}^T_{3\times 1}$$

$$\mathbf{A} = \begin{bmatrix} -2R_{sa} & R_{sb} & R_{sc} & -2\left(S_{a1} - S_{a2}\right) & \left(S_{b1} - S_{b2}\right) & \left(S_{c1} - S_{c2}\right) \\ R_{sa} & -2R_{sb} & R_{sc} & \left(S_{a1} - S_{a2}\right) & -2\left(S_{b1} - S_{b2}\right) & \left(S_{c1} - S_{c2}\right) \\ R_{sa} & R_{sb} & -2R_{sc} & \left(S_{a1} - S_{a2}\right) & \left(S_{b1} - S_{b2}\right) & -2\left(S_{c1} - S_{c2}\right) \end{bmatrix}_{3\times 6}$$

$$\mathbf{x} = \begin{bmatrix} i_{sa} & i_{sb} & i_{sc} & v_{ca} & v_{cb} & v_{cc} \end{bmatrix}^T_{6\times 1}$$

$$\mathbf{B} = \begin{bmatrix} 2 & -1 & -1 \\ -1 & 2 & -1 \\ -1 & -1 & 2 \end{bmatrix}_{3\times 3}$$

$$\mathbf{u} = \begin{bmatrix} v_{sa} & v_{sb} & v_{sc} \end{bmatrix}^T_{3\times 1}$$

Moreover, applying KCL in each SM yields the capacitor voltage dynamics as

$$C_j \frac{dv_{cj}}{dt} = \left(S_{j1} - S_{j2}\right) i_{sj} - \frac{v_{cj}}{R_{mj}} \tag{5.159}$$

Extending (5.159) into the *abc*-frame gives

$$\begin{cases} C_a \dfrac{dv_{ca}}{dt} = \left(S_{a1} - S_{a2}\right) i_{sa} - \dfrac{v_{ca}}{R_{ma}} \\ C_b \dfrac{dv_{cb}}{dt} = \left(S_{b1} - S_{b2}\right) i_{sb} - \dfrac{v_{cb}}{R_{mb}} \\ C_c \dfrac{dv_{cc}}{dt} = \left(S_{c1} - S_{c2}\right) i_{sc} - \dfrac{v_{cc}}{R_{mc}} \end{cases} \tag{5.160}$$

Since **L** is prone to have singularity issue, the modification should be made by considering the total current relationship as

$$i_{sa} + i_{sb} + i_{sc} = 0 \tag{5.161}$$

Substituting phase *c* current as $i_{sc} = -i_{sa} - i_{sb}$ into (5.158) yields

$$\mathbf{L}_r \frac{d\mathbf{i}_r}{dt} = \mathbf{A}_r \mathbf{x}_r + \mathbf{B}_r \mathbf{u} \tag{5.162}$$

where

$$\mathbf{L}_r = \begin{bmatrix} 2L_{sa} + L_{sc} & -L_{sb} + L_{sc} \\ -L_{sa} + L_{sc} & 2L_{sb} + L_{sc} \end{bmatrix}_{2\times 2}$$

$$\mathbf{i}_r = \begin{bmatrix} i_{sa} & i_{sb} \end{bmatrix}^T_{2\times 1}$$

$$\mathbf{A}_r = \begin{bmatrix} -2R_{sa} - R_{sc} & R_{sb} - R_{sc} & -2\left(S_{a1} - S_{a2}\right) & S_{b1} - S_{b2} & S_{c1} - S_{c2} \\ R_{sa} - R_{sc} & -2R_{sb} - R_{sc} & S_{a1} - S_{a2} & -2\left(S_{b1} - S_{b2}\right) & S_{c1} - S_{c2} \end{bmatrix}_{2\times 5}$$

$$\mathbf{x}_r = \begin{bmatrix} i_{sa} & i_{sb} & v_{ca} & v_{cb} & v_{cc} \end{bmatrix}^T_{5\times 1}$$

$$\mathbf{B}_r = \begin{bmatrix} 2 & -1 & -1 \\ -1 & 2 & -1 \end{bmatrix}_{2\times 3}$$

Moreover, (5.160) can be modified into

$$C_c \frac{dv_{cc}}{dt} = -\left(S_{c1} - S_{c2}\right) i_{sa} - \left(S_{c1} - S_{c2}\right) i_{sb} - \frac{v_{cc}}{R_{mc}} \tag{5.163}$$

Hence, incorporating (5.162) and (5.163) into a matrix form results in

$$\dot{\mathbf{x}}_r = \mathbf{F}\mathbf{x}_r + \mathbf{G}\mathbf{u} \tag{5.164}$$

where

$$\mathbf{F} = \begin{bmatrix} & & \mathbf{L}_r^{-1}\mathbf{A}_r & & \\ \frac{S_{a1}-S_{a2}}{C_a} & 0 & -\frac{1}{R_{ma}C_a} & 0 & 0 \\ 0 & \frac{S_{b1}-S_{b2}}{C_b} & 0 & -\frac{1}{R_{mb}C_b} & 0 \\ -\frac{(S_{c1}-S_{c2})}{C_c} & -\frac{(S_{c1}-S_{c2})}{C_b} & 0 & 0 & -\frac{1}{R_{mc}C_c} \end{bmatrix}_{5\times 5}$$

$$\mathbf{G} = \begin{bmatrix} \mathbf{L}_r^{-1}\mathbf{B}_r \\ 0 \\ 0 \\ 0 \end{bmatrix}_{5\times 3}$$

Now, let us extend the three-level CML-VSC into l-level by generalizing (5.156) as

$$\begin{cases} L_{sa}\frac{di_{sa}}{dt} = -R_{sa}i_{sa} - \mathbf{S_a}\mathbf{v_{ca}} + v_{sa} - V_{Hm} \\ L_{sb}\frac{di_{sb}}{dt} = -R_{sb}i_{sb} - \mathbf{S_b}\mathbf{v_{cb}} + v_{sb} - V_{Hm} \\ L_{sc}\frac{di_{sc}}{dt} = -R_{sc}i_{sc} - \mathbf{S_c}\mathbf{v_{cc}} + v_{sc} - V_{Hm} \end{cases} \tag{5.165}$$

where

$$\mathbf{v_{ca}} = \begin{bmatrix} v_{ca1} & \cdots & v_{caN} \end{bmatrix}^T_{N\times 1}$$
$$\mathbf{v_{cb}} = \begin{bmatrix} v_{cb1} & \cdots & v_{cbN} \end{bmatrix}^T_{N\times 1}$$
$$\mathbf{v_{cc}} = \begin{bmatrix} v_{cc1} & \cdots & v_{ccN} \end{bmatrix}^T_{N\times 1}$$
$$\mathbf{S_a} = \begin{bmatrix} (S_{a1} - S_{a2}) & \cdots & (S_{a2N-1} - S_{a2N}) \end{bmatrix}_{1\times N}$$
$$\mathbf{S_b} = \begin{bmatrix} (S_{b1} - S_{b2}) & \cdots & (S_{b2N-1} - S_{b2N}) \end{bmatrix}_{1\times N}$$
$$\mathbf{S_c} = \begin{bmatrix} (S_{c1} - S_{c2}) & \cdots & (S_{c2N-1} - S_{c2N}) \end{bmatrix}_{1\times N}$$

The dynamics of capacitor voltage in each SM can be expressed as

$$\begin{cases} C_{ak}\frac{dv_{cak}}{dt} = \left(S_{a2k-1} - S_{a2k}\right) i_{sa} - \frac{v_{cak}}{R_{mak}} \\ C_{bk}\frac{dv_{cbk}}{dt} = \left(S_{b2k-1} - S_{b2k}\right) i_{sb} - \frac{v_{cbk}}{R_{mbk}} \\ C_{ck}\frac{dv_{cck}}{dt} = \left(S_{c2k-1} - S_{c2k}\right) i_{sc} - \frac{v_{cck}}{R_{mck}} \end{cases} \tag{5.166}$$

where $k \in [1, N]$.

Extending the three-level CML-VSC to l-level CML-VSC, (5.162) can be modified into

$$\mathbf{L}_r\frac{d\mathbf{i}_r}{dt} = \mathbf{A}_r\mathbf{x}_r + \mathbf{B}_r\mathbf{u} \tag{5.167}$$

where

$$\mathbf{L}_r = \begin{bmatrix} 2L_{sa} + L_{sc} & -L_{sb} + L_{sc} \\ -L_{sa} + L_{sc} & 2L_{sb} + L_{sc} \end{bmatrix}_{2\times2}$$

$$\mathbf{A}_r = \begin{bmatrix} -2R_{sa} - R_{sc} & R_{sb} - R_{sc} & -2\mathbf{S_a} & \mathbf{S_b} & \mathbf{S_c} \\ R_{sa} - R_{sc} & -2R_{sb} - R_{sc} & \mathbf{S_a} & -2\mathbf{S_b} & \mathbf{S_c} \end{bmatrix}_{2\times3N+2}$$

$$\mathbf{x}_r = \begin{bmatrix} i_{sa} & i_{sb} & v_{ca} & v_{cb} & v_{cc} \end{bmatrix}^T_{3N+2\times1}$$

$$\mathbf{B}_r = \begin{bmatrix} 2 & -1 & -1 \\ -1 & 2 & -1 \end{bmatrix}_{2\times3}$$

The phase c capacitor voltage dynamics is also modified into

$$C_{ck}\frac{dv_{cck}}{dt} = -\left(S_{c2k-1} - S_{c2k}\right) i_{sa} - \left(S_{c2k-1} - S_{c2k}\right) i_{sb} - \frac{v_{cck}}{R_{mck}} \tag{5.168}$$

Finally, the dynamics of CML-VSC can be expressed as

$$\dot{\mathbf{x}}_r = \mathbf{F}\mathbf{x}_r + \mathbf{G}\mathbf{u} \tag{5.169}$$

where

$$\mathbf{F} = \begin{bmatrix} & & \mathbf{L}_r^{-1}\mathbf{A}_r & & \\ \frac{\mathbf{S_a^T}}{C_a} & 0 & -\frac{1}{R_{ma}C_a} & 0 & 0 \\ 0 & \frac{\mathbf{S_b^T}}{C_b} & 0 & -\frac{1}{R_{mb}C_b} & 0 \\ -\frac{\mathbf{S_c^T}}{C_c} & -\frac{\mathbf{S_c^T}}{C_b} & 0 & 0 & -\frac{1}{R_{mc}C_c} \end{bmatrix}_{3N+2\times5}$$

$$\mathbf{G} = \begin{bmatrix} \mathbf{L}_r^{-1}\mathbf{B}_r \\ 0 \\ 0 \\ 0 \end{bmatrix}_{3N+2\times3}$$

Example 5.1

A three-phase, seven-level CML-VSC, whose parameters are listed in Table 5.8, is driven by PS-PWM. The frequency modulation index is 27 and the modulating signals expressed in symmetrical component have the magnitudes of 0.8 and firing angle of 2.19°. Simulate the CML-VSC and plot the evolution of current waveform if the RMS value of three-phase voltage sources are set to 114.5 V (V_{rms-LL}).

Solutions:

The parameters listed in Table 5.8 are substituted into (5.169). Since the level of CML-VSC is 7, the switching matrices can be defined as

$$\begin{aligned} S_a &= \left[\left(S_{a1} - S_{a2}\right) \ \left(S_{a3} - S_{a4}\right) \ \left(S_{a5} - S_{a6}\right)\right] \\ S_b &= \left[\left(S_{b1} - S_{b2}\right) \ \left(S_{b3} - S_{b4}\right) \ \left(S_{b5} - S_{b6}\right)\right] \\ S_c &= \left[\left(S_{c1} - S_{c2}\right) \ \left(S_{c3} - S_{c4}\right) \ \left(S_{c5} - S_{c6}\right)\right] \end{aligned}$$

The SM capacitors are given by

$$\begin{aligned} v_{ca} &= \begin{bmatrix} v_{ca1} & v_{ca2} & v_{ca3} \end{bmatrix} \\ v_{cb} &= \begin{bmatrix} v_{cb1} & v_{cb2} & v_{cb3} \end{bmatrix} \\ v_{cc} &= \begin{bmatrix} v_{cc1} & v_{cc2} & v_{cc3} \end{bmatrix} \end{aligned}$$

Table 5.8 Cascaded Multilevel Converter Parameters.

Components	Symbol	Value	Unit
Inductor	L_{sa}	2.5	mH
	L_{sb}	0.5	mH
	L_{sc}	1	mH
Resistor	R_{sa}	0.1	Ω
	R_{sb}	0.15	Ω
	R_{sc}	0.2	Ω
Submodule Capacitor	C_{a1}	3600	μF
	C_{a2}	3000	μF
	C_{a3}	2800	μF
	C_{b1}	3000	μF
	C_{b2}	2500	μF
	C_{b3}	2000	μF
	C_{c1}	2800	μF
	C_{c2}	3200	μF
	C_{c3}	3600	μF
Submodule Resistor	R_m	10^9	Ω

Hence, the dimension of $\mathbf{A}_r \in \mathbb{R}^{2\times11}$ and $\mathbf{F} \in \mathbb{R}^{11\times11}$

To begin the simulation, the initialization is required to specify the converter parameters and modulation technique.

```
%% Initialization
level=7;                                    % The number of level
freq=60;                                    % Fundamental frequency
T=1/freq;                                   % Period
Ts=1;                                       % Total simulation Time
w=2*pi*freq;                                % Omega (in rad/s)
%% Converter Parameters
Vs=114.5/sqrt(3);                           % AC Source (rms value)
phi=[0 -2*pi/3 2*pi/3];                     % Phase angle
N=(level-1)/2;                              % Number of Submodules
R=[0.1; 0.15; 0.2];                         % [Ra Rb Rc]
L=[2.5e-3; 0.5e-3; 1e-3];                   % [La Lb Lc]
C=[3600e-6, 2400e-6, 1200e-6;               % [Ca1   Cb1    Cc1]
     3000e-6, 2500e-6, 2000e-6;             % [Ca2   Cb2    Cc2]
     2800e-6, 3200e-6, 3600e-6];            % [Ca3   Cb3    Cc3]
Rm=[1e9 1e9 1e9;                            % [Rma1  Rmb1   Rmc1]
     1e9 1e9 1e9;                           % [Rma2  Rmb2   Rmc2]
     1e9 1e9 1e9];                          % [Rma3  Rmb3   Rmc3]
Rsw=0.005*2*N;                              % Internal resistance of switches
R=[R(1)+Rsw; R(2)+Rsw; R(3)+Rsw];           % Equivalent resistance
```

Since the converter has three SMs, the number of level is 7. `Vs` denotes the ac source voltage in line to neutral. Thus, the line-to-line voltage, 114.5 V, needs to be divided by `sqrt(3)`. `phi` is a column vector whose elements indicate the phase. Some parameters written in row vectors are `R`, `L`, representing the resistance and inductance, respectively. As the number of capacitors and internal resistance in the converter are 9, C and Rm are described by a 3 × 3 matrix. The internal resistance in switching devices is Rsw. If the parasitic components in switching devices are not omitted, the value of internal switching resistance for total SMs in each phase is

$$R_{sw} = 2Nr_{sw} \tag{5.170}$$

where r_{sw} is the internal resistance of each switching device. The modulating signal parameters are written in the following code.

```
%% Modulating Signal
Mags=[0 0.8 0]';                          % [zero positive negative]
phases=[0 2.19 0]'*pi/180;                % [zero positive negative] in rad
modsig=rec2pol(sc2abc([Mags phases]));    % rectangular to polar form
Mag=[modsig(1,1) modsig(2,1) modsig(3,1) -modsig(1,1) -modsig(2,1) ...
 -modsig(3,1)];                           % Magnitude of modulation signal
delta=[modsig(1,2) modsig(2,2) modsig(3,2) modsig(1,2) modsig(2,2) ...
modsig(3,2)];                             % Phase angle of modulation signal
mf=27;                                    % frequency modulation index
```

As can be seen in the aforementioned code, there is a function used to convert the symmetrical components into *abc* components written as sc2abc. Note that the arguments of this function are the magnitude and phase of modulating signal in symmetrical components. Variable Mags describes the row vector of magnitude, while phases denotes the row vector of phase. It is worth to mention that the Mags and phases have to be written in zero-, positive-, and negative-sequence components respectively.

```
function [fabc] = sc2abc(fa012)
% This function transforms the symmetrical components into its original
% abc phasors.
% The symmetrical components can be expressed in a one column array in
% rectangular complex form or in a two column array in polar form, with
% 1st column magnitude and 2nd column phase angle in degree.
% First, second and third rows are Zero-, positives- and negative-sequence
% components respectively.
% The function  returns the abc phsors in rectangular form.
rankf012=length(fa012(1,:));
if rankf012 == 2
    mag= fa012(:,1); ang=fa012(:,2);
    fa012r=mag.*(cos(ang)+j*sin(ang));
elseif rankf012 ==1
    fa012r=fa012;
end
a=cos(2*pi/3)+j*sin(2*pi/3);
A = [1   1   1; 1 a^2 a; 1 a  a^2];
fabc= A*fa012r;
fabc0=fa012r(1)*[1; 1;  1];
fabc1=fa012r(2)*[1;  a^2; a];
fabc2=fa012r(3)*[1;  a;  a^2];
```

The sc2abc function will return the values in rectangular form. To convert the rectangular form into the polar form, the function rec2pol is designed. The function rec2pol is written in the following way:

```
function Fpol = rec2pol(Frec)
Fpol = [abs(Frec)  angle(Frec)];
```

The switching time and switching pattern can be obtained using the following function:

```
[SST,SSseq] = else_sw_time_CMC(level,Mag,delta,freq,mf,Ts);
```

The function `else_sw_time_CMC` can be written in the following way:

```
function[SST,SSseq] = else_sw_time_CMC(level,Mag,delta,freq,mf,Ts)
%% The function to obtain the switching time and switching pattern using...
%% Phase-shifted modulation

N=(level-1)/2;
th2=360/(level-1);
for i=1:N
    th(1,i)=th2*(i-1);
end
th=[th(1,:); th(1,:); th(1,:); th(1,:); th(1,:); th(1,:)];
Fs=1e5;
SN=Ts*Fs;
tt=linspace(0,Ts,SN);

%% Find switching time and switching pattern
ST=zeros(length(tt),1+6*N);
k=1;
for m=1:length(Mag)
    for n=1:N
        mag=Mag(m);
        delta1=delta(m);
        trith=th(m,n);
        for ii=1:length(tt)
            ST(ii,1+N*(m-1)+n) = fST(tt(ii),mag,freq,delta1,mf,trith);
            ST(ii,1)=tt(ii);
        end
        for iii=1:length(tt)-1
            if (ST(iii,1+N*(m-1)+n)>=0 && ST(iii+1,1+N*(m-1)+n)<=0) || ...
                    (ST(iii,1+N*(m-1)+n)<=0 && ST(iii+1,1+N*(m-1)+n)>=0)
                SSX(k,N*(m-1)+n)=fzero(@fST,[tt(iii) tt(iii+1)], ...
                    optimset('Display','off'),mag,freq,delta1,mf,trith);
                k=k+1;
            end
        end
        k=1;
    end
end

SSS=[];
for i=1:size(SSX,2)
    SSS=[SSS;SSX(1:size(SSX,1),i)];
end

SSS=[SSS;1/60];

SSS=sort(SSS);
A=find(SSS==0);
SSS=SSS(length(A)+1:length(SSS),1);
%%
```

```
for m=1:length(Mag)
        for n=1:N
            mag=Mag(m);
            delta1=delta(m);
            trith=th(m,n);
            for ii=1:length(SSS)
            if fST(SSS(ii,1),mag,freq,delta1,mf,trith)>0
               SSS(ii,1+N*(m-1)+n) = 1;
            else
                SSS(ii,1+N*(m-1)+n) = 0;
            end
            end
        end
end
%%
delta_t=1e-15;
delta_T=[-delta_t delta_t];
for t=1:2
    dt=delta_T(t);
    for m=1:length(Mag)
        for n=1:N
            mag=Mag(m);
            delta1=delta(m);
            trith=th(m,n);
            for ii=1:length(SSS)
                if t==1
                    SSSold(ii,1) = SSS(ii)+dt;
                    if fST(SSSold(ii,1),mag,freq,delta1,mf,trith)>0
                        SSSold(ii,1+N*(m-1)+n) = 1;
                    else
                        SSSold(ii,1+N*(m-1)+n) = 0;
                    end
                else
                    SSSnew(ii,1) = SSS(ii)+dt;
                    if fST(SSSnew(ii,1),mag,freq,delta1,mf,trith)>0
                        SSSnew(ii,1+N*(m-1)+n) = 1;
                    else
                        SSSnew(ii,1+N*(m-1)+n) = 0;
                    end
                end
            end
        end
    end
end

SSSEND=[SSSold;SSSnew];
SSSEND=sortrows(SSSEND,1);
%%
SST=SSS(:,1);
SSseq=SSSold(:,2:size(SSS,2));
```

The fST is introduced as a function to compare carrier signal and modulating signal, which can be written as

```
function y = fST(x,mag,f,delta1,mf,trith)
if trith<0
    trith=trith+360;
end
if trith<=180
    y = mag*cos(2*pi*f*x+delta1) - sawtooth(2*pi*f*mf*(x+1/(mf*f*...
```

```
        (360/trith))),0.5);
else
    trithnew=trith-360;
    y = mag*cos(2*pi*f*x+delta1) - sawtooth(2*pi*f*mf*(x+1/(mf*f*...
        (360/trithnew))),0.5);
end
```

To obtain the evolution of current and voltage of the converter, the function `CMC_waveform` is presented in the following way:

```
[iabc, vcap, vsabc] = CMC_waveform(Vs,phi,w,N,R,L,C,Rm,SST,SSseq);
```

Basically, the function `CMC_waveform` contains the model of converter and the exponential matrix calculation to obtain the solution in the time domain. The details of the function `CMC_waveform` is listed as follows:

```
function[iabc, vcap, vsabc] = CMC_waveform(Vs,phi,w,N,R,L,C,Rm,SST,SSseq)
%% The function for simulating the evolution of voltage and current of CMC

%% Extracting phase angle
phi_a=phi(1); phi_b=phi(2); phi_c=phi(3);

%% Switching Time and Switching Pattern
taus_2=SST(:,1);
S_2=SSseq(:,1:size(SSseq,2));
%% Classifiying Switching Pattern into
Sa_up=S_2(1,0*N+1:1*N);
Sb_up=S_2(1,1*N+1:2*N);
Sc_up=S_2(1,2*N+1:3*N);
Sa_low=S_2(1,3*N+1:4*N);
Sb_low=S_2(1,4*N+1:5*N);
Sc_low=S_2(1,5*N+1:6*N);
%% Defining the vector of output variables
iabc=[];
vcap=[];
vsabc=[];
%% Inductor matrix
HH=[((2/3)*L(1))+((1/3)*L(3)) ((-1/3)*L(2))+((1/3)*L(3));
    ((-1/3)*L(1)+(1/3)*L(3)) ((2/3)*L(2))+((1/3)*L(3))];
%% Constructing matrix elements
Fa=zeros(1,N);
Fb=zeros(1,N);
Fc=zeros(1,N);
Pa=zeros(N,1);
Pb=zeros(N,1);
Pc=zeros(N,1);
W1=zeros(N,N);
W2=zeros(N,N);
W3=zeros(N,N);
for i=1:N
    Fa(1,i)=(Sa_up(1,i)-Sa_low(1,i));
    Fb(1,i)=(Sb_up(1,i)-Sb_low(1,i));
    Fc(1,i)=(Sc_up(1,i)-Sc_low(1,i));
    Pa(i,1)=(Sa_up(1,i)-Sa_low(1,i))/C(1,i);
    Pb(i,1)=(Sb_up(1,i)-Sb_low(1,i))/C(2,i);
    Pc(i,1)=(Sc_up(1,i)-Sc_low(1,i))/C(3,i);
    W1(i,i)=-1/(Rm(1,i)*C(1,i));
    W2(i,i)=-1/(Rm(2,i)*C(2,i));
```

```
    W3(i,i)=-1/(Rm(3,i)*C(3,i));
end


F=[ (-2/3)*Fa, (1/3)*Fb, (1/3)*Fc;
    (1/3)*Fa, (-2/3)*Fb, (1/3)*Fc];
F_Hinv=HH\F;

G=[((-2/3)*R(1))-((1/3)*R(3)) ((1/3)*R(2))-((1/3)*R(3));
    ((1/3)*R(1))-((1/3)*R(3)) ((-2/3)*R(2))-((1/3)*R(3))];

G_Hinv=HH\G;

P1=blkdiag(Pa,Pb);
P=[P1; -1*Pc, -1*Pc];

W=blkdiag(W1,W2,W3);

X=[ Vs*sqrt(2)*cos(phi_a);
    Vs*sqrt(2)*sin(phi_a);
    Vs*sqrt(2)*cos(phi_b);
    Vs*sqrt(2)*sin(phi_b);
    Vs*sqrt(2)*cos(phi_c);
    Vs*sqrt(2)*sin(phi_c)];

A_Max=[G_Hinv, F_Hinv;
    P, W];
[VA_Max,HA_Max]=size(A_Max);
B_M=[2/3, 0, -1/3, 0, -1/3, 0;
    -1/3, 0, 2/3, 0, -1/3, 0];
B_Minv=HH\B_M;

B_Max=[B_Minv;
    zeros(3*N,6)];
[VB_Max,HB_Max]=size(B_Max);

omega2=[0 -1*w; 1*w 0];
D_Max=blkdiag(omega2,omega2,omega2);
[VD_Max,HD_Max]=size(D_Max);

Axxx_I=[A_Max B_Max;
    zeros(VD_Max,HA_Max),D_Max];
%% Obtaining the solution via exponential matrix
PHI=expm(Axxx_I*taus_2(1))*[zeros(VA_Max,1); X];

iabc=[iabc,PHI(1:2,1)];
vcap=[vcap,PHI(3:VA_Max,1)];
vsabc=[vsabc,PHI(VA_Max+1:2:VA_Max+5,1)];

for iii=2:length(taus_2)
    Sa_up=S_2(iii,0*N+1:1*N);
    Sb_up=S_2(iii,1*N+1:2*N);
    Sc_up=S_2(iii,2*N+1:3*N);
    Sa_low=S_2(iii,3*N+1:4*N);
    Sb_low=S_2(iii,4*N+1:5*N);
    Sc_low=S_2(iii,5*N+1:6*N);

    %% Constructing matrix elements
    Fa=zeros(1,N);
    Fb=zeros(1,N);
    Fc=zeros(1,N);
```

```
        Pa=zeros(N,1);
        Pb=zeros(N,1);
        Pc=zeros(N,1);
        for i=1:N
            Fa(1,i)=(Sa_up(1,i)-Sa_low(1,i));
            Fb(1,i)=(Sb_up(1,i)-Sb_low(1,i));
            Fc(1,i)=(Sc_up(1,i)-Sc_low(1,i));
            Pa(i,1)=(Sa_up(1,i)-Sa_low(1,i))/C(1,i);
            Pb(i,1)=(Sb_up(1,i)-Sb_low(1,i))/C(2,i);
            Pc(i,1)=(Sc_up(1,i)-Sc_low(1,i))/C(3,i);
        end
        F=[ (-2/3)*Fa, (1/3)*Fb, (1/3)*Fc;
            (1/3)*Fa, (-2/3)*Fb, (1/3)*Fc];
        F_Hinv=HH\F;

        P1=blkdiag(Pa,Pb);
        P=[P1; -1*Pc, -1*Pc];

        A_Max=[G_Hinv, F_Hinv;
            P, W];

        Axxx_I(1:VA_Max,1:HA_Max)=A_Max;

        %% Obtaining the solution via exponential matrix
        PHI=expm(Axxx_I*(taus_2(iii)-taus_2(iii-1)))*PHI;

        %% Updating the output variables
        iabc=[iabc,PHI(1:2,1)];
        vcap=[vcap,PHI(3:VA_Max,1)];
        vsabc=[vsabc,PHI(VA_Max+1:2:VA_Max+5,1)];

    end
```

To verify the accuracy of the presented method, the simulation in PSCAD/EMTDC is conducted as a benchmark. The results of the presented method are shown in Figs. 5.31–5.33 representing i_{sa}, i_{sb}, and v_{ca3}, respectively. Note that the initial conditions for PSCAD/EMTDC and the presented method are arbitrarily set to zeros. It can be clearly seen that the simulation results of proposed method and PSCAD/EMTDC are in good agreement.

5.3.5 Modular Multi-Level VSC

The modular multi-level VSC (MML-VSC) was introduced in [64]. The configuration of MML-VSC is similar to that of the CML-VSC. The only difference between them is the placement of the DC source. Fig. 5.34 shows the simplest form of MML-VSC, known as the three-level MML-VSC. As shown in the figure, MML-VSC only requires a single DC source whereas CML-VSC has many isolated DC sources. Hence, MML-VSC is more suitable for an HVDC system. The common configuration of MML-VSC's SM is half-bridge converters, comprised of a dc capacitor and two active switching components working in a complementary fashion. To simplify the analysis, let us derive the simple three-level MML-VSC model, which later can be extended into l-level.

Three-level MML-VSC

Each phase of MML-VSC is composed of an upper arm (u) and a lower arm (l), respectively. Each of these arms contains an SM, an arm inductor (L), and an arm resistor (R).

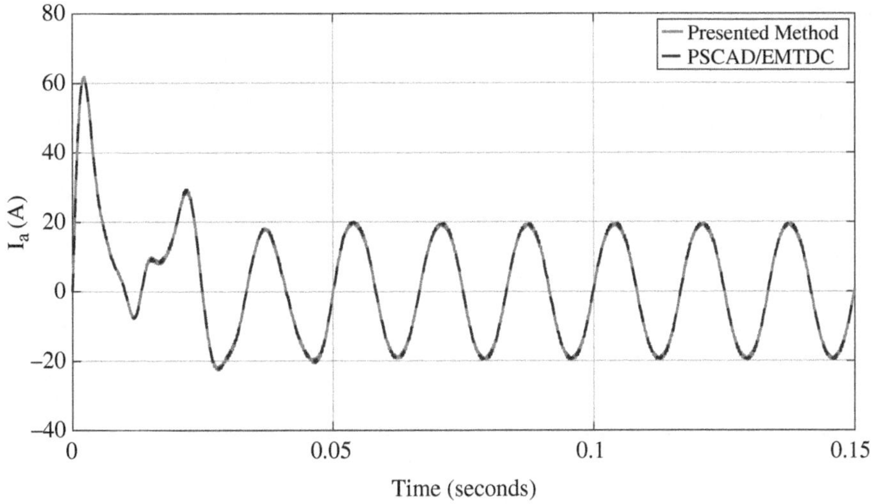

Figure 5.31 Evolution of i_{sa}.

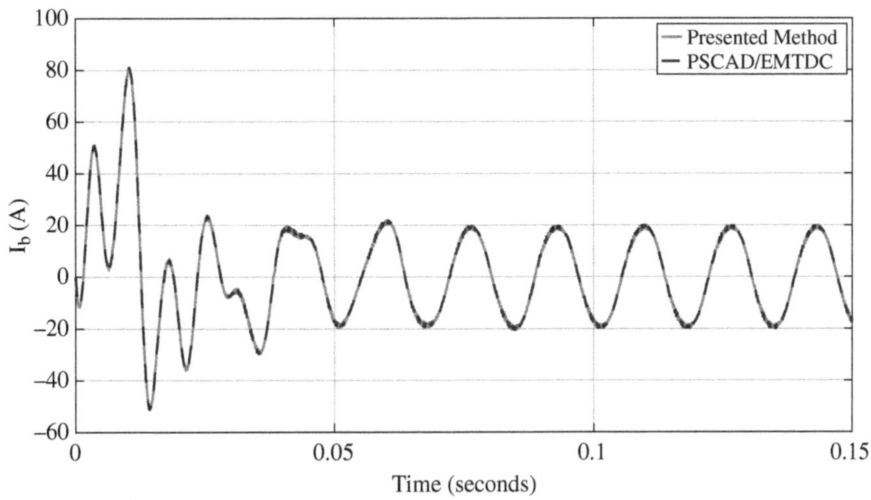

Figure 5.32 Evolution of i_{sb}.

The single-phase MML-VSC is used for analysis as depicted in Fig. 5.35. By employing KVL, the dynamics of Loops 1 and 2 can be obtained as

$$v_{dc} + V_{Hm} = S_j^u v_{cj}^u + R_j^u i_j^u + L_j^u \frac{d}{dt} i_j^u + L_{sj} \frac{d}{dt} i_{sj} + R_{sj} i_{sj} + v_{sj} \tag{5.171}$$

$$-V_{Hm} = S_j^l v_{cj}^l + R_j^l i_j^l + L_j^l \frac{d}{dt} i_j^l - L_{sj} \frac{d}{dt} i_{sj} - R_{sj} i_{sj} - v_{sj} \tag{5.172}$$

where v_{sj} is the source voltage of phase j ($j \in \{a, b, c\}$), i_{sj} is the phase current, and v_{dc} is the dc-link voltage connected to the upper and lower arms. i_j^u, v_{cj}^u and i_j^l, v_{cj}^l denote the current and voltage of the

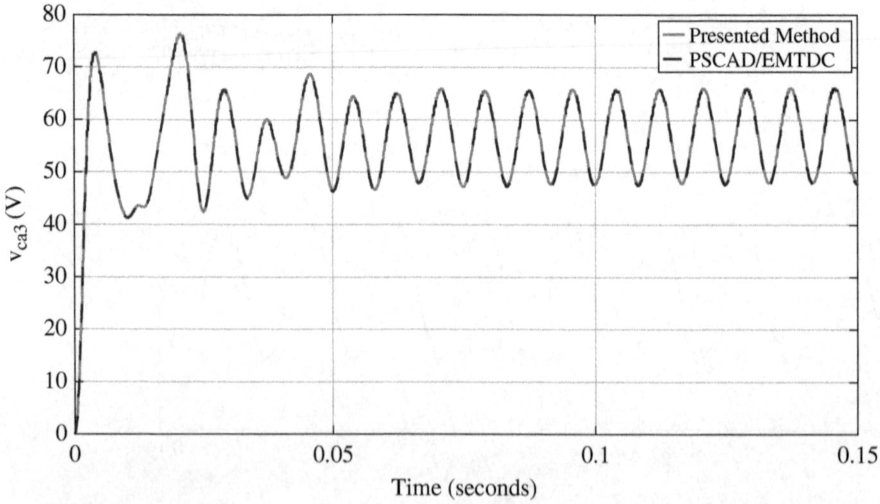

Figure 5.33 Evolution of v_{ca3}.

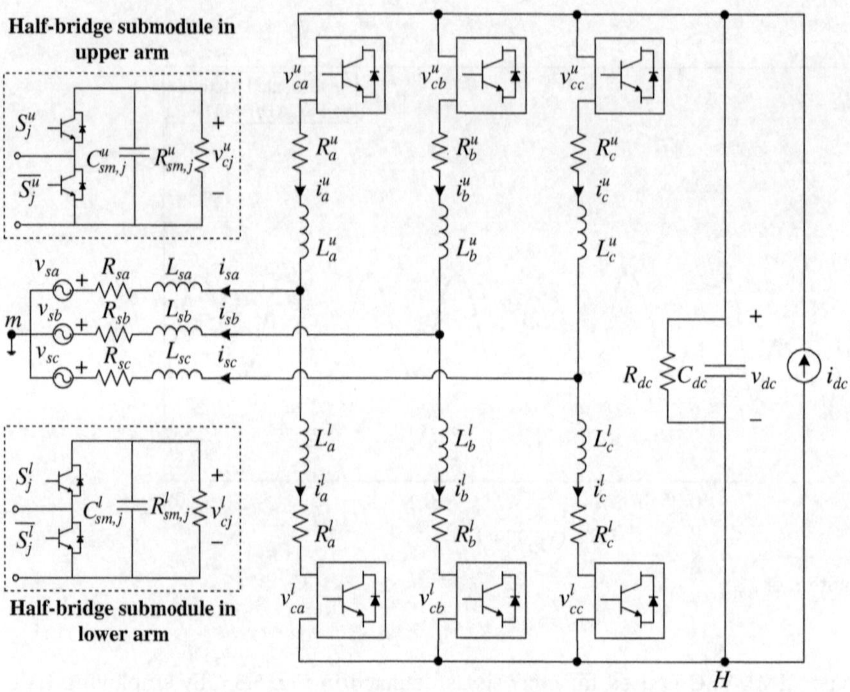

Figure 5.34 Schematic diagram of a three-level MML-VSC.

upper and lower arms, respectively. S_j^u and S_j^l represent switches position for the upper and lower arms, individually.

By defining the circulation current of each phase as $i_{cir,j} = \frac{i_j^u + i_j^l}{2}$ and the phase current as $i_j = i_j^u - i_j^l$ then substituting (5.171) into (5.172) yields

$$v_{dc} = S_j^u v_{cj}^u + S_j^l v_{cj}^l + \frac{L_j^d}{2}\frac{d}{dt}i_{sj} + \frac{R_j^d}{2}i_{sj} + L_j^z\frac{d}{dt}i_{cir,j} + R_j^z i_{cir,j} \tag{5.173}$$

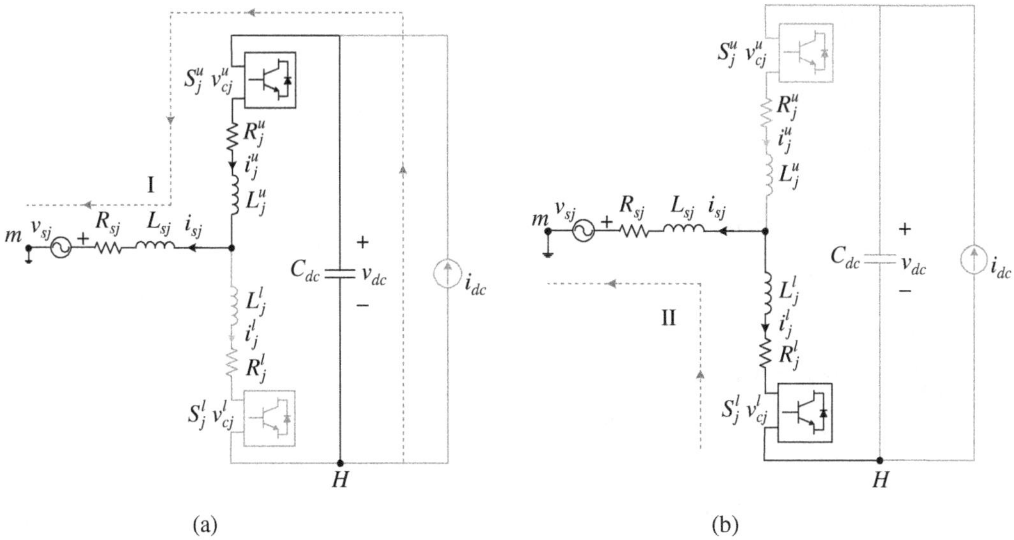

Figure 5.35 Equivalent circuit of single-phase MML-VSC: (a) Loop 1, (b) Loop 2.

where $L_j^d = L_j^u - L_j^l$, $R_j^d = R_j^u - R_j^l$, $L_j^z = L_j^u + L_j^l$,and $R_j^z = R_j^u + R_j^l$.

In addition, subtracting (5.172) from (5.171) results in

$$v_{dc} + 2V_{Hm} = S_j^u v_{cj}^u - S_j^l v_{cj}^l + L'_{sj}\frac{d}{dt}i_{sj} + R'_{sj}i_{sj} + L_j^d\frac{d}{dt}i_{cir,j} + R_j^d i_{cir,j} + 2v_{sj} \tag{5.174}$$

where $L'_{sj} = \frac{L_j^z}{2} + 2L_{sj}$ and $R'_{sj} = \frac{R_j^z}{2} + 2R_{sj}$. Note that (5.174) can be generalized for all $j \in \{a, b, c\}$. Hence, V_{Hm} can be derived as

$$2V_{Hm} = \frac{1}{3}\sum_{j=a}^{c}\left(L'_{sj}\frac{d}{dt}i_{sj} + L_j^d\frac{d}{dt}i_{cir,j} + R'_{sj}i_{sj} + R_j^d i_{cir,j} + S_j^u v_{cj}^u - S_j^l v_{cj}^l + 2v_{sj}\right) - v_{dc} \tag{5.175}$$

Referring to the relationship of phase current as $i_{sa} + i_{sb} + i_{sc} = 0$, the phase c current can be represented by (5.176).

$$i_{sc} = -(i_{sa} + i_{sb}) \tag{5.176}$$

Substituting (5.176) and (5.175) into (5.174), the dynamics of current, including phase current and the circulating current of each phase, can be defined as follow

$$\begin{bmatrix}\mathbf{M_1} & \mathbf{M_2}\\ \mathbf{M_3} & \mathbf{M_4}\end{bmatrix}\frac{d}{dt}\begin{bmatrix}\mathbf{i}\\ \mathbf{i_{cir}}\end{bmatrix} = \begin{bmatrix}\mathbf{M_5} & \mathbf{M_6}\\ \mathbf{M_7} & \mathbf{M_8}\end{bmatrix}\begin{bmatrix}\mathbf{i}\\ \mathbf{i_{cir}}\end{bmatrix} + \mathbf{F_1}\begin{bmatrix}\mathbf{v_c^u}\\ \mathbf{v_c^l}\end{bmatrix} + \mathbf{B_1}\begin{bmatrix}\mathbf{v_s}\\ v_{dc}\end{bmatrix} \tag{5.177}$$

where

$$\mathbf{i} = \begin{bmatrix}i_{sa} & i_{sb}\end{bmatrix}_{2\times1}^T$$
$$\mathbf{i_{cir}} = \begin{bmatrix}i_{cir,a} & i_{cir,b} & i_{cir,c}\end{bmatrix}_{3\times1}^T$$
$$\mathbf{v_c^u} = \begin{bmatrix}v_{ca}^u & v_{cb}^u & v_{cc}^u\end{bmatrix}_{3\times1}^T$$
$$\mathbf{v_c^l} = \begin{bmatrix}v_{ca}^l & v_{cb}^l & v_{cc}^l\end{bmatrix}_{3\times1}^T$$
$$\mathbf{v_s} = \begin{bmatrix}v_{sa} & v_{sb} & v_{sc}\end{bmatrix}_{3\times1}^T$$

$$\mathbf{M}_1 = \begin{bmatrix} 2L'_{sa} + L'_{sc} & -L'_{sb} + L'_{sc} \\ -L'_{sa} + L'_{sc} & 2L'_{sb} + L'_{sc} \end{bmatrix}_{2\times 2}$$

$$\mathbf{M}_2 = \begin{bmatrix} 2L_a^d & -L_b^d & -L_c^d \\ -L_a^d & 2L_b^d & -L_c^d \end{bmatrix}_{2\times 3}$$

$$\mathbf{M}_3 = \begin{bmatrix} \frac{L_a^d}{2} & 0 & \frac{-L_c^d}{2} \\ 0 & \frac{L_b^d}{2} & \frac{-L_c^d}{2} \end{bmatrix}^T_{3\times 2}$$

$$\mathbf{M}_4 = diag\left(L_a^z, L_b^z, L_c^z\right)_{3\times 3}$$

$$\mathbf{M}_5 = \begin{bmatrix} -2R'_{sa} - R'_{sc} & R'_{sb} - R'_{sc} \\ R'_{sa} - R'_{sc} & -2R'_{sb} - R'_{sc} \end{bmatrix}_{2\times 2}$$

$$\mathbf{M}_6 = \begin{bmatrix} -2R_a^d & R_b^d & R_c^d \\ R_a^d & -2R_b^d & R_c^d \end{bmatrix}_{2\times 3}$$

$$\mathbf{M}_7 = \begin{bmatrix} \frac{-R_a^d}{2} & 0 & \frac{R_c^d}{2} \\ 0 & \frac{-R_b^d}{2} & \frac{R_c^d}{2} \end{bmatrix}^T_{3\times 2}$$

$$\mathbf{M}_8 = -diag\left(R_a^z, R_b^z, R_c^z\right)_{3\times 3}$$

$$\mathbf{F}_1 = \begin{bmatrix} -2S_a^u & S_b^u & S_c^u & 2S_a^l & -S_b^l & -S_c^l \\ S_a^u & -2S_b^u & S_c^u & -S_a^l & 2S_b^l & -S_c^l \\ -S_a^u & 0 & 0 & -S_a^l & 0 & 0 \\ 0 & -S_b^u & 0 & 0 & -S_b^l & 0 \\ 0 & 0 & -S_c^u & 0 & 0 & -S_c^l \end{bmatrix}_{5\times 6}$$

$$\mathbf{B}_1 = \begin{bmatrix} -4 & 2 & 2 & 0 \\ 2 & -4 & 2 & 0 \\ 0 & 0 & 0 & 1 \\ 0 & 0 & 0 & 1 \\ 0 & 0 & 0 & 1 \end{bmatrix}_{5\times 4}$$

The capacitor voltage of each SM in the upper and lower arms can be obtained by using KCL as

$$\frac{d}{dt}v_{cj}^u = \frac{S_j^u}{2C_{sm,j}^u}i_{sj} + \frac{S_j^u}{C_{sm,j}^u}i_{cir,j} - \frac{1}{R_{sm,j}^u C_{sm,j}^u}v_{cj}^u \tag{5.178}$$

$$\frac{d}{dt}v_{cj}^l = -\frac{S_j^l}{2C_{sm,j}^l}i_{sj} + \frac{S_j^l}{C_{sm,j}^l}i_{cir,j} - \frac{1}{R_{sm,j}^l C_{sm,j}^l}v_{cj}^l \tag{5.179}$$

where $C_{sm,j}^u$, $C_{sm,j}^l$ and $R_{sm,j}^u$, $R_{sm,j}^l$ denote the capacitor and resistor of SM in the upper and lower arms, respectively. In the same way, the dynamics of the dc-link capacitor voltage of an MML-VSC via KCL is obtained as

$$\frac{d}{dt}v_{dc} = -\frac{1}{C_{dc}}\sum_{j=a}^{c} i_{cir,j} - \frac{1}{R_{dc}C_{dc}}v_{dc} + \frac{1}{C_{dc}}i_{dc} \tag{5.180}$$

where C_{dc} and R_{dc} represent dc-link capacitor and resistor, respectively.

An *l*-level MML-VSC

The schematic circuit of an *l*-level MML-VSC is depicted in Fig. 5.36. The *N*-SMs are added in each arm in series connection via *L* and *R*. Note that "*n*" in Fig. 5.36 represents the *n*-th SM where $n \in [1, N]$. It is also worth to mention that the number of level "*l*" in MML-VSC determines the number of *N*-SMs through (5.181)

$$N = \frac{l-1}{2} \tag{5.181}$$

where *l* is strictly an odd number. The state-space regular form of converter is represented as (5.182)

$$\dot{\mathbf{x}} = \mathbf{A}\mathbf{x} + \mathbf{B}\mathbf{u} \tag{5.182}$$

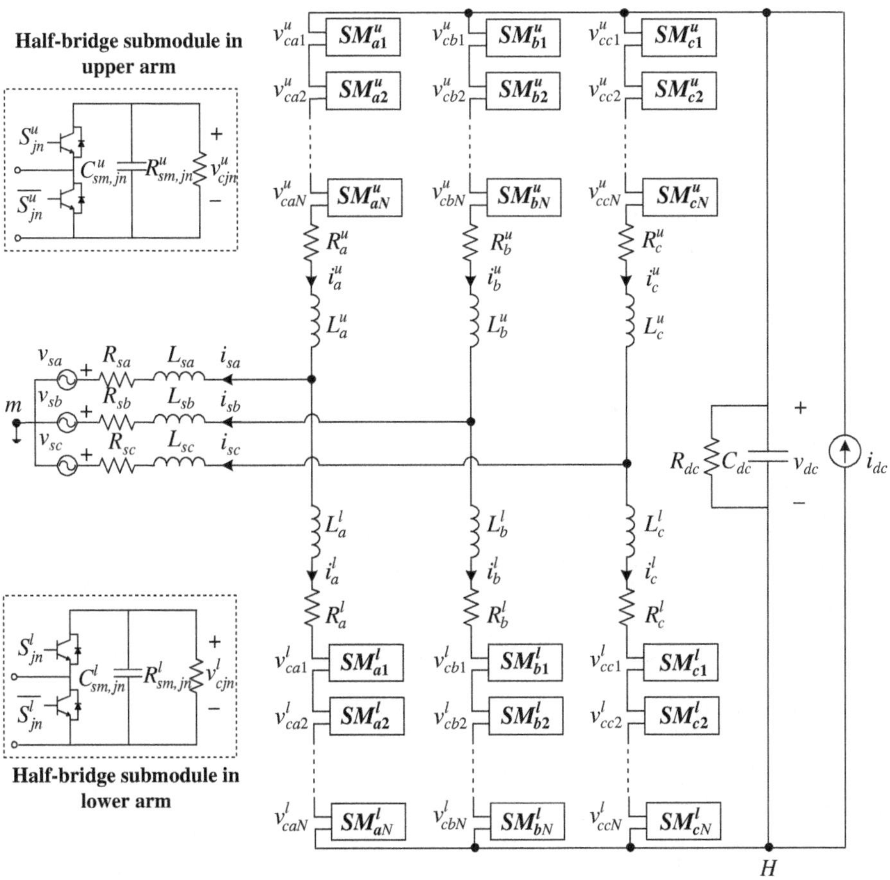

Figure 5.36 Schematic diagram of an *l*-level MML-VSC.

where

$$\mathbf{x} = \begin{bmatrix}\mathbf{i} & \mathbf{i_{cir}} & \mathbf{v_c} & v_{dc}\end{bmatrix}^T_{6N+6\times1}$$

$$\mathbf{u} = \begin{bmatrix}\mathbf{v_{sa}} & \mathbf{v_{sb}} & \mathbf{v_{sc}} & \mathbf{i_{dc}}\end{bmatrix}^T_{8\times1}$$

$$\mathbf{A} = \mathbf{A_1^{-1}A_2}$$

$$\mathbf{B} = \mathbf{A_1^{-1}B_2}$$

$$\mathbf{\Omega} = \begin{bmatrix}\begin{bmatrix}0 & -\omega\\ \omega & 0\end{bmatrix} & 0 & 0\\ 0 & \ddots & 0\\ 0 & 0 & \begin{bmatrix}0 & -\omega\\ \omega & 0\end{bmatrix}\end{bmatrix}_{8\times8}$$

$$\mathbf{v_{cj}} = \begin{bmatrix}v^u_{ca1} & v^u_{cb1} & v^u_{cc1} & v^l_{ca1} & v^l_{cb1} & v^l_{cc1} & \dots & v^u_{caN} & v^u_{cbN} & v^u_{ccN} & v^l_{caN} & v^l_{cbN} & v^l_{ccN}\end{bmatrix}^T_{6N\times1}$$

$$\mathbf{A_1} = \begin{bmatrix}\mathbf{M_1} & \mathbf{M_2} & 0 & 0\\ \mathbf{M_3} & \mathbf{M_4} & 0 & 0\\ 0 & 0 & \mathbf{T_1} & 0\\ 0 & 0 & 0 & C_{dc}\end{bmatrix}_{6N+6\times6N+6}$$

$$\mathbf{A_2} = \begin{bmatrix}\mathbf{M_5} & \mathbf{M_6} & \mathbf{F_2} & 0\\ \mathbf{M_7} & \mathbf{M_8} & \mathbf{F_3} & \mathbf{W_1}^{\mathbf{T}}\\ \mathbf{P_1} & -\mathbf{F_3^T} & -\mathbf{T_2} & 0\\ 0 & -\mathbf{W_1} & 0 & -1/R_{dc}\end{bmatrix}_{6N+6\times6N+6}$$

$$\mathbf{B_2} = \begin{bmatrix}-4 & 0 & 2 & 0 & 2 & 0 & 0 & 0\\ 2 & 0 & -4 & 0 & 2 & 0 & 0 & 0\\ 0 & 0 & 0 & 0 & 0 & 0 & 0 & 0\\ \vdots & \vdots & \vdots & \vdots & \vdots & \vdots & \vdots & \vdots\\ 0 & 0 & 0 & 0 & 0 & 0 & 1 & 0\end{bmatrix}_{6N+6\times8}$$

$$\mathbf{W_1} = \begin{bmatrix}1 & 1 & 1\end{bmatrix}_{3\times1}$$

$$\mathbf{F_2} = \begin{bmatrix}\mathbf{F_{up}}(1) & \mathbf{F_{up}}(2) & \dots & \mathbf{F_{up}}(N)\end{bmatrix}_{2\times6N}$$

$$\mathbf{F_3} = \begin{bmatrix}\mathbf{F_{low}}(1) & \mathbf{F_{low}}(2) & \dots & \mathbf{F_{low}}(N)\end{bmatrix}_{3\times6N}$$

$$\mathbf{F_{up}}(n) = \begin{bmatrix}-2S^u_{an} & S^u_{bn} & S^u_{cn} & 2S^l_{an} & -S^l_{bn} & -S^l_{cn}\\ S^u_{an} & -2S^u_{bn} & S^u_{cn} & -S^l_{an} & 2S^l_{bn} & -S^l_{cn}\end{bmatrix}_{2\times6}$$

$$\mathbf{F_{low}}(n) = \begin{bmatrix}-S^u_{an} & 0 & 0 & -S^l_{an} & 0 & 0\\ 0 & -S^u_{bn} & 0 & 0 & -S^l_{bn} & 0\\ 0 & 0 & -S^u_{cn} & 0 & 0 & -S^l_{cn}\end{bmatrix}_{3\times6}$$

$$\mathbf{P_1} = \begin{bmatrix}\mathbf{P_2}(1) & \mathbf{P_2}(2) & \dots & \mathbf{P_2}(N)\end{bmatrix}^T_{6N\times2}$$

$$\mathbf{P_2}(n) = \begin{bmatrix}\frac{S^u_{an}}{2} & 0 & \frac{-S^u_{cn}}{2} & \frac{-S^l_{an}}{2} & 0 & \frac{S^l_{cn}}{2}\\ 0 & \frac{S^u_{bn}}{2} & \frac{-S^u_{cn}}{2} & 0 & \frac{-S^l_{bn}}{2} & \frac{S^l_{cn}}{2}\end{bmatrix}_{2\times6}$$

$$\mathbf{T_1} = \begin{bmatrix}\mathbf{T_3}(1) & 0 & 0\\ 0 & \ddots & 0\\ 0 & 0 & \mathbf{T_3}(N)\end{bmatrix}_{6N\times6N}$$

$$\mathbf{T}_2 = \begin{bmatrix} \mathbf{T}_4(1) & 0 & 0 \\ 0 & \ddots & 0 \\ 0 & 0 & \mathbf{T}_4(N) \end{bmatrix}_{6N\times 6N}$$

$$\mathbf{T}_3(n) = diag\left(C^u_{sm,an}, C^u_{sm,bn}, C^u_{sm,cn}, C^l_{sm,an}, C^l_{sm,bn}, C^l_{sm,cn}\right)_{6\times 6}$$

$$\mathbf{T}_4(n) = diag\left(\frac{1}{R^u_{sm,an}}, \frac{1}{R^u_{sm,bn}}, \frac{1}{R^u_{sm,cn}}, \frac{1}{R^l_{sm,an}}, \frac{1}{R^l_{sm,bn}}, \frac{1}{R^l_{sm,cn}}\right)_{6\times 6}$$

6

Frequency Coupling Matrices

"Pull one hair and the whole body moves"

– Zizhen Gong (1792–1841)

One of the most common equations used to model electrical elements in the frequency domain or harmonic domain is

$$\mathbf{I} = \mathbf{YV}, \tag{6.1}$$

where $\mathbf{V}$ is the vector containing voltage harmonics at each node, $\mathbf{I}$ is the vector of current harmonics, and $\mathbf{Y}$ is the harmonic admittance matrix. To employ the formulation of (6.1), the harmonic admittance matrix must be found.

For linear time-invariant (LTI) components such as resistors (R), inductors (L), and capacitors (C), these admittance matrices are analytical and can be easily derived via phasor analysis. In addition, the structure of these admittance matrices is diagonal, which means that there is no interaction between harmonic voltages and currents of different orders.

In contrast, an analytical expression for $\mathbf{Y}$ is in general not attainable for power electronic systems. However, if the switching times are known a priori (i.e. the open-loop case), the power electronic systems reduce from nonlinear time-varying to linear time-varying (LTV). In this case, analytical expressions for the admittance matrix are obtainable. Unlike the LTI components, off-diagonal elements exist in the admittance matrices of the power electronic devices, indicating that voltage and current harmonics with different orders are "cross-coupled." Hence, the name "Frequency Coupling Matrix (FCM)" [13, 65] is used in the literature to specifically distinguish it from the admittances of the LTI components. Fig. 6.1 shows the FCM of a two-level VSC whereas Fig. 6.2 shows the admittance matrix of an inductor.

The main objective of this chapter is to show how the FCM of a power converter can be obtained by means of frequency-domain and time-domain methods.

6.1 Construction of FCM in the Harmonic Domain

To convert the differential equation $\dot{\mathbf{x}} = \mathbf{Ax} + \mathbf{Bu}$ into an equivalent algebraic equation in the harmonic domain, one needs to first understand the concept of matrices of differentiation and Toeplitz.

Consider a linear inductor whose voltage and current relation is

$$v(t) = L\frac{di(t)}{dt}. \tag{6.2}$$

Harmonic Modeling of Voltage Source Converters using Basic Numerical Methods, First Edition.
Ryan Kuo-Lung Lian, Ramadhani Kurniawan Subroto, Victor Andrean, and Bing Hao Lin.

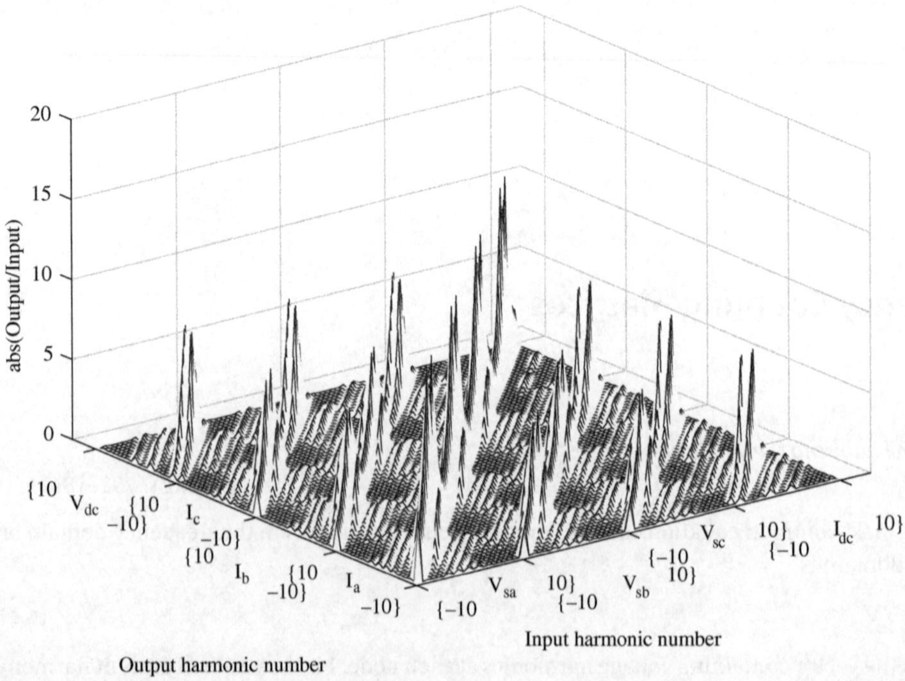

Figure 6.1 The FCM of a two-level VSC.

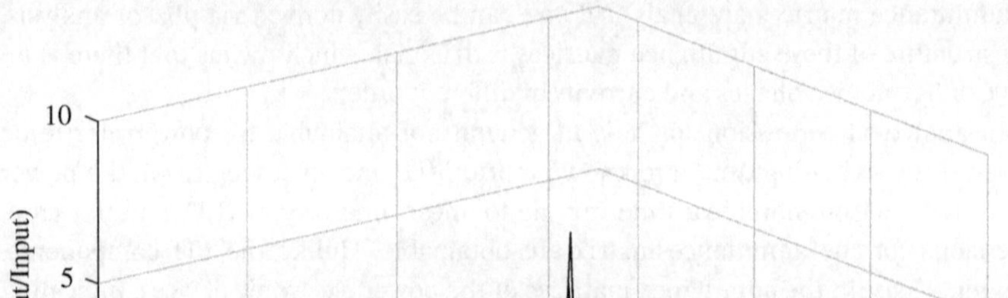

Figure 6.2 The admittance matrix of an LTI component.

Assume $v(t)$ and $i(t)$ are periodic, which can be expressed in terms of their Fourier series:

$$v(t) = \sum_{n=-\infty}^{n=\infty} V_n e^{jnw_0 t} \tag{6.3}$$

$$i(t) = \sum_{n=-\infty}^{n=\infty} I_n e^{jnw_0 t}. \tag{6.4}$$

Hence, (6.2) can be written as

$$\sum_{n=-\infty}^{n=\infty} V_n e^{jn\omega_0 t} = L\frac{d}{dt}\sum_{n=-\infty}^{n=\infty} I_n e^{jn\omega_0 t} = L\sum_{n=-\infty}^{n=\infty} jn\omega_0 I_n e^{jn\omega_0 t}. \tag{6.5}$$

Therefore, (6.2) in the harmonic domain is

$$\mathbf{V} = L\mathbf{D}\mathbf{I}, \tag{6.6}$$

where $\mathbf{V} = \begin{bmatrix} \cdots & V_{-2} & V_{-1} & V_0 & V_1 & V_2 & \cdots \end{bmatrix}^T$, $\mathbf{I} = \begin{bmatrix} \cdots & I_{-2} & I_{-1} & I_0 & I_1 & I_2 & \cdots \end{bmatrix}^T$, and

$$\mathbf{D} = \begin{bmatrix} \ddots & & & & & & \\ & -j2\omega_0 & & & & & \\ & & -j\omega_0 & & & & \\ & & & 0 & & & \\ & & & & j\omega_0 & & \\ & & & & & j2\omega_0 & \\ & & & & & & \ddots \end{bmatrix},$$

which is called the matrix of differentiation.

To understand the role of Toeplitz matrices played in the harmonic domain, consider the following case [12]:

Assume that v and i only contains the fundamental and dc terms:

$$v(t) = \sum_{n=-1}^{1} V_n e^{jn\omega t} \tag{6.7}$$

$$i(t) = \sum_{n=-1}^{1} I_n e^{jn\omega t}. \tag{6.8}$$

Their product, $p(t)$,

$$p(t) = v(t)i(t) = \sum_{n=-1}^{1} V_n e^{jn\omega t} \sum_{n=-1}^{1} I_n e^{jn\omega t} \tag{6.9}$$

$$= \left(V_{-1}I_{-1}\right)e^{-j2\omega t} + \left(V_{-1}I_0 + V_0 I_{-1}\right)e^{-j\omega t} + \left(V_{-1}I_1 + V_0 I_0 + V_1 I_{-1}\right) + \left(V_0 I_1 + V_1 I_0\right)e^{j\omega t} + \left(V_1 I_1\right)e^{j2\omega t}. \tag{6.10}$$

The frequency domain representation of equation (6.10) is

$$\mathbf{P} = \begin{bmatrix} P_{-2} \\ P_{-1} \\ P_0 \\ P_1 \\ P_2 \end{bmatrix} = \begin{bmatrix} V_{-1}I_{-1} \\ V_{-1}I_0 + V_0 I_{-1} \\ V_1 I_{-1} + V_0 I_0 + V_{-1}I_1 \\ V_1 I_0 + V_0 I_1 \\ V_1 I_1 \end{bmatrix} = \mathbf{V_{TP}}\mathbf{I}, \tag{6.11}$$

where

$$\mathbf{V_{TP}} = \begin{bmatrix} V_0 & V_{-1} & 0 & 0 & 0 \\ V_1 & V_0 & V_{-1} & 0 & 0 \\ 0 & V_1 & V_0 & V_{-1} & 0 \\ 0 & 0 & V_1 & V_0 & V_{-1} \\ 0 & 0 & 0 & V_1 & V_0 \end{bmatrix}, \mathbf{I} = \begin{bmatrix} 0 \\ I_{-1} \\ I_0 \\ I_1 \\ 0 \end{bmatrix}.$$

Thus, the convolution between two vectors can be easily calculated via matrix multiplication if one of the vectors is converted into the form of $\mathbf{V_{TP}}$, which is called the Toeplitz matrix. A matrix is Toeplitz, also called a convolution matrix [66], if every diagonal from top left to bottom right has the same element. Mathematically, an $n \times n$ Toeplitz matrix, T_n, can be written as

$$\begin{bmatrix} a_0 & a_{-1} & a_{-2} & \cdots & a_{-(n-1)} \\ a_1 & a_0 & a_{-1} & & \\ a_2 & a_1 & a_0 & \ddots & \vdots \\ \vdots & \ddots & a_{-1} & & \\ a_{n-1} & & \cdots & & a_0 \end{bmatrix}. \tag{6.12}$$

The convolution between vectors **V** and **I** is evaluated using (6.13)

$$\mathbf{V} \otimes \mathbf{I} = \mathbf{V_{TP}I}. \tag{6.13}$$

To see how these two operators are used, we can see how a nonlinear inductor is modeled in the harmonic domain. Assume that a nonlinear inductor has the following relation between its current and flux:

$$i = f(\psi) = 0.001\psi + 0.05\psi^3. \tag{6.14}$$

This can be expressed in the harmonic domain as

$$\mathbf{I} = a\mathbf{\Psi} + b\mathbf{\Psi} \otimes \mathbf{\Psi} \otimes \mathbf{\Psi}, \tag{6.15}$$

where a and b are constants.

Moreover, according to Faraday's law, the voltage across the inductor is

$$v = \frac{d\psi}{dt}. \tag{6.16}$$

This in turn can be replaced with its frequency domain counterpart by means of the differential operator, leading to

$$\mathbf{V} = \mathbf{D\Psi}. \tag{6.17}$$

Substituting $\mathbf{\Psi} = \mathbf{D}^{-1}\mathbf{V}$ into (6.15), we can obtain the relation between inductor current and voltage.

Similarly, one can replace the differential operator of differential equation with differential matrix **D**, and converting the harmonic vectors of the switching functions into their corresponding Toeplitz forms, one can easily build the FCM of the VSC. As an example, the state-space model of a balanced two-level VSC is stated in (6.18).

$$\frac{d}{dt}\begin{bmatrix} i_a \\ i_b \\ i_c \\ v_{dc} \end{bmatrix} = \begin{bmatrix} -\frac{R}{L} & 0 & 0 & \frac{-2S_a + S_b + S_c}{3L} \\ 0 & -\frac{R}{L} & 0 & \frac{S_a - 2S_b + S_c}{3L} \\ 0 & 0 & -\frac{R}{L} & \frac{S_a + S_b - 2S_c}{3L} \\ \frac{S_a}{C} & \frac{S_b}{C} & \frac{S_c}{C} & -\frac{1}{R_{dc}C} \end{bmatrix} \begin{bmatrix} i_a \\ i_b \\ i_c \\ v_{dc} \end{bmatrix} + \begin{bmatrix} \frac{1}{L} & 0 & 0 & 0 \\ 0 & \frac{1}{L} & 0 & 0 \\ 0 & 0 & \frac{1}{L} & 0 \\ 0 & 0 & 0 & \frac{1}{C} \end{bmatrix} \begin{bmatrix} v_{sa} \\ v_{sb} \\ v_{sc} \\ i_{dc} \end{bmatrix} \tag{6.18}$$

Let us look at phase a of the ac side:

$$\frac{d}{dt}i_a = -\frac{R}{L}i_a + \frac{-2S_a + S_b + S_c}{3L}v_{dc} + \frac{1}{L}v_{sa}. \tag{6.19}$$

Converting (6.19) into frequency domain, (6.20) results.

$$\mathbf{I_a} = (LD + R)^{-1}\left[-\frac{2}{3}\mathbf{S_a} \otimes \mathbf{V_{dc}} + \frac{1}{3}\mathbf{S_b} \otimes \mathbf{V_{dc}} + \frac{1}{3}\mathbf{S_c} \otimes \mathbf{V_{dc}} + \frac{1}{L}\mathbf{V_{sa}}\right] \tag{6.20}$$

Applying (6.13) to (6.20), one obtains

$$\mathbf{I_a} = (LD + R)^{-1}\left[-\frac{2}{3}\mathbf{S_{aTP}}\mathbf{V_{dc}} + \frac{1}{3}\mathbf{S_{bTP}}\mathbf{V_{dc}} + \frac{1}{3}\mathbf{S_{cTP}}\mathbf{V_{dc}} + \frac{1}{L}\mathbf{V_{sa}}\right]. \tag{6.21}$$

Similar expression can be obtained for phases b and c.

On the other hand, on the dc side:

$$\frac{d}{dt}v_{dc} = \frac{1}{C}S_a i_a + \frac{1}{C}S_b i_b + \frac{1}{C}S_c i_c - \frac{1}{R_{dc}C}v_{dc}. \tag{6.22}$$

Converting (6.22) into frequency domain, one obtains

$$\mathbf{V_{dc}} = \frac{1}{C}\mathbf{D}^{-1}\mathbf{S_{aTP}}\mathbf{I_a} + \frac{1}{C}\mathbf{D}^{-1}\mathbf{S_{bTP}}\mathbf{I_b} + \frac{1}{C}\mathbf{D}^{-1}\mathbf{S_{cTP}}\mathbf{I_c} - \frac{1}{R_{dc}C}\mathbf{D}^{-1}\mathbf{V_{dc}}. \tag{6.23}$$

The FCM consequently can be built by arranging (6.21) (together with equations for phases b and c) and (6.23) in a matrix form as shown in (6.24).

$$\begin{bmatrix} \mathbf{I_a} \\ \mathbf{I_b} \\ \mathbf{I_c} \\ \mathbf{V_{dc}} \end{bmatrix} = \mathbf{F} \begin{bmatrix} \mathbf{V_{sa}} \\ \mathbf{V_{sb}} \\ \mathbf{V_{sc}} \\ \mathbf{I_{dc}} \end{bmatrix} \tag{6.24}$$

Note that the FCM, $\mathbf{F}$ of equation (6.24) can be partitioned as

$$\mathbf{F} = \begin{bmatrix} \mathbf{A_{vsc}} & \mathbf{B_{vsc}} \\ \mathbf{C_{vsc}} & \mathbf{D_{vsc}} \end{bmatrix}, \tag{6.25}$$

where $\mathbf{A_{vsc}}$, and $\mathbf{D_{vsc}}$ are the ac admittance matrix and dc impedance matrix of the VSC, respectively. $\mathbf{B_{vsc}}$, and $\mathbf{C_{vsc}}$ are the multiplication factors linking the ac current to dc current harmonics, and dc voltage to ac voltage harmonics, respectively.

Note that the Fourier coefficients of the switching functions can be either obtained analytically [67] or numerically (via FFT). The analytical expression of the PWM switching function for a two-level VSC can be obtained by the method presented in [68] and [69]. In [69], it has been shown that a pulse train can be obtained either by a 2-D model (Fig. 6.3(a)) or 3-D model (Fig. 6.3(b)).

For the 2-D model, the modulating signal (v_m) is compared with the carrier waveform (v_c). The intersection points form the switching edge of the pulse train. In the 3-D model, v_m with a unit height in the z-direction is repeatedly drawn on the x-y plane, running parallel to the y-axis. The repetition period in both x ($x = \omega_c t$, where ω_c is the carrier frequency) and y ($y = \omega_m t$, where ω_m is the modulating frequency) direction is 2π. To generate the PWM signal, a plane with slope of $\frac{\omega_c}{\omega_m}$ is traced perpendicularly to the x-y plane, and is intersected with the modulating-signal wall. The intersection points coincide with those generated in the 2-D model, as shown in Fig. 6.3. Since

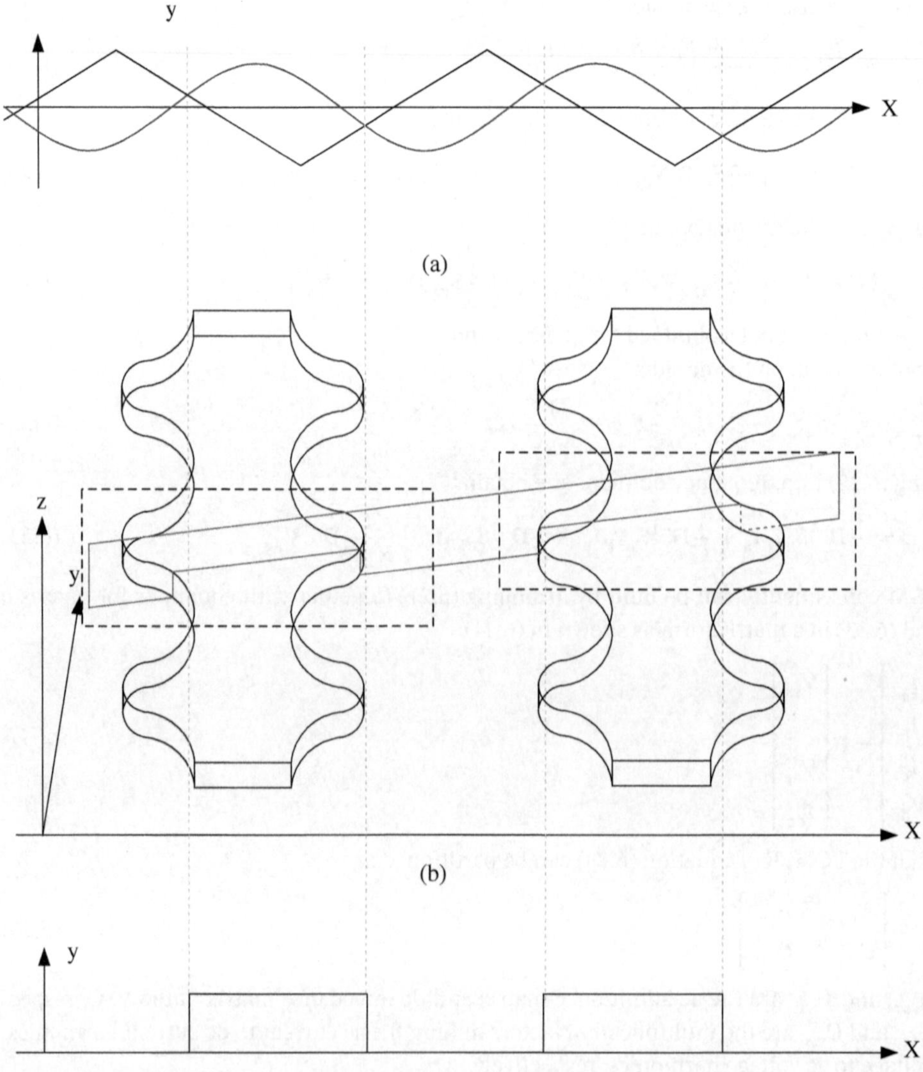

Figure 6.3 (a) 2-D model of the PWM modulation; (b) 3-D model of the PWM modulation.

the resulting signal has a period of 2π in both x and y directions, it can be represented as a double Fourier series of the form [70]:

$$F(x,y) = \frac{A_{00}}{2} + \sum_{n=1}^{\infty} A_{0n} \cos(ny) + B_{0n} \sin(ny) + \sum_{m=1}^{\infty} A_{m0} \cos(mx)$$
$$+ B_{m0} \sin(mx) + \sum_{m=1}^{\infty} \sum_{n=\pm\infty}^{\pm 1} \left[A_{mn}(mx + ny) + B_{mn} \sin(mx + ny)\right] \quad (6.26)$$

The first term of (6.26) corresponds to the dc offset, the second term to the frequency components of the modulating signal, the third term to the carrier frequency and its harmonics, and the last term is the ensemble of all possible harmonics frequency formed by taking the sum and difference between the modulating and carrier waveforms [70].

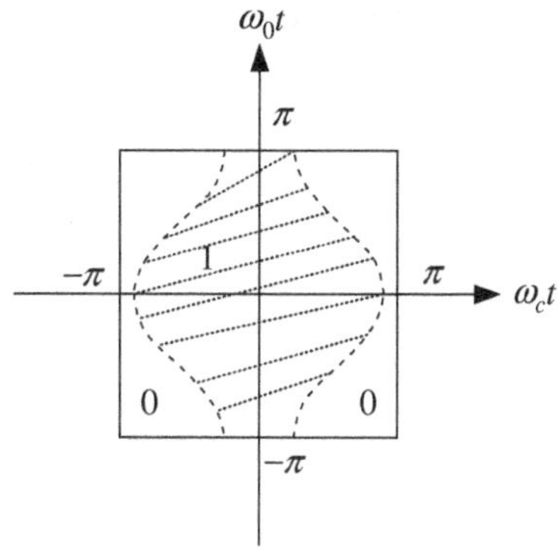

Figure 6.4 The unit cell of the modulating-signal wall.

To find out the Fourier coefficients of (6.26), one can look at the unit cell, indicated by the dashed box in Fig. 6.3, which is shown in Fig. 6.4, and conclude that $F(x, y)$ changes from 0 to 1 when

$$x = 2\pi p - \frac{\pi}{2}\left(1 + M\cos\omega_m t\right) \quad p = 0, 1, 2 \dots \infty, \tag{6.27}$$

and $F(x, y)$ changes from 1 to 0 when

$$x = 2\pi p + \frac{\pi}{2}\left(1 + M\cos\omega_m t\right) \quad p = 0, 1, 2 \dots \infty, \tag{6.28}$$

where M is the modulation index [11].

Under the integration limits defined by (6.27) and (6.28), the Fourier coefficients can be obtained [67, 70]

$$A_{mn} + jB_{mn} = \frac{1}{\pi^2}\int_{-\pi}^{\pi}\int_{-\frac{\pi}{2}(1+M\cos y)}^{\frac{\pi}{2}(1+M\cos y)} e^{j(mx+ny)}dxdy. \tag{6.29}$$

By applying trigonometric identities and Jacob-Anger expansion [67], one can obtain the compact form of analytical expansion of the switching functions of the PWM as shown in (6.30).

$$\begin{aligned} S(t) = 0.5 + 0.5M\cos\left(w_0 t + \theta_0\right) + \frac{2}{\pi}\sum_{m=1}^{\infty}\frac{1}{m}J_0\left(m\frac{\pi}{2}M\right)\cdot \\ \sin\left(m\frac{2}{\pi}\right)\cos\left(m\left(w_c t + \theta_c\right)\right) + \frac{2}{\pi}\sum_{m=1}^{\infty}\sum_{\substack{n=-\infty \\ (n\neq 0)}}^{\infty}\frac{1}{m}\cdot \\ J_n\left(m\frac{\pi}{2}M\right)\sin\left[(m+n)\frac{\pi}{2}\right]\cos\left[m\left(\omega_c t + \theta_c\right) + n\left(\omega_0 t + \theta_0\right)\right] \end{aligned} \tag{6.30}$$

6.2 Construction of FCM in the Time Domain

To formulate the FCM directly in the time domain, we need a bridge connecting between time and frequency domains. This can be done by converting the Fourier transform integral to its differential form.

The Fourier coefficients of any periodic signal, $\psi(t)$, can be obtained by evaluating (6.31)

$$\Psi_n = \frac{1}{T}\int_0^T \psi(t)e^{-jn\omega t}dt. \tag{6.31}$$

Consider the following differential equation [71],

$$\frac{dx_h}{dt} = wx_h + h\psi(t) \tag{6.32}$$

whose solution is

$$x_h(T) = e^{wT}x_h(0) + \int_0^T e^{w(T-t)} h\psi(t)dt. \tag{6.33}$$

Comparing (6.31) and (6.33), one can immediately conclude that:

$$x_h(T) = \Psi_n$$

$$w = jn\omega$$

$$h = \frac{1}{T}e^{-jn\omega T}$$

provided that $x_h(0) = 0$.

Therefore, harmonics of interest can be represented by the differential equation (6.32). The advantage of using (6.32) is that it allows one to go across from time domain to frequency or harmonic domain directly. One can immediately obtain the Fourier coefficients of interest by the end of integration, provided that the system is at steady-state.[1]

Combining the above two concepts, the following generalized equation can be written for a system with a sinusoidal exciting forcing function.

$$\frac{d\hat{\mathbf{x}}}{dt} = \begin{bmatrix} \mathbf{A} & \mathbf{B} & 0 \\ 0 & \mathbf{\Omega} & 0 \\ h & 0 & \mathbf{\Omega} \end{bmatrix} \hat{\mathbf{x}}, \tag{6.34}$$

where $\mathbf{\Omega} = \begin{bmatrix} 0 & -\omega \\ \omega & 0 \end{bmatrix}$ and $\hat{\mathbf{x}}$ contains the system state $\mathbf{x}$, input state $\mathbf{u}$, and harmonic state x_h. That is, $\hat{\mathbf{x}} = \begin{bmatrix} \mathbf{x} & \mathbf{u} & x_h \end{bmatrix}^T$.

Now, we can proceed to construct the FCM in the time domain. When formulating the FCM of a power converter, it is much more convenient to express it in the **ABCD** form. Take a two-level VSC as an example, the harmonic injections caused by external networks can be modeled as voltage source on the ac side and current sources on the dc side. Fig. 6.5 shows the equivalent system.

The general harmonic domain formulation of this system is

$$\begin{bmatrix} \mathbf{I_{ac}} \\ \mathbf{V_{dc}} \end{bmatrix} = \mathbf{F}\left(\begin{bmatrix} \mathbf{V_{ac}} \\ \mathbf{I_{dc}} \end{bmatrix}, \mathbf{t_{sw}} \right). \tag{6.35}$$

For a given set of switching times, these input ac voltage and dc current harmonics can be linked to the output ac current and dc voltage harmonics via (6.24).

Assume that there are $(2n+1)$ ac voltage stimuli per phase, and $(2n+1)$ dc current stimuli, applied to the VSC. The stimuli are assumed to take the form $A_k e^{j(k\omega t+\theta_k)}$, where $k = -n$, $-(n-1), ..0, ..., n-1, n$, and $n =$ *integer*.

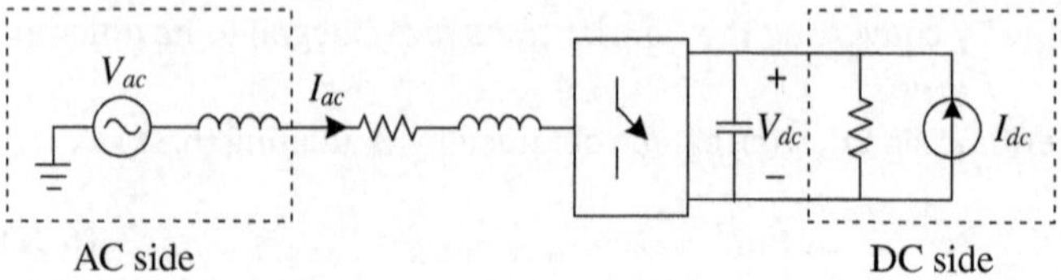

Figure 6.5 External networks of the ac-dc converter.

---

1 $\psi(t)$ needs to be periodic for (6.31) to be valid.

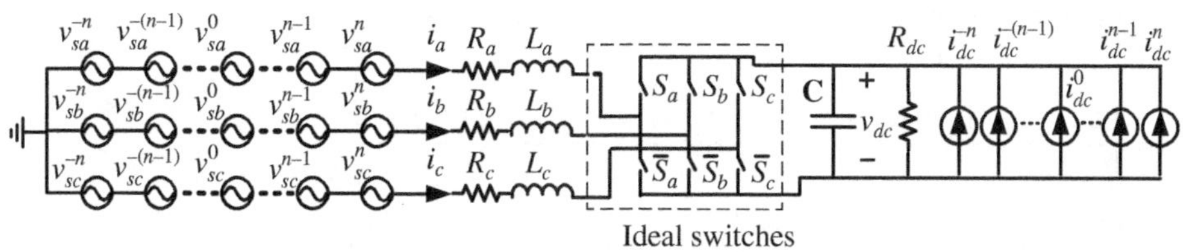

Figure 6.6 Schematic diagram of a two-level VSC with its ac and dc stimuli.

The differential equations describing Fig. 6.6 are:

$$\frac{d}{dt}\begin{bmatrix}\mathbf{x_p}\\ \mathbf{z}\end{bmatrix} = \begin{bmatrix}\mathbf{A_{S_aS_bS_c}} & \mathbf{N}\\ \mathbf{0} & \mathbf{\Omega}\end{bmatrix}\begin{bmatrix}\mathbf{x_p}\\ \mathbf{z}\end{bmatrix}, \tag{6.36}$$

where

$$\mathbf{x_p} = [\ i_a \quad i_b \quad i_c \quad v_{dc}\ ]^T,$$

$$\mathbf{z} = [\ \mathbf{v}_{sa} \quad \mathbf{v}_{sb} \quad \mathbf{v}_{sc} \quad \mathbf{i}_{dc}\ ]^T,$$

$$\mathbf{A_{S_aS_bS_c}} = \begin{bmatrix} -\dfrac{R_{aa}}{L_t} & -\dfrac{R_{ab}}{L_t} & -\dfrac{R_{ac}}{L_t} & \dfrac{S'_a}{L_t} \\ -\dfrac{R_{ba}}{L_t} & -\dfrac{R_{bb}}{L_t} & -\dfrac{R_{bc}}{L_t} & \dfrac{S'_b}{L_t} \\ -\dfrac{R_{ca}}{L_t} & -\dfrac{R_{ca}}{L_t} & -\dfrac{R_{cc}}{L_t} & \dfrac{S'_c}{L_t} \\ \dfrac{S_a}{C} & \dfrac{S_b}{C} & \dfrac{S_c}{C} & \dfrac{-1}{R_{dc}C} \end{bmatrix},$$

$$L_t = L_aL_b + L_bL_c + L_aL_c$$

$$R_{aa} = R_a\left(L_b + L_c\right) + R_bL_c$$

$$R_{ab} = 0$$

$$R_{ac} = R_bL_c - R_cL_b$$

$$R_{ba} = R_cL_a - R_aL_c$$

$$R_{bb} = R_b\left(L_a + L_c\right) + R_cL_a$$

$$R_{bc} = 0$$

$$R_{ca} = 0$$

$$R_{cb} = R_aL_b - R_bL_a$$

$$R_{cc} = R_c\left(L_a + L_b\right) + R_aL_b$$

$$S'_a = -\left(L_b + L_c\right)S_a + L_cS_b + L_bS_c$$

$$S'_b = -\left(L_a + L_c\right)S_b + L_cS_a + L_aS_c$$

$$S'_c = -\left(L_a + L_b\right)S_c + L_bS_a + L_aS_b$$

$$\mathbf{N} = \begin{bmatrix} \mathbf{N_{aa}} & \mathbf{N_{ab}} & \mathbf{N_{ac}} & \mathbf{0} \\ \mathbf{N_{ba}} & \mathbf{N_{bb}} & \mathbf{N_{bc}} & \mathbf{0} \\ \mathbf{N_{ca}} & \mathbf{N_{cb}} & \mathbf{N_{cc}} & \mathbf{0} \\ \mathbf{0} & \mathbf{0} & \mathbf{0} & \mathbf{N_l} \end{bmatrix},$$

$$\mathbf{N_{aa}} = \begin{bmatrix} \frac{L_b+L_c}{L_t} & \frac{L_b+L_c}{L_t} & \cdots & \frac{L_b+L_c}{L_t} & \cdots & \frac{L_b+L_c}{L_t} & \frac{L_b+L_c}{L_t} \end{bmatrix},$$

$$\mathbf{N_{bb}} = \begin{bmatrix} \frac{L_a+L_c}{L_t} & \frac{L_a+L_c}{L_t} & \cdots & \frac{L_a+L_c}{L_t} & \cdots & \frac{L_a+L_c}{L_t} & \frac{L_a+L_c}{L_t} \end{bmatrix},$$

$$\mathbf{N_{cc}} = \begin{bmatrix} \frac{L_a+L_b}{L_t} & \frac{L_a+L_b}{L_t} & \cdots & \frac{L_a+L_b}{L_t} & \cdots & \frac{L_a+L_b}{L_t} & \frac{L_a+L_b}{L_t} \end{bmatrix},$$

$$\mathbf{N_{ab}} = \begin{bmatrix} -\frac{L_c}{L_t} & -\frac{L_c}{L_t} & \cdots & -\frac{L_c}{L_t} & \cdots & -\frac{L_c}{L_t} & -\frac{L_c}{L_t} \end{bmatrix},$$

$$\mathbf{N_{ac}} = \begin{bmatrix} -\frac{L_b}{L_t} & -\frac{L_b}{L_t} & \cdots & -\frac{L_b}{L_t} & \cdots & -\frac{L_b}{L_t} & -\frac{L_b}{L_t} \end{bmatrix},$$

$$\mathbf{N_{bc}} = \begin{bmatrix} -\frac{L_a}{L_t} & -\frac{L_a}{L_t} & \cdots & -\frac{L_a}{L_t} & \cdots & -\frac{L_a}{L_t} & -\frac{L_a}{L_t} \end{bmatrix},$$

$$\mathbf{N_{ba}} = \mathbf{N_{ab}}, \mathbf{N_{ca}} = \mathbf{N_{ac}}, \mathbf{N_{cb}} = \mathbf{N_{bc}},$$

$$\mathbf{N_l} = \begin{bmatrix} \frac{1}{C} & \frac{1}{C} & \cdots & \frac{1}{C} & \cdots & \frac{1}{C} & \frac{1}{C} \end{bmatrix},$$

$$\mathbf{\Omega} = diag(\mathbf{\Omega_1}, \mathbf{\Omega_1}, \mathbf{\Omega_1}, \mathbf{\Omega_1}),$$

$$\mathbf{\Omega_1} = diag(-jn\omega, -j(n-1)\omega, \ldots, 0, \ldots, j(n-1)\omega, jn\omega).$$

In the above formulation, the amplitudes (A_k) and phase angles (θ_k) of the ac and dc stimuli are uniquely related to the initial conditions, $\mathbf{z}(0)$. In addition, although the switching functions, S_a, S_b, and S_c are time-varying, they are periodic with frequency ω. Therefore, Fourier series coefficients exist for the steady-state solutions of (6.36).

Augmenting (6.32) to (6.36) yields:

$$\frac{d}{dt}\hat{\mathbf{x}} = \hat{\mathbf{A}}\hat{\mathbf{x}}, \tag{6.37}$$

where

$$\hat{\mathbf{x}} = [\ \mathbf{x}\ \ \mathbf{z}\ \ \mathbf{x_h}\ ]^T,$$

$$\mathbf{x_h} = \begin{bmatrix} \mathbf{I_a} & \mathbf{I_b} & \mathbf{I_c} & \mathbf{V_{dc}} \end{bmatrix}^T,$$

$$\mathbf{I_a} = \begin{bmatrix} I_a^{-n} & I_a^{-(n-1)} & \cdots & I_a^0 & \cdots & I_a^{n-1} & I_a^n \end{bmatrix}^T,$$

$$\mathbf{I_b} = \begin{bmatrix} I_b^{-n} & I_b^{-(n-1)} & \cdots & I_b^0 & \cdots & I_b^{n-1} & I_b^n \end{bmatrix}^T,$$

$$\mathbf{I_c} = \begin{bmatrix} I_c^{-n} & I_c^{-(n-1)} & \cdots & I_c^0 & \cdots & I_c^{n-1} & I_c^n \end{bmatrix}^T,$$

$$\mathbf{V_{dc}} = \begin{bmatrix} V_{dc}^{-n} & V_{dc}^{-(n-1)} & \cdots & V_{dc}^0 & \cdots & V_{dc}^{n-1} & V_{dc}^n \end{bmatrix}^T,$$

$$\hat{\mathbf{A}} = \begin{bmatrix} \mathbf{A}_{S_aS_bS_c} & \mathbf{N} & \mathbf{0} \\ \mathbf{0} & \mathbf{\Omega} & \mathbf{0} \\ \mathbf{H} & \mathbf{0} & \mathbf{W} \end{bmatrix},$$

$$\mathbf{W} = diag(\mathbf{W_1}, \mathbf{W_1}, \mathbf{W_1}, \mathbf{W_1}),$$

$$\mathbf{W_1} = diag(-jn\omega, -j(n-1)\omega, \ldots, 0, \ldots, j(n-1)\omega, jn\omega),$$

$$\mathbf{H} = \left[\begin{array}{cccc}
\frac{e^{jn\omega T}}{T} & 0 & 0 & 0 \\
\frac{e^{j(n-1)\omega T}}{T} & 0 & 0 & 0 \\
\vdots & \vdots & \vdots & \vdots \\
\frac{1}{T} & 0 & 0 & 0 \\
\vdots & \vdots & \vdots & \vdots \\
\frac{e^{-j(n-1)\omega T}}{T} & 0 & 0 & 0 \\
\frac{e^{-jn\omega T}}{T} & 0 & 0 & 0 \\
\hline
0 & \frac{e^{jn\omega T}}{T} & 0 & 0 \\
0 & \frac{e^{j(n-1)\omega T}}{T} & 0 & 0 \\
\vdots & \vdots & \vdots & \vdots \\
0 & \frac{1}{T} & 0 & 0 \\
\vdots & \vdots & \vdots & \vdots \\
0 & \frac{e^{-j(n-1)\omega T}}{T} & 0 & 0 \\
0 & \frac{e^{-jn\omega T}}{T} & 0 & 0 \\
\hline
0 & 0 & \frac{e^{jn\omega T}}{T} & 0 \\
0 & 0 & \frac{e^{j(n-1)\omega T}}{T} & 0 \\
\vdots & \vdots & \vdots & \vdots \\
0 & 0 & \frac{1}{T} & 0 \\
\vdots & \vdots & \vdots & \vdots \\
0 & 0 & \frac{e^{-j(n-1)\omega T}}{T} & 0 \\
0 & 0 & \frac{e^{-jn\omega T}}{T} & 0 \\
\hline
0 & 0 & 0 & \frac{e^{jn\omega T}}{T} \\
0 & 0 & 0 & \frac{e^{j(n-1)\omega T}}{T} \\
\vdots & \vdots & \vdots & \vdots \\
0 & 0 & 0 & \frac{1}{T} \\
\vdots & \vdots & \vdots & \vdots \\
0 & 0 & 0 & \frac{e^{-j(n-1)\omega T}}{T} \\
0 & 0 & 0 & \frac{e^{-jn\omega T}}{T}
\end{array}\right].$$

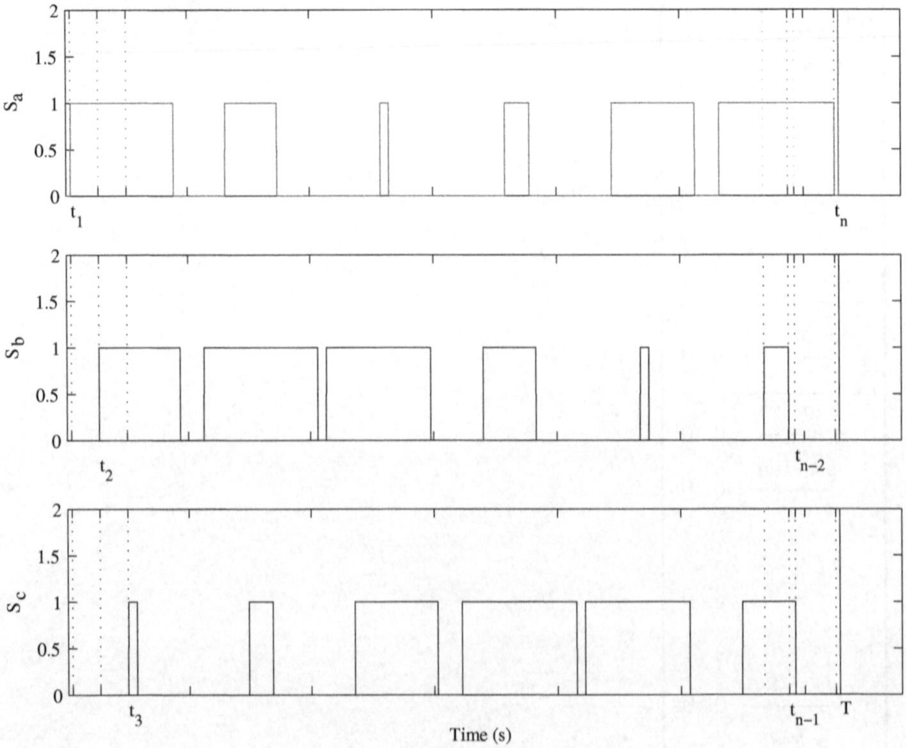

Figure 6.7 A typical PWM switching function waveform.

Fig. 6.7 shows typical PWM switching functions of a two-level VSC over one period, T. For the given switching times (as in Fig. 6.7), the solution of the state, $\hat{\mathbf{x}}$ over T, is given by (6.38).

$$\hat{\mathbf{x}}(T) = \hat{\mathbf{\Phi}}\hat{\mathbf{x}}(0) \tag{6.38}$$

$\hat{\mathbf{\Phi}}$ is calculated from:

$$\hat{\mathbf{\Phi}} = e^{\hat{\mathbf{A}}_{t_{n+1}}(T-t_n)} \dots e^{\hat{\mathbf{A}}_{t2}(t_2-t_1)} e^{\hat{\mathbf{A}}_{t_1}(t_1-0)}, \tag{6.39}$$

where $\hat{\mathbf{A}}_{t_k}$ represents $\hat{\mathbf{A}}$ between t_{k-1} and t_k. Fig. 6.8 shows the sparsity pattern of matrix $\hat{\mathbf{A}}$. It is evident that the matrix, $\hat{\mathbf{A}}$, is very sparse, and is dominantly diagonal. Therefore, one can evaluate $\hat{\mathbf{\Phi}}\hat{\mathbf{x}}(0)$ by means of Krylov subspace method, described in Chapter 4, to speed up the calculation time. This will be further illustrated in Example 6.2. $\hat{\mathbf{x}}$ can be obtained if we substitute the steady-state $\mathbf{x}(0)$, which can be produced by solving the multiplication of exponential matrix $\bar{\mathbf{A}}$, $\mathbf{z}(0)$, and $\mathbf{x_h} = 0$ into (6.38).

Alternatively, one can observe that $\hat{\mathbf{\Phi}}$ invariably has the following unique structure,

$$\hat{\mathbf{\Phi}} \triangleq \begin{bmatrix} \mathbf{\Phi} & \mathbf{\Gamma} & 0 \\ 0 & \mathbf{\Omega}_p & 0 \\ \mathbf{H}_p & \mathbf{Q}_p & \mathbf{W}_p \end{bmatrix}, \tag{6.40}$$

where subscript $\mathbf{p}$ represents "the product" term.

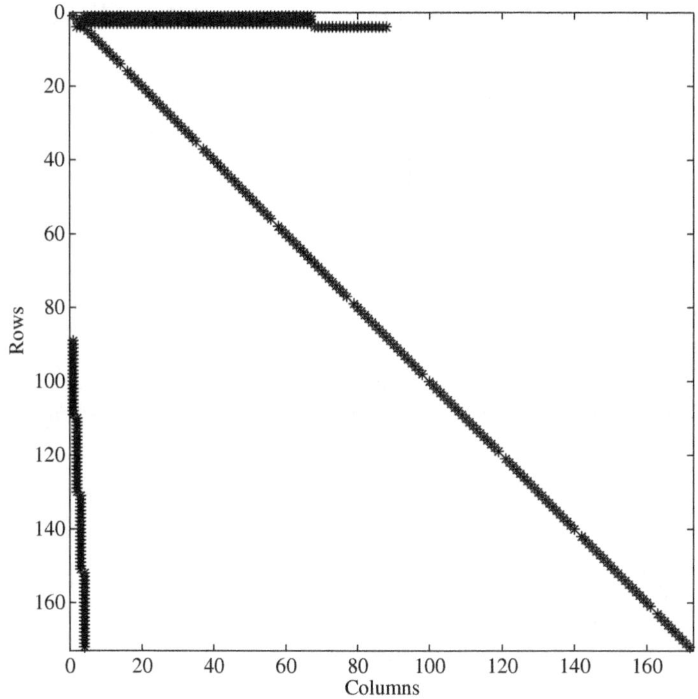

Figure 6.8 A typical sparsity pattern of matrix $\hat{\mathbf{A}}$.

The following provides a simple numerical proof for (6.40). The product of the two matrices, $\hat{\mathbf{A}}_1$ and $\hat{\mathbf{A}}_2$, is:

$$\begin{bmatrix} \mathbf{A}_1 & \mathbf{N} & \mathbf{0} \\ 0 & \boldsymbol{\Omega} & \mathbf{0} \\ \mathbf{H} & \mathbf{0} & \mathbf{W} \end{bmatrix} \begin{bmatrix} \mathbf{A}_2 & \mathbf{N} & \mathbf{0} \\ 0 & \boldsymbol{\Omega} & \mathbf{0} \\ \mathbf{H} & \mathbf{0} & \mathbf{W} \end{bmatrix} = \begin{bmatrix} \mathbf{A}_1\mathbf{A}_2 & \mathbf{A}_1\mathbf{N} + \mathbf{N}\boldsymbol{\Omega} & \mathbf{0} \\ \mathbf{0} & \boldsymbol{\Omega}^2 & \mathbf{0} \\ \mathbf{H}\mathbf{A}_2 + \mathbf{W}\mathbf{H} & \mathbf{H}\mathbf{N} & \mathbf{W}^2 \end{bmatrix}, \tag{6.41}$$

The product of 3 matrices, $\hat{\mathbf{A}}_1$, $\hat{\mathbf{A}}_2$, and $\hat{\mathbf{A}}_3$, is

$$\begin{bmatrix} \mathbf{A}_1\mathbf{A}_2 & \mathbf{A}_1\mathbf{N} + \mathbf{N}\boldsymbol{\Omega} & \mathbf{0} \\ \mathbf{0} & \boldsymbol{\Omega}^2 & \mathbf{0} \\ \mathbf{H}\mathbf{A}_2 + \mathbf{W}\mathbf{H} & \mathbf{H}\mathbf{N} & \mathbf{W}^2 \end{bmatrix} \begin{bmatrix} \mathbf{A}_3 & \mathbf{N} & \mathbf{0} \\ 0 & \boldsymbol{\Omega} & \mathbf{0} \\ \mathbf{H} & \mathbf{0} & \mathbf{W} \end{bmatrix}, \tag{6.42}$$

which has the form of

$$\begin{bmatrix} \alpha & \beta & \mathbf{0} \\ \mathbf{0} & \gamma & \mathbf{0} \\ \eta & \xi & \varsigma \end{bmatrix}. \tag{6.43}$$

By extension, matrix, $e^{\hat{\mathbf{A}}\mathbf{t}}$ will also have the form of (6.43) because of the fact that

$$e^{\hat{\mathbf{A}}t} = I + \hat{\mathbf{A}}t + \frac{\hat{\mathbf{A}}^2 t^2}{2!} + \frac{\hat{\mathbf{A}}^3 t^3}{3!} + \cdots + \frac{\hat{\mathbf{A}}^n t^n}{n!}. \tag{6.44}$$

The augmented state transition matrix, $\hat{\mathbf{\Phi}}$, will have the form of (6.43) as well. This is because

$$\hat{\mathbf{\Phi}} = e^{\hat{\mathbf{A}}_{t_{n+1}}(T-t_n)} \ldots e^{\hat{\mathbf{A}}_{t2}(t_2-t_1)} e^{\hat{\mathbf{A}}_{t_1}(t_1-0)}$$

$$= \begin{bmatrix} \alpha_{\mathbf{n+1}} & \beta_{\mathbf{n+1}} & \mathbf{0} \\ \mathbf{0} & \gamma_{\mathbf{n+1}} & \mathbf{0} \\ \eta_{\mathbf{n+1}} & \xi_{\mathbf{n+1}} & \varsigma_{\mathbf{n+1}} \end{bmatrix} \cdots \begin{bmatrix} \alpha_2 & \beta_2 & \mathbf{0} \\ \mathbf{0} & \gamma_2 & \mathbf{0} \\ \eta_2 & \xi_2 & \varsigma_2 \end{bmatrix} \begin{bmatrix} \alpha_1 & \beta_1 & \mathbf{0} \\ \mathbf{0} & \gamma_1 & \mathbf{0} \\ \eta_1 & \xi_1 & \varsigma_1 \end{bmatrix}. \tag{6.45}$$

Therefore, we can conclude $\hat{\mathbf{\Phi}}$ has the form of

$$\hat{\mathbf{\Phi}} = \begin{bmatrix} \mathbf{\Phi} & \mathbf{\Gamma} & \mathbf{0} \\ \mathbf{0} & \mathbf{\Omega_p} & \mathbf{0} \\ \mathbf{H_p} & \mathbf{Q_p} & \mathbf{W_p} \end{bmatrix}.$$

In general, analytical expressions for $\mathbf{\Phi}$, $\mathbf{\Gamma}$, $\mathbf{H_p}$, $\mathbf{Q_p}$ are not required as these matrices may be extracted from the calculated $\hat{\mathbf{\Phi}}$ matrix.

Equation (6.38) can then be re-expressed as (6.46):

$$\begin{bmatrix} \mathbf{x_p}(T) \\ \mathbf{z}(T) \\ \mathbf{x_h}(T) \end{bmatrix} = \begin{bmatrix} \mathbf{\Phi} & \mathbf{\Gamma} & \mathbf{0} \\ \mathbf{0} & \mathbf{\Omega_p} & \mathbf{0} \\ \mathbf{H_p} & \mathbf{Q_p} & \mathbf{W_p} \end{bmatrix} \begin{bmatrix} \mathbf{x_p}(0) \\ \mathbf{z}(0) \\ \mathbf{x_h}(0) \end{bmatrix}. \tag{6.46}$$

The third row of (6.46) yields the solution to the Fourier integrals:

$$\mathbf{x_h}(T) = \mathbf{H_p}\mathbf{x_p}(0) + \mathbf{Q_p}\mathbf{z}(0). \tag{6.47}$$

The steady-state initial condition, $\mathbf{x}(0)$, related to $\mathbf{z}(0)$, is restated in (6.48).

$$\mathbf{x_p}(0) = (\mathbb{I} - \mathbf{\Phi})^{-1}\mathbf{\Gamma}\mathbf{z}(0), \tag{6.48}$$

Substituting (6.48) to (6.47) yields the steady-state output harmonics related to the initial conditions, $\mathbf{z}(0)$, stated in (6.49):

$$\mathbf{x_h}(T) = \{\mathbf{H_p}[(\mathbb{I} - \mathbf{\Phi})^{-1}\mathbf{\Gamma}] + \mathbf{Q_p}\}\mathbf{z}(0). \tag{6.49}$$

Note that merely by computing $\hat{\mathbf{\Phi}}$, all submatrices necessary to solve (6.49) are available. An alternative to (6.49) is described on pg. 168.

The FCM in (6.24) relates the output harmonics to input harmonic phasors, while (6.49) instead relates them to input initial conditions. To derive the FCM from (6.49), a one-to-one relation between input harmonic phasor and input initial conditions must be obtained.

As stated, stimuli take the form of $A_k e^{j(k\omega t+\theta_k)}$. Therefore, the initial condition for the stimulus's differential equations is simply equal to the harmonic phasor of that stimulus:

$$
\mathbf{z}(0) = \begin{bmatrix} V_{sa}^{-n} \\ \vdots \\ V_{sa}^{0} \\ \vdots \\ V_{sa}^{n} \\ V_{sb}^{-n} \\ \vdots \\ V_{sb}^{0} \\ \vdots \\ V_{sb}^{n} \\ V_{sc}^{-n} \\ \vdots \\ V_{sc}^{0} \\ \vdots \\ V_{sc}^{n} \\ I_{dc}^{-n} \\ \vdots \\ I_{dc}^{0} \\ \vdots \\ I_{dc}^{n} \end{bmatrix}.
$$

Consequently, (6.49) directly relates input to output harmonics. This is explicitly shown in (6.50).

$$
\begin{bmatrix} I_{a}^{-n} \\ \vdots \\ I_{a}^{0} \\ \vdots \\ I_{a}^{n} \\ I_{b}^{-n} \\ \vdots \\ I_{b}^{0} \\ \vdots \\ I_{b}^{n} \\ I_{c}^{-n} \\ \vdots \\ I_{c}^{0} \\ \vdots \\ I_{c}^{n} \\ V_{dc}^{-n} \\ \vdots \\ V_{dc}^{0} \\ \vdots \\ V_{dc}^{n} \end{bmatrix} = \mathbf{F} \begin{bmatrix} V_{sa}^{-n} \\ \vdots \\ V_{sa}^{0} \\ \vdots \\ V_{sa}^{n} \\ V_{sb}^{-n} \\ \vdots \\ V_{sb}^{0} \\ \vdots \\ V_{sb}^{n} \\ V_{sc}^{-n} \\ \vdots \\ V_{sc}^{0} \\ \vdots \\ V_{sc}^{n} \\ I_{dc}^{-n} \\ \vdots \\ I_{dc}^{0} \\ \vdots \\ I_{dc}^{n} \end{bmatrix}, \tag{6.50}
$$

where

$$\mathbf{F} = \{\mathbf{H_p}[(\mathbb{I} - \mathbf{\Phi})^{-1}\mathbf{\Gamma}] + \mathbf{Q_p}\}.$$

Note that since the stimuli have the form of $A_k e^{j(k\omega t + \theta_k)}$, it takes both positive and negative harmonics to constitute one sinusoid. However, this complex Fourier frequency domain is widely employed to analyze LTV systems because of its convenience for calculating the convolution in the frequency domain. To aid comparison with the frequency domain FCM, the time domain FCM in this chapter is derived based on complex Fourier series. An equivalent FCM may be derived based on the trigonometric Fourier series. In that case, the FCM needs to be expressed in terms of real numbers to relate input and output harmonics as in (6.51).

$$\begin{bmatrix} \mathrm{Re}\{I_a^0\} \\ \mathrm{Im}\{I_a^0\} \\ \vdots \\ \mathrm{Re}\{I_a^n\} \\ \mathrm{Im}\{I_a^n\} \\ \mathrm{Re}\{I_b^0\} \\ \mathrm{Im}\{I_b^0\} \\ \vdots \\ \mathrm{Re}\{I_b^n\} \\ \mathrm{Im}\{I_b^n\} \\ \mathrm{Re}\{I_c^0\} \\ \mathrm{Im}\{I_c^0\} \\ \vdots \\ \mathrm{Re}\{I_c^n\} \\ \mathrm{Im}\{I_c^n\} \\ \mathrm{Re}\{V_{dc}^0\} \\ \mathrm{Im}\{V_{dc}^0\} \\ \vdots \\ \mathrm{Re}\{V_{dc}^n\} \\ \mathrm{Im}\{V_{dc}^n\} \end{bmatrix} = \mathbf{F_{re}} \begin{bmatrix} \mathrm{Re}\{V_{sa}^0\} \\ \mathrm{Im}\{V_{sa}^0\} \\ \vdots \\ \mathrm{Re}\{V_{sa}^n\} \\ \mathrm{Im}\{V_{sa}^n\} \\ \mathrm{Re}\{V_{sb}^0\} \\ \mathrm{Im}\{V_{sb}^0\} \\ \vdots \\ \mathrm{Re}\{V_{sb}^n\} \\ \mathrm{Im}\{V_{sb}^n\} \\ \mathrm{Re}\{V_{sc}^0\} \\ \mathrm{Im}\{V_{sc}^0\} \\ \vdots \\ \mathrm{Re}\{V_{sc}^n\} \\ \mathrm{Im}\{V_{sc}^n\} \\ \mathrm{Re}\{I_{dc}^0\} \\ \mathrm{Im}\{I_{dc}^0\} \\ \vdots \\ \mathrm{Re}\{I_{dc}^n\} \\ \mathrm{Im}\{I_{dc}^n\} \end{bmatrix}. \tag{6.51}$$

The following code is a sample code to evaluate the FCM for a CML-VSC. The input of this function is similar to the function for modeling the CML-VSC, `CMC_waveform`, in **Example 5.1**, with the additional variables `Hh` and `Hh2`, denoting the current and capacitor voltage harmonics of interests, respectively. The difference between the sample code in **Example 5.1** and the following code is the existence of augmented system matrix $\hat{\mathbf{A}}$, which is represented by `Axxx_II`. The multiplication of exponential matrix over the switching time intervals are performed in Line 175, while the FCM is evaluated in Line 186.

```
1 function [FCM] = CML_VSC_FCM(Vs,phi,w,N,R,L,C,Rm,SST,SSseq,Hh,Hh2)
2 %% The function for evaluating FCM of Cascaded Multi-Level VSC
3
4 %% Extracting phase angle
5 phi_a=phi(1); phi_b=phi(2); phi_c=phi(3);
6
```

```
%% Switching Time and Switching Pattern
taus_2=SST(:,1);
S_2=SSseq(:,1:size(SSseq,2));
%% Classifying Switching Pattern into
Sa_up=S_2(1,0*N+1:1*N);
Sb_up=S_2(1,1*N+1:2*N);
Sc_up=S_2(1,2*N+1:3*N);
Sa_low=S_2(1,3*N+1:4*N);
Sb_low=S_2(1,4*N+1:5*N);
Sc_low=S_2(1,5*N+1:6*N);

%% Inductor matrix
HH=[((2/3)*L(1))+((1/3)*L(3)) ((-1/3)*L(2))+((1/3)*L(3));
    ((-1/3)*L(1)+(1/3)*L(3)) ((2/3)*L(2))+((1/3)*L(3))];
%% Constructing matrix elements
Fa=zeros(1,N);
Fb=zeros(1,N);
Fc=zeros(1,N);
Pa=zeros(N,1);
Pb=zeros(N,1);
Pc=zeros(N,1);
W1=zeros(N,N);
W2=zeros(N,N);
W3=zeros(N,N);
for i=1:N
    Fa(1,i)=(Sa_up(1,i)-Sa_low(1,i));
    Fb(1,i)=(Sb_up(1,i)-Sb_low(1,i));
    Fc(1,i)=(Sc_up(1,i)-Sc_low(1,i));
    Pa(i,1)=(Sa_up(1,i)-Sa_low(1,i))/C(1,i);
    Pb(i,1)=(Sb_up(1,i)-Sb_low(1,i))/C(2,i);
    Pc(i,1)=(Sc_up(1,i)-Sc_low(1,i))/C(3,i);
    W1(i,i)=-1/(Rm(1,i)*C(1,i));
    W2(i,i)=-1/(Rm(2,i)*C(2,i));
    W3(i,i)=-1/(Rm(3,i)*C(3,i));
end


F=[ (-2/3)*Fa, (1/3)*Fb, (1/3)*Fc;
    (1/3)*Fa, (-2/3)*Fb, (1/3)*Fc];
F_Hinv=HH\F;

G=[((-2/3)*R(1))-((1/3)*R(3)) ((1/3)*R(2))-((1/3)*R(3));
    ((1/3)*R(1))-((1/3)*R(3)) ((-2/3)*R(2))-((1/3)*R(3))];

G_Hinv=HH\G;

P1=blkdiag(Pa,Pb);
P=[P1; -1*Pc, -1*Pc];

W=blkdiag(W1,W2,W3);

z0=[ Vs*sqrt(2)*cos(phi_a);      % Input states
    Vs*sqrt(2)*sin(phi_a);
    Vs*sqrt(2)*cos(phi_b);
    Vs*sqrt(2)*sin(phi_b);
    Vs*sqrt(2)*cos(phi_c);
    Vs*sqrt(2)*sin(phi_c)];

A_Max=[G_Hinv, F_Hinv;
    P, W];
[VA_Max,HA_Max]=size(A_Max);
```

```
B_M=[2/3, 0, -1/3, 0, -1/3, 0;
    -1/3, 0, 2/3, 0, -1/3, 0];
B_Minv=HH\B_M;

B_Max=[B_Minv;
    zeros(3*N,6)];
[VB_Max,HB_Max]=size(B_Max);

omega2=[0 -1*w; 1*w 0];
D_Max=blkdiag(omega2,omega2,omega2);
[VD_Max,HD_Max]=size(D_Max);

Axxx_I=[A_Max B_Max;
    zeros(VD_Max,HA_Max),D_Max];
[VAxxx_I,HAxxx_I]=size(Axxx_I);
Ip=eye(VA_Max,HA_Max);

%% Constructing H matrix
for i=1:3*N+2                          %Iabc + Vcap
    if i<3                             % Iabc
        for ii=1:length(Hh)
        H(length(Hh)*(i-1)+ii,:)=[Ip(i,:)*1/T*exp(-j*Hh(ii)*w*T)];
        end
    else                               % Vcap
        for ii=1:length(Hh2)
        H(2*length(Hh)+length(Hh2)*(i-3)+ii,:)=[Ip(i,:)*1/T*exp(-j*Hh2(ii)
        *w*T)];
        end
    end
end
[barH,kolH]=size(H);
Hz = zeros(barH,HB_Max);
Hnew = [H,Hz];
[VH_Max,HH_Max]=size(Hnew);

%% Constructing M matrix
[barHh,kolHh]=size(Hh);
[barHh2,kolHh2]=size(Hh2);
M1=[Hh,Hh];
M2=zeros(1,3*N*kolHh2);
for i=1:3*N
    M2(1,(i-1)*kolHh2+1:i*kolHh2)=Hh2;
end
M=j*w*diag([M1, M2]);

[VM_Max,HM_Max]=size(M);
Z=zeros(VAxxx_I,length(Hh)*2+length(Hh2)*(3*N));     % zeros matrices

%% Constructing augmented matrix (system, input, and harmonic states)
Axxx_II=[Axxx_I,Z;
    Hnew,M];

%% Obtaining the solution via exponential matrix
PHI=expm(Axxx_II*(taus_2(1)));                  % Computing exponential matrix

if taus_2(length(taus_2))>T
    i=1;
    while taus_2(i)<=T
        cycle=i;
        i=i+1;
    end
```

```
    tau=[taus_2(1:cycle,1); T];
else
    cycle=length(else_SST);
    tau=[else_SST];

    S_2=else_SSseq(:,1:size(else_SSseq,2));
end

for iii=2:length(tau)
    Sa_up=S_2(iii,0*N+1:1*N);
    Sb_up=S_2(iii,1*N+1:2*N);
    Sc_up=S_2(iii,2*N+1:3*N);
    Sa_low=S_2(iii,3*N+1:4*N);
    Sb_low=S_2(iii,4*N+1:5*N);
    Sc_low=S_2(iii,5*N+1:6*N);

    %% Constructing matrix elements
    Fa=zeros(1,N);
    Fb=zeros(1,N);
    Fc=zeros(1,N);
    Pa=zeros(N,1);
    Pb=zeros(N,1);
    Pc=zeros(N,1);
    for i=1:N
        Fa(1,i)=(Sa_up(1,i)-Sa_low(1,i));
        Fb(1,i)=(Sb_up(1,i)-Sb_low(1,i));
        Fc(1,i)=(Sc_up(1,i)-Sc_low(1,i));
        Pa(i,1)=(Sa_up(1,i)-Sa_low(1,i))/C(1,i);
        Pb(i,1)=(Sb_up(1,i)-Sb_low(1,i))/C(2,i);
        Pc(i,1)=(Sc_up(1,i)-Sc_low(1,i))/C(3,i);
    end
    F=[ (-2/3)*Fa, (1/3)*Fb, (1/3)*Fc;
        (1/3)*Fa, (-2/3)*Fb, (1/3)*Fc];
    F_Hinv=HH\F;

    P1=blkdiag(Pa,Pb);
    P=[P1; -1*Pc, -1*Pc];

    A_Max=[G_Hinv, F_Hinv;
        P, W];

    Axxx_I(1:VA_Max,1:HA_Max)=A_Max;

    Axxx_II=[Axxx_I,Z;Hnew,M]; % Augmented matrix

    %% Obtaining the solution via exponential matrix
    PHI=expm(Axxx_II*(tau(iii)-tau(iii-1)))*PHI;
    [VPHI_Max,HPHI_Max]=size(PHI);
end

%% Partitioning matrix produced by multiplication of exponential matrix
Ap=PHI(1:VA_Max,1:HA_Max);
Np=PHI(1:VA_Max,HA_Max+1:HA_Max+HB_Max);
Hp=PHI(VPHI_Max-VH_Max+1:VPHI_Max,1:HA_Max);
Qp=PHI(VPHI_Max-VH_Max+1:VPHI_Max,HA_Max+1:HA_Max+HB_Max);

%% Computing the Frequency Coupling Matrix (FCM)
FCM=[(Hp*(inv(Ip-Ap)*Np)+Qp)*z0];
```

As one may see, evaluating the exponential of `Axxx_II` over the switching time intervals could be expensive, particularly if the dimension of the harmonics state is large. To efficiently compute the evaluation of exponential matrix multiplication, Krylov subspace technique is employed, as described in Chapter 4. The following procedures describe the application of Krylov subspace technique to evaluate the harmonics as presented in [72].

1. Obtain the state transition matrix of the augmented system and input matrices $\left(\bar{\mathbf{A}}\right)$ by solving (6.52)

$$\bar{\boldsymbol{\Phi}} = \begin{bmatrix} \boldsymbol{\Phi} & \boldsymbol{\Gamma} \\ \mathbf{0} & \boldsymbol{\Omega}_{\mathbf{p}} \end{bmatrix} = \prod_{i=n_{sw}}^{1} e^{\bar{\mathbf{A}}_i(t_i - t_{i-1})} = e^{\bar{\mathbf{A}}_{n_{sw}}\left(T - t_{(n_{sw}-1)}\right)} \ldots e^{\bar{\mathbf{A}}_2(t_2-t_1)} e^{\bar{\mathbf{A}}_1(t_1-0)}. \tag{6.52}$$

2. According to the state and input periodicity constraints, the steady-state initial condition vector can be obtained by

$$\begin{bmatrix} \mathbf{x}(T) \\ \mathbf{u}(T) \end{bmatrix} = \begin{bmatrix} \mathbf{x}(0) \\ \mathbf{u}(0) \end{bmatrix} = \bar{\boldsymbol{\Phi}} \begin{bmatrix} \mathbf{x}(0) \\ \mathbf{u}(0) \end{bmatrix}. \tag{6.53}$$

Hence,

$$\left(\mathbb{I} - \bar{\boldsymbol{\Phi}}\right) \begin{bmatrix} \mathbf{x}(0) \\ \mathbf{u}(0) \end{bmatrix} = \begin{bmatrix} 0 \\ 0 \end{bmatrix}. \tag{6.54}$$

Partitioning (6.54) yields

$$\begin{bmatrix} \mathbf{E}_1 & \mathbf{E}_2 \\ \mathbf{E}_3 & \mathbf{E}_4 \end{bmatrix} \begin{bmatrix} \mathbf{x}(0) \\ \mathbf{u}(0) \end{bmatrix} = \begin{bmatrix} 0 \\ 0 \end{bmatrix}, \tag{6.55}$$

where $\mathbf{E}_1$, $\mathbf{E}_2$, $\mathbf{E}_3$, and $\mathbf{E}_4$ are submatrices of $\left(\mathbb{I} - \bar{\boldsymbol{\Phi}}\right)$. Consequently, the initial condition vector can be obtained as

$$\mathbf{x}(0) = -\mathbf{E}_1^{-1}\mathbf{E}_2\mathbf{u}(0). \tag{6.56}$$

3. As the augmented initial condition vector $(\hat{x}(0))$, consisting of $\mathbf{x}(0)$, $\mathbf{z}(0)$, and $\mathbf{x}_h(0)$, are known, the augmented state transition matrix can be directly solved by using Krylov subspace technique described in Chapter 4. The following code is the sample code for evaluating harmonics of interest of CML-VSC via Krylov subspace technique.

```
function[Xx_harm] = CML_KRYLOV_FCM(Vs,phi,w,N,R,L,C,Rm,SST,SSseq,Hh,Hh2)
%% The function for evaluating harmonics of interests of Cascaded Multi-Level
% VSC

%% Extracting phase angle
phi_a=phi(1); phi_b=phi(2); phi_c=phi(3);

%% Switching Time and Switching Pattern
taus_2=SST(:,1);
S_2=SSseq(:,1:size(SSseq,2));
%% Classifying Switching Pattern into
Sa_up=S_2(1,0*N+1:1*N);
Sb_up=S_2(1,1*N+1:2*N);
Sc_up=S_2(1,2*N+1:3*N);
Sa_low=S_2(1,3*N+1:4*N);
Sb_low=S_2(1,4*N+1:5*N);
Sc_low=S_2(1,5*N+1:6*N);

```

```
%% Inductor matrix
HH=[((2/3)*L(1))+((1/3)*L(3)) ((-1/3)*L(2))+((1/3)*L(3));
    ((-1/3)*L(1)+(1/3)*L(3)) ((2/3)*L(2))+((1/3)*L(3))];
%% Constructing matrix elements
Fa=zeros(1,N);
Fb=zeros(1,N);
Fc=zeros(1,N);
Pa=zeros(N,1);
Pb=zeros(N,1);
Pc=zeros(N,1);
W1=zeros(N,N);
W2=zeros(N,N);
W3=zeros(N,N);
for i=1:N
    Fa(1,i)=(Sa_up(1,i)-Sa_low(1,i));
    Fb(1,i)=(Sb_up(1,i)-Sb_low(1,i));
    Fc(1,i)=(Sc_up(1,i)-Sc_low(1,i));
    Pa(i,1)=(Sa_up(1,i)-Sa_low(1,i))/C(1,i);
    Pb(i,1)=(Sb_up(1,i)-Sb_low(1,i))/C(2,i);
    Pc(i,1)=(Sc_up(1,i)-Sc_low(1,i))/C(3,i);
    W1(i,i)=-1/(Rm(1,i)*C(1,i));
    W2(i,i)=-1/(Rm(2,i)*C(2,i));
    W3(i,i)=-1/(Rm(3,i)*C(3,i));
end


F=[ (-2/3)*Fa, (1/3)*Fb, (1/3)*Fc;
    (1/3)*Fa, (-2/3)*Fb, (1/3)*Fc];
F_Hinv=HH\F;

G=[((-2/3)*R(1))-((1/3)*R(3)) ((1/3)*R(2))-((1/3)*R(3));
    ((1/3)*R(1))-((1/3)*R(3)) ((-2/3)*R(2))-((1/3)*R(3))];

G_Hinv=HH\G;

P1=blkdiag(Pa,Pb);
P=[P1; -1*Pc, -1*Pc];

W=blkdiag(W1,W2,W3);

z0=[ Vs*sqrt(2)*cos(phi_a);      % Input states
    Vs*sqrt(2)*sin(phi_a);
    Vs*sqrt(2)*cos(phi_b);
    Vs*sqrt(2)*sin(phi_b);
    Vs*sqrt(2)*cos(phi_c);
    Vs*sqrt(2)*sin(phi_c)];

A_Max=[G_Hinv, F_Hinv;
    P, W];
[VA_Max,HA_Max]=size(A_Max);
B_M=[2/3, 0, -1/3, 0, -1/3, 0;
    -1/3, 0, 2/3, 0, -1/3, 0];
B_Minv=HH\B_M;

B_Max=[B_Minv;
    zeros(3*N,6)];
[VB_Max,HB_Max]=size(B_Max);

omega2=[0 -1*w; 1*w 0];
D_Max=blkdiag(omega2,omega2,omega2);
[VD_Max,HD_Max]=size(D_Max);
```

```
Axxx_I(:,:,1)=[A_Max B_Max;
    zeros(VD_Max,HA_Max),D_Max];
[VAxxx_I,HAxxx_I]=size(Axxx_I);
Ip=eye(VA_Max,HA_Max);

%% Constructing H matrix
for i=1:3*N+2                          %Iabc + Vcap
    if i<3                             % Iabc
        for ii=1:length(Hh)
        H(length(Hh)*(i-1)+ii,:)=[Ip(i,:)*1/T*exp(-j*Hh(ii)*w*T)];
        end
    else                               % Vcap
        for ii=1:length(Hh2)
        H(2*length(Hh)+length(Hh2)*(i-3)+ii,:)=[Ip(i,:)*1/T*exp(-j*Hh2(ii)
        *w*T)];
        end
    end
end
[barH,kolH]=size(H);
Hz = zeros(barH,HB_Max);
Hnew = [H,Hz];
[VH_Max,HH_Max]=size(Hnew);

%% Constructing M matrix
[barHh,kolHh]=size(Hh);
[barHh2,kolHh2]=size(Hh2);
M1=[Hh,Hh];
M2=zeros(1,3*N*kolHh2);
for i=1:3*N
    M2(1,(i-1)*kolHh2+1:i*kolHh2)=Hh2;
end
M=j*w*diag([M1, M2]);

[VM_Max,HM_Max]=size(M);
Z=zeros(VAxxx_I,length(Hh)*2+length(Hh2)*(3*N));      % zeros matrices

if taus_2(length(taus_2))>T
    i=1;
    while taus_2(i)<=T
        cycle=i;
        i=i+1;
    end
    tau=[taus_2(1:cycle,1); T];
else
    cycle=length(else_SST);
    tau=[else_SST];

    S_2=else_SSseq(:,1:size(else_SSseq,2));
end
PHI_bar=expm(Axxx_I(:,:,1)*(tau(1)));

for iii=2:length(tau)
    Sa_up=S_2(iii,0*N+1:1*N);
    Sb_up=S_2(iii,1*N+1:2*N);
    Sc_up=S_2(iii,2*N+1:3*N);
    Sa_low=S_2(iii,3*N+1:4*N);
    Sb_low=S_2(iii,4*N+1:5*N);
    Sc_low=S_2(iii,5*N+1:6*N);

    %% Constructing matrix elements
```

```
141      Fa=zeros(1,N);
142      Fb=zeros(1,N);
143      Fc=zeros(1,N);
144      Pa=zeros(N,1);
145      Pb=zeros(N,1);
146      Pc=zeros(N,1);
147      for i=1:N
148          Fa(1,i)=(Sa_up(1,i)-Sa_low(1,i));
149          Fb(1,i)=(Sb_up(1,i)-Sb_low(1,i));
150          Fc(1,i)=(Sc_up(1,i)-Sc_low(1,i));
151          Pa(i,1)=(Sa_up(1,i)-Sa_low(1,i))/C(1,i);
152          Pb(i,1)=(Sb_up(1,i)-Sb_low(1,i))/C(2,i);
153          Pc(i,1)=(Sc_up(1,i)-Sc_low(1,i))/C(3,i);
154      end
155      F=[ (-2/3)*Fa, (1/3)*Fb, (1/3)*Fc;
156          (1/3)*Fa, (-2/3)*Fb, (1/3)*Fc];
157      F_Hinv=HH\F;
158
159      P1=blkdiag(Pa,Pb);
160      P=[P1; -1*Pc, -1*Pc];
161
162      A_Max=[G_Hinv, F_Hinv;
163          P, W];
164
165      Axxx_I(:,:,iii)=A_Max;
166      PHI_bar=expm(Axxx_I(:,:,iii)*(tau(iii)-tau(iii-1)))*PHI_bar;
167  end
168
169  %% Obtaining initial conditions vector
170  PHI_I=PHI_bar-eye(size(PHI_bar));
171  E2=PHI_I(1:3*N+2,3*N+2+1:3*N+2+1+5);
172  E1=PHI_I(1:3*N+2,1:3*N+2);
173  Xabc0=-inv(E1)*E2*z0;                % initial condition of ia,ib,vca,vcb,vcc
174  xx0=[Xabc0; z0; zeros(barH,1)];      % initial condition of augmented system
175
176  %% Computing the harmonics of interests using Krylov subspace
177  Axxx_II(:,:,1)=[Axxx_I(:,:,1),Z;Hnew,M];
178  PHI=expv(tau(1),Axxx_II,xx0);
179
180  for iii=2:length(tau)
181
182      Axxx_II(:,:,iii)=[Axxx_I(:,:,iii),Z;Hnew,M];
183      PHI=expv(tau(iii)-tau(iii-1),Axxx_II(:,:,iii),PHI);
184
185  end
186  [VPHI13_Max,HPHI13_Max]=size(PHI);
187
188  Xx_harm=PHI(3*N+8:VPHI13_Max);
```

Note that the function expv in Lines 178 and 183 is the Krylov subspace technique to compute the multiplication of exponential matrix and initial condition vector, provided in Chapter 4.

Example 6.1 In this example, and the following two examples, we will study and compare the FCMs constructed in the frequency domain and the time domain. The main limitation of the frequency domain analysis is that it suffers from the harmonic truncation error if insufficient harmonic terms are included in the calculation. The only case where the frequency domain does not

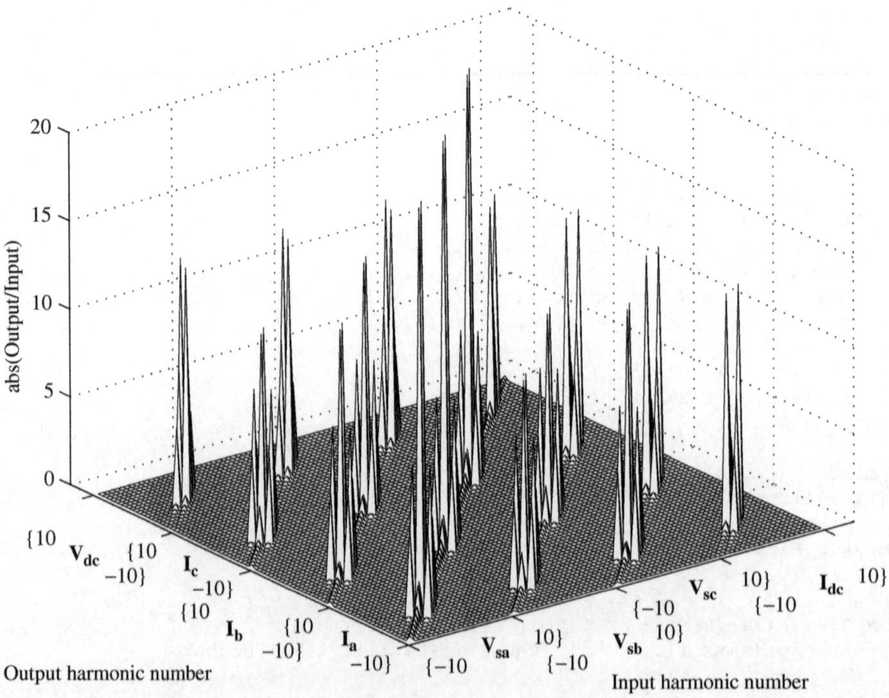

Figure 6.9 FCM of a VSC with $m_f = \infty$ based on frequency domain analysis under balanced system parameter condition.

suffer from this problem is when the ratio between switching and modulating frequency, m_f, is set to ∞. In this case, balanced switching functions can be simply represented as:

$$\begin{aligned} S_a &= \frac{m_a}{2} cos(\omega t + \delta) + \frac{1}{2} \\ S_b &= \frac{m_a}{2} cos(\omega t - 2\pi/3 + \delta) + \frac{1}{2} \\ S_c &= \frac{m_a}{2} cos(\omega t - 4\pi/3 + \delta) + \frac{1}{2}. \end{aligned} \tag{6.57}$$

This case can serve as a benchmark to validate the derived FCM in the time domain when m_f is set sufficiently large in the time domain.

Figs. 6.9 and 6.10 show the absolute values of the FCM based on frequency domain (for $m_f = \infty$) and time domain analysis (for $m_f = 111$) for harmonic numbers from −10 to 10. As can be seen, there is hardly any distinction between these two figures. Table 6.1 lists the normalized parameters being used in the simulation.

A sample set of the input stimuli, as given in Table 6.2, is applied to the two FCMs. This allows the low order harmonics produced by the two methods to be compared. Fig. 6.11 shows the resulting ac current and dc voltage harmonics for frequency domain, fast time domain, and brute force time domain (PSCAD/EMTDC) simulations. Good agreement among these three exists.

Example 6.2 Continued from **Example** 6.1 when m_f is reduced to 3, the errors due to harmonic truncation are apparent. Table 6.3 shows phase a current under the excitations of Table 6.2 when only harmonic numbers from −3 to 3 are included in the time and frequency domain calculations. There exist some discrepancies between frequency and time domains. As more harmonics are included in the frequency domain calculation, frequency domain results approach those of the

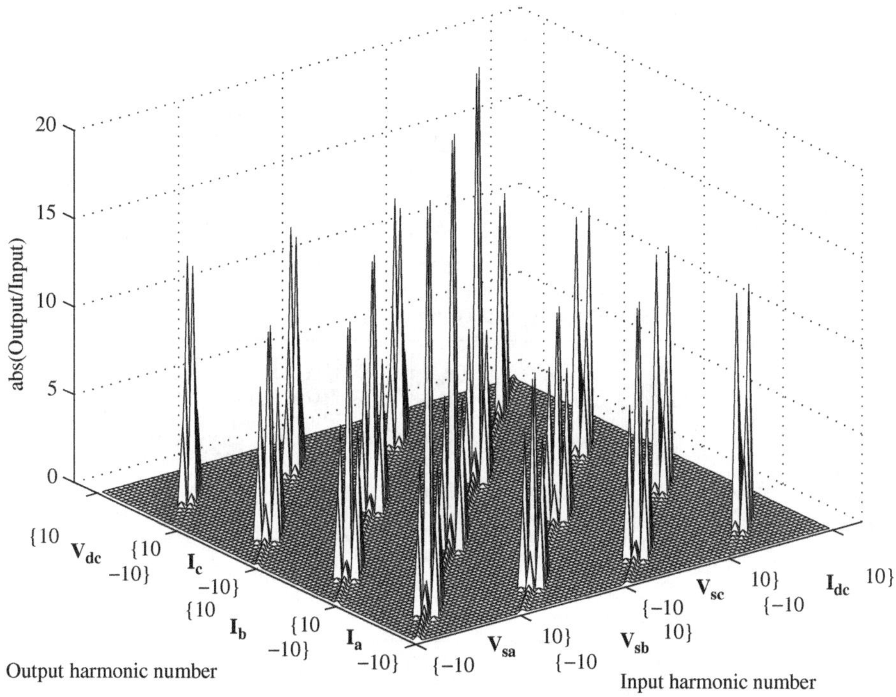

Figure 6.10 FCM of a VSC with $m_f = 111$ based on time domain analysis under balanced system parameter condition.

Table 6.1 Table of parameter list.

Parameter list	$R_a = R_b = R_c = 0.02\Omega$
	$L_a = L_b = L_c = 0.2H$
	$C = 0.5F, \omega = 1rad/sec$
	$m_a = 0.9, \delta = -1.2°$
	$R_{dc} = \infty$

Table 6.2 Table of input stimuli.

AC Input stimuli	$V_{sa}^{-3} = 0.04V, V_{sa}^{3} = 0.04V$
	$V_{sa}^{-2} = 0.04V, V_{sa}^{2} = 0.04V$
	$V_{sa}^{-1} = 0.5V, V_{sa}^{1} = 0.5V$
	$V_{sb}^{-3} = -0.020 + j\frac{\sqrt{3}}{50}V, V_{sb}^{3} = -0.020 - j\frac{\sqrt{3}}{50}V$
	$V_{sb}^{-2} = -0.020 + j\frac{\sqrt{3}}{50}V, V_{sb}^{2} = -0.020 - j\frac{\sqrt{3}}{50}V$
	$V_{sb}^{-1} = -0.25 + j\frac{\sqrt{3}}{4}V, V_{sb}^{1} = -0.25 - j\frac{\sqrt{3}}{4}V$
	$V_{sc}^{-3} = -0.020 - j\frac{\sqrt{3}}{50}V, V_{sc}^{3} = -0.020 + j\frac{\sqrt{3}}{50}V$
	$V_{sc}^{-2} = -0.020 - j\frac{\sqrt{3}}{50}V, V_{sc}^{2} = -0.020 + j\frac{\sqrt{3}}{50}V$
	$V_{sc}^{-1} = -0.25 - j\frac{\sqrt{3}}{4}V, V_{sc}^{1} = -0.25 + j\frac{\sqrt{3}}{4}V$
DC Input stimuli	$I_{dc}^{0} = 0.05A$

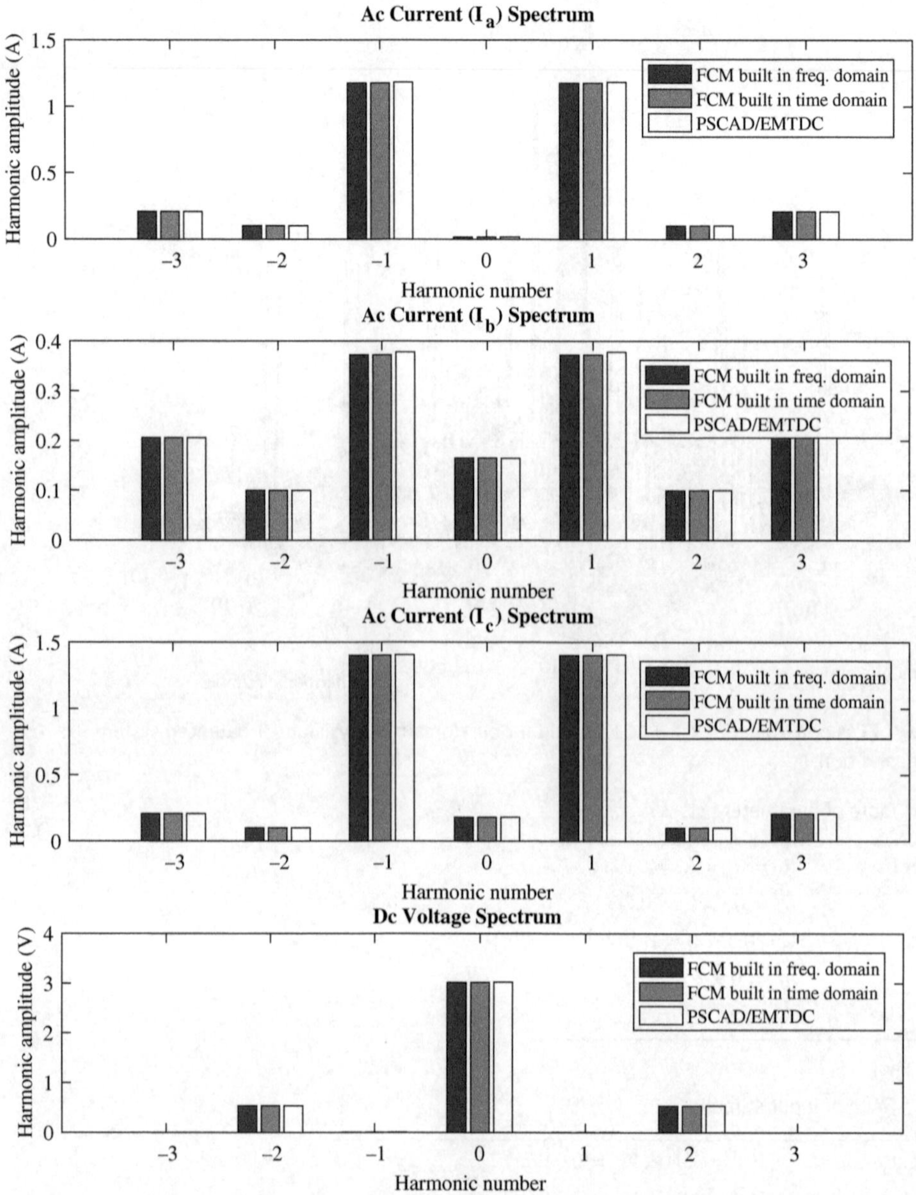

Figure 6.11 Harmonic spectra under balanced system parameters where m_f is very large.

time domain solution. This fact can be observed from Table 6.4 when harmonic numbers from −20 to 20 are included in the harmonic domain calculation. Such a truncation phenomenon does not exist for the time domain solution as may be noted by comparing Tables 6.3 and 6.4.

Example 6.3 Continued from **Example 6.2**, R_b, R_c, and L_b, L_c are kept the same; however, R_a and L_a are increased by 10% to make the system unbalanced. In addition, the dc capacitance, C, is reduced to 0.05 F. The FCMs of the two domains are shown in Figs. 6.12 and 6.13, respectively, when harmonic numbers from −10 to 10 are included in the calculation.

Table 6.3 First four current harmonics for phase *a* when $h = -3$ to 3 under balanced system parameter condition.

	FCM built in frequency domain	FCM built in time domain
I_a^{-3}	$0.0087 + j0.0237A$	$0.0078 + j0.0244A$
I_a^{-2}	$0.0101 + j0.1002A$	$0.0096 + j0.0976A$
I_a^{-1}	$-0.0169 - j0.6009A$	$-0.0152 - j0.6152A$
I_a^{0}	$-0.0145A$	$-0.0142A$
I_a^{1}	$-0.0169 + j0.6009A$	$-0.0152 + j0.6152A$
I_a^{2}	$0.0101 - j0.1002A$	$0.0096 - j0.0976A$
I_a^{3}	$0.0087 - j0.0237A$	$0.0078 - j0.0244A$

Table 6.4 First four current harmonics for phase *a* when $h = -20$ to 20 under balanced system parameter condition.

	FCM built in frequency domain	FCM built in time domain
I_a^{-3}	$0.0079 + j0.0245A$	$0.0078 + j0.0244A$
I_a^{-2}	$0.0099 + j0.0987A$	$0.0096 + j0.0976A$
I_a^{-1}	$-0.0154 - j0.6052A$	$-0.0152 - j0.6152A$
I_a^{0}	$-0.0140A$	$-0.0142A$
I_a^{1}	$-0.0154 + j0.6052A$	$-0.0152 + j0.6152A$
I_a^{2}	$0.0099 - j0.0987A$	$0.0096 - j0.0976A$
I_a^{3}	$0.0079 - j0.0245A$	$0.0078 - j0.0244A$

As can be noted in the figures, the FCMs differ because insufficient harmonic terms are included in the frequency domain. Consequently, when the sample set of input stimuli (see Table 6.2) are applied to the resulting FCMs, the output currents differ significantly (Table 6.5). Tables 6.5 through 6.7 show the resulting phase *a* current for the first four harmonics.

Once again, as the number of included harmonics increases, the frequency domain results approach those of the time domain. However, 50 harmonic terms are needed in this case to obtain the same degree of accuracy as in the balanced case. Nevertheless, it is to be noted that the situations described in **Examples** 6.2 and 6.3 rarely happen. They merely are demonstrate the advantage of computing FCM in time domain. Also, the reader needs to be aware of the importance of selecting the right numerical method in evaluating the matrix exponential to avoid numerical truncation errors or round-off errors.

FCM Computational Time of Various Exponential Matrix Methods

Fig. 6.14 depicts the computational time of various exponential methods for computing FCM of *m*-level MML-VSC. Note that the simulation is carried out in Intel Core i5-6500 with RAM 16 GB. In this case, *m* is varied from 3, 51, 101, 151, 201, 251, and 401. Four methods are evaluated, i.e.

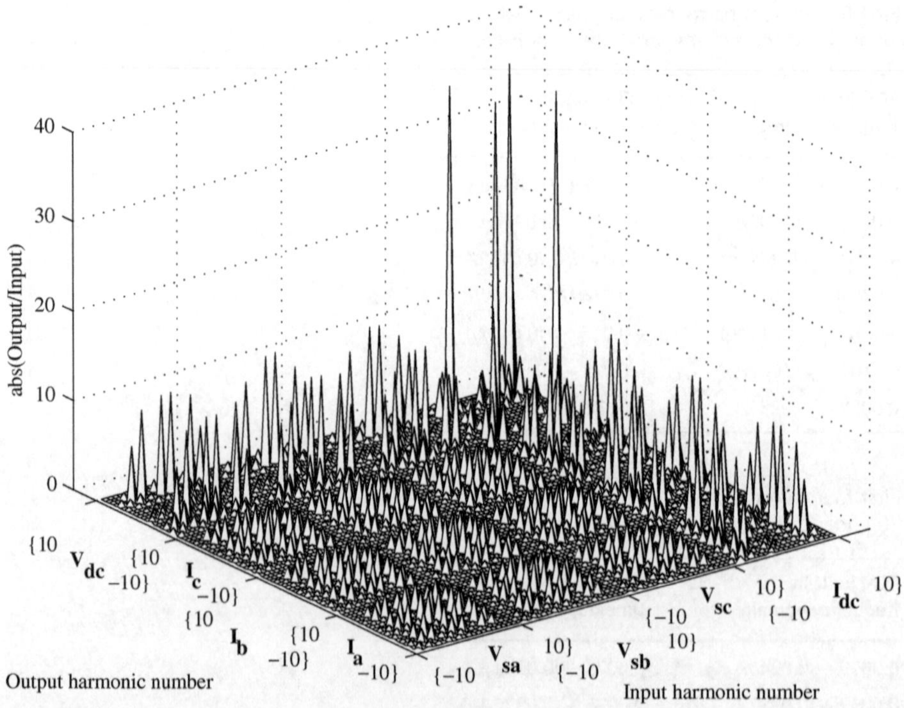

Figure 6.12 Absolute values of **F** of a VSC with $m_f = 3$ based on frequency domain analysis under unbalanced system parameter condition.

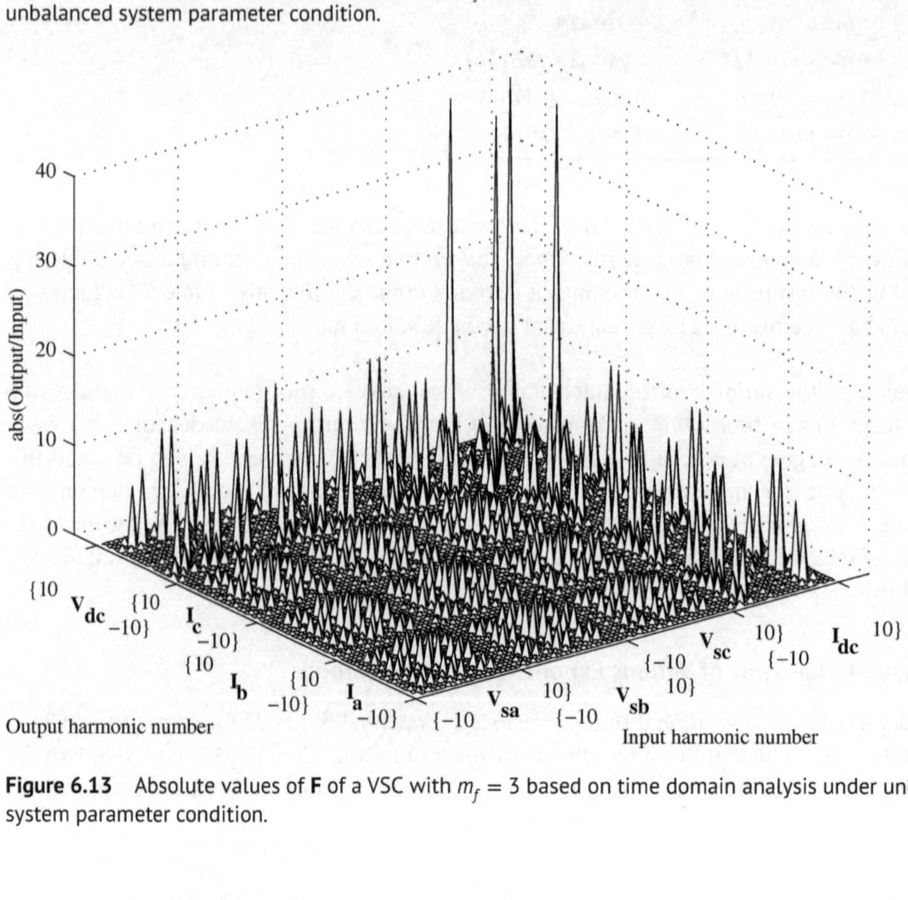

Figure 6.13 Absolute values of **F** of a VSC with $m_f = 3$ based on time domain analysis under unbalanced system parameter condition.

Table 6.5 First four current harmonics for phase a when $h = -10$ to 10 under unbalanced system parameter condition.

	FCM built in frequency domain	**FCM built in time domain**
I_a^{-3}	$0.0153 + j0.0579A$	$0.0168 + j0.0443A$
I_a^{-2}	$0.0102 - j0.0329A$	$0.0081 - j0.0126A$
I_a^{-1}	$-0.0400 - j0.1828A$	$-0.0385 - j0.1548A$
I_a^{0}	$-0.0147A$	$-0.0127A$
I_a^{1}	$-0.0400 + j0.1828A$	$-0.0385 + j0.1548A$
I_a^{2}	$0.0102 + j0.0329A$	$0.0081 + j0.0126A$
I_a^{3}	$0.0153 - j0.0579A$	$0.0168 - j0.0443A$

Table 6.6 First four current harmonics for phase a when $h = -25$ to 25 under unbalanced system parameter condition.

	FCM built in frequency domain	**FCM built in time domain**
I_a^{-3}	$0.0158 + j0.0436A$	$0.0168 + j0.0443A$
I_a^{-2}	$0.0086 - j0.0164A$	$0.0081 - j0.0126A$
I_a^{-1}	$-0.0379 - j0.1456A$	$-0.0385 - j0.1548A$
I_a^{0}	$-0.0134A$	$-0.0127A$
I_a^{1}	$-0.0379 + j0.1456A$	$-0.0385 + j0.1548A$
I_a^{2}	$0.0086 + j0.0164A$	$0.0081 + j0.0126A$
I_a^{3}	$0.0158 - j0.0436A$	$0.0168 - j0.0443A$

Table 6.7 First four current harmonics for phase a when $h = -50$ to 50 under unbalanced system parameter condition.

	FCM built in frequency domain	**FCM built in time domain**
I_a^{-3}	$0.0169 + j0.0445A$	$0.0168 + j0.0443A$
I_a^{-2}	$0.0081 - j0.0130A$	$0.0081 - j0.0126A$
I_a^{-1}	$-0.0385 - j0.1519A$	$-0.0385 - j0.1548A$
I_a^{0}	$-0.0128A$	$-0.0127A$
I_a^{1}	$-0.0385 + j0.1519A$	$-0.0385 + j0.1548A$
I_a^{2}	$0.0081 + j0.0130A$	$0.0081 + j0.0126A$
I_a^{3}	$0.0169 - j0.0445A$	$0.0168 - j0.0443A$

expm (MATLAB toolbox), scaling and squaring with truncated Taylor [73], KSM (Section 4.3), and KSM with restarting technique (Section 4.4). The modulation frequency index (m_f) is set to be the same, which is 27. In general, the larger (m_f) results in the larger number of switching times. This problem becomes worse as the level of MML-VSC is larger, which leads to a significantly large system matrix. Hence, the computational time can be extensive. Fig. 6.14 justifies that the KSM

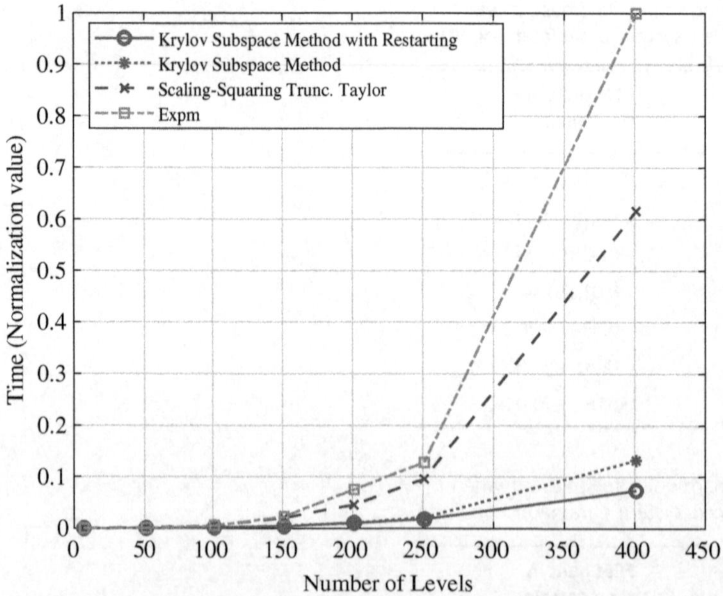

Figure 6.14 The normalized computational time of exponential matrix methods.

with restarting technique shows the superiority compared to other methods. As can be seen, the required computational time for $m = 401$ is almost equal to those of $m = 201$ of `expm` and $m = 251$ of scaling and squaring with truncated Taylor. Moreover, the significant difference appears when $m > 251$ between the KSM and KSM with restarting technique. This concludes that KSM with restarting technique has superior performance for evaluating FCM of large size matrix.

7

General Control Approaches of a VSC

"One man's magic is another man's engineering. Supernatural is a null word."

– Robert A. Heinlein

In this chapter, we will describe some of the control structures commonly used in a VSC. Since there are many books devoted to control structures and designs, this section is only intended as a brief introduction to the readers.

7.1 Reference Frame

For a three-phase system, the voltages and currents can be described in various reference frames. Depending on which reference frame the power converter is chosen to be modeled, different types of the controllers will result. Thus, a brief introduction to different reference frames is given in this section.

7.1.1 Stationary-*abc* Frame

The three phases a, b, c in the natural frame can be regarded as spatially distributed away from each other by $\frac{2\pi}{3}$, as shown in Fig. 7.1. It is drawn to be consistent with the phase sequence *abc*. Space vector at a particular time t or phase θ can be expressed as a vector according to the values spatially distributed on the corresponding coordinates. As shown in Fig. 7.1, the space phasor $\mathbf{f}(t)$ rotates in the counterclockwise direction at a speed ω_0 relative to a-axis in a stationary-*abc* frame. The functions $f_a(t)$, $f_b(t)$, and $f_c(t)$ are acquired through casting the space phasor on the a, b, and c axes, respectively.

A three-phase sinusoidal function in balanced condition is expressed as follows:

$$f_a(t) = f_m cos(\omega_0 t + \theta_0) \tag{7.1a}$$

$$f_b(t) = f_m cos(\omega_0 t + \theta_0 - \frac{2\pi}{3}) \tag{7.1b}$$

$$f_c(t) = f_m cos(\omega_0 t + \theta_0 - \frac{4\pi}{3}), \tag{7.1c}$$

where f_m, ω_0, and θ_0 denote the amplitude, the angular frequency, and the initial phase angle of the function, respectively. (7.1) can be replaced via a space phasor in the stationary-*abc* frame, as

Harmonic Modeling of Voltage Source Converters using Basic Numerical Methods, First Edition.
Ryan Kuo-Lung Lian, Ramadhani Kurniawan Subroto, Victor Andrean, and Bing Hao Lin.

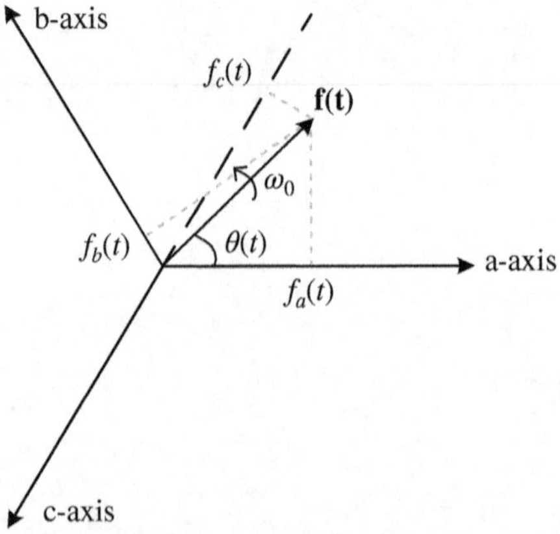

Figure 7.1 A space phasor in the *abc*-frame.

stated in (7.2).

$$\mathbf{f}(t) = \frac{2}{3}\left[e^{j0}f_a(t) + e^{j\frac{2\pi}{3}}f_b(t) + e^{j\frac{4\pi}{3}}f_c(t)\right] \tag{7.2}$$

The phase angle of the space phasor is defined by (7.3)

$$\theta(t) = \theta_0 + \int_0^t \omega_0(\tau)d\tau. \tag{7.3}$$

7.1.2 Stationary-$\alpha\beta$ Frame

The space phasor in the stationary-*abc* frame can be separated into real and imaginary components. This can be easily achieved by projecting the space phasor into the real and imaginary axes, which are referred to as the alpha-beta ($\alpha\beta$) axes, as shown in Fig. 7.2.

This transformation can be algebraically done via the Clarke transformation matrix, which is

$$\begin{bmatrix} f_\alpha \\ f_\beta \\ f_0 \end{bmatrix} = \frac{2}{3}\begin{bmatrix} 1 & -\frac{1}{2} & -\frac{1}{2} \\ 0 & \frac{\sqrt{3}}{2} & -\frac{\sqrt{3}}{2} \\ \frac{1}{2} & \frac{1}{2} & \frac{1}{2} \end{bmatrix}\begin{bmatrix} f_a \\ f_b \\ f_c \end{bmatrix} \tag{7.4}$$

Fig. 7.2 demonstrates that a space phasor in the $\alpha\beta$-frame can be represented by (7.5).

$$\mathbf{f}(t) = f_\alpha(t) + jf_\beta(t), \tag{7.5}$$

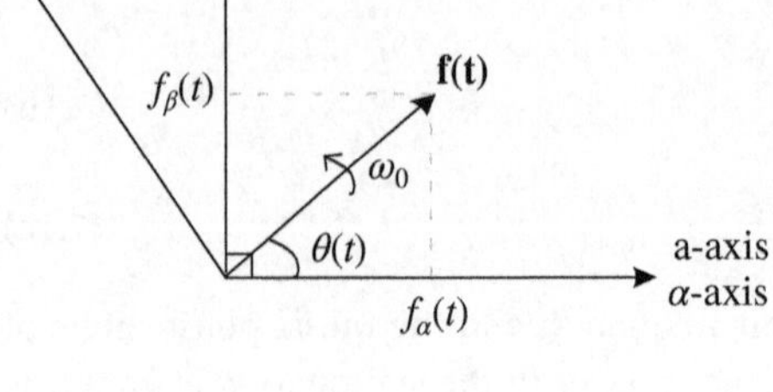

Figure 7.2 A space phasor in the $\alpha\beta$-frame.

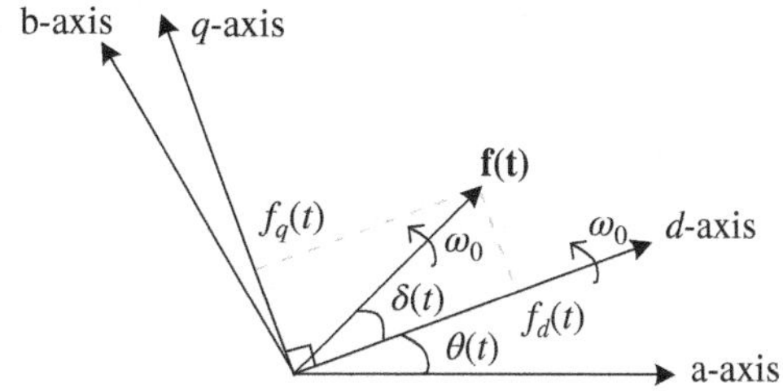

Figure 7.3 A space phasor in the *dq*-frame.

where $f_\alpha(t)$ and $f_\beta(t)$ are acquired through casting the space phasor on the α and β axis, respectively. The space phasor rotates at a speed of ω_0 with respect to the α-axis. $\theta(t)$ can be obtained via (7.6) from the relationship between the real and imaginary components of $f_\alpha(t)$ and $f_\beta(t)$.

$$\theta(t) = \tan^{-1}\left[\frac{f_\beta(t)}{f_\alpha(t)}\right] \tag{7.6}$$

7.1.3 Synchronous-*dq* Frame

The space phasor in the stationary-*abc* frame can be converted to the rotating components in the synchronous reference frame. This frame is also called the *dq*0-frame where the axes of *d* and *q* denote direct and quadrature, respectively.

As shown in Fig. 7.3, the two axes are perpendicular to each other and they rotate at an angular speed of ω_0 with respect to the *a*-axis. Unlike the stationary frame (i.e. *abc*-frame and $\alpha\beta$-frame) which is ac in nature, the *dq*-frame is dc in nature. There is an angle $\theta(t)$, defined in (7.3), in the *dq*-frame with respect to the *abc*-frame. A space phasor in the *dq*-frame can be represented by (7.7).

$$\mathbf{f}(t) = f_d(t) + jf_q(t) \tag{7.7}$$

$\delta(t)$ can be acquired via (7.8) from the relationship with $f_d(t)$ and $f_q(t)$.

$$\delta(t) = \tan^{-1}\left[\frac{f_q(t)}{f_d(t)}\right] \tag{7.8}$$

The corresponding space vector in the *dq* frame is

$$\mathbf{f} = f_d + jf_q = \left(f_\alpha + jf_\beta\right)e^{-j\omega_0 t} = \underbrace{\begin{bmatrix}\cos\left(\omega_0 t\right) & \sin\left(\omega_0 t\right)\\ -\sin\left(\omega_0 t\right) & \cos\left(\omega_0 t\right)\end{bmatrix}}_{\mathbf{P}_M}\begin{bmatrix}f_\alpha\\ f_\beta\end{bmatrix}. \tag{7.9}$$

Thus, the Park transformation matrix, converting a three-phase system from the *abc*- to the *dq*-frame is stated in (7.10), and the inverse Park transform matrix is expressed in (7.11).

$$\begin{bmatrix}f_d(t)\\ f_q(t)\\ f_0(t)\end{bmatrix} = \frac{2}{3}\begin{bmatrix}\cos\left(\omega_0 t\right) & \cos\left(\omega_0 t - \frac{2\pi}{3}\right) & \cos\left(\omega_0 t - \frac{4\pi}{3}\right)\\ -\sin\left(\omega_0 t\right) & -\sin\left(\omega_0 t - \frac{2\pi}{3}\right) & -\sin\left(\omega_0 t - \frac{4\pi}{3}\right)\\ \frac{1}{2} & \frac{1}{2} & \frac{1}{2}\end{bmatrix}\begin{bmatrix}f_a(t)\\ f_b(t)\\ f_c(t)\end{bmatrix}, \tag{7.10}$$

where $f_0(t)$ represents the zero component of the two-axis system in this frame.

$$\begin{bmatrix} f_a(t) \\ f_b(t) \\ f_c(t) \end{bmatrix} = \begin{bmatrix} \cos(\omega_0 t) & -\sin(\omega_0 t) & 1 \\ \cos\left(\omega_0 t - \frac{2\pi}{3}\right) & -\sin\left(\omega_0 t - \frac{2\pi}{3}\right) & 1 \\ \cos\left(\omega_0 t - \frac{4\pi}{3}\right) & -\sin\left(\omega_0 t - \frac{4\pi}{3}\right) & 1 \end{bmatrix} \begin{bmatrix} f_d(t) \\ f_q(t) \\ f_0(t) \end{bmatrix} \tag{7.11}$$

7.1.4 Phase-Locked Loop

The grid synchronization is very important for power converters. The fundamental magnitude, frequency, and phase angle of the grid voltage are important information for grid-connected or grid-following converters to generate the reference for converter output current as well as real and reactive powers. Some of the grid synchronization techniques include the zero-crossings detection, the weighted least square estimation, the adaptive notch filtering, and the phase-locked loop (PLL) techniques. Among them, the PLL-based synchronization techniques are probably the most commonly adopted for converter control due to its easy implementation and good accuracy [74]. The synchronous reference frame PLL (SRF-PLL) [75] is the most frequently used PLL for a balanced grid. The PLL uses the *dq*-frame grid voltage to synchronize the inverter with the grid phase by a PI controller, as shown in Fig. 7.4.

The fundamental component of the grid voltage is forced to be aligned with the *d*-axis, meaning that the voltage in *q*-axis, v_{sq} is set to zero. The difference between v_{sqref} and v_{sq} is fed into the proportional integrator (PI) controller to obtain the difference of the angular frequency ω_c. Then, ω_c is added to the grid angular frequency ($\omega_0 = 2\pi f$) and the output value is integrated to obtain an output angle $\hat{\theta}$. The relationship between the angular frequency and phase angle can be formulated as

$$\hat{\theta} = \int \hat{\omega} dt, \tag{7.12}$$

where the output angle $\hat{\theta}$ must be in the range of $[0, 2\pi]$. Afterwards, the output angle $\hat{\theta}$ is sent back to the Park transform matrix to satisfy the closed-loop control. When the difference between v_{sqref} and v_{sq} reaches zero, the PLL output angle $\hat{\theta}$ is the estimated grid phase angle θ.

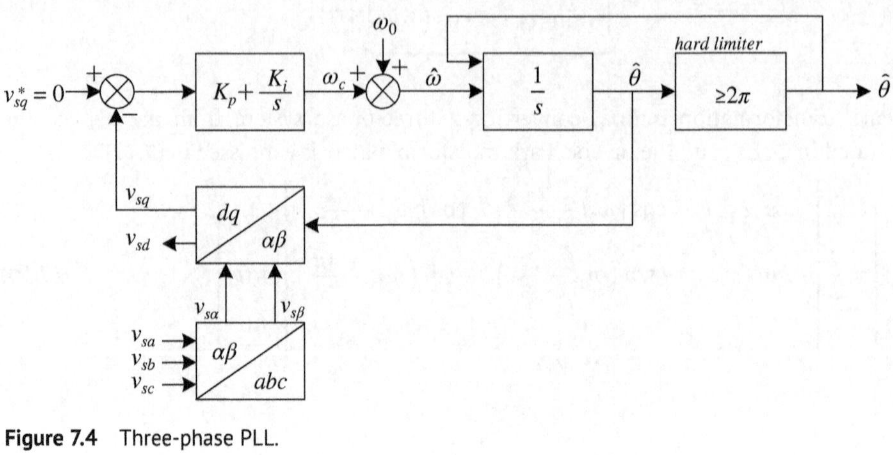

Figure 7.4 Three-phase PLL.

Note that the SRF-PLL cannot track the phase angle if the grid voltage is distorted or unbalanced. Therefore, sophisticated PLLs such as the enhanced PLL proposed in [76] are used to track the phase angle of the fundamental component of the grid voltage despite the presence of harmonics.

7.2 Control Strategies

The active and reactive power controllers for a VSC can be categorized into vector-current controller (VCC) and direct power controller (DPC). The main difference of these strategies rely on the existence of PLL and Park transformation. DPC can eliminate the use of PLL and Park transformation by replacing them with power calculator. Moreover, the models derived for control design for both strategies are also different.

7.2.1 Vector-Current Controller

For a grid-connected VSC, it should resemble a current source. Therefore, the output currents of a VSC need to be well regulated. Depending on the reference frame chosen, a different forms of controllers can be designed. For example, in the *dq* reference frame, PI controllers are used. In the $\alpha\beta$ or *abc* reference frames, proportional resonant (PR) controllers are used. In this section, we only discuss the *dq* current controllers, which are the most widely used ones.

To design a PI controller, we need to derive the small-signal model of the VSC. A simplified equivalent circuit of a VSC is shown in Fig. 7.5. Note that v_s and v_t represent the source and converter voltages. Moreover, the equivalent resistor, R_t, is the summation of R_s and R_f, and the equivalent inductor, L_t, is the summation of L_s and L_f.

The differential equation describing Fig. 7.5 is stated as

$$\mathbf{L}_t \frac{d}{dt}\mathbf{i}_t = -\mathbf{R}_t\mathbf{i}_t + \mathbf{v}_s - \mathbf{v}_t, \tag{7.13}$$

$$L_t \frac{d}{dt}\begin{bmatrix} i_{ta} \\ i_{tb} \\ i_{tc} \end{bmatrix} = -R_t \begin{bmatrix} i_{ta} \\ i_{tb} \\ i_{tc} \end{bmatrix} - \begin{bmatrix} v_{ta} \\ v_{tb} \\ v_{tc} \end{bmatrix} + \begin{bmatrix} v_{sa} \\ v_{sb} \\ v_{sc} \end{bmatrix} \tag{7.14}$$

The Clarke transformation is used to convert (7.14) to its $\alpha\beta$-frame form as stated in (7.15) with the zero component ignored because it is a three-phase, three-wire system.

$$L_t \frac{d}{dt}\begin{bmatrix} i_{t\alpha} \\ i_{t\beta} \end{bmatrix} = -R_t \begin{bmatrix} i_{t\alpha} \\ i_{t\beta} \end{bmatrix} + \begin{bmatrix} v_{s\alpha} \\ v_{s\beta} \end{bmatrix} - \begin{bmatrix} v_{t\alpha} \\ v_{t\beta} \end{bmatrix} \tag{7.15}$$

(7.15) can in turn be written as

$$L_t \frac{d}{dt}\left(\mathbf{P}_M^{-1}\begin{bmatrix} i_{td} \\ i_{tq} \end{bmatrix}\right) = -R_t\mathbf{P}_M^{-1}\begin{bmatrix} i_{td} \\ i_{tq} \end{bmatrix} - \mathbf{P}_M^{-1}\begin{bmatrix} v_{td} \\ v_{tq} \end{bmatrix} + \mathbf{P}_M^{-1}\begin{bmatrix} v_{sd} \\ v_{sq} \end{bmatrix}. \tag{7.16}$$

Furthermore, multiplying both sides of (7.16) with $\mathbf{P}_M$

$$L_t\mathbf{P}_M \frac{d}{dt}\mathbf{P}_M^{-1}\begin{bmatrix} i_{td} \\ i_{tq} \end{bmatrix} + L_t \frac{d}{dt}\begin{bmatrix} i_{td} \\ i_{tq} \end{bmatrix} = -R_t\begin{bmatrix} i_{td} \\ i_{tq} \end{bmatrix} - \begin{bmatrix} v_{td} \\ v_{tq} \end{bmatrix} + \begin{bmatrix} v_{sd} \\ v_{sq} \end{bmatrix}. \tag{7.17}$$

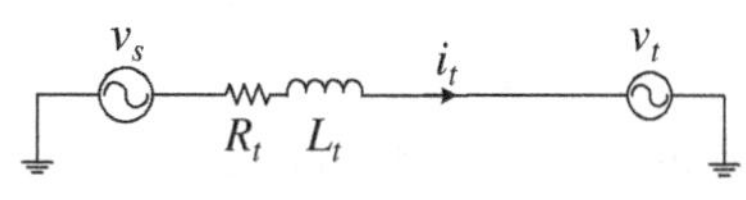

Figure 7.5 Equivalent circuit of the VSC.

Note that the term $\mathbf{P}_M \frac{d}{dt}\mathbf{P}_M^{-1}$ can be rewritten in (7.18).

$$\mathbf{P}_M \frac{d}{dt}\mathbf{P}_M^{-1} = \begin{bmatrix} \cos(\omega_0 t) & \sin(\omega_0 t) \\ -\sin(\omega_0 t) & \cos(\omega_0 t) \end{bmatrix} \omega_0 \begin{bmatrix} \cos(\omega_0 t) & -\sin(\omega_0 t) \\ \sin(\omega_0 t) & \cos(\omega_0 t) \end{bmatrix} = \begin{bmatrix} 0 & -\omega_0 \\ \omega_0 & 0 \end{bmatrix} \tag{7.18}$$

Thus, incorporating (7.18) into (7.17) produces (7.19).

$$L_t \frac{d}{dt}\begin{bmatrix} i_{td} \\ i_{tq} \end{bmatrix} = \begin{bmatrix} -R_t & \omega_0 L_t \\ -\omega_0 L_t & -R_t \end{bmatrix}\begin{bmatrix} i_{td} \\ i_{tq} \end{bmatrix} - \begin{bmatrix} v_{td} \\ v_{tq} \end{bmatrix} + \begin{bmatrix} v_{sd} \\ v_{sq} \end{bmatrix} \tag{7.19}$$

v_{td1} and v_{tq1} are defined in (7.20) to decouple the coupling term in (7.19).

$$v_{td1} = v_{sd} + \omega_0 L_t i_{tq} - v_{td} \tag{7.20a}$$

$$v_{tq1} = v_{sq} - \omega_0 L_t i_{td} - v_{tq} \tag{7.20b}$$

Therefore, (7.19) can be rewritten as (7.21).

$$\begin{bmatrix} v_{td1} \\ v_{tq1} \end{bmatrix} = L_t \frac{d}{dt}\begin{bmatrix} i_{td} \\ i_{tq} \end{bmatrix} + \begin{bmatrix} R_t & 0 \\ 0 & R_t \end{bmatrix}\begin{bmatrix} i_{td} \\ i_{tq} \end{bmatrix} \tag{7.21}$$

A PI controller can be designed based on (7.22), which is derived via Laplace transform to (7.21).

$$i_{td} = \frac{\frac{1}{L_t}}{s + \frac{R_t}{L_t}} v_{td1} \tag{7.22a}$$

$$i_{tq} = \frac{\frac{1}{L_t}}{s + \frac{R_t}{L_t}} v_{tq1} \tag{7.22b}$$

Let the PI controller structures be defined as

$$\begin{aligned} v_{td1} &= K_p\left(i_{tdref} - i_{td}\right) + K_i \int \left(i_{tdref} - i_{td}\right) dt \\ v_{tq1} &= K_p\left(i_{tqref} - i_{tq}\right) + K_i \int \left(i_{tqref} - i_{tq}\right) dt, \end{aligned} \tag{7.23}$$

where K_p and K_i represent the proportional and integral constants.

Consequently, the closed-loop transfer function can be achieved by substituting (7.23) into (7.21) and taking Laplace transform as

$$\begin{aligned} \frac{i_{td}(s)}{i_{tdref}(s)} &= \frac{\frac{K_p}{L_t}s + \frac{K_i}{L_t}}{s^2 + \left(\frac{R_t}{L_t} + \frac{K_p}{L_t}\right)s + \frac{K_i}{L_t}} \\ \frac{i_{tq}(s)}{i_{tqref}(s)} &= \frac{\frac{K_p}{L_t}s + \frac{K_i}{L_t}}{s^2 + \left(\frac{R_t}{L_t} + \frac{K_p}{L_t}\right)s + \frac{K_i}{L_t}}. \end{aligned} \tag{7.24}$$

As one may see, the closed-loop transfer functions with PI controllers lead to second-order system. The block diagram representation of (7.20) is depicted in Fig. 7.6.

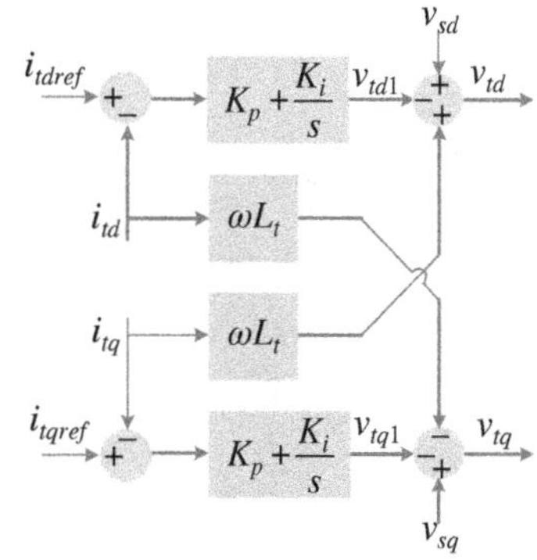

Figure 7.6 Block diagram of VCC with PI controller.

The question may arise how to determine the proper i_{tdref} and i_{tqref}. To answer this problem, the derived i_{tdref} and i_{tqref} are the products of active power reference $\left(P_{ref}\right)$ and reactive power reference $\left(Q_{ref}\right)$, simply obtained from the active and reactive powers in dq frame as (7.25).

$$P_{ref} = \frac{3}{2}\left(v_{sd}i_{tdref} + v_{sq}i_{tqref}\right) \tag{7.25a}$$

$$Q_{ref} = \frac{3}{2}\left(v_{sq}i_{tdref} - v_{sd}i_{tqref}\right) \tag{7.25b}$$

Since v_{sq} is zero for a balanced condition, (7.25) can be rewritten as (7.26).

$$i_{tdref} = \frac{2P_{ref}}{3v_{sd}} \tag{7.23}$$

$$i_{tqref} = -\frac{2Q_{ref}}{3v_{sd}} \tag{7.26a}$$

Fig. 7.7 shows a general control diagram for a VSC, incorporating the aforementioned controls.

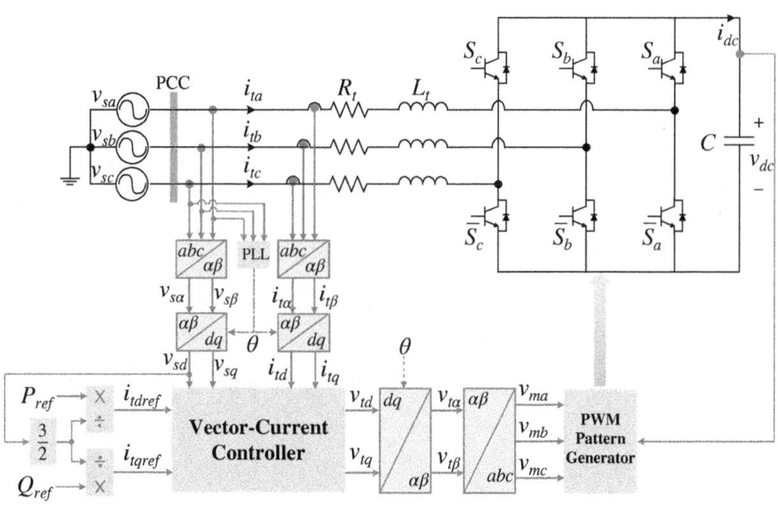

Figure 7.7 Control configuration of VCC for a VSC.

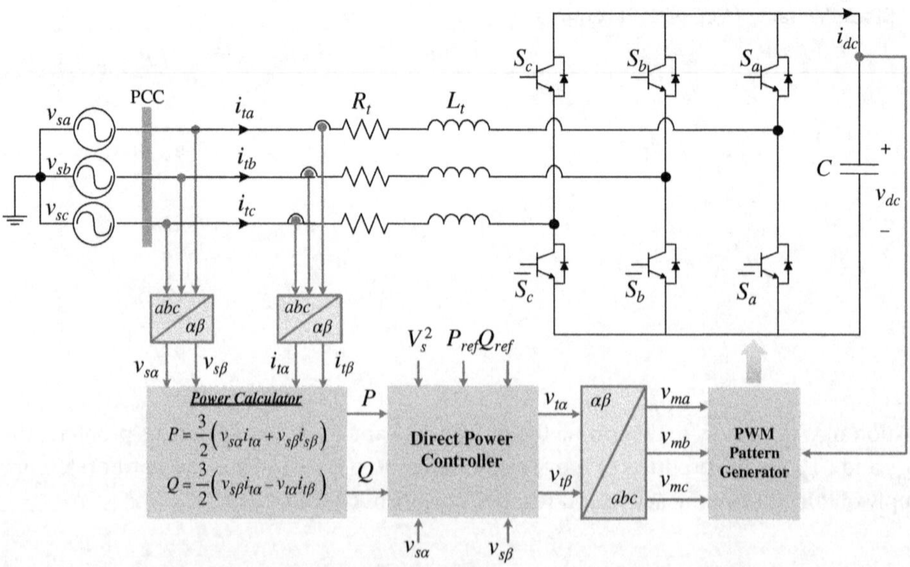

Figure 7.8 Control configuration of DPC for a VSC.

7.2.2 Direct Power Controller

The main concept of a direct power controller (DPC) is motivated by [77], [78], [79]. Compared to VCC, this control strategy offers some benefits [78]:

1. It does not require the coordinate transformation and angular information of grid voltage or virtual flux.
2. It is able to improve transient performance.

Fig. 7.8 illustrates the block diagram of DPC strategy for a three-phase two-level VSC. As can be seen, the active and reactive powers are directly regulated via the feedback of active and reactive powers. Hence, it is important to define the instantaneous power in the stationary $\alpha\beta$-frame as

$$P = \frac{3}{2}\left(v_{s\alpha} i_{t\alpha} + v_{s\beta} i_{t\beta}\right)$$
$$Q = \frac{3}{2}\left(v_{s\beta} i_{t\alpha} - v_{s\alpha} i_{t\beta}\right). \tag{7.27}$$

Defining the grid voltage in the $\alpha\beta$-frame as

$$v_{s\alpha} = V_s \cos(\omega t)$$
$$v_{s\beta} = V_s \sin(\omega t). \tag{7.28}$$

Taking the derivative of (7.28) with respect to time yields

$$\dot{v}_{s\alpha} = -\omega V_s \sin(\omega t) = -\omega v_{s\beta}$$
$$\dot{v}_{s\beta} = \omega V_s \cos(\omega t) = \omega v_{s\alpha}. \tag{7.29}$$

The dynamics of active and reactive power can also be obtained by taking the time derivative of (7.27) as

$$\dot{P} = \frac{3}{2}\left(v_{s\alpha} \dot{i}_{t\alpha} + \dot{v}_{s\alpha} i_{t\alpha} + v_{s\beta} \dot{i}_{t\beta} + \dot{v}_{s\beta} i_{t\beta}\right)$$
$$\dot{Q} = \frac{3}{2}\left(v_{s\beta} \dot{i}_{t\alpha} + \dot{v}_{s\beta} i_{t\alpha} - v_{s\alpha} \dot{i}_{t\beta} - \dot{v}_{s\alpha} i_{t\beta}\right). \tag{7.30}$$

Incorporating (7.15) and (7.29) into (7.30) gives

$$\dot{P} = -\frac{R_t}{L_t}P - \omega Q + \frac{3}{2L_t}\left(V_s^2 - v_{s\alpha}v_{t\alpha} - v_{s\beta}v_{t\beta}\right) \tag{7.31}$$

$$\dot{Q} = \omega P - \frac{R_t}{L_t}Q + \frac{3}{2L_t}\left(-v_{s\beta}v_{t\alpha} + v_{s\alpha}v_{t\beta}\right). \tag{7.32}$$

One may observe that (7.31) and (7.32) consist of the multiplication of control signal and grid voltage. To simplify the analysis, let us represent v_P and v_Q as

$$\begin{aligned} v_P &= -v_{s\alpha}v_{t\alpha} - v_{s\beta}v_{t\beta} \\ v_Q &= -v_{s\beta}v_{t\alpha} + v_{s\alpha}v_{t\beta}. \end{aligned} \tag{7.33}$$

Hence, (7.33) can be rewritten into matrix form as

$$\begin{bmatrix} v_P \\ v_Q \end{bmatrix} = \begin{bmatrix} -v_{s\alpha} & -v_{s\beta} \\ -v_{s\beta} & v_{s\alpha} \end{bmatrix}\begin{bmatrix} v_{t\alpha} \\ v_{t\beta} \end{bmatrix} = \begin{bmatrix} -V_s\cos(\omega t) & -V_s\sin(\omega t) \\ -V_s\sin(\omega t) & V_s\cos(\omega t) \end{bmatrix}\begin{bmatrix} v_{t\alpha} \\ v_{t\beta} \end{bmatrix} \tag{7.34}$$

and the inverse of (7.34) is obtained as

$$\begin{bmatrix} v_{t\alpha} \\ v_{t\beta} \end{bmatrix} = -\frac{1}{V_s^2}\begin{bmatrix} v_{s\alpha} & v_{s\beta} \\ v_{s\beta} & -v_{s\alpha} \end{bmatrix}\begin{bmatrix} v_P \\ v_Q \end{bmatrix} \tag{7.35}$$

Consequently, the dynamics of VSC in terms of active and reactive power can be modeled as

$$\begin{bmatrix} \dot{P} \\ \dot{Q} \end{bmatrix} = \begin{bmatrix} -\frac{R_t}{L_t} & -\omega \\ \omega & -\frac{R_t}{L_t} \end{bmatrix}\begin{bmatrix} P \\ Q \end{bmatrix} + \begin{bmatrix} \frac{3}{2L_t} & 0 \\ 0 & \frac{3}{2L_t} \end{bmatrix}\begin{bmatrix} v_P \\ v_Q \end{bmatrix} + \begin{bmatrix} \frac{3}{2L_t} \\ 0 \end{bmatrix}V_s^2. \tag{7.36}$$

To independently control the active and reactive powers, the decoupling terms are introduced as

$$\begin{aligned} v_P &= \frac{2L_t}{3}\omega Q - V_s^2 + \frac{2L_t}{3}v_{P1} \\ v_Q &= -\frac{2L_t}{3}\omega P + \frac{2L_t}{3}v_{Q1}, \end{aligned} \tag{7.37}$$

where v_{P1} and v_{Q1} denote the control signal from controllers for P and Q, respectively.

Substituting (7.37) into (7.36) results in

$$\begin{bmatrix} \dot{P} \\ \dot{Q} \end{bmatrix} = \begin{bmatrix} -\frac{R_t}{L_t} & 0 \\ 0 & -\frac{R_t}{L_t} \end{bmatrix}\begin{bmatrix} P \\ Q \end{bmatrix} + \begin{bmatrix} v_{P1} \\ v_{Q1} \end{bmatrix}. \tag{7.38}$$

Taking the Laplace transform of (7.38) yields

$$\begin{aligned} P(s) &= \frac{1}{s + \frac{R_t}{L_t}}v_{P1}(s) \\ Q(s) &= \frac{1}{s + \frac{R_t}{L_t}}v_{Q1}(s). \end{aligned} \tag{7.39}$$

If PI controllers are employed to regulate the active and reactive powers, v_{P1} and v_{Q1} are represented as

$$\begin{aligned} v_{P1} &= K_p\left(P_{ref} - P\right) + K_i\int\left(P_{ref} - P\right)dt \\ v_{Q1} &= K_p\left(Q_{ref} - Q\right) + K_i\int\left(Q_{ref} - Q\right)dt, \end{aligned} \tag{7.40}$$

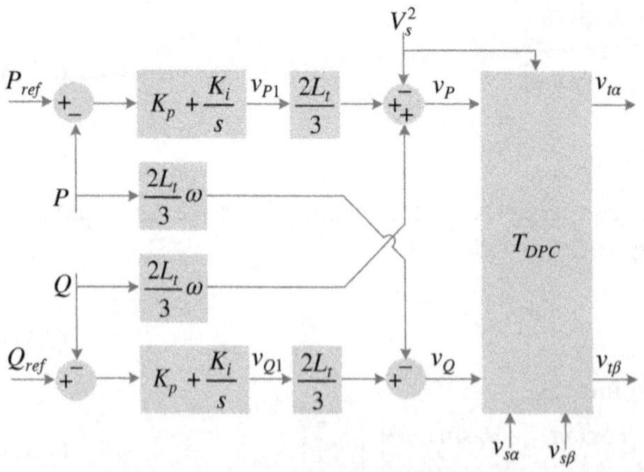

Figure 7.9 Block diagram of DPC with PI controller.

where P_{ref} and Q_{ref} define the active and reactive powers, respectively. The proportional and integral constants are denoted as K_p and K_i, respectively.

Finally, the closed-loop transfer function can be achieved by substituting (7.40) to (7.38)

$$\frac{P(s)}{P_{ref}(s)} = \frac{K_p s + K_i}{s^2 + \left(\frac{R_t}{L_t} + K_p\right) s + K_i}$$

$$\frac{Q(s)}{Q_{ref}(s)} = \frac{K_p s + K_i}{s^2 + \left(\frac{R_t}{L_t} + K_p\right) s + K_i} \tag{7.41}$$

From (7.41), one may notice that the closed-loop system with PI controller forms a second-order system. The block diagram of DPC with PI controller is depicted in Fig. 7.9. Note that $\mathbf{T}_{DPC}$ represents the transformation matrix to generate the modulating signal, as stated in (7.35).

7.2.3 DC-bus Voltage Controller

DC-bus voltage regulation is very prominent for the majority of back-to-back converter topology, as a constant voltage is the fundamental requirement for the power converter. Fig. 7.10 shows the typical control strategy to stabilize the dc-bus voltage. As can be seen, the control strategy consists of two loops. The inner loop is responsible for current control, while the outer loop is aimed at stabilizing the dc-bus voltage. To successfully realize this configuration, the inner loop bandwidth has to be greater than the outer loop one. The closed-loop bandwidth of the outer loop controller is typically selected as one-tenth of grid frequency [80]. Consequently, the inner loop controller acts faster than the outer loop controller.

As demonstrated in Fig. 7.10, the d-axis current reference, i_{tdref}, can also be produced by the voltage controller. The reference of dc-bus voltage, v_{dcref}, is compared with v_{dc}, and their difference will be fed to the PI regulator to yield i_{tdref}. In general, it can be expressed as

$$i_{tdref} = K_{pv}\left(v_{dcref} - v_{dc}\right) + K_{iv}\int \left(v_{dcref} - v_{dc}\right) dt, \tag{7.42}$$

where K_{pv} and K_{iv} are the proportional and integral constants for v_{dc} controller.

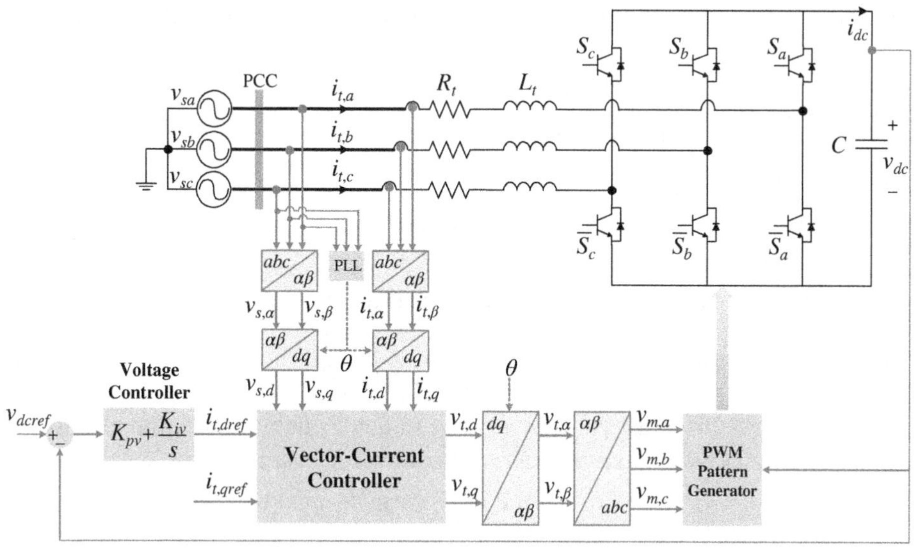

Figure 7.10 Block diagram of dc voltage control with PI controller.

7.2.4 Circulating Current Controller

The above-mentioned PLL, current, dc voltage, and power controllers are the basic control structures applicable to both two-level and multilevel VSCs. Nevertheless, there are controllers specifically required for a multilevel converter. For example, a voltage balancer between the converter submodules is usually needed for multilevel converters. If no corrective control is employed, voltages of the capacitors of the submodules will drift from their nominal values. This drift occurs due to the asymmetries of the PWM gating pulses and tolerances of the converter components [30]. Voltage balancing can be achieved via control implementation or PWM strategies. Readers who are interested in this topic can refer to [30], [32], [81], and [82].

For an MML-VSC, another technical issue is the circulating currents flowing among the three legs of the converter. Generation of the circulating current is due to the voltage difference between the upper and lower arms of the MML-VSC. It will raise power losses and the capacitor voltage ripple in the submodules. In order to eliminate the circulating current, the control strategy on the synchronous reference frame is widely used. Rewrite (5.173) to obtain the formulas of the circulating currents in the balanced condition as follows:

$$L_{arm}\frac{d}{dt}i_{cirj} = -R_{arm}i_{cirj} - \frac{1}{2}(S_j^u v_{cj}^u + S_j^l v_{cj}^l) + \frac{1}{2}v_{dc}, \tag{7.43}$$

where $L_{arm} = \frac{1}{3}\sum_{j=a}^{c}\left(\frac{L_j^u + L_j^l}{2}\right)$ and $R_{arm} = \frac{1}{3}\sum_{j=a}^{c}\left(\frac{R_j^u + R_j^l}{2}\right)$. To simplify the analysis, the voltage terms can be replaced by $v_{cir,j} = \frac{1}{2}(v_{dc} - (S_j^u v_{cj}^u + S_j^l v_{cj}^l))$, and (7.44) is resulted.

$$L_{arm}\frac{d}{dt}\begin{bmatrix} i_{cira} \\ i_{cirb} \\ i_{circ} \end{bmatrix} = -R_{arm}\begin{bmatrix} i_{cira} \\ i_{cirb} \\ i_{circ} \end{bmatrix} + \begin{bmatrix} v_{cira} \\ v_{cirb} \\ v_{circ} \end{bmatrix} \tag{7.44}$$

The Clarke transformation is used to convert (7.44) to its $\alpha\beta$-frame form as stated in (7.45) with the zero component ignored.

$$L_{arm}\frac{d}{dt}\begin{bmatrix} i_{cir\alpha} \\ i_{cir\beta} \end{bmatrix} = -R_{arm}\begin{bmatrix} i_{cir\alpha} \\ i_{cir\beta} \end{bmatrix} + \begin{bmatrix} v_{cir\alpha} \\ v_{cir\beta} \end{bmatrix} \tag{7.45}$$

(7.45) can be converted to dq-frame as

$$L_{arm}\frac{d}{dt}\left(\mathbf{P}_M^{-1}\begin{bmatrix} i_{cird} \\ i_{cirq} \end{bmatrix}\right) = -R_{arm}\mathbf{P}_M^{-1}\begin{bmatrix} i_{cird} \\ i_{cirq} \end{bmatrix} + \mathbf{P}_M^{-1}\begin{bmatrix} v_{cird} \\ v_{cirq} \end{bmatrix} \tag{7.46}$$

Multiplying both sides of (7.46) with $\mathbf{P}_M$ to get the dq-frame equation

$$L_{arm}\mathbf{P}_M\frac{d}{dt}\mathbf{P}_M^{-1}\begin{bmatrix} i_{cird} \\ i_{cirq} \end{bmatrix} + L_{arm}\frac{d}{dt}\begin{bmatrix} i_{cird} \\ i_{cirq} \end{bmatrix} = -R_{arm}\begin{bmatrix} i_{cird} \\ i_{cirq} \end{bmatrix} + \begin{bmatrix} v_{cird} \\ v_{cirq} \end{bmatrix} \tag{7.47}$$

For circulating current case, the transformation from the stationary frame into dq-frame must use $-2\omega_0$ term, as the circulating current is mainly formed by the second harmonic frequency negative sequence component [83], [84]. Hence, the term $\mathbf{P}_M\frac{d}{dt}\mathbf{P}_M^{-1}$ can be rewritten in (7.48).

$$\begin{aligned}\mathbf{P}_M\frac{d}{dt}\mathbf{P}_M^{-1} &= \begin{bmatrix} \cos(2\omega_0 t) & -\sin(2\omega_0 t) \\ \sin(2\omega_0 t) & \cos(2\omega_0 t) \end{bmatrix} 2\omega_0 \begin{bmatrix} -\sin(2\omega_0 t) & \cos(2\omega_0 t) \\ -\cos(2\omega_0 t) & -\sin(2\omega_0 t) \end{bmatrix} \\ &= \begin{bmatrix} 0 & 2\omega_0 \\ -2\omega_0 & 0 \end{bmatrix}\end{aligned} \tag{7.48}$$

Therefore, incorporating (7.48) into (7.47) produces (7.49).

$$L_{arm}\frac{d}{dt}\begin{bmatrix} i_{cird} \\ i_{cirq} \end{bmatrix} = \begin{bmatrix} -R_{arm} & -2\omega_0 L_{arm} \\ 2\omega_0 L_{arm} & -R_{arm} \end{bmatrix}\begin{bmatrix} i_{cird} \\ i_{cirq} \end{bmatrix} + \begin{bmatrix} v_{cird} \\ v_{cirq} \end{bmatrix} \tag{7.49}$$

v_{cird1} and v_{cirq1} are defined in (7.50) to decouple the coupling term in (7.49).

$$v_{cird1} = v_{cird} - 2\omega_0 L_{arm} i_{cirq} \tag{7.50a}$$

$$v_{cirq1} = v_{cirq} + 2\omega_0 L_{arm} i_{cird} \tag{7.50b}$$

Therefore, (7.49) can be rewritten as (7.51).

$$\begin{bmatrix} v_{cird1} \\ v_{cirq1} \end{bmatrix} = L_{arm}\frac{d}{dt}\begin{bmatrix} i_{cird} \\ i_{cirq} \end{bmatrix} + \begin{bmatrix} R_{arm} & 0 \\ 0 & R_{arm} \end{bmatrix}\begin{bmatrix} i_{cird} \\ i_{cirq} \end{bmatrix} \tag{7.51}$$

A PI controller can be designed based on (7.52), which is derived via the Laplace transform of (7.51). The block diagram representation of (7.49) is depicted in Fig. 7.11.

$$i_{cird} = \frac{\frac{1}{L_{arm}}}{s + \frac{R_{arm}}{L_{arm}}} v_{cird1} \tag{7.52a}$$

$$i_{cirq} = \frac{\frac{1}{L_{arm}}}{s + \frac{R_{arm}}{L_{arm}}} v_{cirq1} \tag{7.52b}$$

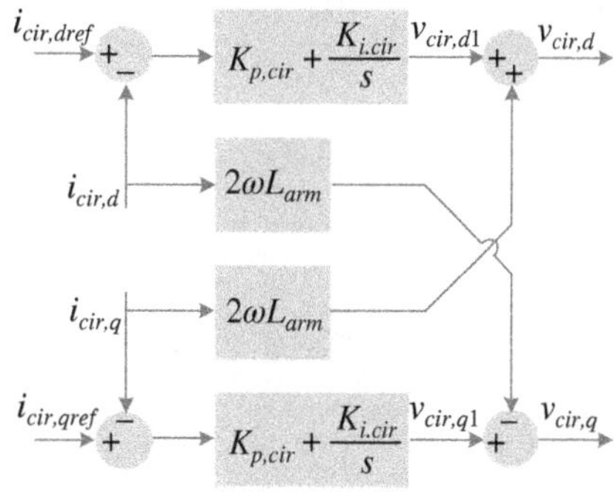

Figure 7.11 Block diagram of circulating current control designed in *dq*-frame.

PI controllers for circulating current controllers can be expressed as

$$\begin{aligned} v_{cird1} &= K_{p,cir}\left(i_{cir,dref} - i_{cir,d}\right) + K_{i,cir}\int \left(i_{cir,dref} - i_{cir,d}\right) dt \\ v_{cirq1} &= K_{p,cir}\left(i_{cir,qref} - i_{cir,q}\right) + K_{i,cir}\int \left(i_{cir,qref} - i_{cir,q}\right) dt, \end{aligned} \tag{7.53}$$

where $K_{p,cir}$ and $K_{i,cir}$ denote proportional and integral constants, respectively.

8

Generalized Steady-State Solution Procedure for Closed-Loop Converter Systems

"One who tied the bell knot shall be the one who unknots it."

– Hong Hui (1071–1128)

8.1 Introduction

Mathematically, a closed-loop power converter can be described by transcendental equations, which do not have closed-form solutions. This is evident from observing Fig. 8.1. As the figure indicates, the state variables, $\mathbf{x}$, are a function of the switching times, $\mathbf{t}_{sw}$, and the input,[1] $\mathbf{u}$. In the feedforward direction the relationship is given by:

$$\mathbf{x} = \mathbf{f}(\mathbf{t}_{sw}, \mathbf{u}). \tag{8.1}$$

However, $\mathbf{t}_{sw}$ itself is also a function of both states and input, due to the feedback path:

$$\mathbf{t}_{sw} = \mathbf{g}(\mathbf{x}, \mathbf{u}) \tag{8.2}$$

Combining (8.1) and (8.2), one will obtain

$$\mathbf{x} = \mathbf{f}(\mathbf{g}(\mathbf{x}, \mathbf{u})). \tag{8.3}$$

Equation (8.3) in general can only be solved by iterative methods. Thus, the objectives of this chapter are:

1. To present a unified solution procedure linking previous steady-state methods [85, 86]. By following the solution procedure, one can easily solve for a closed-loop power converter or in fact any closed-loop system.
2. To show how one may able to derive a new method from the proposed solution procedure.

8.2 Generalized Procedure

Fig. 8.2 describes the generalized solution procedure of solving a closed-loop converter system. As indicated in the figure, the procedure consists of five steps, which are described in detail as follows:

1 In the most general case the input $\mathbf{u}$ in Fig. 8.1 may itself be an unknown variable that depends on the behavior of neighboring converter systems.

Harmonic Modeling of Voltage Source Converters using Basic Numerical Methods, First Edition.
Ryan Kuo-Lung Lian, Ramadhani Kurniawan Subroto, Victor Andrean, and Bing Hao Lin.

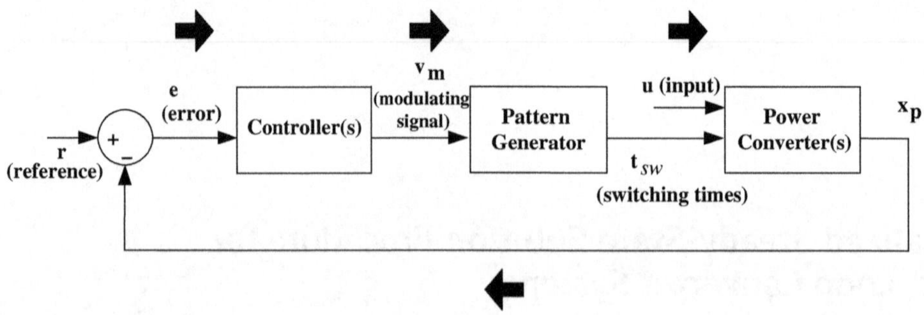

Figure 8.1 Closed-loop converter system.

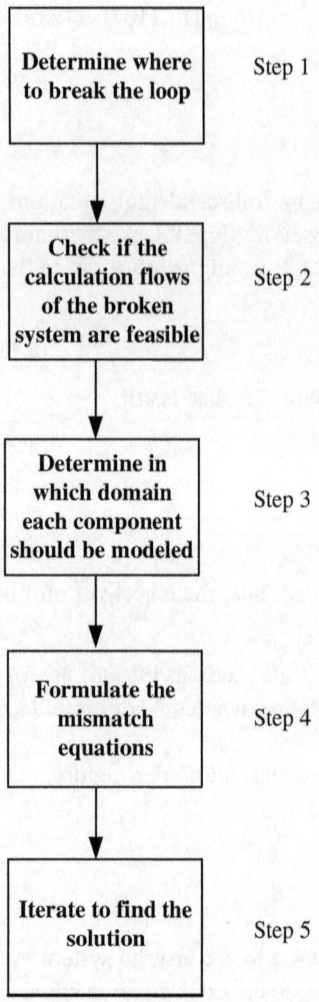

Figure 8.2 Generalized solution procedure diagram for solving a closed-loop power converter system.

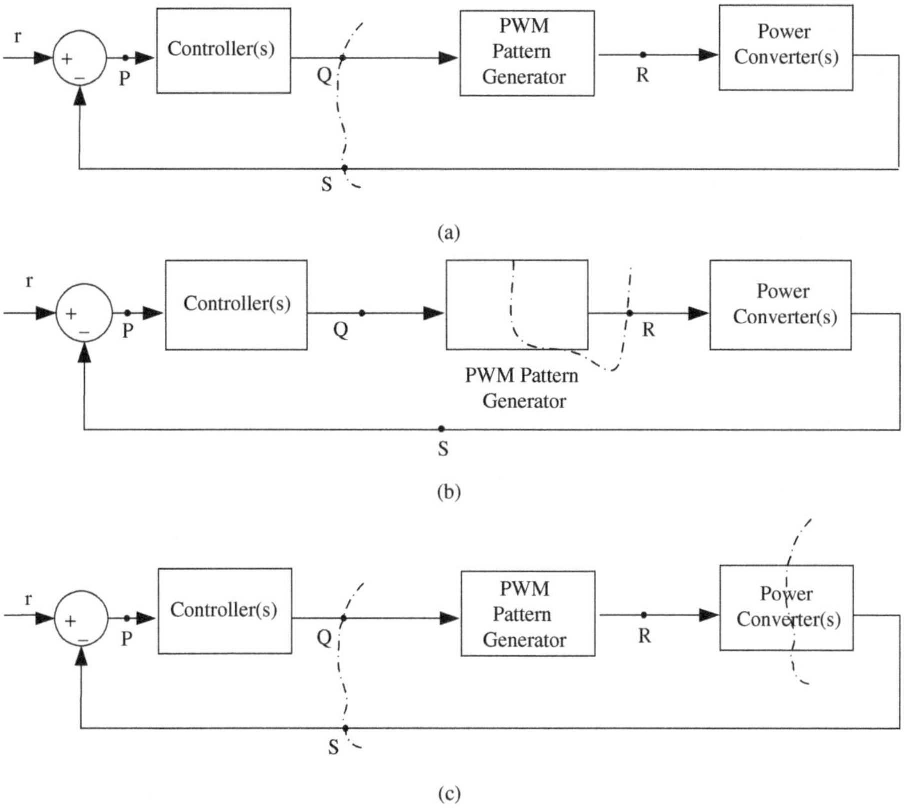

Figure 8.3 Various loop-breaking strategies: (a) loop-breaking 1 method; (b) loop-breaking 2 method; (c) loop-breaking 3 method.

8.2.1 Step 1: Determine How and Where to Break the Loop

The first step in formulating an iterative algorithm, one needs to first find a set of mismatch equations [87]. The solution is obtained when the mismatch equations are driven to zero (or below a specified tolerance). The simplest manner to formulate the mismatch equations of a closed-loop system is by means of breaking the loop. There are, in general, three methods of breaking loops.

1. Loop-Breaking 1 (LB 1) Method: break at the connection point between blocks (see Fig. 8.3(a) as an example).
2. Loop-Breaking 2 (LB 2) Method: break within the individual block (see Fig. 8.3(b) as an example).
3. Loop-Breaking 3 (LB 3) Method: have multiple break points both within and between components (see Fig. 8.3(c) as an example).

Note that the "breaking" location is the place where the mismatch equations are formulated. In addition to formulating the mismatch equations, it has been observed that breaking the loop can, in some cases, also avoid the non-convergence problem of the feedback system [88].

8.2.2 Step 2: Check if the Calculation Flows of the Broken System are Feasible

As mentioned, three methods of breaking a closed-loop converter system exist and, for each method, numerous break locations can be selected. However, regardless of how or where one may

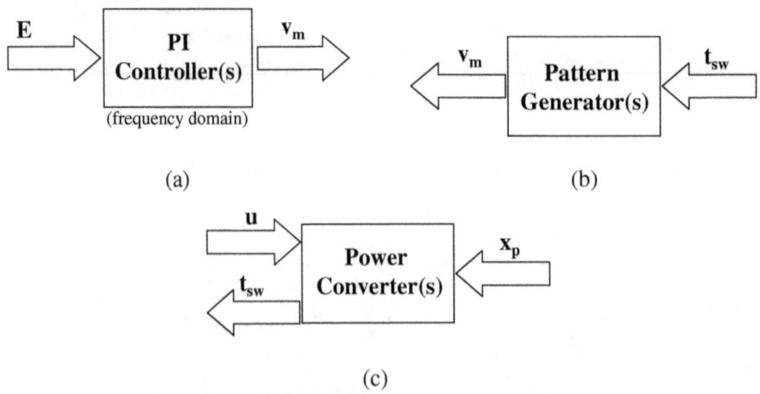

Figure 8.4 Infeasible calculation paths.

break the loop, the resulting calculation flows need to be feasible. Since a closed-loop converter normally consists of three components—a power converter, a PWM pattern generator, and a controller—there are three infeasible cases pertaining to these components. They are identified as follows:

Case I: If a PI controller is modeled in the frequency domain, the calculation flow indicated in Fig. 8.4(a) is not feasible because the error signal **E**[2] may contain a dc component. However, the reverse calculation path (back calculation from output to input of the PI controller) is feasible since the integrator behaves as a differentiator with zero dc gain. In contrast, if a PI controller is modeled in the time domain, the forward path is feasible.

Case II: The PWM pattern generator is a nonlinear system that offers a one-to-one mapping of input (the modulating signal $\mathbf{v_m}$) to output (the switching time $\mathbf{t_{sw}}$) only in the forward direction as indicated in Fig. 8.1. Thus, the calculation path flow in Fig. 8.4(b) is not feasible.

Case III: For a given set of input **u** and switching times (or switching functions), it is feasible to obtain the states of the converter system, $\mathbf{x_p}$. However, the reversed path as shown in Fig. 8.4(c) is infeasible. This can be seen from the solution of $\mathbf{x_p}$ at time t_k, which is restated in (8.4).

$$\mathbf{x_p}(t_k) = e^{\mathbf{A_k}(t_k - t_{k-1})}\mathbf{x_p}(t_{k-1}) + \int_{t_{k-1}}^{t_k} e^{\mathbf{A_k}(t_k-\tau)}\mathbf{B_k}\mathbf{u}(\tau)d\tau. \tag{8.4}$$

where $\mathbf{A_k}$ and $\mathbf{B_k}$ are the state space and input matrices, respectively, for the interval between t_{k-1} and t_k. One can see that solving for switching times requires taking the logarithm of (8.4). If $x(t_{k-1}) = 0$, taking the logarithm of (8.4) results in blow-up of the solution.

8.2.3 Step 3: Determine What Domain of Each Component in the System Should be Modeled

Once the loop is broken in a correct manner, one may easily determine what domain is most suitable for modeling each of the components. Also, one should determine whether the secondary or parasitic effects should be added in the model. Although adding such effects will increase the accuracy of the results, simulation time may increase.

As a general rule, Fourier and inverse Fourier transforms are needed in cases where all components are not modeled in the same domain. As an exception to this rule, the concept of harmonic

2 Here, "**E**" represents the error signal $\mathbf{e}(t)$ in the frequency domain.

state space, discussed in Chapter 6, may be used to link the time domain variables with frequency domain variables without using direct application of Fourier and inverse Fourier transforms.

8.2.4 Step 4: Formulate the Mismatch Equations

As mentioned previously, to solve for the steady-state, a set of mismatch equations must be driven to zero (or below a specified tolerance). For closed-loop converter systems, one set of mismatch equations will be associated with the LB that is carried out in **Step 1**. These will be referred to as the LB mismatch equations. Through driving the LB mismatch equations to zero, we effectively close the feedback loop that is broken in **Step 1**. Additional mismatch equations may also exist. A prevalent example of additional mismatch equations would be the periodicity constraints associated with a time domain system model, i.e. $\mathbf{x}(T) = \mathbf{x}(0)$.

8.2.5 Step 5: Iterate to Find the Solution

Most of the iterative solvers are based on Newton-type methods due to their excellent convergence rate [87]. However, the convergence of the Newton-type algorithm is highly dependent on the initial conditions. Poor initialization is not only detrimental to the speed of the simulation but can, in extreme cases, even lead to overall convergence problems. For most closed-loop power converter systems, a good seed value can be found from open-loop system analysis [13] or from the values obtained from the first iteration of a positive-sequence model [87].

8.3 Previously Proposed Methods Derived from the Proposed Solution Procedures

In this section, it is shown how several previously proposed methods can all be derived from the proposed general solution procedure.

8.3.1 Steady-State Methods Derived from Loop-Breaking 1 Method

LB 1 method breaks the loop at the connection point between the components. The steady-state method proposed in [86] can be viewed as having the loop breaking at the connection point between the controller and the PWM pattern generator as shown in Fig. 8.3. Knowing that Fig. 8.4(b) is not a feasible calculation path, one will conclude that the calculation flow indicated in black arrows in Fig. 8.5 is the only valid manner to formulate the mismatch equations. **Block 1**, consisting of the PWM pattern generator and the power converter, is then chosen to be modeled in the time domain, whereas the controller **Block 2** is modeled in the frequency domain. The unknown variable required to be solved is the modulating signal, $\mathbf{V}_{\mathbf{m}}$. It is initialized (as indicated by white arrows) at point **Q**.

It is supplied forward to **Block 1** to produce $\mathbf{\Gamma}_1$, and supplied backward to **Block 2** to produce $\mathbf{\Gamma}_2$, as shown in the figure. To close the system, one needs to impose the following mismatch equation:

$$\mathbf{\Gamma}_1 - \mathbf{\Gamma}_2 = 0. \tag{8.5}$$

Since the power converter is modeled in the time domain, one needs to include a periodicity constraint in the mismatch equations. The corresponding unknown variables due to this constraint

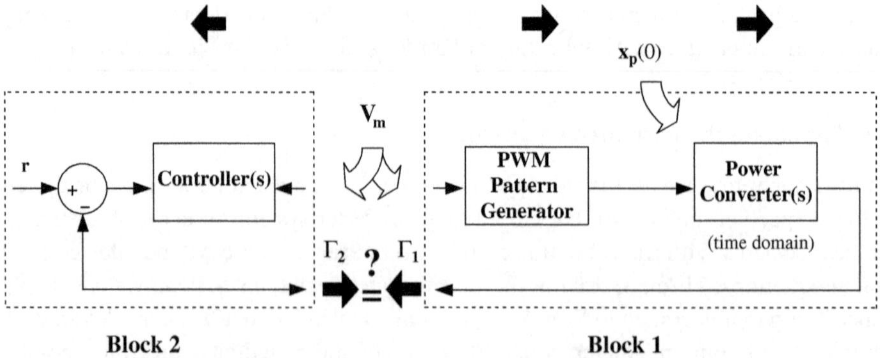

Figure 8.5 Formulating the mismatch equations for Loop-Breaking 1 Method when the converter is modeled in the time domain.

are the initial conditions of the power converter states, $\mathbf{x_p}(0)$, as shown in Fig. 8.5. Therefore, the overall iterative formula to be solved by Newton's method are written in (8.6).

$$\zeta^{(k+1)} = \zeta^{(k)} - [\mathbf{J}^{(k)}]^{-1}\mathbf{M}^{(k)}, \tag{8.6}$$

where $\zeta = \begin{bmatrix} \mathbf{V_m} \\ \mathbf{x_p}(0) \end{bmatrix}$, $\mathbf{M} = \begin{bmatrix} \mathbf{\Gamma}_1 - \mathbf{\Gamma}_2 \\ \mathbf{x_p}(T) - \mathbf{x_p}(0) \end{bmatrix}$, and $\mathbf{J} = \left[\frac{\partial \mathbf{M}}{\partial \zeta}\right]$. Equation (8.6) coincides with the formulation of the hybrid method that [86] proposed to solve a closed-loop VSC.

Moreover, other new methods not yet presented in the literature can also be deduced from LB 1 method. For example, one may choose to model the power converters in the frequency domain in Fig. 8.5, leading to a new iterative formula.

$$\zeta^{(k+1)} = \zeta^{(k)} - [\mathbf{J}^{(k)}]^{-1}\mathbf{M}^{(k)}, \tag{8.7}$$

where $\zeta = [\mathbf{V_m}]$, $\mathbf{M} = [\mathbf{\Gamma}_1 - \mathbf{\Gamma}_2]$, and $\mathbf{J} = \left[\frac{\partial \mathbf{M}}{\partial \zeta}\right]$.

Application of (8.7) would allow frequency domain converter models, such as those proposed in [89], to be employed for closed-loop steady-state computation.

8.3.2 Steady-State Methods Derived from Loop-Breaking 2 Method

The shooting method, presented in Chapter 3, is used to solve a simple dc power supply, which only contains a nonlinear component—the diode. Moreover, the relationship between voltage and current of the diode is given by (3.86). The dc power supply can be completely modeled by continuous-time state space equations as given by (3.87). Hence, the steady-state solution can be solved by the shooting method as done in Section 3.4.5. Generally, Aprille and Trick's formulation [38] only works for continuous-time networks or switched networks whose switching times are independent of changes in the network solution [90]. Nevertheless, in many situations, the diodes and other semiconductor switches are simply modeled as ideal switches because the detailed switching behaviors are not main concerns and do not affect the steady-state solution of the power converters. Consequently, precisely determining the turn-on and turn-off times of these ideal switches becomes important because erroneous solutions or divergence will result when switching time variations are not taken into consideration or falsely determined [90–92]. Grötzbach and von Lutz overcame this deficiency by introducing switching time constraints [93] that determine the transition from one topological mode to another. Therefore, the problem of finding solutions to power electronic converters can be recast as satisfying the following boundary constraints simultaneously:

$$\mathbf{x_p}(T) - \mathbf{x_p}(0) = 0 \tag{8.8}$$

$$\mu \mathbf{v_{cont}} - \mathbf{g}(\mathbf{x_p}(0), \mathbf{u}, \mathbf{t_{sw}}) = 0, \tag{8.9}$$

where μ is a scalar, $\mathbf{v_{cont}}$ is the control voltage, $\mathbf{g}$ contains the switching time constraints, and $\mathbf{u}$ is a vector of the input sources. (8.8) is the periodicity constraint whereas (8.9) represents the switching time constraints. Note that the switching time constraints are usually formulated in terms of threshold conditions, and these threshold conditions can be further classified into internal and external ones. An internal threshold condition causes a diode or a thyristor to change state and depends only on the internal circuit parameters. In this case, μ in (8.9) is zero. On the other hand, an external threshold condition controls the turn-on or turn-off instances of the switches by means of some control methods. In this case, μ is non-zero.

If one looks closely at Grötzbach and von Lutz's approach, the system can be viewed as having the loop broken within the PWM pattern generator component, as shown in Fig. 8.3(b). How the PWM pattern generator is broken depends on which pattern generation strategy is chosen. For the purpose of illustration, sinusoidal pulse-width modulation (SPWM) is used as an example to show how one may "break" the PWM pattern generator. Other PWM pattern generator strategies, such as the SVM, can be easily deduced from the presented example. The PWM signal is obtained by comparing the modulating signals with a triangular carrier waveform. The turn-on and turn-off times are the times at which the modulating waveform intersects with the triangular carrier waveforms, as shown in Fig. 8.6. While Fig. 8.6 shows how we implement SPWM in practice, mathematically we may view the process in an alternate manner, as follows. Knowing the switching times, samples of the carrier signal at these switching times (see Fig. 8.7(a)) should align with samples of the modulating signals at the corresponding switching times (see Fig. 8.7(b)).

Fig. 8.8 shows the feasible calculation flow (shown with black arrows) for this configuration, together with the unknown variables that must be initialized (shown with white arrows) to begin the iteration. The corresponding iterative formula is therefore stated in (8.10).

$$\zeta^{(k+1)} = \zeta^{(k)} - [\mathbf{J}^{(k)}]^{-1}\mathbf{M}^{(k)}, \tag{8.10}$$

where $\zeta = \begin{bmatrix} \mathbf{t_{sw}} \\ \mathbf{x_p}(0) \\ \mathbf{x_c}(0) \end{bmatrix}$, $\mathbf{M} = \begin{bmatrix} \mathbf{v_m}(\mathbf{t_{sw}}) - \mathbf{v_{carr}}(\mathbf{t}_{sw}) \\ \mathbf{x_p}(T) - \mathbf{x_p}(0) \\ \mathbf{x_c}(T) - \mathbf{x_c}(0) \end{bmatrix}$,

$\mathbf{v_m}$ is the modulating waveform, $\mathbf{v_{carr}}$ is the carrier waveform, $\mathbf{x_p}$ is the state variable of the power converter, $\mathbf{x_c}$ is the state variable of the controller(s), and $\mathbf{J} = \left[\frac{\partial \mathbf{M}}{\partial \zeta}\right]$.

Equation (8.10) precisely coincides with the formulation proposed in [93] (i.e. (8.8) and (8.9)). Chapter 10 shows in detail how this method can be used to solve a closed-loop, two-level VSC-based static synchronous compensator (STATCOM).

Derivation of (8.10) requires modeling all the components in Fig. 8.1. Nevertheless, the concept of the LB 2 method can be effectively applied if one wants to model components in the frequency domain. The frequency domain methods proposed in [85, 94, 95] can be viewed as a special case of the LB 2 method. Fig. 8.9 shows how a mismatch equation may be formed for a STATCOM [94, 96]. The power converter itself is broken into two parts as indicated in the figure. Thus, the mismatch equation needed to be driven to zero is

$$\mathbf{I_{dc1}} + \mathbf{I_{dc2}} = 0, \tag{8.11}$$

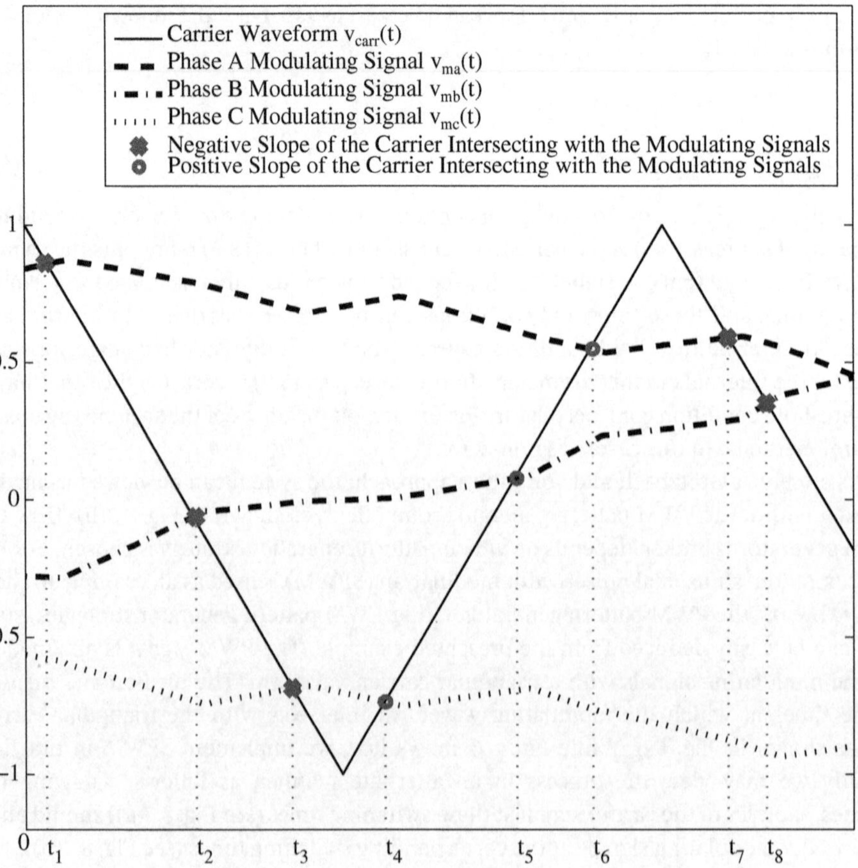

Figure 8.6 Triangular carrier waveform together with the modulating signals.

and the unknown variable is the dc voltage, $\mathbf{V_{dc}}$. Note that [94] and [96] choose to break the circuit at the dc side of the converter as shown in Fig. 8.9. Alternatively, it is also possible to break the circuit on the ac side of the converter.

It is important to note that (8.11) is only valid for an open-loop converter system. To include the controller, control constraints also need to be included in the mismatch equations. This can be done without modeling the control dynamics [85, 94, 95].

8.4 The Loop-Breaking 3 Method

The LB 3 method can be considered as the most general method, which encompasses LB 1 and 2 methods. It involves multiple break points between and within components, and enables solution of more complicated closed-loop systems. For example, multi-converter networks may be analyzed where the input voltage to the power converters are unknown. A converter-fed microgrid would fall into this class of system, as would various shipboard and aircraft power systems. Fig. 8.3(c) shows how one may apply the LB 3 method to solve a closed-loop system. The loop is broken at point **S** and within the converter itself. Several example systems are presented in Chapter 11 to illustrate how one may solve a system consisting of multiple converters by means of the LB 3 method.

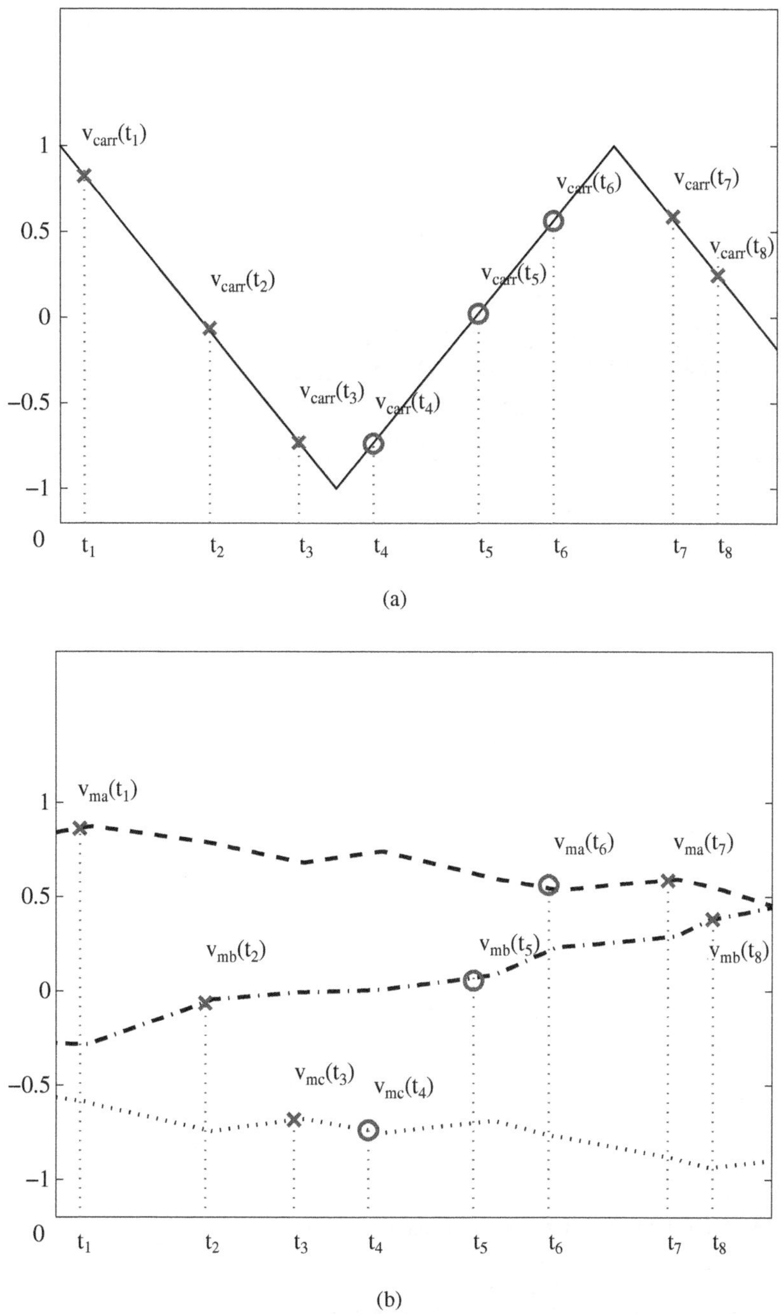

Figure 8.7 (a) The sample points of the carrier signal; (b) the sample points of the modulating signal.

Nevertheless, the LB 3 method can be equally applied to a single converter system such as Fig. 8.10, although such a system is preferably solved by the LB 1 and 2 methods.

To solve the system of Fig. 8.10 by means of loop-breaking 3 method, one may first break the system at the point, **P**, into three blocks (**Blocks 1**, **2**, and **3**) as shown in Fig. 8.11. **Block 2** is further broken into **Blocks 4** and **5**, as shown in Fig. 8.12.

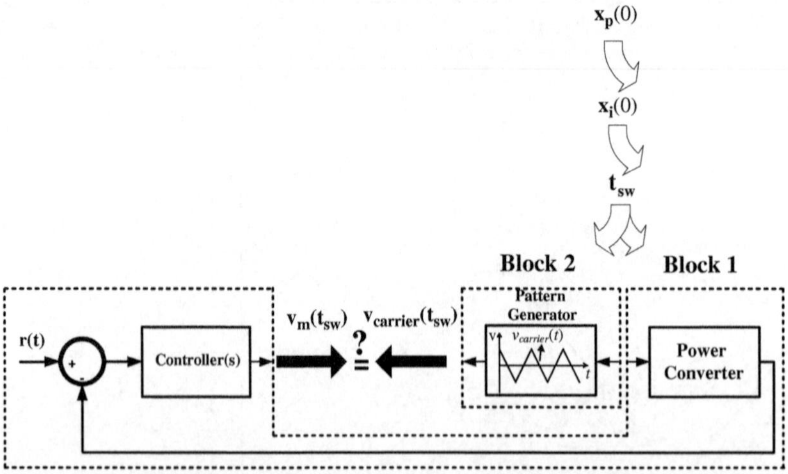

Figure 8.8 Formulating the mismatch equations for Loop-Breaking 2.

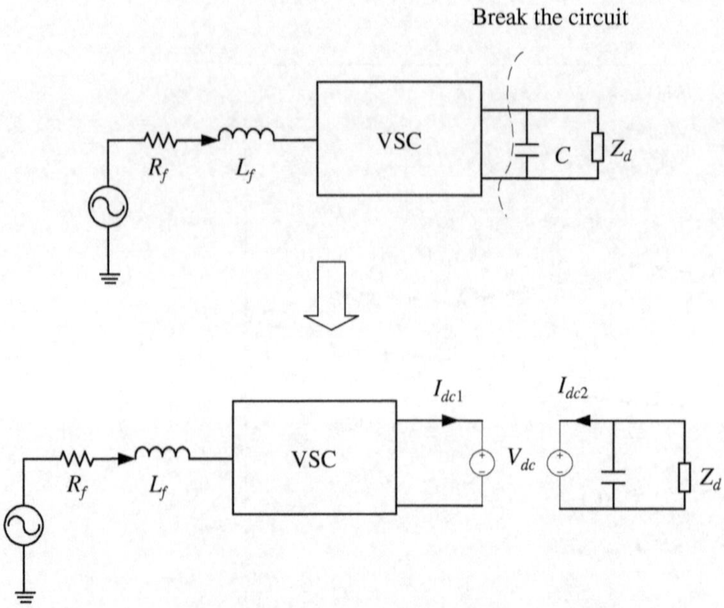

Figure 8.9 How a STATCOM is modeled in the frequency domain.

As an example, we may choose **Blocks 1**, **3**, and **5** to be modeled in the frequency domain, and **Block 4** is modeled in the time domain. The unknown variables are the nodal voltage **V**, the modulating signal of VSC, $\mathbf{V_m}$, and the initial conditions of the state variables of VSC ($\mathbf{x_p}(0)$). Note that one may choose to solve **V** first as an outer loop and then solve for $\mathbf{V_m}$ and $\mathbf{x_p}$ (0) in the inner loop. Chapter 11 illustrates how the outer loop can be solved by the Gauss–Seidel method and the inner loop can be solved by the Newton's method. [97] shows how the outer and inner loops are solved by Newton's method. Alternatively, one can also choose to solve all the unknown variables simultaneously using, for instance, Newton's method.

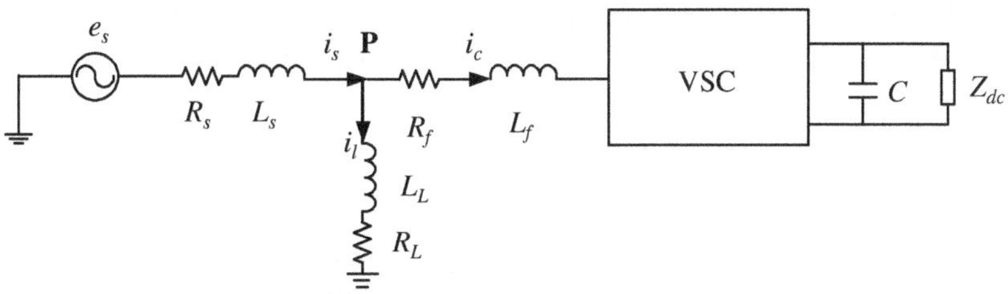

Figure 8.10 The single-line diagram of the D-STATCOM.

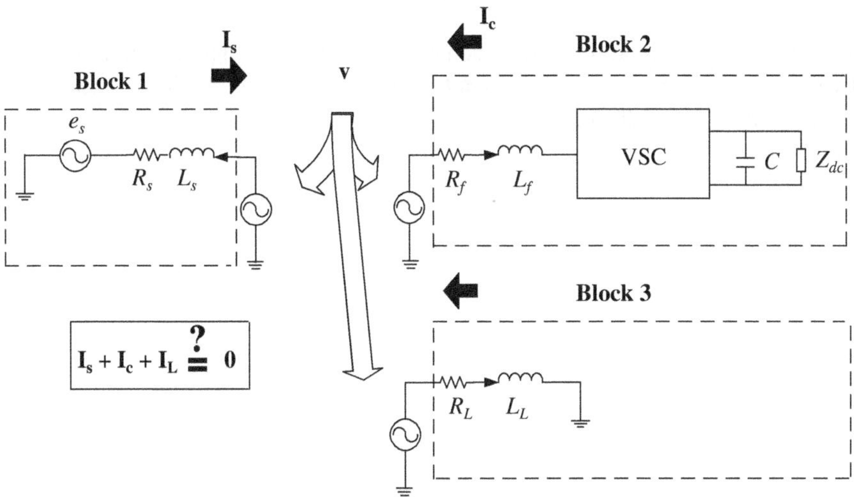

Figure 8.11 Breaking the converter system into three blocks.

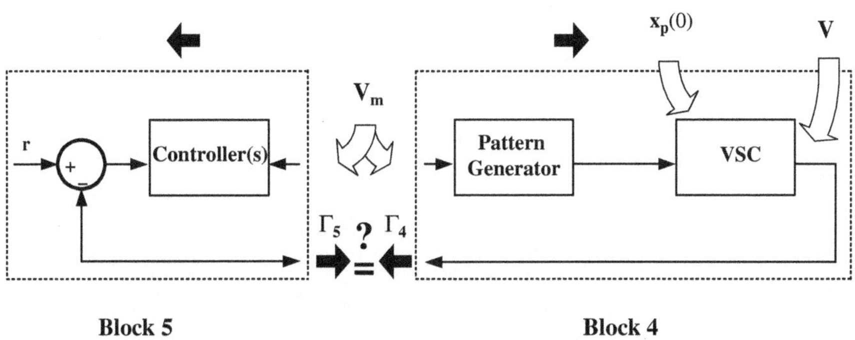

Figure 8.12 Breaking the VSC into two blocks.

The mismatch equations needed to be driven to zero are

$$\mathbf{M}_1 = [\ \boldsymbol{\Gamma}_4 - \boldsymbol{\Gamma}_5\ ] \tag{8.12}$$

$$\mathbf{M}_2 = [\ \mathbf{x}_\mathbf{p}(T) - \mathbf{x}_\mathbf{p}(0)\ ] \tag{8.13}$$

$$\mathbf{M}_3 = [\ \mathbf{I}_\mathbf{s} + \mathbf{I}_\mathbf{c} + \mathbf{I}_\mathbf{L}\ ] \tag{8.14}$$

Essentially, **V**, $\mathbf{V_m}$, and $\mathbf{x_p}(0)$ are the unknown variables to be solved, and need to be initialized. Then, **V** is passed to **Block 1** to produce $\mathbf{I_S}$. **V**, $\mathbf{V_m}$, and $\mathbf{x_p}(0)$ are passed to **Block 2** (consisting of **Block 4** and **Block 5**) to check if $\mathbf{M_1}$, $\mathbf{M_2}$, and $\mathbf{M_3}$ are close to zero.

Newton's formulation to solve (8.12) to (8.14) can be stated as

$$\eta^{(i+1)} = \eta^{(i)} - [\mathbf{J}^{(i)}]^{-1}\mathbf{M}^{(i)}, \tag{8.15}$$

where $\eta = [\mathbf{V}]$, where $\eta = \begin{bmatrix} \mathbf{V} & \mathbf{V_m} & \mathbf{X_p}(0) \end{bmatrix}^T$ and $\mathbf{J} = \begin{bmatrix} \frac{\partial \mathbf{M_1}}{\partial \mathbf{V}} & \frac{\partial \mathbf{M_1}}{\partial \mathbf{V_m}} & \frac{\partial \mathbf{M_1}}{\partial \mathbf{X_{vsc}}(0)} \\ \frac{\partial \mathbf{M_2}}{\partial \mathbf{V}} & \frac{\partial \mathbf{M_2}}{\partial \mathbf{V_m}} & \frac{\partial \mathbf{M_2}}{\partial \mathbf{X_{vsc}}(0)} \\ \frac{\partial \mathbf{M_3}}{\partial \mathbf{V}} & \frac{\partial \mathbf{M_3}}{\partial \mathbf{V_m}} & \frac{\partial \mathbf{M_3}}{\partial \mathbf{X_{vsc}}(0)} \end{bmatrix}$.

In particular, $\frac{\partial \mathbf{M_3}}{\partial \mathbf{V}} = \frac{\partial \mathbf{I_s}}{\partial \mathbf{V}} + \frac{\partial \mathbf{I_c}}{\partial \mathbf{V}} + \frac{\partial \mathbf{I_L}}{\partial \mathbf{V}}$, and $\frac{\partial \mathbf{I_c}}{\partial \mathbf{V}}$ can be considered as the small-signal harmonic model of the two-level VSC, which will be described in detail in Chapter 12. Also, the initial conditions need to be chosen carefully to avoid divergence.

9

Loop-Breaking 1 Method

"Divide the fire and you will soon put it out."

– Greek Proverb

The purpose of this chapter is to show in a stepwise manner how a three-phase two-level VSC can be modeled by the loop-breaking 1 (LB 1) method. Readers can then follow these steps to model other types of closed-loop VSCs. The LB 1 method requires the closed-loop system at the connection point between blocks to be broken. According to the feasibility analysis done in Chapter 8, Fig. 9.1 is the legitimate way for LB 1 method. As can be seen from the figure, the power converter circuit is "separated" from the controller(s). Therefore, the open-loop equation will only contain the model of the VSC power circuit, which is LTI for each switching time interval. Note that the PWM generator is not part of the open-loop system equation because its modeling is simply obtained by solving algebraic equations (as in the case of regular sampled PWM or SVM) or transcendental equations (as in the case of the naturally sampled PWM), and it does not involve any state-space differential equations. Furthermore, since the controller(s) are described in terms of transfer functions, they can be modeled in the frequency domain so that only algebraic calculations are involved.

The modeling procedure of the LB 1 method is described as follows:

1. Initialize modulating signal harmonics $\mathbf{V_m}$ as shown in Fig. 9.1.
2. Supply $\mathbf{V_m}$ to Block 1, which consists of the pattern generator and the model of the power converter. Note that Block 1 is modeled in the time domain, so $\mathbf{V_m}$ needs to be converted to time domain signals, $\mathbf{v_m(t)}$.
3. Supply $\mathbf{V_m}$ to Block 2, which contains the controllers modeled in the frequency domain.
4. Check if the feedback voltage/current harmonics $\mathbf{X_1}$ and $\mathbf{X_2}$ are the same at the point of $\mathbf{R}$. If this condition holds true, then the loop is closed and the closed-loop steady state solution is obtained.

Note that going backwards through the controller block (Block 2) resolves the singularity introduced by the PI controllers, which has a pole at zero frequency due to the integrator.

9.1 A Typical Two-Level VSC with AC Current Control and DC Voltage Control

Fig. 9.2 shows a general grid-connected VSC circuit, where the dc current injection i_l may represent either a source or a load current. Fig. 9.2 also shows a typical VSC controller, employing the direct

Harmonic Modeling of Voltage Source Converters using Basic Numerical Methods, First Edition.
Ryan Kuo-Lung Lian, Ramadhani Kurniawan Subroto, Victor Andrean, and Bing Hao Lin.

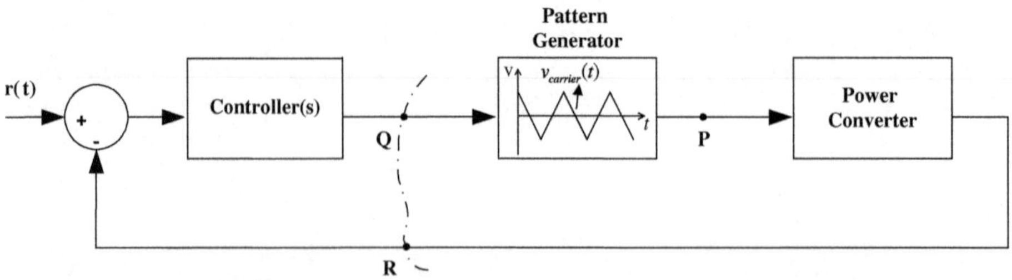

Figure 9.1 A general block diagram of a closed-loop converter system for the LB 1 method.

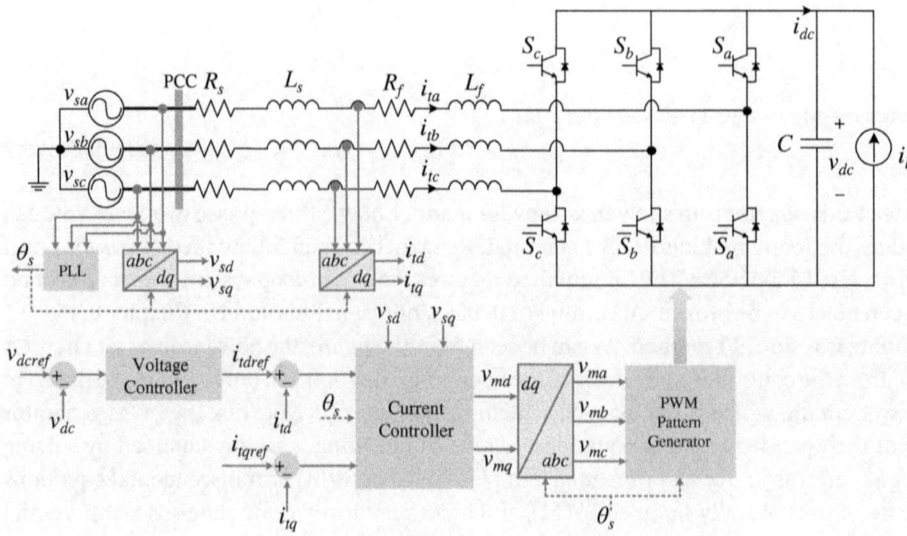

Figure 9.2 The schematic diagram of a VSC with its internal controllers.

control scheme as discussed in Chapter 7. Current regulators are controlled in the synchronous *dq* reference frame to exploit the tracking properties of PI regulators. Since the *d*-axis of the converter controller is aligned with the grid voltage, the *d*-axis current determines the active power exchanged with the grid. A dc voltage regulator assigns the *d*-axis reference current. The *q*-axis current determines the reactive power exchanged with the grid. Its reference value may be freely assigned to meet grid requirements. Without loss of generality, the following sections consider the VSC operated as a STATCOM, where the dc current injection is neglected.

9.2 Loop-Breaking 1 Method for a Two-Level VSC

Fig. 9.1 presents the basic modeling procedure for the LB 1 method, assuming that the controller only contains one PI controller. As a typical VSC has three PI controllers (1 dc voltage controller and 2 ac current controllers), such a system must be subdivided into three solution blocks as shown in Fig. 9.3.

Block 1 consists of the pattern generator and the model of the VSC. Though not shown in Fig. 9.3, Block 1 may also contain feedforward or auxiliary controllers provided that they do not include integral terms. Block 2 is the "Current Control Block," which contains the PI regulator for the ac currents. Block 3 is the "Voltage Control Block," which contains the PI regulator for dc bus voltage.

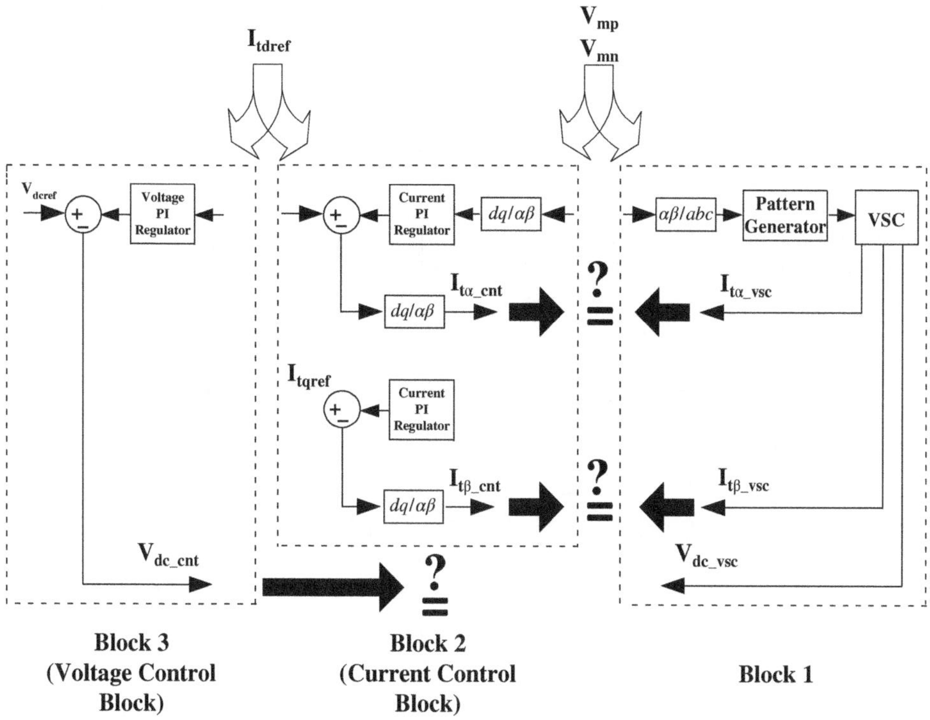

Figure 9.3 The calculation flow diagram of the hybrid method for simulating a closed-loop VSC.

The following describes the modeling procedures:

1. Initialize modulating signal harmonics in terms of the sequence components, $\mathbf{V_{mp}}$ and $\mathbf{V_{mn}}$, and active current reference, $\mathbf{I_{tdref}}$. Note that modulating voltage signals initialized in terms of the sequence components can simplify initialization and improve convergence under unbalanced conditions [87].
2. $\mathbf{V_{mp}}$ and $\mathbf{V_{mn}}$ are supplied to Block 1 to calculate the ac current and dc voltage harmonics predicted by the VSC ($\mathbf{I_{t\alpha\_vsc}}$, $\mathbf{I_{t\beta\_vsc}}$, and $\mathbf{V_{dc\_vsc}}$). Note that the currents are expressed in terms of α and β components rather than *abc* to reduce the dimensions of the mismatch equations.
3. $\mathbf{V_{mp}}$ and $\mathbf{V_{mn}}$ are also supplied to the Current Controller Block. In addition, the initialized active current reference, $\mathbf{I_{tdref}}$, is passed forward to the Current Control Block to calculate the ac current harmonics predicted by the Current Controller Block ($\mathbf{I_{t\alpha\_cnt}}$ and $\mathbf{I_{t\beta\_cnt}}$).
4. $\mathbf{I_{tdref}}$ is passed to the Voltage Controller Block for the prediction of $\mathbf{V_{dc\_cnt}}$.
5. Check if

$$\mathbf{I_{t\alpha\_vsc}} - \mathbf{I_{t\alpha\_cnt}} = 0 \tag{9.1}$$

$$\mathbf{I_{t\beta\_vsc}} - \mathbf{I_{t\beta\_cnt}} = 0 \tag{9.2}$$

$$\mathbf{V_{dc\_vsc}} - \mathbf{V_{dc\_cnt}} = 0 \tag{9.3}$$

If (9.1), (9.2), and (9.3) hold true, then the predicted modulating signal harmonics and active current reference harmonics are correct, and the loop is effectively closed; otherwise, the modulating signal harmonics and active current reference harmonics must be iterated until convergence is achieved. The modelings of Block 1, the Current Control Block, and the Voltage Control Block in the LB 1 method are briefly described as follows:

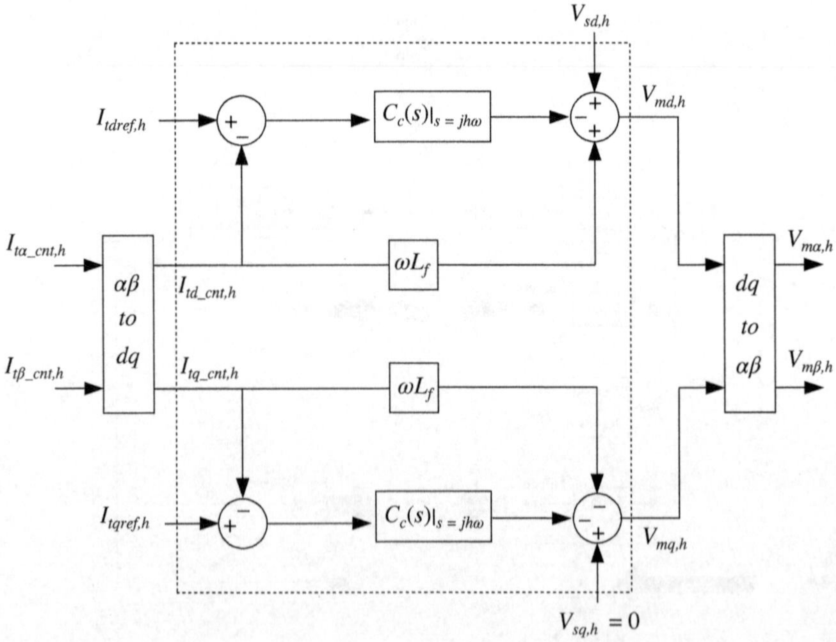

Figure 9.4 Block diagram of the current controllers in the frequency domain.

9.2.1 Block 1

Both the pattern generator and the VSC power circuit are modeled in the time domain. Since the modulating harmonics are initialized, they need to be converted into time domain signals before being fed to the pattern generator. The pattern generator then produces the corresponding switching times, $\mathbf{t_{sw}}$, which in turn are fed into the VSC.

In Chapter 5, we demonstrated how ac current and dc voltage harmonics can be obtained via the FCM for given switching times. Therefore, the modeling of the VSC is simply the FCM.

9.2.2 Current Controller Block

Fig. 9.4 shows the current controller of a VSC in the synchronous reference frame, where $C_c(s)|_{s=jhw}$ gives the steady-state response of the PI current controller at harmonic frequencies.

For given V^h_{md} and V^h_{mq}, the ac current harmonics at the current controllers, $I^h_{td\_cnt}$ and $I^h_{tq\_cnt}$, can be back calculated and found as in (9.4).

$$\begin{bmatrix} I_{td\_cnt,h} \\ I_{tq\_cnt,h} \end{bmatrix} = \begin{bmatrix} \frac{C_c(jh\omega)}{(C_c(jh\omega))^2 + (\omega L_f)^2} & \frac{-\omega L_f}{(C_c(jh\omega))^2 + (\omega L_f)^2} \\ \frac{\omega L_f}{(C_c(jh\omega))^2 + (\omega L_f)^2} & \frac{C_c(jh\omega)}{(C_c(jh\omega))^2 + (\omega L_f)^2} \end{bmatrix} \begin{bmatrix} V_{md,h} \\ V_{mq,h} \end{bmatrix} + \begin{bmatrix} \frac{-C_c(jh\omega)}{(C_c(jh\omega))^2 + (\omega L_f)^2} & \frac{\omega L_f}{(C_c(jh\omega))^2 + (\omega L_f)^2} \\ \frac{-\omega L_f}{(C_c(jh\omega))^2 + (\omega L_f)^2} & \frac{-C_c(jh\omega)}{(C_c(jh\omega))^2 + (\omega L_f)^2} \end{bmatrix} \begin{bmatrix} V_{sd,h} \\ V_{sq,h} \end{bmatrix}$$

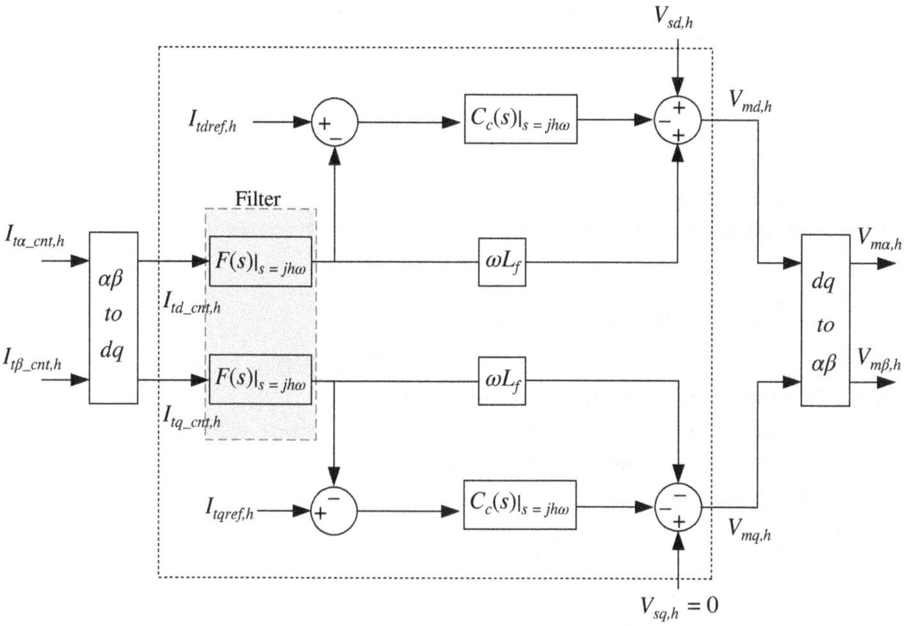

Figure 9.5 The block diagram of the current controllers incorporating current filters in the frequency domain.

$$+\begin{bmatrix} \frac{(C_c(jh\omega))^2}{(C_c(jh\omega))^2+(\omega L_f)^2} & \frac{-C_c(jh\omega)\,\omega L_f}{(C_c(jh\omega))^2+(\omega L_f)^2} \\ \frac{C_c(jh\omega)\,\omega L_f}{(C_c(jh\omega))^2+(\omega L_f)^2} & \frac{(C_c(jh\omega))^2}{(C_c(jh\omega))^2+(\omega L_f)^2} \end{bmatrix} \begin{bmatrix} I_{tdref,h} \\ I_{tqref,h} \end{bmatrix} \tag{9.4}$$

For almost all the applications of the VSC, low-pass current filters are used before the currents enter into the controllers to prevent undesired switching frequency components from going into the pattern generator. Fig. 9.5 shows the block diagram of current controllers in which current filters, $F(s)$, are incorporated. Accordingly, (9.4) can be modified into (9.5).

$$\begin{bmatrix} I_{td\_cnt,h} \\ I_{tq\_cnt,h} \end{bmatrix} = \begin{bmatrix} \frac{C_c(jh\omega)F(jh\omega)}{(C_c(jh\omega))^2+(\omega L_f)^2} & \frac{-\omega L_f F(jh\omega)}{(C_c(jh\omega))^2+(\omega L_f)^2} \\ \frac{\omega L_f F(jh\omega)}{(C_c(jh\omega))^2+(w_0 L_f)^2} & \frac{C_c(jh\omega)F(jh\omega)}{(C_c(jh\omega))^2+(\omega L_f)^2} \end{bmatrix} \begin{bmatrix} V_{md,h} \\ V_{mq,h} \end{bmatrix}$$
$$+\begin{bmatrix} \frac{-C_c(jh\omega)F(jh\omega)}{(C_c(jh\omega))^2+(\omega L_f)^2} & \frac{\omega L_f F(jh\omega)}{(C_c(jh\omega))^2+(\omega L_f)^2} \\ \frac{-\omega L_f F(jh\omega)}{(C_c(jh\omega))^2+(\omega L_f)^2} & \frac{-C_c(jh\omega)F(jh\omega)}{(C_c(jh\omega))^2+(\omega L_f)^2} \end{bmatrix} \begin{bmatrix} V_{sd,h} \\ V_{sq,h} \end{bmatrix}$$
$$+\begin{bmatrix} \frac{(C_c(jh\omega))^2F(jh\omega)}{(C_c(jh\omega))^2+(\omega L_f)^2} & \frac{-C_c(jh\omega)\omega L_f F(jh\omega)}{(C_c(jh\omega))^2+(\omega L_f)^2} \\ \frac{C_c(jh\omega)\omega L_f F(jh\omega)}{(C_c(jh\omega))^2+(\omega L_f)^2} & \frac{(C_c(jh\omega))^2F(jh\omega)}{(C_c(jh\omega))^2+(\omega L_f)^2} \end{bmatrix} \begin{bmatrix} I_{tdref,h} \\ I_{tqref,h} \end{bmatrix} \tag{9.5}$$

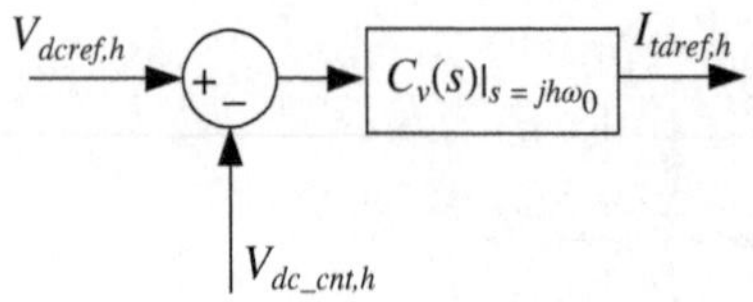

Figure 9.6 The block diagram of the dc voltage controller in the frequency domain.

Note that (9.4) or (9.5) must be evaluated at each harmonic of interest. For multiple harmonics of interest, the corresponding multiple equations of the form of (9.4) can be written in a compact form as (9.6),

$$\mathbf{I_{tdq\_cnt}} = \mathbf{Y_{mdq}V_{mdq}} + \mathbf{Y_{sdq}V_{sdq}} + \mathbf{K_{dqref}I_{tdqref}}, \tag{9.6}$$

where $\mathbf{I_{tdq\_cnt}} = \left[I_{tdq\_cnt,0} \quad I_{tdq\_cnt,1} \quad I_{tdq\_cnt,2} \quad \cdots \quad I_{tdq\_cnt,n}\right]^T$. Other harmonic vectors such as $\mathbf{V_{mdq}}$ and $\mathbf{V_{sdq}}$ contain the same number of harmonic orders as $\mathbf{I_{tdq\_cnt}}$.

To form a mismatch equation between solutions of the current control block and Block 1, (9.6) must be transformed into the $\alpha\beta$-frame. This can be done by multiplying both sides of (9.6) by the connection matrix, $\mathbf{C_{con}}$ [87], [98].

$$\mathbf{I_{t\alpha\beta\_cnt}} = \mathbf{C_{con}Y_{mdq}V_{mdq}} + \mathbf{C_{con}Y_{sdq}V_{sdq}} + \mathbf{C_{con}K_{dqref}I_{tdqref}}, \tag{9.7}$$

where $\mathbf{C_{con}} = \frac{1}{2}\begin{bmatrix} \mathbf{0} & \mathbf{M_j} & \mathbf{0} & \mathbf{0} & \mathbf{0} \\ \mathbf{N_j} & \mathbf{0} & \ddots & \mathbf{0} & \mathbf{0} \\ \mathbf{0} & \ddots & \mathbf{0} & \mathbf{M_j} & \mathbf{0} \\ \mathbf{0} & \mathbf{0} & \mathbf{N_j} & \mathbf{0} & \mathbf{M_j} \\ \mathbf{0} & \mathbf{0} & \mathbf{0} & \mathbf{N_j} & \mathbf{0} \end{bmatrix}$, $\mathbf{M_j} = \begin{bmatrix} 1 & j \\ -j & 1 \end{bmatrix}$, and $\mathbf{N_j} = \begin{bmatrix} 1 & -j \\ j & 1 \end{bmatrix}$.

However, since the modulating signals $\mathbf{V_{m\alpha\beta}}$ are to be initialized in the $\alpha\beta$-frame as indicated in Fig. 9.3, a more convenient form of the equation is given by (9.8).

$$\mathbf{I_{t\alpha\beta\_cnt}} = \mathbf{C_{con}Y_{mdq}C_{con}^{-1}V_{m\alpha\beta}} + \mathbf{C_{con}Y_{sdq}C_{con}^{-1}V_{s\alpha\beta}} + \mathbf{C_{on}K_{dqref}I_{tdqref}} \tag{9.8}$$

9.2.3 Voltage Controller Block

Fig. 9.6 shows the outer dc bus voltage controller that assigns the active current reference (I^h_{tdref}), where $C_v(s)|_{s=jh\omega}$ gives the steady-state response of the PI voltage controller at harmonic frequencies.

Given I^h_{tdref}, the dc voltage harmonics are back calculated to be (9.9).

$$V_{dc\_cnt,h} = V_{dcref,h} - \frac{I_{tdref,h}}{C_v(jh\omega)} \tag{9.9}$$

For each harmonic of interest, an equation in the form of (9.9) is required, yielding the matrix relation (9.10).

$$\mathbf{V_{dc\_cnt}} = \mathbf{V_{dcref}} - (\mathbf{C_v})^{-1}\mathbf{I_{tdref}} \tag{9.10}$$

9.3 Solution Flow Diagram

Fig. 9.7 shows the overall flow diagram of the LB 1's large signal analysis of a closed-loop VSC.

$\mathbf{V_{mp}}$ and $\mathbf{V_{mn}}$, together with $\mathbf{I_{tdref}}$, are passed to Block 1 and the control blocks. In Block 1, the modulating signal harmonics are converted to abc-components and fed to a PWM pattern generator.

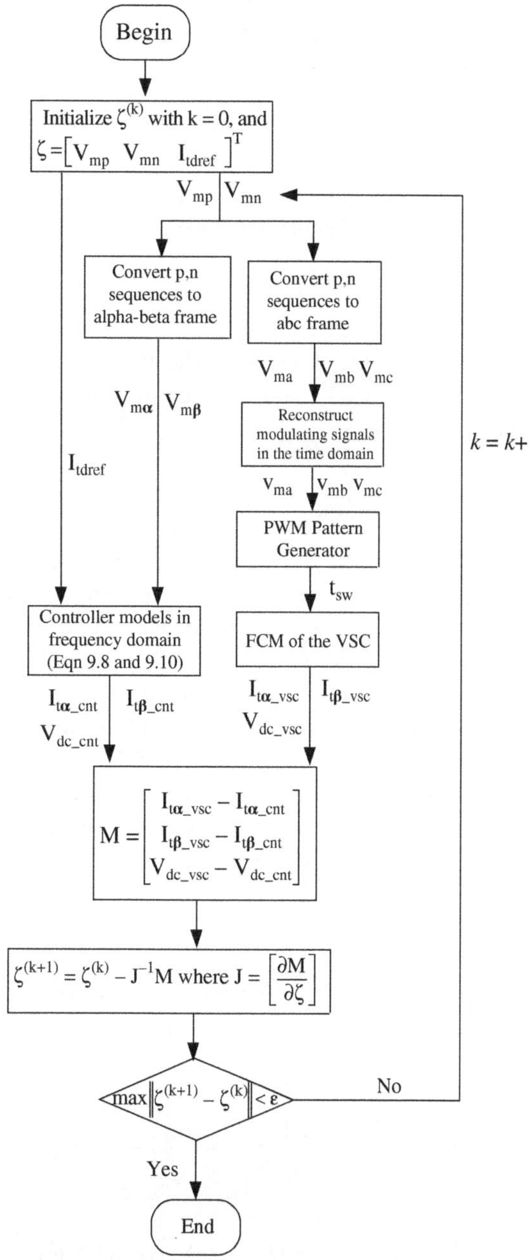

Figure 9.7 Solution flow chart of the LB 1 method for the VSC.

The switching times ($\mathbf{t_{sw}}$) produced by the pattern generator are fed to the VSC model. The VSC model generates ac current harmonics ($\mathbf{I_{t\alpha\_vsc}}$ and $\mathbf{I_{t\beta\_vsc}}$) and dc voltage harmonics ($\mathbf{V_{dc\_vsc}}$). In the control blocks, $\mathbf{V_{mp}}$ and $\mathbf{V_{mn}}$ are converted to the $\alpha\beta$-components $\mathbf{V_{m\alpha}}$ and $\mathbf{V_{m\beta}}$ via (9.52). These voltages, together with $\mathbf{I_{tdref}}$, are substituted into (9.8) and (9.10) to yield $\mathbf{I_{t\alpha\_cnt}}$, and $\mathbf{I_{t\beta\_cnt}}$ and $\mathbf{V_{dc\_cnt}}$. Then, the mismatch equations can be found by (9.11).

$$\mathbf{M} = \begin{bmatrix} \mathbf{I_{t\alpha\_vsc}} - \mathbf{I_{t\alpha\_cnt}} \\ \mathbf{I_{t\beta\_vsc}} - \mathbf{I_{t\beta\_cnt}} \\ \mathbf{V_{dc\_vsc}} - \mathbf{V_{dc\_cnt}} \end{bmatrix} = \left[\mathbf{f_{vsc}} - \mathbf{f_{cnt}}\right] \tag{9.11}$$

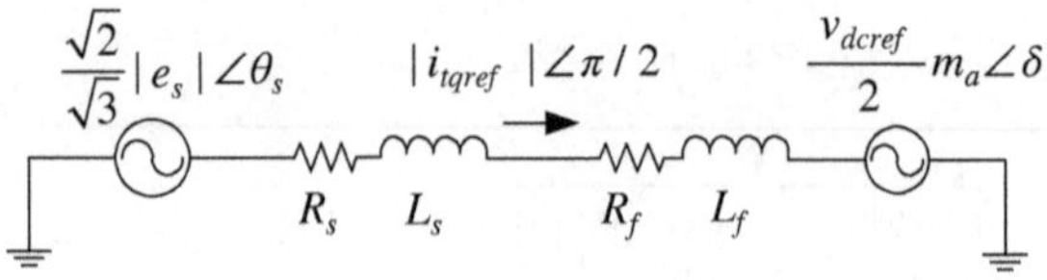

Figure 9.8 The simplified open-loop VSC model with the assumption of infinite switching frequency.

The mismatch equations set the stage for a Newton-type iterative method, which starts with a seed value $\zeta^{(0)}$, and generates the sequence $\zeta^{\infty}_{n=0}$ by (9.12).

$$\zeta^{(k+1)} = \zeta^{(k)} - \mathbf{J}^{-1}\mathbf{M}, \tag{9.12}$$

where $\zeta = \begin{bmatrix}\mathbf{V_{mp}} & \mathbf{V_{mn}} & \mathbf{I_{tdref}}\end{bmatrix}^T$.

Convergence and efficiency of the process are dependent on the initial values, $\zeta^{(0)}$, and the Jacobian, **J**. Sections 9.3.1 and 9.3.2 are devoted to the discussions on initialization and finding the Jacobian matrix.

9.3.1 Initialization

Proper initialization is critical for Newton's method. Poor initialization is not only detrimental to the speed of the simulation but can, in extreme cases, even lead to overall convergence problems. A good seed value must therefore be found for the first iteration. Fortunately, this is not difficult as the dominant harmonic produced by a VSC is the positive sequence fundamental component. A simple phasor solution, based on Fig. 9.8, is therefore used for initialization.

The initialization model assumes infinite switching frequency and neglects both control action and converter losses. The phasor analysis yields modulation index (m_a) and firing angle (δ) associated with the fundamental frequency positive sequence component for a given value of $|e_s|$, θ_s, v_{dcref} and i_{tqref}. All other harmonic components are initialized near zero. Note that initialization of precisely 0 will result in singularity.

9.3.2 Jacobian Matrix

The Jacobian matrix needed for a Newton-type iteration is given by the partial derivative of the mismatch equations with respect to the input vector, $\zeta = \begin{bmatrix}\mathbf{V_{mp}} & \mathbf{V_{mn}} & \mathbf{I_{tdref}}\end{bmatrix}^T$. Consequently,

$$\mathbf{J} = \left[\frac{\partial \mathbf{M}}{\partial \zeta}\right] = \begin{bmatrix} \frac{\partial \mathbf{I}_{t\alpha\_vsc}}{\partial \zeta} - \frac{\partial \mathbf{I}_{t\alpha\_cnt}}{\partial \zeta} \\ \frac{\partial \mathbf{I}_{t\beta\_vsc}}{\partial \zeta} - \frac{\partial \mathbf{I}_{t\beta\_cnt}}{\partial \zeta} \\ \frac{\partial \mathbf{V}_{dc\_vsc}}{\partial \zeta} - \frac{\partial \mathbf{V}_{dc\_cnt}}{\partial \zeta} \end{bmatrix}^T = \left[\frac{\partial \mathbf{f}_{vsc}}{\partial \zeta} - \frac{\partial \mathbf{f}_{cnt}}{\partial \zeta}\right] \tag{9.13}$$

$$\mathbf{J} = \mathbf{J}_{vsc} - \mathbf{J}_{cnt} = \begin{bmatrix}\mathbf{J}_{11\_vsc} & | & \mathbf{J}_{12\_vsc}\end{bmatrix} - \begin{bmatrix}\mathbf{J}_{11\_cnt} & | & \mathbf{J}_{12\_cnt}\end{bmatrix}. \tag{9.14}$$

Note that in the following derivations, all the variables will be only considered up to the fundamental component to avoid excessively large matrix. Jacobian matrices pertaining to variables with higher harmonics can be easily extended by following the same steps.

$$\mathbf{J}_{\mathbf{11\_vsc}} = \begin{bmatrix} \frac{\partial \text{Re}\{I_{t\alpha\_vsc,1}\}}{\partial \text{Re}\{V_{mp,1}\}} & \frac{\partial \text{Re}\{I_{t\alpha\_vsc,1}\}}{\partial \text{Im}\{V_{mp,1}\}} & \frac{\partial \text{Re}\{I_{t\alpha\_vsc,1}\}}{\partial \text{Re}\{V_{mn,1}\}} & \frac{\partial \text{Re}\{I_{t\alpha\_vsc,1}\}}{\partial \text{Im}\{V_{mn,1}\}} \\ \frac{\partial \text{Re}\{I_{t\beta\_vsc,1}\}}{\partial \text{Re}\{V_{mp,1}\}} & \frac{\partial \text{Re}\{I_{t\beta\_vsc,1}\}}{\partial \text{Im}\{V_{mp,1}\}} & \frac{\partial \text{Re}\{I_{t\beta\_vsc,1}\}}{\partial \text{Re}\{V_{mn,1}\}} & \frac{\partial \text{Re}\{I_{t\beta\_vsc,1}\}}{\partial \text{Im}\{V_{mn,1}\}} \\ \frac{\partial \text{Im}\{I_{t\alpha\_vsc,1}\}}{\partial \text{Re}\{V_{mp,1}\}} & \frac{\partial \text{Im}\{I_{t\alpha\_vsc,1}\}}{\partial \text{Im}\{V_{mp,1}\}} & \frac{\partial \text{Im}\{I_{t\alpha\_vsc,1}\}}{\partial \text{Re}\{V_{mn,1}\}} & \frac{\partial \text{Im}\{I_{t\alpha\_vsc,1}\}}{\partial \text{Im}\{V_{mn,1}\}} \\ \frac{\partial \text{Im}\{I_{t\beta\_vsc,1}\}}{\partial \text{Re}\{V_{mp,1}\}} & \frac{\partial \text{Im}\{I_{t\beta\_vsc,1}\}}{\partial \text{Im}\{V_{mp,1}\}} & \frac{\partial \text{Im}\{I_{t\beta\_vsc,1}\}}{\partial \text{Re}\{V_{mn,1}\}} & \frac{\partial \text{Im}\{I_{t\beta\_vsc,1}\}}{\partial \text{Im}\{V_{mn,1}\}} \\ \frac{\partial V_{dc,0}}{\partial \text{Re}\{V_{mp,1}\}} & \frac{\partial V_{dc,0}}{\partial \text{Im}\{V_{mp,1}\}} & \frac{\partial V_{dc,0}}{\partial \text{Re}\{V_{mn,1}\}} & \frac{\partial V_{dc,0}}{\partial \text{Im}\{V_{mn,1}\}} \\ \frac{\partial \text{Re}\{V_{dc,1}\}}{\partial \text{Re}\{V_{mp,1}\}} & \frac{\partial \text{Re}\{V_{dc,1}\}}{\partial \text{Im}\{V_{mp,1}\}} & \frac{\partial \text{Re}\{V_{dc,1}\}}{\partial \text{Re}\{V_{mn,1}\}} & \frac{\partial \text{Re}\{V_{dc,1}\}}{\partial \text{Im}\{V_{mn,1}\}} \\ \frac{\partial \text{Im}\{V_{dc,1}\}}{\partial \text{Re}\{V_{mp,1}\}} & \frac{\partial \text{Im}\{V_{dc,1}\}}{\partial \text{Im}\{V_{mp,1}\}} & \frac{\partial \text{Im}\{V_{dc,1}\}}{\partial \text{Re}\{V_{mn,1}\}} & \frac{\partial \text{Im}\{V_{dc,1}\}}{\partial \text{Im}\{V_{mn,1}\}} \end{bmatrix} \tag{9.15}$$

Since in the open-loop case $\mathbf{I}_{\mathbf{tdref}}$ has no influence over the VSC,

$$\mathbf{J}_{\mathbf{12\_vsc}} = \begin{bmatrix} \frac{\partial \text{Re}\{I_{t\alpha\_vsc,1}\}}{\partial I_{tdref,0}} & \frac{\partial \text{Re}\{I_{t\alpha\_vsc,1}\}}{\partial \text{Re}\{I_{tdref,1}\}} & \frac{\partial \text{Re}\{I_{t\alpha\_vsc,1}\}}{\partial \text{Im}\{I_{tdref,1}\}} \\ \frac{\partial \text{Re}\{I_{t\beta\_vsc,1}\}}{\partial I_{tdref,0}} & \frac{\partial \text{Re}\{I_{t\beta\_vsc,1}\}}{\partial \text{Re}\{I_{tdref,1}\}} & \frac{\partial \text{Re}\{I_{t\beta\_vsc,1}\}}{\partial \text{Im}\{I_{tdref,1}\}} \\ \frac{\partial \text{Im}\{I_{t\alpha\_vsc,1}\}}{\partial I_{tdref,0}} & \frac{\partial \text{Im}\{I_{t\alpha\_vsc,1}\}}{\partial \text{Re}\{I_{tdref,1}\}} & \frac{\partial \text{Im}\{I_{t\alpha\_vsc,1}\}}{\partial \text{Im}\{I_{tdref,1}\}} \\ \frac{\partial \text{Im}\{I_{t\beta\_vsc,1}\}}{\partial I_{tdref,0}} & \frac{\partial \text{Im}\{I_{t\beta\_vsc,1}\}}{\partial \text{Re}\{I_{tdref,1}\}} & \frac{\partial \text{Im}\{I_{t\beta\_vsc,1}\}}{\partial \text{Im}\{I_{tdref,1}\}} \\ \frac{\partial V_{dc,0}}{\partial I_{tdref,0}} & \frac{\partial V_{dc,0}}{\partial \text{Re}\{I_{tdref,1}\}} & \frac{\partial V_{dc,0}}{\partial \text{Re}\{I_{tdref,1}\}} \\ \frac{\partial \text{Re}\{V_{dc,1}\}}{\partial I_{tdref,0}} & \frac{\partial \text{Re}\{V_{dc,1}\}}{\partial \text{Re}\{I_{tdref,1}\}} & \frac{\partial \text{Re}\{V_{dc,1}\}}{\partial \text{Im}\{I_{tdref,1}\}} \\ \frac{\partial \text{Im}\{V_{dc,1}\}}{\partial I_{tdref,0}} & \frac{\partial \text{Im}\{V_{dc,1}\}}{\partial \text{Re}\{I_{tdref,1}\}} & \frac{\partial \text{Im}\{V_{dc,1}\}}{\partial \text{Im}\{I_{tdref,1}\}} \end{bmatrix} = \mathbf{0}. \tag{9.16}$$

$$\mathbf{J_{11\_cnt}} = \begin{bmatrix} \frac{\partial \mathrm{Re}\{I_{t\alpha\_cnt,1}\}}{\partial \mathrm{Re}\{V_{mp,1}\}} & \frac{\partial \mathrm{Re}\{I_{t\alpha\_cnt,1}\}}{\partial \mathrm{Im}\{V_{mp,1}\}} & \frac{\partial \mathrm{Re}\{I_{t\alpha\_cnt,1}\}}{\partial \mathrm{Re}\{V_{mn,1}\}} & \frac{\partial \mathrm{Re}\{I_{t\alpha\_cnt,1}\}}{\partial \mathrm{Im}\{V_{mn,1}\}} \\ \frac{\partial \mathrm{Re}\{I_{t\beta\_cnt,1}\}}{\partial \mathrm{Re}\{V_{mp,1}\}} & \frac{\partial \mathrm{Re}\{I_{t\beta\_cnt,1}\}}{\partial \mathrm{Im}\{V_{mp,1}\}} & \frac{\partial \mathrm{Re}\{I_{t\beta\_cnt,1}\}}{\partial \mathrm{Re}\{V_{mn,1}\}} & \frac{\partial \mathrm{Re}\{I_{t\beta\_cnt,1}\}}{\partial \mathrm{Im}\{V_{mn,1}\}} \\ \frac{\partial \mathrm{Im}\{I_{t\alpha\_cnt,1}\}}{\partial \mathrm{Re}\{V_{mp,1}\}} & \frac{\partial \mathrm{Im}\{I_{t\alpha\_cnt,1}\}}{\partial \mathrm{Im}\{V_{mp,1}\}} & \frac{\partial \mathrm{Im}\{I_{t\alpha\_cnt,1}\}}{\partial \mathrm{Re}\{V_{mn,1}\}} & \frac{\partial \mathrm{Im}\{I_{t\alpha\_cnt,1}\}}{\partial \mathrm{Im}\{V_{mn,1}\}} \\ \frac{\partial \mathrm{Im}\{I_{t\beta\_cnt,1}\}}{\partial \mathrm{Re}\{V_{mp,1}\}} & \frac{\partial \mathrm{Im}\{I_{t\beta\_cnt,1}\}}{\partial \mathrm{Im}\{V_{mp,1}\}} & \frac{\partial \mathrm{Im}\{I_{t\beta\_cnt,1}\}}{\partial \mathrm{Re}\{V_{mn,1}\}} & \frac{\partial \mathrm{Im}\{I_{t\beta\_cnt,1}\}}{\partial \mathrm{Im}\{V_{mn,1}\}} \\ \frac{\partial V_{dc,0}}{\partial \mathrm{Re}\{V_{mp,1}\}} & \frac{\partial V_{dc,0}}{\partial \mathrm{Im}\{V_{mp,1}\}} & \frac{\partial V_{dc,0}}{\partial \mathrm{Re}\{V_{mn,1}\}} & \frac{\partial V_{dc,0}}{\partial \mathrm{Im}\{V_{mn,1}\}} \\ \frac{\partial \mathrm{Re}\{V_{dc,1}\}}{\partial \mathrm{Re}\{V_{mp,1}\}} & \frac{\partial \mathrm{Re}\{V_{dc,1}\}}{\partial \mathrm{Im}\{V_{mp,1}\}} & \frac{\partial \mathrm{Re}\{V_{dc,1}\}}{\partial \mathrm{Re}\{V_{mn,1}\}} & \frac{\partial \mathrm{Re}\{V_{dc,1}\}}{\partial \mathrm{Im}\{V_{mn,1}\}} \\ \frac{\partial \mathrm{Im}\{V_{dc,1}\}}{\partial \mathrm{Re}\{V_{mp,1}\}} & \frac{\partial \mathrm{Im}\{V_{dc,1}\}}{\partial \mathrm{Im}\{V_{mp,1}\}} & \frac{\partial \mathrm{Im}\{V_{dc,1}\}}{\partial \mathrm{Re}\{V_{mn,1}\}} & \frac{\partial \mathrm{Im}\{V_{dc,1}\}}{\partial \mathrm{Im}\{V_{mn,1}\}} \end{bmatrix}, \tag{9.17}$$

$$\mathbf{J_{12\_cnt}} = \begin{bmatrix} \frac{\partial \mathrm{Re}\{I_{t\alpha\_cnt,1}\}}{\partial I_{tdref,0}} & \frac{\partial \mathrm{Re}\{I_{t\alpha\_cnt,1}\}}{\partial \mathrm{Re}\{I_{tdref,1}\}} & \frac{\partial \mathrm{Re}\{I_{t\alpha\_cnt,1}\}}{\partial \mathrm{Im}\{I_{tdref,1}\}} \\ \frac{\partial \mathrm{Re}\{I_{t\beta\_cnt,1}\}}{\partial I_{tdref,0}} & \frac{\partial \mathrm{Re}\{I_{t\beta\_cnt,0}\}}{\partial \mathrm{Re}\{I_{tdref,1}\}} & \frac{\partial \mathrm{Re}\{I_{t\beta\_cnt,1}\}}{\partial \mathrm{Im}\{I_{tdref,1}\}} \\ \frac{\partial \mathrm{Im}\{I_{t\alpha\_cnt,1}\}}{\partial I_{tdref,0}} & \frac{\partial \mathrm{Im}\{I_{t\alpha\_cnt,1}\}}{\partial \mathrm{Re}\{I_{tdref,1}\}} & \frac{\partial \mathrm{Im}\{I_{t\alpha\_cnt,1}\}}{\partial \mathrm{Im}\{I_{tdref,1}\}} \\ \frac{\partial \mathrm{Im}\{I_{t\beta\_cnt,1}\}}{\partial I_{tdref,0}} & \frac{\partial \mathrm{Im}\{I_{t\beta\_cnt,1}\}}{\partial \mathrm{Re}\{I_{tdref,1}\}} & \frac{\partial \mathrm{Im}\{I_{t\beta\_cnt,1}\}}{\partial \mathrm{Im}\{I_{tdref,1}\}} \\ \frac{\partial V_{dc,0}}{\partial I_{tdref,0}} & \frac{\partial V_{dc,0}}{\partial \mathrm{Re}\{I_{tdref,1}\}} & \frac{\partial V_{dc,0}}{\partial \mathrm{Re}\{I_{tdref,1}\}} \\ \frac{\partial \mathrm{Re}\{V_{dc,1}\}}{\partial I_{tdref,0}} & \frac{\partial \mathrm{Re}\{V_{dc,1}\}}{\partial \mathrm{Re}\{I_{tdref,1}\}} & \frac{\partial \mathrm{Re}\{V_{dc,1}\}}{\partial \mathrm{Im}\{I_{tdref,1}\}} \\ \frac{\partial \mathrm{Im}\{V_{dc,1}\}}{\partial I_{tdref,0}} & \frac{\partial \mathrm{Im}\{V_{dc,1}\}}{\partial \mathrm{Re}\{I_{tdref,1}\}} & \frac{\partial \mathrm{Im}\{V_{dc,1}\}}{\partial \mathrm{Im}\{I_{tdref,1}\}} \end{bmatrix}, \tag{9.18}$$

(9.15) can be decomposed into:

$$\mathbf{J_{11\_vsc}} = \mathbf{J_1} \cdot \mathbf{J_2} \cdot \mathbf{J_3}, \tag{9.19}$$

where

$$\mathbf{J}_1 = \begin{bmatrix} \frac{\partial \mathrm{Re}\{I_{t\alpha\_vsc,1}\}}{\partial t_1} & \frac{\partial \mathrm{Re}\{I_{t\alpha\_vsc,1}\}}{\partial t_2} & \frac{\partial \mathrm{Re}\{I_{t\alpha\_vsc,1}\}}{\partial t_3} & \frac{\partial \mathrm{Re}\{I_{t\alpha\_vsc,1}\}}{\partial t_4} & \cdots & \frac{\partial \mathrm{Re}\{I_{t\alpha\_vsc,1}\}}{\partial t_{n-1}} & \frac{\partial \mathrm{Re}\{I_{t\alpha\_vsc,1}\}}{\partial t_n} \\ \frac{\partial \mathrm{Re}\{I_{t\beta\_vsc,1}\}}{\partial t_1} & \frac{\partial \mathrm{Re}\{I_{t\beta\_vsc,1}\}}{\partial t_2} & \frac{\partial \mathrm{Re}\{I_{t\beta\_vsc,1}\}}{\partial t_3} & \frac{\partial \mathrm{Re}\{I_{t\beta\_vsc,1}\}}{\partial t_4} & \cdots & \frac{\partial \mathrm{Re}\{I_{t\beta\_vsc,1}\}}{\partial t_{n-1}} & \frac{\partial \mathrm{Re}\{I_{t\beta\_vsc,1}\}}{\partial t_n} \\ \frac{\partial \mathrm{Im}\{I_{t\alpha\_vsc,1}\}}{\partial t_1} & \frac{\partial \mathrm{Im}\{I_{t\alpha\_vsc,1}\}}{\partial t_2} & \frac{\partial \mathrm{Im}\{I_{t\alpha\_vsc,1}\}}{\partial t_3} & \frac{\partial \mathrm{Im}\{I_{t\alpha\_vsc,1}\}}{\partial t_4} & \cdots & \frac{\partial \mathrm{Im}\{I_{t\alpha\_vsc,1}\}}{\partial t_{n-1}} & \frac{\partial \mathrm{Im}\{I_{t\alpha\_vsc,1}\}}{\partial t_n} \\ \frac{\partial \mathrm{Im}\{I_{t\beta\_vsc,1}\}}{\partial t_1} & \frac{\partial \mathrm{Im}\{I_{t\beta\_vsc,1}\}}{\partial t_2} & \frac{\partial \mathrm{Im}\{I_{t\beta\_vsc,1}\}}{\partial t_3} & \frac{\partial \mathrm{Im}\{I_{t\beta\_vsc,1}\}}{\partial t_4} & \cdots & \frac{\partial \mathrm{Im}\{I_{t\beta\_vsc,1}\}}{\partial t_{n-1}} & \frac{\partial \mathrm{Im}\{I_{t\beta\_vsc,1}\}}{\partial t_n} \\ \frac{\partial V_{dc,0}}{\partial t_1} & \frac{\partial V_{dc,0}}{\partial t_2} & \frac{\partial V_{dc,0}}{\partial t_3} & \frac{\partial V_{dc,0}}{\partial t_4} & \cdots & \frac{\partial V_{dc,0}}{\partial t_{n-1}} & \frac{\partial V_{dc,0}}{\partial t_n} \\ \frac{\partial \mathrm{Re}\{V_{dc,1}\}}{\partial t_1} & \frac{\partial \mathrm{Re}\{V_{dc,1}\}}{\partial t_2} & \frac{\partial \mathrm{Re}\{V_{dc,1}\}}{\partial t_3} & \frac{\partial \mathrm{Re}\{V_{dc,1}\}}{\partial t_4} & \cdots & \frac{\partial \mathrm{Re}\{V_{dc,1}\}}{\partial t_{n-1}} & \frac{\partial \mathrm{Re}\{V_{dc,1}\}}{\partial t_n} \\ \frac{\partial \mathrm{Im}\{V_{dc,1}\}}{\partial t_1} & \frac{\partial \mathrm{Im}\{V_{dc,1}\}}{\partial t_2} & \frac{\partial \mathrm{Im}\{V_{dc,1}\}}{\partial t_3} & \frac{\partial \mathrm{Im}\{V_{dc,1}\}}{\partial t_4} & \cdots & \frac{\partial \mathrm{Im}\{V_{dc,1}\}}{\partial t_{n-1}} & \frac{\partial \mathrm{Im}\{V_{dc,1}\}}{\partial t_n} \end{bmatrix}, \tag{9.20}$$

$$\mathbf{J}_2 = \begin{bmatrix} \frac{\partial t_1}{\partial \mathrm{Re}\{V_{ma,1}\}} & \frac{\partial t_1}{\partial \mathrm{Im}\{V_{ma,1}\}} & \frac{\partial t_1}{\partial \mathrm{Re}\{V_{mb,1}\}} & \frac{\partial t_1}{\partial \mathrm{Im}\{V_{mb,1}\}} & \frac{\partial t_1}{\partial \mathrm{Re}\{V_{mc,1}\}} & \frac{\partial t_1}{\partial \mathrm{Im}\{V_{mc,1}\}} \\ \frac{\partial t_2}{\partial \mathrm{Re}\{V_{ma,1}\}} & \frac{\partial t_2}{\partial \mathrm{Im}\{V_{ma,1}\}} & \frac{\partial t_2}{\partial \mathrm{Re}\{V_{mb,1}\}} & \frac{\partial t_2}{\partial \mathrm{Im}\{V_{mb,1}\}} & \frac{\partial t_2}{\partial \mathrm{Re}\{V_{mc,1}\}} & \frac{\partial t_2}{\partial \mathrm{Im}\{V_{mc,1}\}} \\ \frac{\partial t_3}{\partial \mathrm{Re}\{V_{ma,1}\}} & \frac{\partial t_3}{\partial \mathrm{Im}\{V_{ma,1}\}} & \frac{\partial t_3}{\partial \mathrm{Re}\{V_{mb,1}\}} & \frac{\partial t_3}{\partial \mathrm{Im}\{V_{mb,1}\}} & \frac{\partial t_3}{\partial \mathrm{Re}\{V_{mc,1}\}} & \frac{\partial t_3}{\partial \mathrm{Im}\{V_{mc,1}\}} \\ \frac{\partial t_4}{\partial \mathrm{Re}\{V_{ma,1}\}} & \frac{\partial t_4}{\partial \mathrm{Im}\{V_{ma,1}\}} & \frac{\partial t_4}{\partial \mathrm{Re}\{V_{mb,1}\}} & \frac{\partial t_4}{\partial \mathrm{Im}\{V_{mb,1}\}} & \frac{\partial t_4}{\partial \mathrm{Re}\{V_{mc,1}\}} & \frac{\partial t_4}{\partial \mathrm{Im}\{V_{mc,1}\}} \\ \vdots & \vdots & \vdots & \vdots & \vdots & \vdots \\ \frac{\partial t_{n-1}}{\partial \mathrm{Re}\{V_{ma,1}\}} & \frac{\partial t_{n-1}}{\partial \mathrm{Im}\{V_{ma,1}\}} & \frac{\partial t_{n-1}}{\partial \mathrm{Re}\{V_{mb,1}\}} & \frac{\partial t_{n-1}}{\partial \mathrm{Im}\{V_{mb,1}\}} & \frac{\partial t_{n-1}}{\partial \mathrm{Re}\{V_{mc,1}\}} & \frac{\partial t_{n-1}}{\partial \mathrm{Im}\{V_{mc,1}\}} \\ \frac{\partial t_n}{\partial \mathrm{Re}\{V_{ma,1}\}} & \frac{\partial t_n}{\partial \mathrm{Im}\{V_{ma,1}\}} & \frac{\partial t_n}{\partial \mathrm{Re}\{V_{mb,1}\}} & \frac{\partial t_n}{\partial \mathrm{Im}\{V_{mb,1}\}} & \frac{\partial t_n}{\partial \mathrm{Re}\{V_{mc,1}\}} & \frac{\partial t_n}{\partial \mathrm{Im}\{V_{mc,1}\}} \end{bmatrix}, \tag{9.21}$$

$$\mathbf{J}_3 = \begin{bmatrix} \frac{\partial \mathrm{Re}\{V_{ma,1}\}}{\partial \mathrm{Re}\{V_{mp,1}\}} & \frac{\partial \mathrm{Re}\{V_{ma,1}\}}{\partial \mathrm{Im}\{V_{mp,1}\}} & \frac{\partial \mathrm{Re}\{V_{ma,1}\}}{\partial \mathrm{Re}\{V_{mn,1}\}} & \frac{\partial \mathrm{Re}\{V_{ma,1}\}}{\partial \mathrm{Im}\{V_{mn,1}\}} \\ \frac{\partial \mathrm{Im}\{V_{ma,1}\}}{\partial \mathrm{Re}\{V_{mp,1}\}} & \frac{\partial \mathrm{Im}\{V_{ma,1}\}}{\partial \mathrm{Im}\{V_{mp,1}\}} & \frac{\partial \mathrm{Im}\{V_{ma,1}\}}{\partial \mathrm{Re}\{V_{mn,1}\}} & \frac{\partial \mathrm{Im}\{V_{ma,1}\}}{\partial \mathrm{Im}\{V_{mn,1}\}} \\ \frac{\partial \mathrm{Re}\{V_{mb,1}\}}{\partial \mathrm{Re}\{V_{mp,1}\}} & \frac{\partial \mathrm{Re}\{V_{mb,1}\}}{\partial \mathrm{Im}\{V_{mp,1}\}} & \frac{\partial \mathrm{Re}\{V_{mb,1}\}}{\partial \mathrm{Re}\{V_{mn,1}\}} & \frac{\partial \mathrm{Re}\{V_{mb,1}\}}{\partial \mathrm{Im}\{V_{mn,1}\}} \\ \frac{\partial \mathrm{Im}\{V_{mb,1}\}}{\partial \mathrm{Re}\{V_{mp,1}\}} & \frac{\partial \mathrm{Im}\{V_{mb,1}\}}{\partial \mathrm{Im}\{V_{mp,1}\}} & \frac{\partial \mathrm{Im}\{V_{mb,1}\}}{\partial \mathrm{Re}\{V_{mn,1}\}} & \frac{\partial \mathrm{Im}\{V_{mb,1}\}}{\partial \mathrm{Im}\{V_{mn,1}\}} \\ \frac{\partial \mathrm{Re}\{V_{mc,1}\}}{\partial \mathrm{Re}\{V_{mp,1}\}} & \frac{\partial \mathrm{Re}\{V_{mc,1}\}}{\partial \mathrm{Im}\{V_{mp,1}\}} & \frac{\partial \mathrm{Re}\{V_{mc,1}\}}{\partial \mathrm{Re}\{V_{mn,1}\}} & \frac{\partial \mathrm{Re}\{V_{mc,1}\}}{\partial \mathrm{Im}\{V_{mn,1}\}} \\ \frac{\partial \mathrm{Im}\{V_{mc,1}\}}{\partial \mathrm{Re}\{V_{mp,1}\}} & \frac{\partial \mathrm{Im}\{V_{mc,1}\}}{\partial \mathrm{Im}\{V_{mp,1}\}} & \frac{\partial \mathrm{Im}\{V_{mc,1}\}}{\partial \mathrm{Re}\{V_{mn,1}\}} & \frac{\partial \mathrm{Im}\{V_{mc,1}\}}{\partial \mathrm{Im}\{V_{mn,1}\}} \end{bmatrix}. \tag{9.22}$$

The derivations of $\mathbf{J}_1$, $\mathbf{J}_2$, and $\mathbf{J}_3$ are described as follows:

Derivations of $\mathbf{J}_1$

$\mathbf{J}_1$ is obtained by differentiating (6.49) with respect to the switching times.

$$\mathbf{J}_1 = \left\{ \left(\frac{d\mathbf{H_p}}{d\mathbf{t_{sw}}}\right)[\mathbb{I} - \mathbf{\Phi}]^{-1}\mathbf{\Gamma} + \mathbf{H_p}[\mathbb{I} - \mathbf{\Phi}]^{-2}\left(\frac{d\mathbf{\Phi}}{d\mathbf{t_{sw}}}\right)\mathbf{\Gamma} + \mathbf{H_p}[\mathbb{I} - \mathbf{\Phi}]^{-1}\left(\frac{d\mathbf{\Gamma}}{d\mathbf{t_{sw}}}\right) + \left(\frac{d\mathbf{Q_p}}{d\mathbf{t_{sw}}}\right) \right\} \mathbf{z}(0). \tag{9.23}$$

Note that $\frac{d\mathbf{H_p}}{d\mathbf{t_{sw}}}$, $\frac{d\mathbf{\Phi}}{d\mathbf{t_{sw}}}$, $\frac{d\mathbf{\Gamma}}{d\mathbf{t_{sw}}}$ and $\frac{d\mathbf{Q_p}}{d\mathbf{t_{sw}}}$ can be easily extracted from $\frac{d\Phi}{d\mathbf{t_{sw}}}$, whose structure is

$$\frac{d\Phi}{d\mathbf{t_{sw}}} = \begin{bmatrix} \frac{d\mathbf{\Phi}}{d\mathbf{t_{sw}}} & \frac{d\mathbf{\Gamma}}{d\mathbf{t_{sw}}} & \mathbf{0} \\ \mathbf{0} & \frac{d\mathbf{\Omega_p}}{d\mathbf{t_{sw}}} & \mathbf{0} \\ \frac{d\mathbf{H_p}}{d\mathbf{t_{sw}}} & \frac{d\mathbf{Q_p}}{d\mathbf{t_{sw}}} & \frac{d\mathbf{M_p}}{d\mathbf{t_{sw}}} \end{bmatrix}. \tag{9.24}$$

The following lists the MATLAB code for $\mathbf{J}_1$:

```
len=length(Axxx_II); %Axxx_II represents state matrix augmented with the
% harmonics
PP_II=[];
%when n=1;
pp_xx_II=eye(len);
for i=2:length(tau)
     pp_xx_II=expm(Axxx_II(:,:,Sdseq_In(i))*(tau(i)-tau(i-1)))*pp_xx_II;
end
pp_xx_II=pp_xx_II*(-Axxx_II(:,:,Sdseq_In(2))+Axxx_II(:,:,Sdseq_In(1)))*...
expm(Axxx_II(:,:,Sdseq_In(1))*tau(1));
PP_II=[pp_xx_II,PP_II];

```

```
counter=0;
%when n>= 2
for n=2:length(tau)-1
    PP2_II=expm(Axxx_II(:,:,Sdseq_In(1))*tau(1));
    phi22_II=eye(len);
    for k=2:n
        PP2_II=expm(Axxx_II(:,:,Sdseq_In(k))*(tau(k)-tau(k-1)))*PP2_II;
    end
    phi_II=(-Axxx_II(:,:,Sdseq_In(n+1))+Axxx_II(:,:,Sdseq_In(n)))*PP2_II;
    for p=n+1:length(tau)
        phi22_II=expm(Axxx_II(:,:,Sdseq_In(p))*(tau(p)-tau(p-1)))*phi22_II;
    end

    PP_II=[PP_II,phi22_II*phi_II];
end
%%%%%%%%%%%%%%%%%%%%%%%%%%%%%%%%%%%%%%%%%%%%%%%%%%%%%%%%%%%%%%%%%%%%%%%%%%

PPP_xxharm=[];

state_nbr=3; %states are i_alpha, i_beta, and v_dc
input_nbr=2; %input states are v_s\alpha and v_s\beta

for n=1:len:length(PP_II)
    A_partial_II=PP_II(:,n:n+len-1);
    Ap_d=A_partial_II(1:state_nbr,1:state_nbr);
    Np_d=A_partial_II(1:state_nbr,state_nbr+1:state_nbr+input_nbr);
    H1p_d=A_partial_II(1+state_nbr+input_nbr:size(A_partial_II),1:state_nbr);
    H2p_d=A_partial_II(1+state_nbr+input_nbr:size(A_partial_II), ...
    state_nbr+1:state_nbr+input_nbr);
    x_harm2_tau=((H1p_d)*inv(II-Ap)*Np+H1p*(II-Ap)^-2*(Ap_d)*Np+H1p*
    inv(II-Ap)*...
    (Np_d)+H2p_d)*z0;
    PPP_xxharm=[PPP_xxharm x_harm2_tau];

end

len_ac=length(Hh); %ac harmonic number
len_dc=length(fa); %dc harmonic number

Ialpha_sen_real=real(PPP_xxharm(1:len_ac,:));
Ialpha_sen_imag=imag(PPP_xxharm(1:len_ac,:));
Ibeta_sen_real=real(PPP_xxharm(len_ac+1:2*len_ac,:));
Ibeta_sen_imag=imag(PPP_xxharm(len_ac+1:2*len_ac,:));
Vdc_sen_real_0=PPP_xxharm(2*len_ac+1,:);
Vdc_sen_real=real(PPP_xxharm(2*len_ac+2:2*len_ac+len_dc,:));
Vdc_sen_imag=imag(PPP_xxharm(2*len_ac+2:2*len_ac+len_dc,:));

J_1=[Ialpha_sen_real;Ibeta_sen_real;Ialpha_sen_imag;Ibeta_sen_imag; ...
Vdc_sen_real_0;Vdc_sen_real;Vdc_sen_imag];
```

Derivations of J_2

$\mathbf{J_2}$ is the sensitivity of switching times to the harmonics in the modulating signals. This can be obtained via the switching time constraints as listed in (9.25)–(9.27). Note that the derivations of (9.25)–(9.27) are described in Chapter 10.

$$\mathbf{F^a_{sw}} = \pm\frac{4}{T_{sw}}\mathbf{t_a} \mp \mathbf{k_a} - v_{ma}(\mathbf{t_a}) = 0 \tag{9.25}$$

$$\mathbf{F^b_{sw}} = \pm\frac{4}{T_{sw}}\mathbf{t_b} \mp \mathbf{k_b} - v_{mb}(\mathbf{t_b}) = 0 \tag{9.26}$$

$$\mathbf{F}_{\mathbf{sw}}^{\mathbf{c}} = \pm\frac{4}{T_{sw}}\mathbf{t_c} \mp \mathbf{k_c} - v_{mc}(\mathbf{t_c}) = 0 \tag{9.27}$$

$$\mathbf{F}_{\mathbf{sw}}^{\mathbf{a}} = \pm\frac{4}{T_{sw}}\mathbf{t_a} \mp \mathbf{k_a} - \sum_{h=1}^{\infty}\sqrt{(\mathrm{Re}\{V_{ma}^h\})^2 + (\mathrm{Im}\{V_{ma}^h\})^2}\cos\left(h\omega\mathbf{t_a} + \tan^{-1}\left(\frac{\mathrm{Im}\{V_{ma}^h\}}{\mathrm{Re}\{V_{ma}^h\}}\right)\right) = 0 \tag{9.28}$$

$$\mathbf{F}_{\mathbf{sw}}^{\mathbf{b}} = \pm\frac{4}{T_{sw}}\mathbf{t_b} \mp \mathbf{k_b} - \sum_{h=1}^{\infty}\sqrt{(\mathrm{Re}\{V_{mb}^h\})^2 + (\mathrm{Im}\{V_{mb}^h\})^2}\cos\left(h\omega\mathbf{t_b} + \tan^{-1}\left(\frac{\mathrm{Im}\{V_{mb}^h\}}{\mathrm{Re}\{V_{mb}^h\}}\right)\right) = 0 \tag{9.29}$$

$$\mathbf{F}_{\mathbf{sw}}^{\mathbf{c}} = \pm\frac{4}{T_{sw}}\mathbf{t_c} \mp \mathbf{k_c} - \sum_{h=1}^{\infty}\sqrt{(\mathrm{Re}\{V_{mc}^h\})^2 + (\mathrm{Im}\{V_{mc}^h\})^2}\cos\left(h\omega\mathbf{t_c} + \tan^{-1}\left(\frac{\mathrm{Im}\{V_{mc}^h\}}{\mathrm{Re}\{V_{mc}^h\}}\right)\right) = 0 \tag{9.30}$$

Since in this case, only the fundamental of the modulating signal is considered, the following analysis (for phase a) applies:

Differentiate (9.28) with respect to $\mathrm{Re}\{V_{ma}^1\}$:

$$\pm\frac{4}{T_{sw}}\frac{\partial\mathbf{t_a}}{\partial\mathrm{Re}\{V_{ma}^1\}} - \left\{\frac{(\mathrm{Re}\{V_{ma}^1\})}{\sqrt{(\mathrm{Re}\{V_{ma}^1\})^2 + (\mathrm{Im}\{V_{ma}^1\})^2}}\cos\left(\omega\mathbf{t_a} + \tan^{-1}\frac{\mathrm{Im}\{V_{ma}^1\}}{\mathrm{Re}\{V_{ma}^1\}}\right)\right.$$
$$-\sqrt{(\mathrm{Re}\{V_{ma}^1\})^2 + (\mathrm{Im}\{V_{ma}^1\})^2}\sin\left(\omega\mathbf{t_a} + \tan^{-1}\frac{\mathrm{Im}\{V_{ma}^1\}}{\mathrm{Re}\{V_{ma}^1\}}\right)\left(\omega\frac{\partial\mathbf{t_a}}{\partial\mathrm{Re}\{V_{ma}^1\}}\right) -$$
$$\left.\left\{1+\left(\frac{\mathrm{Im}\{V_{ma}^1\}}{\mathrm{Re}\{V_{ma}^1\}}\right)^2\right\}^{-1}\frac{\mathrm{Im}\{V_{ma}^1\}}{(\mathrm{Re}\{V_{ma}^1\})^2}\right\} = 0. \tag{9.31}$$

Differentiate (9.28) with respect to $\mathrm{Im}\{V_{ma}^1\}$:

$$\pm\frac{4}{T_{sw}}\frac{\partial\mathbf{t_a}}{\partial\mathrm{Im}\{V_{ma}^1\}} - \left\{\frac{(\mathrm{Im}\{V_{ma}^1\})}{\sqrt{(\mathrm{Re}\{V_{ma}^1\})^2 + (\mathrm{Im}\{V_{ma}^1\})^2}}\cos\left(\omega\mathbf{t_a} + \tan^{-1}\frac{\mathrm{Im}\{V_{ma}^1\}}{\mathrm{Re}\{V_{ma}^1\}}\right)\right.$$
$$-\sqrt{(\mathrm{Re}\{V_{ma}^1\})^2 + (\mathrm{Im}\{V_{ma}^1\})^2}\sin\left(\omega\mathbf{t_a} + \tan^{-1}\frac{\mathrm{Im}\{V_{ma}^1\}}{\mathrm{Re}\{V_{ma}^1\}}\right)\left(\omega\frac{\partial\mathbf{t_a}}{\partial\mathrm{Im}\{V_{ma}^1\}}\right) +$$
$$\left.\left\{1+\left(\frac{\mathrm{Im}\{V_{ma}^1\}}{\mathrm{Re}\{V_{ma}^1\}}\right)^2\right\}^{-1}\frac{1}{\mathrm{Re}\{V_{ma}^1\}}\right\} = 0. \tag{9.32}$$

Differentiate (9.28) with respect to $\mathrm{Re}\{V_{mb}^1\}$:

$$\pm\frac{4}{T_{sw}}\frac{\partial\mathbf{t_a}}{\partial\mathrm{Re}\{V_{ma}^1\}} + \sqrt{(\mathrm{Re}\{V_{ma}^1\})^2 + (\mathrm{Im}\{V_{ma}^1\})^2}\sin\left(\omega\mathbf{t_a} + \tan^{-1}\frac{\mathrm{Im}\{V_{ma}^1\}}{\mathrm{Re}\{V_{ma}^1\}}\right)\omega\frac{\partial\mathbf{t_a}}{\partial\mathrm{Re}\{V_{mb}^1\}} = 0. \tag{9.33}$$

Differentiate (9.28) with respect to $\mathrm{Im}\{V_{mb}^1\}$:

$$\pm\frac{4}{T_{sw}}\frac{\partial\mathbf{t_a}}{\partial\mathrm{Im}\{V_{mb}^1\}} + \sqrt{(\mathrm{Re}\{V_{ma}^1\})^2 + (\mathrm{Im}\{V_{ma}^1\})^2}\sin\left(\omega\mathbf{t_a} + \tan^{-1}\frac{\mathrm{Im}\{V_{ma}^1\}}{\mathrm{Re}\{V_{ma}^1\}}\right)\omega\frac{\partial\mathbf{t_a}}{\partial\mathrm{Im}\{V_{mb}^1\}} = 0. \tag{9.34}$$

Differentiate (9.28) with respect to $\text{Re}\{V_{mc}^1\}$:

$$\pm\frac{4}{T_{sw}}\frac{\partial \mathbf{t_a}}{\partial \text{Re}\{V_{mc}^1\}}+\sqrt{(\text{Re}\{V_{ma}^1\})^2+(\text{Im}\{V_{ma}^1\})^2}\sin\left(\omega\mathbf{t_a}+\tan^{-1}\frac{\text{Im}\{V_{ma}^1\}}{\text{Re}\{V_{ma}^1\}}\right)\omega\frac{\partial \mathbf{t_a}}{\partial \text{Re}\{V_{mc}^1\}}=0. \tag{9.35}$$

Differentiate (9.28) with respect to $\text{Im}\{V_{mc}^1\}$:

$$\pm\frac{4}{T_{sw}}\frac{\partial \mathbf{t_a}}{\partial \text{Im}\{V_{mc}^1\}}+\sqrt{(\text{Re}\{V_{ma}^1\})^2+(\text{Im}\{V_{ma}^1\})^2}\sin\left(\omega\mathbf{t_a}+\tan^{-1}\frac{\text{Im}\{V_{ma}^1\}}{\text{Re}\{V_{ma}^1\}}\right)\omega\frac{\partial \mathbf{t_a}}{\partial \text{Im}\{V_{mc}^1\}}=0. \tag{9.36}$$

Rearranging equation (9.31)–(9.36), one can obtain

$$\frac{\partial \mathbf{t_a}}{\partial \text{Re}\{V_{ma}^1\}}=\left\{\pm\frac{4}{T_{sw}}+\sqrt{(\text{Re}\{V_{ma}^1\})^2+(\text{Im}\{V_{ma}^1\})^2}\sin\left(\omega\mathbf{t_a}+\tan^{-1}\frac{\text{Im}\{V_{ma}^1\}}{\text{Re}\{V_{ma}^1\}}\right)\omega\right\}^{-1}$$
$$\left\{\sqrt{(\text{Re}\{V_{ma}^1\})^2+(\text{Im}\{V_{ma}^1\})^2}\sin\left(\omega\mathbf{t_a}+\tan^{-1}\frac{\text{Im}\{V_{ma}^1\}}{\text{Re}\{V_{ma}^1\}}\right)\left\{1+\left(\frac{\text{Im}\{V_{ma}^1\}}{\text{Re}\{V_{ma}^1\}}\right)^2\right\}^{-1}\cdot\right.$$
$$\left.\frac{\text{Im}\{V_{ma}^1\}}{(\text{Re}\{V_{ma}^1\})^2}+\frac{(\text{Re}\{V_{ma}^1\})}{\sqrt{(\text{Re}\{V_{ma}^1\})^2+(\text{Im}\{V_{ma}^1\})^2}}\cos\left(\omega\mathbf{t_a}+\tan^{-1}\frac{\text{Im}\{V_{ma}^1\}}{\text{Re}\{V_{ma}^1\}}\right)\right\} \tag{9.37}$$

$$\frac{\partial \mathbf{t_a}}{\partial \text{Im}\{V_{ma}^1\}}=\left\{\pm\frac{4}{T_{sw}}-\sqrt{(\text{Re}\{V_{ma}^1\})^2+(\text{Im}\{V_{ma}^1\})^2}\sin\left(\omega\mathbf{t_a}+\tan^{-1}\frac{\text{Im}\{V_{ma}^1\}}{\text{Re}\{V_{ma}^1\}}\right)\omega\right\}^{-1}$$
$$\left\{-\sqrt{(\text{Re}\{V_{ma}^1\})^2+(\text{Im}\{V_{ma}^1\})^2}\sin\left(\omega\mathbf{t_a}+\tan^{-1}\frac{\text{Im}\{V_{ma}^1\}}{\text{Re}\{V_{ma}^1\}}\right)\left\{1+\left(\frac{\text{Im}\{V_{ma}^1\}}{\text{Re}\{V_{ma}^1\}}\right)^2\right\}^{-1}\cdot\right.$$
$$\left.\frac{1}{\text{Re}\{V_{ma}^1\}}+\frac{(\text{Re}\{V_{ma}^1\})}{\sqrt{(\text{Re}\{V_{ma}^1\})^2+(\text{Im}\{V_{ma}^1\})^2}}\cos\left(\omega\mathbf{t_a}+\tan^{-1}\frac{\text{Im}\{V_{ma}^1\}}{\text{Re}\{V_{ma}^1\}}\right)\right\} \tag{9.38}$$

$$\frac{\partial \mathbf{t_a}}{\partial \text{Re}\{V_{mb}^1\}}=0 \tag{9.39}$$

$$\frac{\partial \mathbf{t_a}}{\partial \text{Im}\{V_{mb}^1\}}=0 \tag{9.40}$$

$$\frac{\partial \mathbf{t_a}}{\partial \text{Re}\{V_{mc}^1\}}=0 \tag{9.41}$$

$$\frac{\partial \mathbf{t_a}}{\partial \text{Im}\{V_{mc}^1\}}=0. \tag{9.42}$$

Similar analysis can be applied to phase *b* and *c*.

The following lists the MATLAB code for $\mathbf{J}_2$:

```
%The matrix converting the 120 to the alpha beta and 0 components
B=[1,1,0;-j, j,0;0,0,sqrt(2)];
% The inverse Clarke Transformation
C_inv=[1 0 1/sqrt(2);-1/2 sqrt(3)/2 1/sqrt(2);-1/2 -sqrt(3)/2 1/sqrt(2)];

%The modulating signal are initialized by the positive sequences xp+jyp ...
%and negative sequences xn+jyn (note multiple of harmonics)
%The following is to simply obtain the equivalent abc-frame counterparts,
%which are xa+jya, xb+jyb, and xc+jyc

%By the same token, the three modulation indices for ...
%the three phases (ma,mb,mc) and firing angles(phi_a,phi_b,phi_c) can...
%also be obtained.

for k=1:hlenth
    valp_bet=B*[xp(k)+yp(k)*j;xn(k)+yn(k)*j;0];
    vtabc=1*C_inv*valp_bet;
    ma(k)=abs(vtabc(1));
    phi_a(k)=angle(vtabc(1));
    xa(k)=real(vtabc(1));
    ya(k)=imag(vtabc(1));
    mb(k)=abs(vtabc(2));
    phi_b(k)=angle(vtabc(2));
    xb(k)=real(vtabc(2));
    yb(k)=imag(vtabc(2));
    mc(k)=abs(vtabc(3));
    phi_c(k)=angle(vtabc(3));
    xc(k)=real(vtabc(3));
    yc(k)=imag(vtabc(3));
end

%switching frequency
Tsw=1/f/mf;
Jt=[];
%%%%%%%%%%%%%%%%%%%%%%%%%%%%%%

%The first positive slope.
[Jaa]=myfun_sin_a_p(xa,ya,Tsw,Hh,f,swt_t(1));
Jt=[Jt;Jaa];

for i=2:length(swt_t)
    % positive slope
    if seq_phase_all(i) == 11
         [JJ]=myfun_sin_a_swT_p(xa,ya,Tsw,Hh,f,swt_t(i));
    end
    if seq_phase_all(i) == 21
        [JJ]=myfun_sin_b_swT_p(xb,yb,Tsw,Hh,f,swt_t(i));
    end
    if seq_phase_all(i) == 31
        [JJ]=myfun_sin_c_swT_p(xc,yc,Tsw,Hh,f,swt_t(i));
    end
    % negative slope
    if seq_phase_all(i) == 12
     [JJ]=myfun_sin_a_swT_n(xa,ya,Tsw,Hh,f,swt_t(i));
    end
    if seq_phase_all(i) == 22
     [JJ]=myfun_sin_b_swT_n(xb,yb,Tsw,Hh,f,swt_t(i));
    end
    if seq_phase_all(i) == 32
     [JJ]=myfun_sin_c_swT_n(xc,yc,Tsw,Hh,f,swt_t(i));
    end
Jt=[Jt;JJ];

end
```

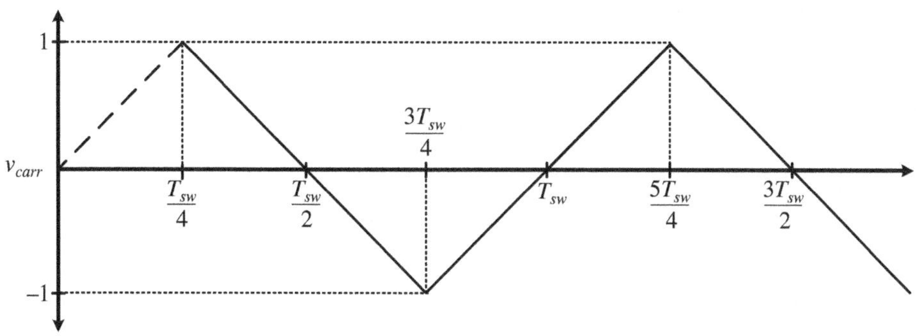

Figure 9.9 The carrier waveform.

Note that the function `seq_phase_all` is used to inform us which phase of the modulating signal intersects with what slope of the carrier waveform (positive or negative). For instance, "`11`" means phase a of the modulating signal intersects with the positive slope of the carrier waveform, whereas "`12`" means phase a of the modulating signal intersects with the negative slope of the carrier waveform. Also note that, we assume the carrier waveform takes the form shown in Fig. 9.9.

The first positive slope as indicated by dashed-line is accounted by the function `myfun_sin_a_p`. The other positive slopes are accounted by `myfun_sin_a_swT_p`, `myfun_sin_b_swT_p`, and `myfun_sin _c_swT_p`. The negative slopes are accounted by `myfun_sin_a_swT_n`, `myfun_sin_b_swT_n`, and `myfun_sin_c_swT_n`. The MATLAB codes for all these functions are listed below.

```
function [Jaa]=myfun_sin_a_p(xa,ya,Tsw,Hh,f,tt)
%F=(4/Tsw)*(x-(t2))-1-(ma)*cos(2*pi*f*x+phia)-(ma3)*cos(2*pi*3*f*x+phia3);

w=2*pi*f;
temp=4/Tsw;

for k=1:length(Hh)
    temp=sqrt(xa(k)*xa(k)+ya(k)*ya(k))*Hh(k)*sin(w*Hh(k)*tt+...
    atan2(ya(k),xa(k)))*w+temp;
end


Jaa=[];
for k=1:length(Hh)
    d_t_ya_temp=(xa(k)*xa(k)+ya(k)*ya(k))^(-0.5)*ya(k)*cos(w*Hh(k)*tt+...
    atan2(ya(k),xa(k)))-sqrt(xa(k)*xa(k)+ya(k)*ya(k))*sin(w*Hh(k)*tt+...
    atan2(ya(k),xa(k)))*(1/(1+(ya(k)/xa(k))^2))*(1/xa(k));

    d_t_xa_temp=(xa(k)*xa(k)+ya(k)*ya(k))^(-0.5)*xa(k)*cos(w*Hh(k)*tt+...
    atan2(ya(k),xa(k)))+sqrt(xa(k)*xa(k)+ya(k)*ya(k))*sin(w*Hh(k)*tt+...
    atan2(ya(k),xa(k)))*(1/(1+(ya(k)/xa(k))^2))*(ya(k)/(xa(k)*xa(k)));

    d_t_xa=inv(temp)*d_t_xa_temp;
    d_t_ya=inv(temp)*d_t_ya_temp;

    Jaa=[Jaa,d_t_xa d_t_ya 0 0 0 0 ];
end
```

```
function [Ja]=myfun_sin_a_swT_p(xa,ya,Tsw,Hh,f,tt)

w=2*pi*f;
temp=4/Tsw;

for k=1:length(Hh)
    temp=sqrt(xa(k)*xa(k)+ya(k)*ya(k))*Hh(k)*sin(w*Hh(k)*tt+...
    atan2(ya(k),xa(k)))*w+temp;
end


Ja=[];
for k=1:length(Hh)
    d_t_ya_temp=(xa(k)*xa(k)+ya(k)*ya(k))^(-0.5)*ya(k)*cos(w*Hh(k)*tt+...
    atan2(ya(k),xa(k)))-sqrt(xa(k)*xa(k)+ya(k)*ya(k))*sin(w*Hh(k)*tt+...
    atan2(ya(k),xa(k)))*(1/(1+(ya(k)/xa(k))^2))*(1/xa(k));

    d_t_xa_temp=(xa(k)*xa(k)+ya(k)*ya(k))^(-0.5)*xa(k)*cos(w*Hh(k)*tt+...
    atan2(ya(k),xa(k)))+sqrt(xa(k)*xa(k)+ya(k)*ya(k))*sin(w*Hh(k)*tt+...
    atan2(ya(k),xa(k)))*(1/(1+(ya(k)/xa(k))^2))*(ya(k)/(xa(k)*xa(k)));

    d_t_xa=inv(temp)*d_t_xa_temp;
    d_t_ya=inv(temp)*d_t_ya_temp;

    Ja=[Ja,d_t_xa d_t_ya 0 0 0 0];
end
```

```
function [Jb]=myfun_sin_b_swT_p(xb,yb,Tsw,Hh,f,tt)

w=2*pi*f;

temp=4/Tsw;

for k=1:length(Hh)
    temp=sqrt(xb(k)*xb(k)+yb(k)*yb(k))*Hh(k)*sin(w*Hh(k)*tt+...
    atan2(yb(k),xb(k)))*w+temp;
end


Jb=[];
for k=1:length(Hh)
    d_t_yb_temp=(xb(k)*xb(k)+yb(k)*yb(k))^(-0.5)*yb(k)*cos(w*Hh(k)*tt+...
    atan2(yb(k),xb(k)))-sqrt(xb(k)*xb(k)+yb(k)*yb(k))*sin(w*Hh(k)*tt+...
    atan2(yb(k),xb(k)))*(1/(1+(yb(k)/xb(k))^2))*(1/xb(k));

    d_t_xb_temp=(xb(k)*xb(k)+yb(k)*yb(k))^(-0.5)*xb(k)*cos(w*Hh(k)*tt+...
    atan2(yb(k),xb(k)))+sqrt(xb(k)*xb(k)+yb(k)*yb(k))*sin(w*Hh(k)*tt+...
    atan2(yb(k),xb(k)))*(1/(1+(yb(k)/xb(k))^2))*(yb(k)/(xb(k)*xb(k)));

    d_t_xb=inv(temp)*d_t_xb_temp;
    d_t_yb=inv(temp)*d_t_yb_temp;

    Jb=[Jb,0 0 d_t_xb d_t_yb 0 0];
end
```

```
function [Jc]=myfun_sin_c_swT_p(xc,yc,Tsw,Hh,f,tt)

w=2*pi*f;

temp=4/Tsw;


for k=1:length(Hh)
    temp=sqrt(xc(k)*xc(k)+yc(k)*yc(k))*Hh(k)*sin(w*Hh(k)*tt+...
    atan2(yc(k),xc(k)))*w+temp;
end

Jc=[];
for k=1:length(Hh)
    d_t_yc_temp=(xc(k)*xc(k)+yc(k)*yc(k))^(-0.5)*yc(k)*cos(w*Hh(k)*tt+...
    atan2(yc(k),xc(k)))-sqrt(xc(k)*xc(k)+yc(k)*yc(k))*sin(w*Hh(k)*tt+...
    atan2(yc(k),xc(k)))*(1/(1+(yc(k)/xc(k))^2))*(1/xc(k));

    d_t_xc_temp=(xc(k)*xc(k)+yc(k)*yc(k))^(-0.5)*xc(k)*cos(w*Hh(k)*tt+...
    atan2(yc(k),xc(k)))+sqrt(xc(k)*xc(k)+yc(k)*yc(k))*sin(w*Hh(k)*tt+...
    atan2(yc(k),xc(k)))*(1/(1+(yc(k)/xc(k))^2))*(yc(k)/(xc(k)*xc(k)));

    d_t_xc=inv(temp)*d_t_xc_temp;
    d_t_yc=inv(temp)*d_t_yc_temp;

    Jc=[Jc,0 0 0 0 d_t_xc d_t_yc];
end
```

```
function [Ja]=myfun_sin_a_swT_n(xa,ya,Tsw,Hh,f,tt)

w=2*pi*f;
temp=-4/Tsw;

for k=1:length(Hh)
    temp=sqrt(xa(k)*xa(k)+ya(k)*ya(k))*Hh(k)*sin(w*Hh(k)*tt+...
    atan2(ya(k),xa(k)))*w+temp;
end


Ja=[];
for k=1:length(Hh)
    d_t_ya_temp=(xa(k)*xa(k)+ya(k)*ya(k))^(-0.5)*ya(k)*cos(w*Hh(k)*tt+...
    atan2(ya(k),xa(k)))-sqrt(xa(k)*xa(k)+ya(k)*ya(k))*sin(w*Hh(k)*tt+...
    atan2(ya(k),xa(k)))*(1/(1+(ya(k)/xa(k))^2))*(1/xa(k));

    d_t_xa_temp=(xa(k)*xa(k)+ya(k)*ya(k))^(-0.5)*xa(k)*cos(w*Hh(k)*tt+...
    atan2(ya(k),xa(k)))+sqrt(xa(k)*xa(k)+ya(k)*ya(k))*sin(w*Hh(k)*tt+...
    atan2(ya(k),xa(k)))*(1/(1+(ya(k)/xa(k))^2))*(ya(k)/(xa(k)*xa(k)));

    d_t_xa=inv(temp)*d_t_xa_temp;
    d_t_ya=inv(temp)*d_t_ya_temp;

    Ja=[Ja,d_t_xa d_t_ya 0 0 0 0];
end
```

```
function [Jb]=myfun_sin_b_swT_n(xb,yb,Tsw,Hh,f,tt)
w=2*pi*f;

temp=-4/Tsw;

for k=1:length(Hh)
    temp=sqrt(xb(k)*xb(k)+yb(k)*yb(k))*Hh(k)*sin(w*Hh(k)*tt+...
    atan2(yb(k),xb(k)))*w+temp;
end


Jb=[];
for k=1:length(Hh)
    d_t_yb_temp=(xb(k)*xb(k)+yb(k)*yb(k))^(-0.5)*yb(k)*cos(w*Hh(k)*tt+...
    atan2(yb(k),xb(k)))-sqrt(xb(k)*xb(k)+yb(k)*yb(k))*sin(w*Hh(k)*tt+...
    atan2(yb(k),xb(k)))*(1/(1+(yb(k)/xb(k))^2))*(1/xb(k));

    d_t_xb_temp=(xb(k)*xb(k)+yb(k)*yb(k))^(-0.5)*xb(k)*cos(w*Hh(k)*tt+...
    atan2(yb(k),xb(k)))+sqrt(xb(k)*xb(k)+yb(k)*yb(k))*sin(w*Hh(k)*tt+...
    atan2(yb(k),xb(k)))*(1/(1+(yb(k)/xb(k))^2))*(yb(k)/(xb(k)*xb(k)));

    d_t_xb=inv(temp)*d_t_xb_temp;
    d_t_yb=inv(temp)*d_t_yb_temp;

    Jb=[Jb,0 0 d_t_xb d_t_yb 0 0];
end
```

```
function [Jc]=myfun_sin_c_swT_n(xc,yc,Tsw,Hh,f,tt)
w=2*pi*f;

temp=-4/Tsw;

for k=1:length(Hh)
    temp=sqrt(xc(k)*xc(k)+yc(k)*yc(k))*Hh(k)*sin(w*Hh(k)*tt+...
    atan2(yc(k),xc(k)))*w+temp;
end

Jc=[];
for k=1:length(Hh)
    d_t_yc_temp=(xc(k)*xc(k)+yc(k)*yc(k))^(-0.5)*yc(k)*cos(w*Hh(k)*tt+...
    atan2(yc(k),xc(k)))-sqrt(xc(k)*xc(k)+yc(k)*yc(k))*sin(w*Hh(k)*tt+...
    atan2(yc(k),xc(k)))*(1/(1+(yc(k)/xc(k))^2))*(1/xc(k));

    d_t_xc_temp=(xc(k)*xc(k)+yc(k)*yc(k))^(-0.5)*xc(k)*cos(w*Hh(k)*tt+...
    atan2(yc(k),xc(k)))+sqrt(xc(k)*xc(k)+yc(k)*yc(k))*sin(w*Hh(k)*tt+...
    atan2(yc(k),xc(k)))*(1/(1+(yc(k)/xc(k))^2))*(yc(k)/(xc(k)*xc(k)));

    d_t_xc=inv(temp)*d_t_xc_temp;
    d_t_yc=inv(temp)*d_t_yc_temp;

    Jc=[Jc,0 0 0 0 d_t_xc d_t_yc];
end
```

Derivations of $\mathbf{J}_3$

Since

$$\begin{bmatrix} V_{ma} \\ V_{mb} \\ V_{mc} \end{bmatrix} = \begin{bmatrix} 1 & 1 & 1 \\ \frac{-1}{2} - j\frac{\sqrt{3}}{2} & \frac{-1}{2} + j\frac{\sqrt{3}}{2} & 1 \\ \frac{-1}{2} + j\frac{\sqrt{3}}{2} & \frac{-1}{2} - j\frac{\sqrt{3}}{2} & 1 \end{bmatrix} \begin{bmatrix} V_{mp} \\ V_{mn} \\ V_{m0} \end{bmatrix}, \tag{9.43}$$

$$\mathbf{J}_3 = \begin{bmatrix} 1 & 0 & 1 & 0 \\ 0 & 1 & 0 & 1 \\ \frac{-1}{2} & \frac{\sqrt{3}}{2} & \frac{-1}{2} & \frac{-\sqrt{3}}{2} \\ \frac{-\sqrt{3}}{2} & \frac{-1}{2} & \frac{\sqrt{3}}{2} & \frac{-1}{2} \\ \frac{-1}{2} & \frac{-\sqrt{3}}{2} & \frac{-1}{2} & \frac{\sqrt{3}}{2} \\ \frac{\sqrt{3}}{2} & \frac{-1}{2} & \frac{-\sqrt{3}}{2} & \frac{-1}{2} \end{bmatrix}. \tag{9.44}$$

The MATLAB code listing for $\mathbf{J}_3$ is as follows:

```
AE=[1 0 1 0;0 1 0 1;-1/2 sqrt(3)/2 -1/2 -sqrt(3)/2; ...
-sqrt(3)/2 -1/2 sqrt(3)/2 -1/2; ...
-1/2 -sqrt(3)/2 -1/2 sqrt(3)/2;sqrt(3)/2 -1/2 -sqrt(3)/2 -1/2];

 J3=AE;

 for k=1:length(Hh)-1
     J3=blkdiag(tempAE,AE);
 end
```

Derivations of $\mathbf{J}_{11\_cnt}$

$\mathbf{J}_{11\_cnt}$ is first partitioned into

$$\mathbf{J}_{11\_cnt} = \begin{bmatrix} \mathbf{J}_a \\ \hline \mathbf{J}_b \\ \hline \mathbf{J}_c \end{bmatrix}, \tag{9.45}$$

where

$$\mathbf{J_a} = \begin{bmatrix} \frac{\partial \mathrm{Re}\{I^1_{t\alpha\_cnt}\}}{\partial \mathrm{Re}\{V^1_{mp}\}} & \frac{\partial \mathrm{Re}\{I^1_{t\alpha\_cnt}\}}{\partial \mathrm{Im}\{V^1_{mp}\}} & \frac{\partial \mathrm{Re}\{I^1_{t\alpha\_cnt}\}}{\partial \mathrm{Re}\{V^1_{mn}\}} & \frac{\partial \mathrm{Re}\{I^1_{t\alpha\_cnt}\}}{\partial \mathrm{Im}\{V^1_{mn}\}} \\ \frac{\partial \mathrm{Re}\{I^1_{t\beta\_cnt}\}}{\partial \mathrm{Re}\{V^1_{mp}\}} & \frac{\partial \mathrm{Re}\{I^1_{t\beta\_cnt}\}}{\partial \mathrm{Im}\{V^1_{mp}\}} & \frac{\partial \mathrm{Re}\{I^1_{t\beta\_cnt}\}}{\partial \mathrm{Re}\{V^1_{mn}\}} & \frac{\partial \mathrm{Re}\{I^1_{t\beta\_cnt}\}}{\partial \mathrm{Im}\{V^1_{mn}\}} \\ \frac{\partial \mathrm{Im}\{I^1_{t\alpha\_cnt}\}}{\partial \mathrm{Re}\{V^1_{mp}\}} & \frac{\partial \mathrm{Im}\{I^1_{t\alpha\_cnt}\}}{\partial \mathrm{Im}\{V^1_{mp}\}} & \frac{\partial \mathrm{Im}\{I^1_{t\alpha\_cnt}\}}{\partial \mathrm{Re}\{V^1_{mn}\}} & \frac{\partial \mathrm{Im}\{I^1_{t\alpha\_cnt}\}}{\partial \mathrm{Im}\{V^1_{mn}\}} \\ \frac{\partial \mathrm{Im}\{I^1_{t\beta\_cnt}\}}{\partial \mathrm{Re}\{V^1_{mp}\}} & \frac{\partial \mathrm{Im}\{I^1_{t\beta\_cnt}\}}{\partial \mathrm{Im}\{V^1_{mp}\}} & \frac{\partial \mathrm{Im}\{I^1_{t\beta\_cnt}\}}{\partial \mathrm{Re}\{V^1_{mn}\}} & \frac{\partial \mathrm{Im}\{I^1_{t\beta\_cnt}\}}{\partial \mathrm{Im}\{V^1_{mn}\}} \end{bmatrix}, \tag{9.46}$$

$$\mathbf{J_b} = \begin{bmatrix} \frac{\partial V^0_{dc}}{\partial \mathrm{Re}\{V^1_{mp}\}} & \frac{\partial V^0_{dc}}{\partial \mathrm{Im}\{V^1_{mp}\}} & \frac{\partial V^0_{dc}}{\partial \mathrm{Re}\{V^1_{mn}\}} & \frac{\partial V^0_{dc}}{\partial \mathrm{Im}\{V^1_{mn}\}} \end{bmatrix} = \mathbf{0}, \tag{9.47}$$

$$\mathbf{J_c} = \begin{bmatrix} \frac{\partial \mathrm{Re}\{V^1_{dc}\}}{\partial \mathrm{Re}\{V^1_{mp}\}} & \frac{\partial \mathrm{Re}\{V^1_{dc}\}}{\partial \mathrm{Im}\{V^1_{mp}\}} & \frac{\partial \mathrm{Re}\{V^1_{dc}\}}{\partial \mathrm{Re}\{V^1_{mn}\}} & \frac{\partial \mathrm{Re}\{V^1_{dc}\}}{\partial \mathrm{Im}\{V^1_{mn}\}} \\ \frac{\partial \mathrm{Im}\{V^1_{dc}\}}{\partial \mathrm{Re}\{V^1_{mp}\}} & \frac{\partial \mathrm{Im}\{V^1_{dc}\}}{\partial \mathrm{Im}\{V^1_{mp}\}} & \frac{\partial \mathrm{Im}\{V^1_{dc}\}}{\partial \mathrm{Re}\{V^1_{mn}\}} & \frac{\partial \mathrm{Im}\{V^1_{dc}\}}{\partial \mathrm{Im}\{V^1_{mn}\}} \end{bmatrix} = \mathbf{0}. \tag{9.48}$$

$\mathbf{J_a}$ can be further decomposed into

$$\mathbf{J_a} = \mathbf{J_{a1}} \cdot \mathbf{J_{a2}}, \tag{9.49}$$

where

$$\mathbf{J_{a1}} = \begin{bmatrix} \frac{\partial \mathrm{Re}\{I^1_{t\alpha\_cnt}\}}{\partial \mathrm{Re}\{V^1_{m\alpha}\}} & \frac{\partial \mathrm{Re}\{I^1_{t\alpha\_cnt}\}}{\partial \mathrm{Im}\{V^1_{m\alpha}\}} & \frac{\partial \mathrm{Re}\{I^1_{t\alpha\_cnt}\}}{\partial \mathrm{Re}\{V^1_{m\beta}\}} & \frac{\partial \mathrm{Re}\{I^1_{t\alpha\_cnt}\}}{\partial \mathrm{Im}\{V^1_{m\beta}\}} \\ \frac{\partial \mathrm{Re}\{I^1_{t\beta\_cnt}\}}{\partial \mathrm{Re}\{V^1_{m\alpha}\}} & \frac{\partial \mathrm{Re}\{I^1_{t\beta\_cnt}\}}{\partial \mathrm{Im}\{V^1_{m\alpha}\}} & \frac{\partial \mathrm{Re}\{I^1_{t\beta\_cnt}\}}{\partial \mathrm{Re}\{V^1_{m\beta}\}} & \frac{\partial \mathrm{Re}\{I^1_{t\beta\_cnt}\}}{\partial \mathrm{Im}\{V^1_{m\beta}\}} \\ \frac{\partial \mathrm{Im}\{I^1_{t\alpha\_cnt}\}}{\partial \mathrm{Re}\{V^1_{m\alpha}\}} & \frac{\partial \mathrm{Im}\{I^1_{t\alpha\_cnt}\}}{\partial \mathrm{Im}\{V^1_{m\alpha}\}} & \frac{\partial \mathrm{Im}\{I^1_{t\alpha\_cnt}\}}{\partial \mathrm{Re}\{V^1_{m\beta}\}} & \frac{\partial \mathrm{Im}\{I^1_{t\alpha\_cnt}\}}{\partial \mathrm{Im}\{V^1_{m\beta}\}} \\ \frac{\partial \mathrm{Im}\{I^1_{t\beta\_cnt}\}}{\partial \mathrm{Re}\{V^1_{m\alpha}\}} & \frac{\partial \mathrm{Im}\{I^1_{t\beta\_cnt}\}}{\partial \mathrm{Im}\{V^1_{m\alpha}\}} & \frac{\partial \mathrm{Im}\{I^1_{t\beta\_cnt}\}}{\partial \mathrm{Re}\{V^1_{m\beta}\}} & \frac{\partial \mathrm{Im}\{I^1_{t\beta\_cnt}\}}{\partial \mathrm{Im}\{V^1_{m\beta}\}} \end{bmatrix}, \tag{9.50}$$

$$\mathbf{J_{a2}} = \begin{bmatrix} \frac{\partial \mathrm{Re}\{V^1_{m\alpha}\}}{\partial \mathrm{Re}\{V^1_{mp}\}} & \frac{\partial \mathrm{Re}\{V^1_{m\alpha}\}}{\partial \mathrm{Im}\{V^1_{mp}\}} & \frac{\partial \mathrm{Re}\{V^1_{m\alpha}\}}{\partial \mathrm{Re}\{V^1_{mn}\}} & \frac{\partial \mathrm{Re}\{V^1_{m\alpha}\}}{\partial \mathrm{Im}\{V^1_{mn}\}} \\ \frac{\partial \mathrm{Re}\{V^1_{m\beta}\}}{\partial \mathrm{Re}\{V^1_{mp}\}} & \frac{\partial \mathrm{Re}\{V^1_{m\beta}\}}{\partial \mathrm{Im}\{V^1_{mp}\}} & \frac{\partial \mathrm{Re}\{V^1_{m\beta}\}}{\partial \mathrm{Re}\{V^1_{mn}\}} & \frac{\partial \mathrm{Re}\{V^1_{m\beta}\}}{\partial \mathrm{Im}\{V^1_{mn}\}} \\ \frac{\partial \mathrm{Im}\{V^1_{m\alpha}\}}{\partial \mathrm{Re}\{V^1_{mp}\}} & \frac{\partial \mathrm{Im}\{V^1_{m\alpha}\}}{\partial \mathrm{Im}\{V^1_{mp}\}} & \frac{\partial \mathrm{Im}\{V^1_{m\alpha}\}}{\partial \mathrm{Re}\{V^1_{mn}\}} & \frac{\partial \mathrm{Im}\{V^1_{m\alpha}\}}{\partial \mathrm{Im}\{V^1_{mn}\}} \\ \frac{\partial \mathrm{Im}\{V^1_{m\beta}\}}{\partial \mathrm{Re}\{V^1_{mp}\}} & \frac{\partial \mathrm{Im}\{V^1_{m\beta}\}}{\partial \mathrm{Im}\{V^1_{mp}\}} & \frac{\partial \mathrm{Im}\{V^1_{m\beta}\}}{\partial \mathrm{Re}\{V^1_{mn}\}} & \frac{\partial \mathrm{Im}\{V^1_{m\beta}\}}{\partial \mathrm{Im}\{V^1_{mn}\}} \end{bmatrix}. \tag{9.51}$$

$\mathbf{J_{a2}}$ can be easily found by referring to (9.52), and its expression is given in (9.53).

$$\begin{bmatrix} X_\alpha \\ X_\beta \\ X_0 \end{bmatrix} = \begin{bmatrix} 1 & 1 & 0 \\ -j & j & 0 \\ 0 & 0 & \sqrt{2} \end{bmatrix} \begin{bmatrix} X_p \\ X_n \\ X_0 \end{bmatrix}, \tag{9.52}$$

$$\mathbf{J_{a2}} = \begin{bmatrix} 1 & 0 & 1 & 0 \\ 0 & 1 & 0 & 1 \\ 0 & 1 & 0 & -1 \\ -1 & 0 & 1 & 0 \end{bmatrix}. \tag{9.53}$$

To find $\mathbf{J_{a1}}$, the following observations can be noted by studying (9.8):

1. $\mathbf{C_{con}Y_{mdq}C_{con}^{-1}}$ is the only term remaining after (9.8) is differentiated with respect to $\mathbf{V_{m\alpha}}$ and $\mathbf{V_{m\beta}}$.
2. $\mathbf{C_{con}Y_{mdq}C_{con}^{-1}}$ has the structure of

$$\begin{bmatrix} x + jy & a + jb \\ -a - jb & x + jy \end{bmatrix}, \tag{9.54}$$

where x, y, a, and b are real numbers.
3. Hence,

$$\mathbf{J_{a1}} = \begin{bmatrix} \frac{\partial \mathrm{Re}\{I^1_{t\alpha\_cnt}\}}{\partial \mathrm{Re}\{V^1_{m\alpha}\}} & \frac{\partial \mathrm{Re}\{I^1_{t\alpha\_cnt}\}}{\partial \mathrm{Im}\{V^1_{m\alpha}\}} & \frac{\partial \mathrm{Re}\{I^1_{t\alpha\_cnt}\}}{\partial \mathrm{Re}\{V^1_{m\beta}\}} & \frac{\partial \mathrm{Re}\{I^1_{t\alpha\_cnt}\}}{\partial \mathrm{Im}\{V^1_{m\beta}\}} \\ \frac{\partial \mathrm{Re}\{I^1_{t\beta\_cnt}\}}{\partial \mathrm{Re}\{V^1_{m\alpha}\}} & \frac{\partial \mathrm{Re}\{I^1_{t\beta\_cnt}\}}{\partial \mathrm{Im}\{V^1_{m\alpha}\}} & \frac{\partial \mathrm{Re}\{I^1_{t\beta\_cnt}\}}{\partial \mathrm{Re}\{V^1_{m\beta}\}} & \frac{\partial \mathrm{Re}\{I^1_{t\beta\_cnt}\}}{\partial \mathrm{Im}\{V^1_{m\beta}\}} \\ \frac{\partial \mathrm{Im}\{I^1_{t\alpha\_cnt}\}}{\partial \mathrm{Re}\{V^1_{m\alpha}\}} & \frac{\partial \mathrm{Im}\{I^1_{t\alpha\_cnt}\}}{\partial \mathrm{Im}\{V^1_{m\alpha}\}} & \frac{\partial \mathrm{Im}\{I^1_{t\alpha\_cnt}\}}{\partial \mathrm{Re}\{V^1_{m\beta}\}} & \frac{\partial \mathrm{Im}\{I^1_{t\alpha\_cnt}\}}{\partial \mathrm{Im}\{V^1_{m\beta}\}} \\ \frac{\partial \mathrm{Im}\{I^1_{t\beta\_cnt}\}}{\partial \mathrm{Re}\{V^1_{m\alpha}\}} & \frac{\partial \mathrm{Im}\{I^1_{t\beta\_cnt}\}}{\partial \mathrm{Im}\{V^1_{m\alpha}\}} & \frac{\partial \mathrm{Im}\{I^1_{t\beta\_cnt}\}}{\partial \mathrm{Re}\{V^1_{m\beta}\}} & \frac{\partial \mathrm{Im}\{I^1_{t\beta\_cnt}\}}{\partial \mathrm{Im}\{V^1_{m\beta}\}} \end{bmatrix} = \begin{bmatrix} x & -y & a & -b \\ -a & b & x & -y \\ y & x & b & a \\ -b & -a & y & x \end{bmatrix}. \tag{9.55}$$

Derivations of $\mathbf{J_{12\_cnt}}$

Similar to $\mathbf{J_{11\_cnt}}$, $\mathbf{J_{12\_cnt}}$ is partitioned into

$$\mathbf{J_{12\_cnt}} = \begin{bmatrix} \mathbf{J_d} \\ \hline \mathbf{J_e} \\ \hline \mathbf{J_f} \end{bmatrix}, \tag{9.56}$$

where

$$\mathbf{J_d} = \begin{bmatrix} \frac{\partial \mathrm{Re}\{I^1_{t\alpha\_cnt}\}}{\partial I^0_{tdref}} & \frac{\partial \mathrm{Re}\{I^1_{t\alpha\_cnt}\}}{\partial \mathrm{Re}\{I^1_{tdref}\}} & \frac{\partial \mathrm{Re}\{I^1_{t\alpha\_cnt}\}}{\partial \mathrm{Im}\{I^1_{tdref}\}} \\ \frac{\partial \mathrm{Re}\{I^1_{t\beta\_cnt}\}}{\partial I^0_{tdref}} & \frac{\partial \mathrm{Re}\{I^1_{t\beta\_cnt}\}}{\partial \mathrm{Re}\{I^1_{tdref}\}} & \frac{\partial \mathrm{Re}\{I^1_{t\beta\_cnt}\}}{\partial \mathrm{Im}\{I^1_{tdref}\}} \\ \frac{\partial \mathrm{Im}\{I^1_{t\alpha\_cnt}\}}{\partial I^0_{tdref}} & \frac{\partial \mathrm{Im}\{I^1_{t\alpha\_cnt}\}}{\partial \mathrm{Re}\{I^1_{tdref}\}} & \frac{\partial \mathrm{Im}\{I^1_{t\alpha\_cnt}\}}{\partial \mathrm{Im}\{I^1_{tdref}\}} \\ \frac{\partial \mathrm{Im}\{I^1_{t\beta\_cnt}\}}{\partial I^0_{tdref}} & \frac{\partial \mathrm{Im}\{I^1_{t\beta\_cnt}\}}{\partial \mathrm{Re}\{I^1_{tdref}\}} & \frac{\partial \mathrm{Im}\{I^1_{t\beta\_cnt}\}}{\partial \mathrm{Im}\{I^1_{tdref}\}} \end{bmatrix}, \tag{9.57}$$

$$\mathbf{J_e} = \begin{bmatrix} \dfrac{\partial V_{dc}^0}{\partial I_{tdref}^0} & \dfrac{\partial V_{dc}^0}{\partial \mathrm{Re}\{I_{tdref}^1\}} & \dfrac{\partial V_{dc}^0}{\partial \mathrm{Re}\{I_{tdref}^1\}} \end{bmatrix} = \mathbf{0}, \tag{9.58}$$

$$\mathbf{J_f} = \begin{bmatrix} \dfrac{\partial \mathrm{Re}\{V_{dc}^1\}}{\partial I_{tdref}^0} & \dfrac{\partial \mathrm{Re}\{V_{dc}^1\}}{\partial \mathrm{Re}\{I_{tdref}^1\}} & \dfrac{\partial \mathrm{Re}\{V_{dc}^1\}}{\partial \mathrm{Im}\{I_{tdref}^1\}} \\ \dfrac{\partial \mathrm{Im}\{V_{dc}^1\}}{\partial I_{tdref}^0} & \dfrac{\partial \mathrm{Im}\{V_{dc}^1\}}{\partial \mathrm{Re}\{I_{tdref}^1\}} & \dfrac{\partial \mathrm{Im}\{V_{dc}^1\}}{\partial \mathrm{Im}\{I_{tdref}^1\}} \end{bmatrix}. \tag{9.59}$$

$\mathbf{J_d}$ has the same structure as (9.55) because $\mathbf{CK_{dqref}}$ is the only term remaining after (9.8) is differentiated with respect to $\mathbf{I_{tdref}}$, and $\mathbf{CK_{dqref}}$ has the same matrix structure as (9.54).

$\mathbf{J_f}$ is obtained by differentiating (9.10) with respect to $\mathbf{I_{tdref}}$. Hence,

$$\mathbf{J_f} = \begin{bmatrix} 0 & \mathrm{Re}\left\{\dfrac{1}{C_v(j\omega)}\right\} & -\mathrm{Im}\left\{\dfrac{1}{C_v(j\omega)}\right\} \\ 0 & \mathrm{Im}\left\{\dfrac{1}{C_v(j\omega)}\right\} & \mathrm{Re}\left\{\dfrac{1}{C_v(j\omega)}\right\} \end{bmatrix}. \tag{9.60}$$

Note that since the Jacobian matrix for the controller's part is straightforward, the MATLAB code is omitted here. The reader is encouraged to write the corresponding code and combine it with the Jacobian matrix for the VSC part to formulate the overall Jacobian matrix.

9.3.3 Number of Modulating Voltage Harmonics to be Included

While the converter model is not inherently bandwidth limited, only a finite number of harmonics are considered for computation of the controller response. The number of modulating harmonics included is chosen by considering: (i) the diminishing amplitude of PWM harmonics with frequency; (ii) the low-pass nature of the power circuit itself; (iii) the low-pass characteristic of the PI controllers; and (iv) the filters' frequency responses. The following describes the procedure for determining the number of modulating voltage harmonics to be included.

Consider the block diagram of Fig. 9.10. Here, we treat the modulating voltage harmonics as disturbances, $D(s)$. Based on the closed-loop analysis, one can obtain an expression linking the reference, $R(s)$, and disturbance, $D(s)$, to the modulating signals, $M(s)$:

$$\begin{aligned} M(s) &= T_1(s)R(s) + T_2(s)D(s) \\ &= \frac{C(s)}{1 + G(s)P(s)F(s)C(s)}R(s) + \frac{-G(s)P(s)F(s)C(s)}{1 + G(s)P(s)F(s)C(s)}D(s) \end{aligned} \tag{9.61}$$

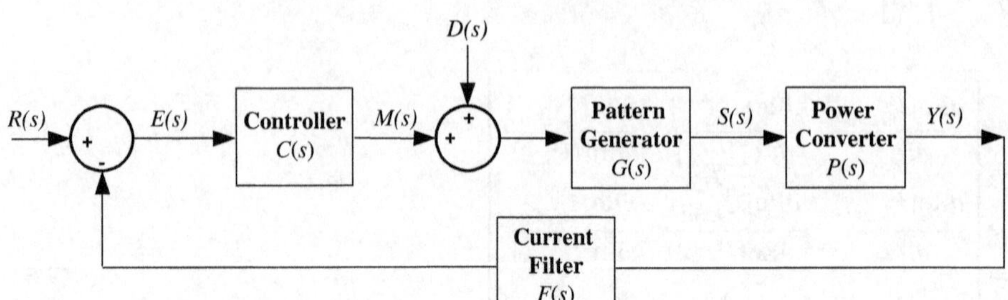

Figure 9.10 The block diagram of the closed-loop VSC.

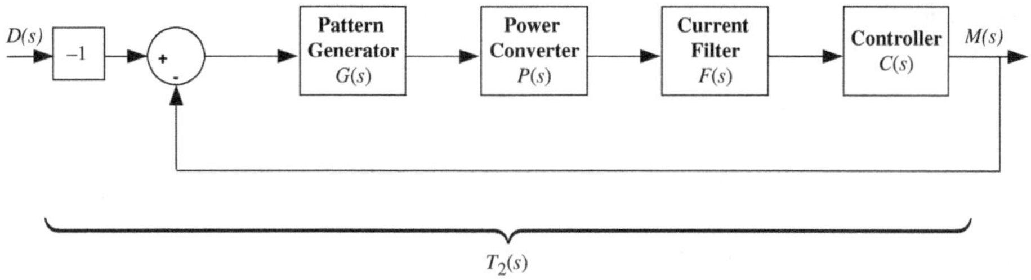

Figure 9.11 The equivalent block diagram of $T_2(s)$.

Note that:

1. $C(s)$ here represents the current controller and $Y(s)$ denotes the phase currents. The effect of the dc voltage controller has been neglected because its bandwidth is much smaller than that of the current controller.
2. Since both controllers and filters are described in the dq-frame, their true transfer functions should be [99]

$$C_s(s) = C(s - j\omega) \tag{9.62}$$

$$F_s(s) = F(s - j\omega), \tag{9.63}$$

where C_s and F_s represent the transfer function of current controllers and filters, respectively in the stationary $\alpha\beta$ reference frame.

It is $T_2(s)$ in (9.61) that is of interest. Fig. 9.11 shows the equivalent block diagram describing $T_2(s)$.

Therefore, the problem can be recast as:

Find the maximum frequency at which the influence of $D(s)$ on $M(s)$ can be considered negligible.

As a rule of thumb, −70dB[1] can be considered as the level below which the modulating harmonics do not affect the switching times produced by the controller.

Moreover, what is important is that the number of mismatch equations depends only on the number of system harmonics that will be affected by the control action. For converter's switching at multiple kHz, this provides an enormous reduction in the number of mismatch equations as compared to the LB 2, which will become apparent in Chapter 10.

Example 9.1 One common approach to design the current controller of a VSC is to use zero-pole cancellation. A PI controller takes the form of $\frac{k(s+k_a)}{s}$. The SSM of a VSC is $\frac{1}{sL_f+R_f}$, where L_f and R_f are the filter inductors and resistors of the VSC. Since the pole of the VSC is stable, k_a can be directly set to $\frac{R_f}{L_f}$. Then, one only needs to determine k, which can be determined by root-locus or frequency response method. Thus, current controllers of the VSC based on the parameters of IEEE benchmark model of D-STATCOM [100] (see Table 9.1) have the form of $\frac{k(s+235.4788)}{s}$. Based on the root-locus analysis, two current controllers, $C_{c\_slow} = \frac{0.04247(s+235.4788)}{s}$ and $C_{c\_fast} = \frac{0.4247(s+235.4788)}{s}$,

1 −70dB is chosen because it is lower than 0.1% (−60dB) — the value of the gain can be considered 0 in practice. A −70dB threshold would also be comparable to the noise floor of a 14-bit A/D converter. Note that both $M(s)$ and $D(s)$ are normalized with respect to the fundamental modulating signal.

Table 9.1 The parameter values used with reference to Fig. 9.2 as compared to those values employed in the IEEE benchmark model of the D-STATCOM.

	IEEE Benchmark	**Fig. 9.2**
Source voltage	$e_s = 112V_{rms}$ (line-line)	$e_s = 112V_{rms}$ (line-line)
Synchronous angle	$\theta_s = 0°$	$\theta_s = 0°$
Source inductance	$L_s = 145\mu H$	$L_s = 145\mu H$
Source resistance	$R_s = 25m\Omega$	$R_s = 25m\Omega$
Filter inductance	$L_f = 0.637mH$	$L_f = 0.637mH$
Filter resistance	$R_f = 0.15\Omega$	$R_f = 0.15\Omega$
DC capacitance	$C = 4820\mu F$	$C = 1000\mu F$
Reactive current reference	$i_{tqref} = 30$A	$i_{tqref} = 30$A
DC voltage reference	$v_{dcref} = 230$V	$v_{dcref} = 230$V
PWM strategy	Natural-sampled SPWM	Natural-sampled SPWM
Switching frequency	$f_s = 1620Hz$	$f_s = 540Hz$
Frequency modulation	$m_f = 27$	$m_f = 9$
Proportional gain	$K = 0.04247$	$K = 0.04247$
Integral constant	$\tau = 0.1$	$\tau = 0.1$
DC proportional gain	$K_p = 0.5$	$K_p = 0.5$
DC integral constant	$\tau_p = 0.04$	$\tau_p = 0.04$
Current filter	$\frac{1}{1 + 0.0019895s}$	$\frac{1}{1 + 0.0019895s}$

are stable controllers. The two current controllers differ by a factor of 10 in their gains. In general, the high gain controller offers high controller bandwidth, and high gain bandwidth also offers fast dynamic response of states. Therefore, the controller with the high gain is referred to as the fast controller, $C_{\_fast}$, and the controller with the low gain as the slow controller, C_{c_slow}.

Fig. 9.12(a) shows the Bode plot of T_2 when $C(s) = C_{c\_slow} = \frac{0.04247(s+235.4788)}{s}$ and $F(s) = \frac{1}{1+0.0019895s}$. As indicated in the figure, the −70dB point is at 3.6659×10^3 rad/sec (around the 10th harmonic). On the other hand, Fig. 9.12(b) shows the Bode plot of T_2 when $C(s)$ is replaced with $C_{c\_fast} = \frac{0.4247(s+235.4788)}{s}$. The corresponding point is at 1.1226×10^4 rad/sec (around the 30th harmonic). Note that the 60 Hz frequency shift has minor effects on the maximum number of relevant harmonics as seen in Fig. 9.12. □

Example 9.2 ***Balanced Two-Level VSC***

To validate the derived LB 1 method, its steady-state solutions are compared with those obtained by brute-force time domain simulation (PSCAD/EMTDC). The system under study in this example is as shown in Fig. 9.2. The parameters and PWM strategy are extracted from the IEEE benchmark model of the D-STATCOM [100]. However, in order to show all harmonics more clearly, the frequency modulation ratio is reduced from the original value of 27 to 9 (i.e. a switching frequency of 540 Hz) and the value of the dc capacitance is reduced from the original value of 4860 μF to 1000 μF. Table 9.1 lists the parameter values used with reference to Fig. 9.2 as compared to those values employed in the IEEE benchmark model of the D-STATCOM.

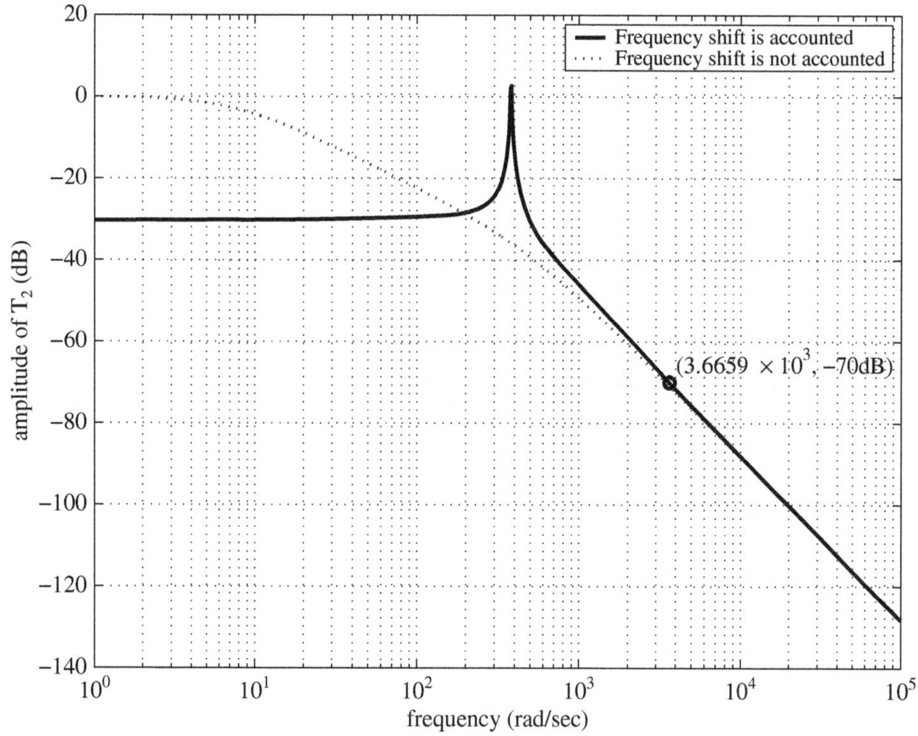

(a)

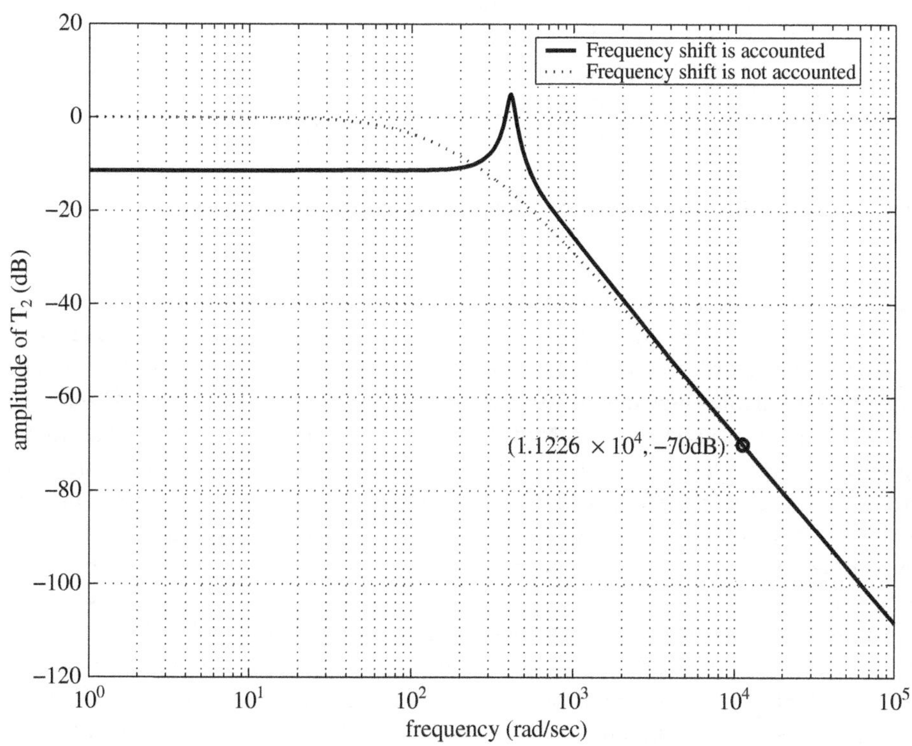

(a)

Figure 9.12 Examination of closed-loop controller transfer function $T_2(s)$ to determine number of relevant harmonics – (a) slow controller; (b) fast controller.

In general, low-order harmonics are expected to be heavily influenced by the control action. To demonstrate how harmonics are affected by the controls, the two current controllers presented in **Example 9.1** are employed. The same outer dc voltage controller $C_v = \frac{0.5(s+50)}{s}$ is employed for both studies.

The VSC is assumed to connect to a perfectly balanced three-phase system, with system voltages given by

$$e_{sa}(t) = |e_s| \cos(wt)$$

$$e_{sb}(t) = |e_s| \cos(wt - 2\pi/3)$$

$$e_{sc}(t) = |e_s| \cos(wt - 4\pi/3).$$

Based on phasor analysis, the positive sequence fundamental component of the modulating signal, $V^1_{mp} = 0.8721 - 0.0458jV$, is obtained. Other harmonics are all arbitrarily initialized to $0.0001 + j0.0001V$.

For balanced operation, only characteristic harmonics are nonzero [11]. Hence, the following harmonics are initialized when the slow controller is employed:

- Modulating signal: $h = 1, 5$, and 7.
- Active current reference: $h = 0, 2, 4, 6$, and 8.

In the case of the fast controller, the following harmonics are initialized:

- Modulating signal: $h = 1, 5, 7, 11, 13, 17, 19, 23, 25$, and 29.
- Active current reference: $h = 0, 2, 4, 6, 8, 10, 12, 14, 16, 18, 20, 22, 24, 26, 28$ and 30.

Since the number of unknown variables is still quite large, it is advised to use Quasi-Newton's method, as discussed in Chapter 3, as the iterative method to avoid inverting a large Jacobian matrix at every iteration. Once the closed-loop modulating signal harmonics are obtained, these signals are passed to the PWM pattern generator to determine the steady state switching times. Applying the switching times to the FCM of the VSC derived in Chapter 5 yields the ac current and dc voltage harmonics. Fig. 9.13 shows the ac current space vector $(I_{t\alpha} + jI_{t\beta})$ harmonic spectrum due to the slow and fast current controllers under balanced operation. The ac current space vector harmonics of PSCAD/EMTDC are obtained by first converting the three-phase datasets, $\mathbf{I_{ta}}$, $\mathbf{I_{tb}}$, and $\mathbf{I_{tc}}$ into the datasets of $\mathbf{I}_{t\alpha}$ and $\mathbf{I}_{t\beta}$ by Clarke transformation. Then, a complex space vector dataset is formed from $\mathbf{I_t} = \mathbf{I}_{t\alpha} + j\mathbf{I}_{t\beta}$. Since the ac current space vectors are complex-valued, the complex conjugate of the positive harmonic components will not be equal to the negative harmonic components, as shown in Fig. 9.13. Therefore, the negative frequencies here mean the negative sequences. The nonzero ac current harmonics shown in the figures (from left to right) are [−53 −47 −41 −35 −29 −23 −17 −11 −5 1 7 13 19 25 31 37 43 49 55]. Note that although the modulating voltage harmonics are only specified up to the converter's bandwidth, the ac current and dc voltage harmonics may be calculated up to much higher frequencies. The assumption is merely that harmonics above the control bandwidth do not propagate through the controller into the modulating signals. When correct switching times are determined, the ac current and dc voltage harmonics may be found for arbitrarily high frequency. Fig. 9.14 shows the dc voltage harmonic spectrum due to the slow and fast current controllers, respectively, under balanced operation. Note that since dc voltage harmonics roll off very rapidly, the magnitude of dc voltage harmonics are plotted using a log scale to allow comparison of small high frequency components. As can be noted from Figs. 9.13 and 9.14, excellent agreement exists between the LB 1 method and the PSCAD/EMTDC simulation results. Hence, the prediction of the converter bandwidth based on the analysis of Section 9.3.3 is accurate. □

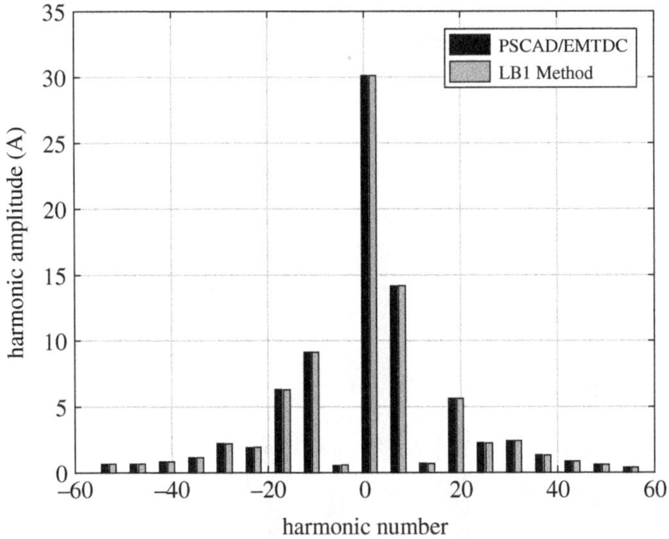

(a)

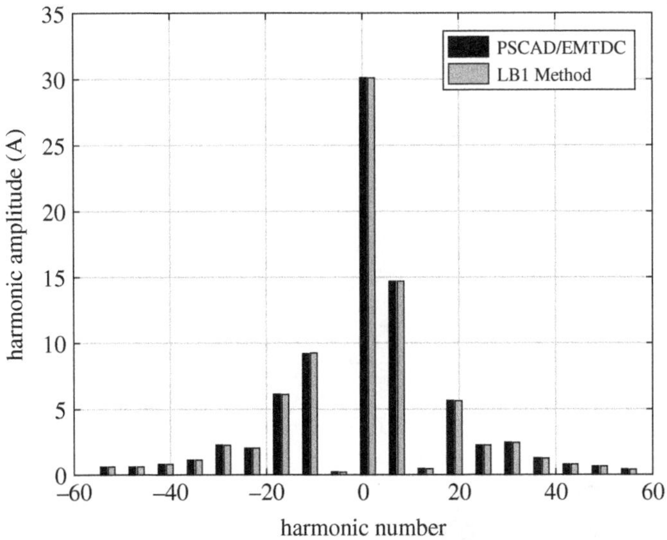

(a)

Figure 9.13 Balanced case, ac current harmonics with ac current filters included – (a) slow controller; (b) fast controller.

Example 9.3 ***Unbalanced Two-Level VSC***
A 10% negative sequence voltage is superimposed on the system voltage to study operation under unbalanced conditions. The system voltage is given by

$$e_{sa}(t) = |e_s| \cos(wt) + 0.1\,|e_s| \cos(wt)$$

$$e_{sb}(t) = |e_s| \cos(wt - 2\pi/3) + 0.1\,|e_s| \cos(wt + 2\pi/3)$$

$$e_{sc}(t) = |e_s| \cos(wt - 4\pi/3) + 0.1\,|e_s| \cos(wt + 4\pi/3).$$

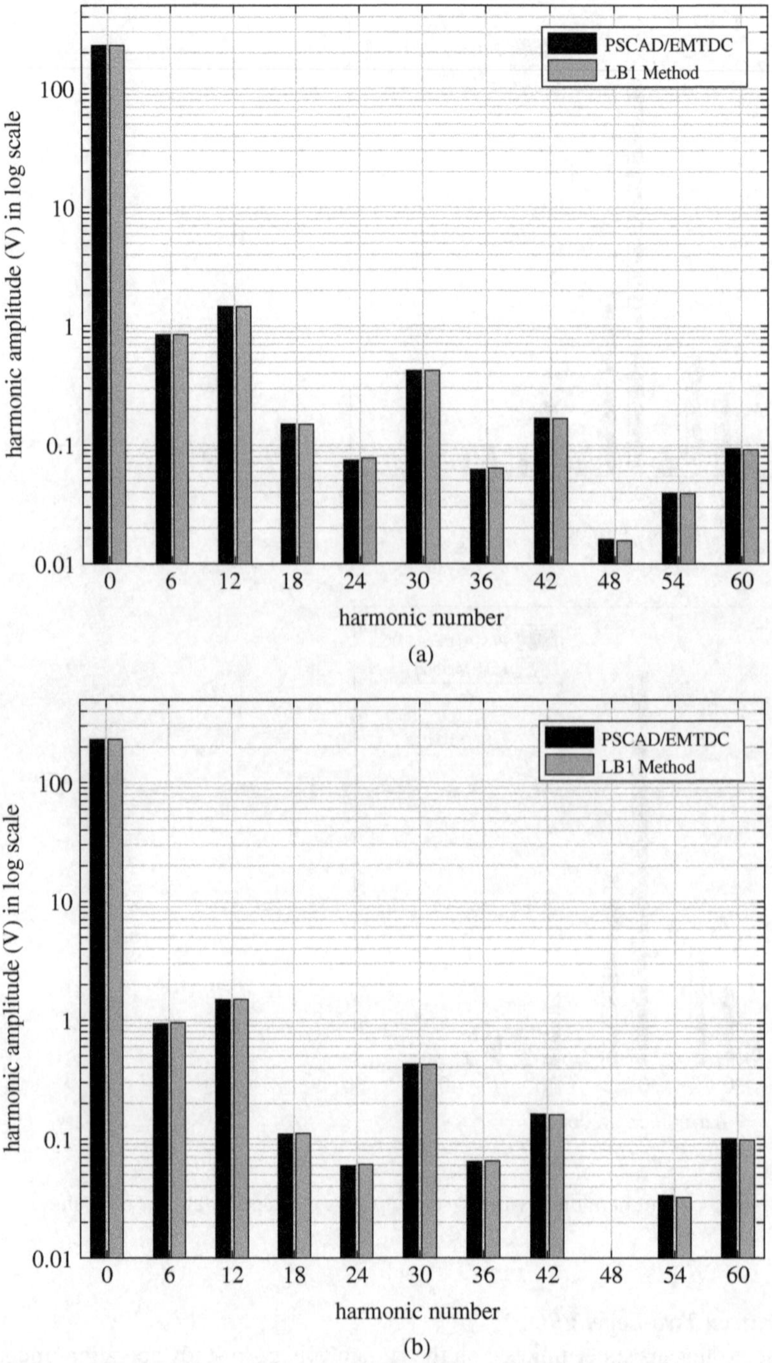

Figure 9.14 Balanced case, dc voltage harmonics with AC current filters included – (a) slow controller; (b) fast controller.

Note that a 10% bus voltage imbalance is selected for visualization purpose only, and that such a high level of imbalance would not ordinarily occur in a power system.

For the unbalanced case considered, the waveform exhibits half-wave symmetry [71], so all odd harmonics up to the maximum harmonic of interest must be included. Hence, the following harmonics are initialized when the slow controller is employed:

- Modulating signal: $h = 1, 3, 5, 7$, and 9.
- Active current reference: $h = 0, 2, 4, 6, 8$, and 10.

In the case of fast controller, on the other hand, the following harmonics are initialized:

- Modulating signal: $h = 1, 3, 5, 7, 9, 11, 13, 15, 17, 19, 21, 23, 25, 27$, and 29.
- Active current reference: $h = 0, 2, 4, 6, 8, 10, 12, 14, 16, 18, 20, 22, 24, 26, 28$, and 30.

Here the initial values obtained by the open-loop phasor analysis were used for this unbalanced case. Note that the number of iterations can be further reduced if the initialized values are set to the converged modulating voltage harmonics and the harmonics of the active current reference computed for the balanced operation.

Fig. 9.15 shows the ac current space vector harmonic spectra for the slow and fast current controllers under unbalanced operation. Comparing Figs. 9.13 and 9.15, one sees that both controllers successfully regulate the positive sequence fundamental current to 30A in both balanced and unbalanced operation. Nevertheless, under unbalanced condition, the current spectra produced by the slow and fast controllers differ significantly, as compared to the balanced condition. We can see that harmonics at the -13th, -9th, -1th, 3rd, 5th, and 9th differ quite noticeably. For instance, the negative sequence fundamental current component is 28A (93.33% of the positive sequence of the fundamental component) for C_{c_slow} while it is only 17A (56.67% of the positive sequence of the fundamental component) for C_{c_fast}. C_{c_fast} can suppress the negative sequence of the fundamental component much more than C_{c_slow}. On the other hand, the 3rd harmonic component is 9A (30% of the positive sequence of the fundamental component) for C_{c_fast} and 16A (53.33%) for C_{c_fast}. C_{c_slow} performs better than C_{c_slow} for the 3rd harmonic component.

Figs. 9.14 and 9.16 show that the dc component of the capacitor voltage V_{dc}^{0} is successfully regulated to its reference value of 230 V in all cases. Similar to the ac current spectra, the voltage spectra difference between the two controllers is more noticeable for the case of the unbalanced case. For instance, the second harmonic component is lower for C_{c_fast} (i.e. 16V), as compared to C_{c_slow} (i.e. 13V). The fourth harmonic component, on the other hand, is higher for C_{c_fast} (i.e. 0.58V), as compared to C_{c_slow} (i.e. 0.28V), though both of them are very small. □

Example 9.4 ***Three-Level MML-VSC with Unsymmetrical Parameter Values***

A three-level MML-VSC with unsymmetrical parameter values is connected to a balanced three-phase ac voltage source as stated in (9.64) and the parameters are listed in Table 9.2

$$e_{sa} = |e_s| \, cos(\omega t) \tag{9.64a}$$

$$e_{sb} = |e_s| \, cos(\omega t - \frac{2\pi}{3}) \tag{9.64b}$$

$$e_{sc} = |e_s| \, cos(\omega t + \frac{2\pi}{3}), \tag{9.64c}$$

where $|e_s| = 4.16$ kV.

To solve such an MML-VSC with output current control, circulating current control, and dc voltage control, one can apply the LB 1 method as shown in Fig. 9.17, which divides the system into five solution blocks. Block 1 is composed of the PWM generator and the model of MML-VSC. Block 2 is

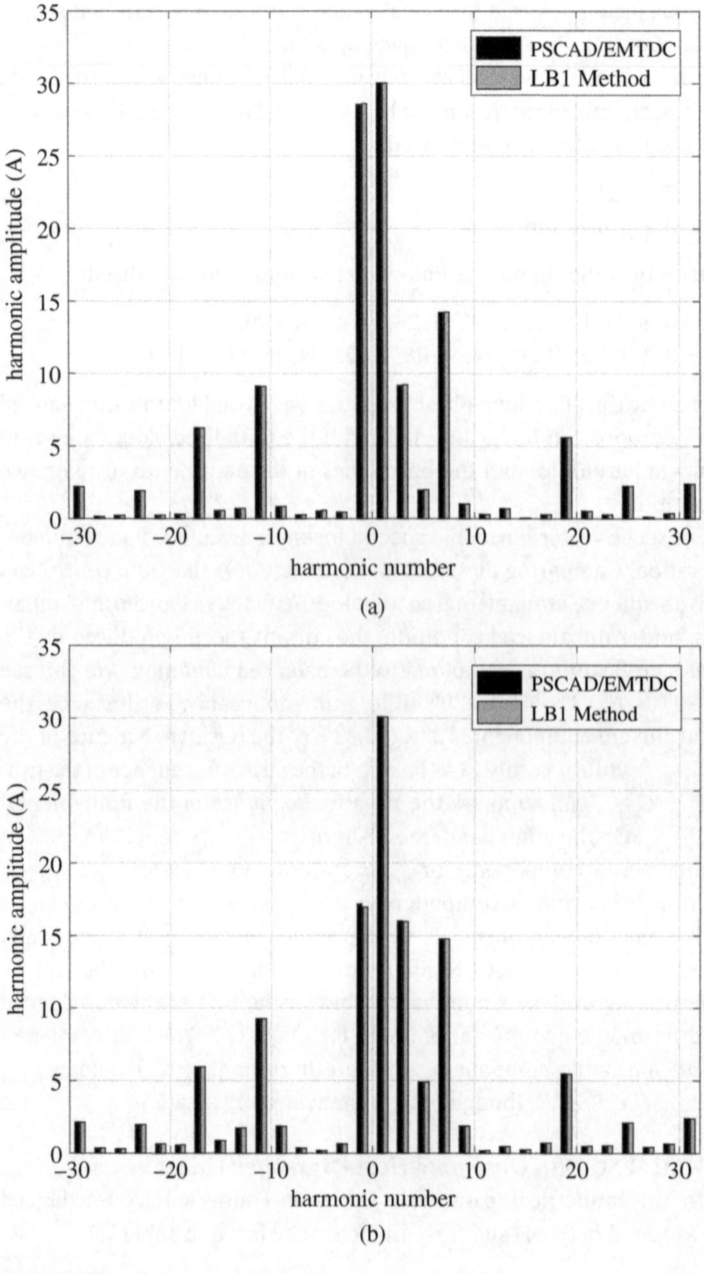

Figure 9.15 Unbalanced case, AC current harmonics with AC current filters included – (a) slow controller; (b) fast controller.

the "Output Current Control Block," which is comprised of PI regulator for the ac current. Block 3 is the "Circulating Current Control Block," which is comprised of PI regulator for suppressing the second-order harmonic in the circulating current. Block 4 is the "Voltage Control Block," which is comprised of PI regulator for the dc bus voltage.

How the output current and dc voltage controllers are modeled in the frequency domain have been shown previously. Here, a brief discussion to the circulating current controller is provided.

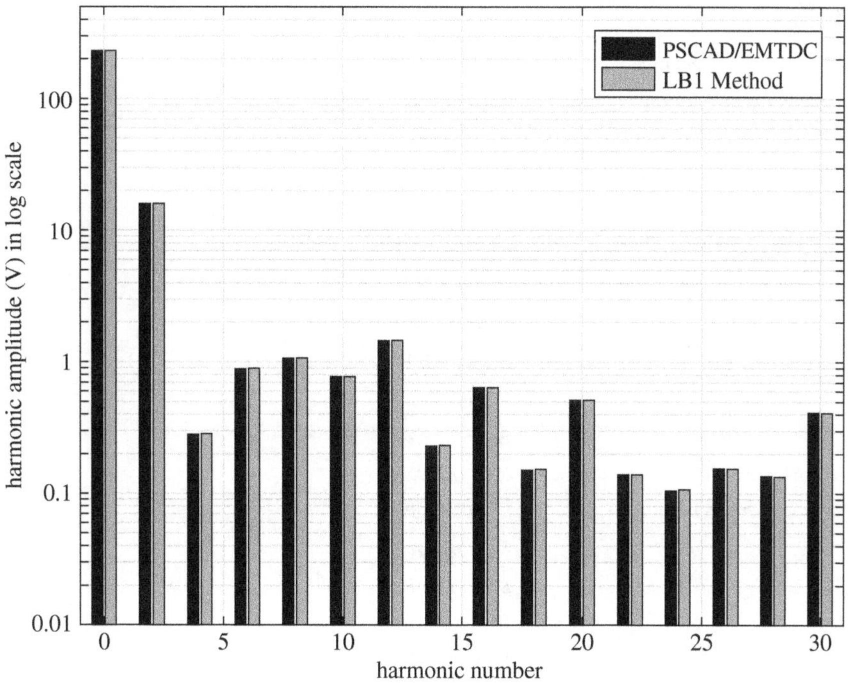

(a)

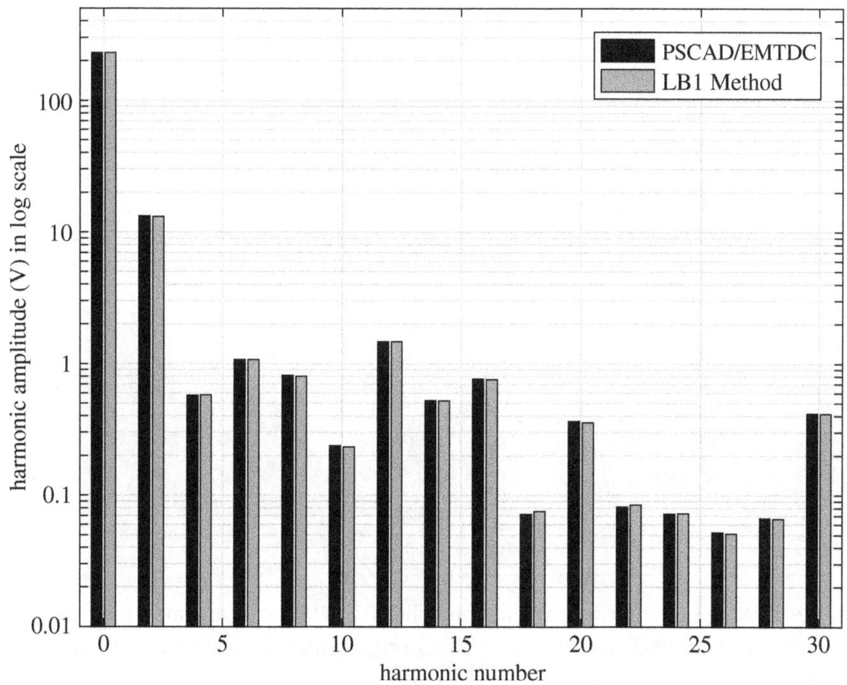

(b)

Figure 9.16 Unbalanced case, dc voltage harmonics with ac current filters included – (a) slow controller; (b) fast controller.

Table 9.2 Three-level MML-VSC parameters.

MML-VSC system	Line resistance	$R_{sa} = 0.15\Omega$
		$R_{sb} = 0.1\Omega$
		$R_{sc} = 0.0255\Omega$
	Line inductance	$L_{sa} = 690\mu H$
		$L_{sb} = 600\mu H$
		$L_{sc} = 145\mu H$
	Arm resistance	$R_a^u = 0.01\Omega$
		$R_b^u = 0.03\Omega$
		$R_c^u = 0.05\Omega$
		$R_a^l = 0.07\Omega$
		$R_b^l = 0.09\Omega$
		$R_c^l = 0.1\Omega$
	Arm inductance	$L_a^u = 690\mu H$
		$L_b^u = 650\mu H$
		$L_c^u = 600\mu H$
		$L_a^l = 700\mu H$
		$L_b^l = 750\mu H$
		$L_c^l = 800\mu H$
	Submodule capacitor	$C_{sm}^{ua} = 5000\mu F$
		$C_{sm}^{ub} = 7500\mu F$
		$C_{sm}^{uc} = 9000\mu F$
		$C_{sm}^{la} = 6000\mu F$
		$C_{sm}^{lb} = 8500\mu F$
		$C_{sm}^{lc} = 10000\mu F$
	Submodule resistor	$R_{sm}^{ua} = 1G\Omega$
		$R_{sm}^{ub} = 2G\Omega$
		$R_{sm}^{uc} = 5G\Omega$
		$R_{sm}^{la} = 10G\Omega$
		$R_{sm}^{lb} = 20G\Omega$
		$R_{sm}^{lc} = 30G\Omega$
	DC-link capacitor	$C_{dc} = 12000\mu F$
	DC-link resistor	$R_{dc} = 1G\Omega$
		$V_{dcref} = 8320\ V$
		$I_{tqref} = 1000\ A$

As presented in Fig. 9.18, the circulating current controllers of MML-VSC are shown in the synchronous reference frame. The angle fed into the $\alpha\beta/dq$ transformation is -2θ, as explained in Section 7.2.4. $C_{cir}(s)|_{s=jh\omega}$ is the transfer function of PI controller.

The modeling procedure for a closed-loop MML-VSC is described as follows:

1. Initialize the modulating signal harmonics, $V_{mt\alpha}$, $V_{mt\beta}$, $V_{mcir\alpha}$, $V_{mcir\beta}$, and the active current reference, I_{tdref}, as shown in Fig. 9.17.

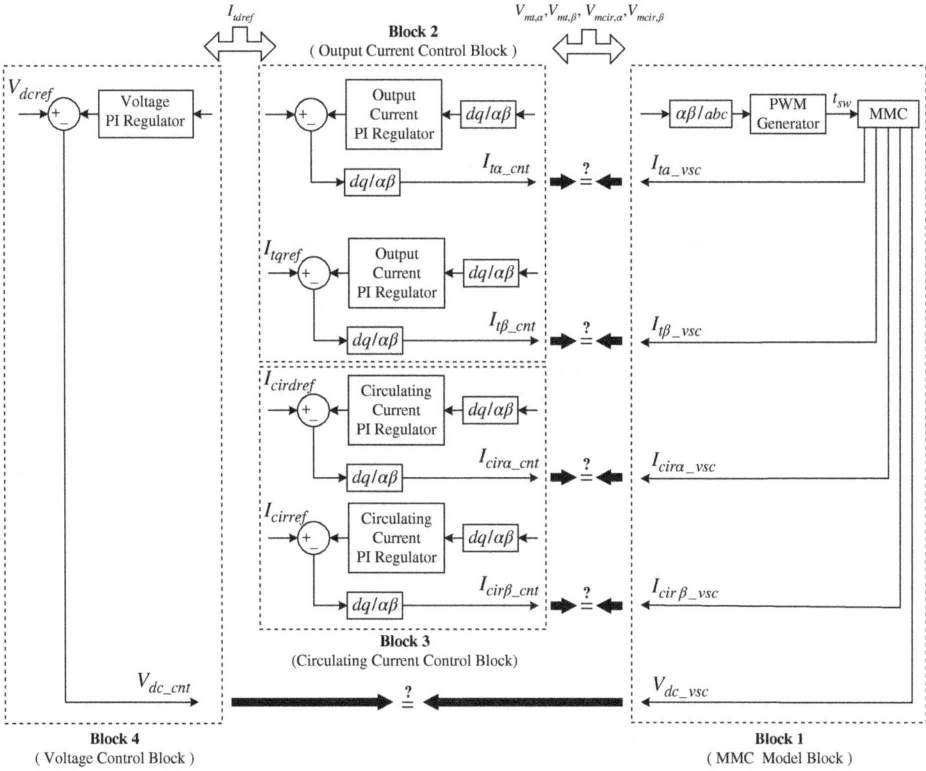

Figure 9.17 A calculation flow diagram of the LB 1 method for simulating a closed-loop MML-VSC.

2. Supply $V_{mt\alpha}$, $V_{mt\beta}$, $V_{mcir\alpha}$, $V_{mcir\beta}$, and I_{tdref} to Block 1 to calculate the output current, circulating current, and dc voltage harmonics predicted by the MML-VSC ($I_{t\alpha\_vsc}, I_{t\beta\_vsc}, I_{cir\alpha\_vsc}, I_{cir\beta\_vsc}$, and V_{dc_vsc}).
3. Supply $V_{cirt\alpha}$ and $V_{cirt\beta}$ to Block 3 for the prediction of circulating current harmonics ($I_{cir\alpha\_cnt}$ and $I_{cir\beta\_cnt}$). Note that I_{tdref} is not only supplied to Voltage Control Block to calculate the dc voltage harmonics by Voltage Control Block (V_{dc_cnt}), but also passed to the Output Current Control Block with $V_{mt\alpha}$ and $V_{mt\beta}$ for the prediction of output current harmonics ($I_{t\alpha\_cnt}$ and $I_{t\beta\_cnt}$).
4. Check whether

$$I_{t\alpha\_vsc} - I_{t\alpha\_cnt} = 0 \tag{9.65}$$

$$I_{t\beta\_vsc} - I_{t\beta\_cnt} = 0 \tag{9.66}$$

$$I_{cir\alpha\_vsc} - I_{cir\alpha\_cnt} = 0 \tag{9.67}$$

$$I_{cir\beta\_vsc} - I_{cir\beta\_cnt} = 0 \tag{9.68}$$

$$V_{dc\_vsc} - V_{dc\_cnt} = 0 \tag{9.69}$$

If (9.65), (9.66), (9.67), (9.68), and (9.69) hold true, the loop is closed and $V_{mt\alpha}$, $V_{mt\beta}$, $V_{mcir\alpha}$, $V_{mcir\beta}$, and I_{tdref} are correct. Otherwise, iterations are continued to search for the correct values of these unknown variables until convergence is reached.

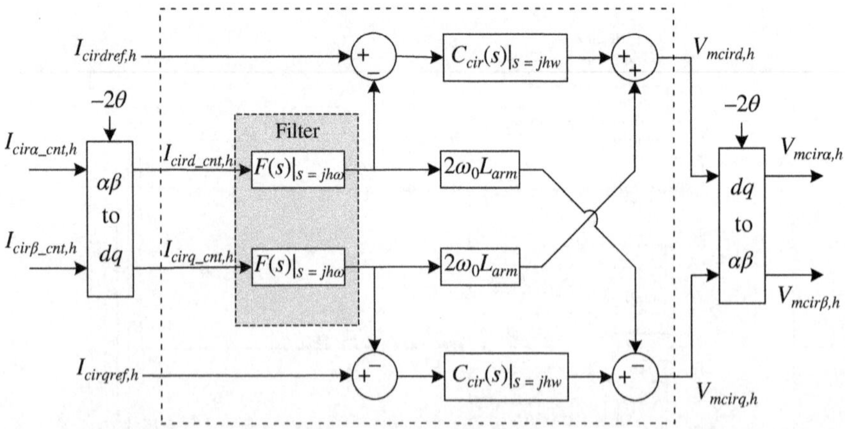

Figure 9.18 Block diagram of the Circulating Current Controllers including the current filters in the frequency domain.

As shown in (9.70), the current harmonics at the circulating current controllers, $I^h_{cird\_cnt}$ and $I^h_{cirq\_cnt}$, can be inversely calculated and obtained for the given $V_{mcird,h}$, $V_{mcirq,h}$, $I_{cirdref,h}$, and $I_{cirqref,h}$.

$$\begin{bmatrix} I_{cird\_cnt,h} \\ I_{cirq\_cnt,h} \end{bmatrix} = \begin{bmatrix} \frac{-C_{cir}(jh\omega)}{F(jh\omega)((C_{cir}(jh\omega))^2 + (2\omega_0 L_{arm})^2)} & \frac{-2\omega_0 L_{arm}}{F(jh\omega)((C_{cir}(jh\omega))^2 + (2\omega_0 L_{arm})^2)} \\ \frac{2\omega_0 L_{arm}}{F(jh\omega)((C_{cir}(jh\omega))^2 + (2\omega_0 L_{arm})^2)} & \frac{-C_{cir}(jh\omega)}{F(jh\omega)((C_{cir}(jh\omega))^2 + (2\omega_0 L_{arm})^2)} \end{bmatrix} \begin{bmatrix} V_{mcird,h} \\ V_{mcirq,h} \end{bmatrix} + \begin{bmatrix} \frac{C_{cir}(jh\omega)}{F(jh\omega)((C_{cir}(jh\omega))^2 + (2\omega_0 L_{arm})^2)} & \frac{2\omega_0 L_{arm}}{F(jh\omega)((C_{cir}(jh\omega))^2 + (2\omega_0 L_{arm})^2)} \\ \frac{-2\omega_0 L_{arm}}{F(jh\omega)((C_{cir}(jh\omega))^2 + (2\omega_0 L_{arm})^2)} & \frac{C_{cir}(jh\omega)}{F(jh\omega)((C_{cir}(jh\omega))^2 + (2\omega_0 L_{arm})^2)} \end{bmatrix} \begin{bmatrix} I_{cirdref,h} \\ I_{cirqref,h} \end{bmatrix} \tag{9.70}$$

Equation (9.70) can be written in compact form as (9.71) for multiple harmonics of concern.

$$\mathbf{I_{cirdq\_cnt}} = \mathbf{Y_{mcir,dq}} \mathbf{V_{mcir,dq}} + \mathbf{K_{cirdqref}} \mathbf{I_{cirdqref}} \tag{9.71}$$

Note that $\mathbf{I_{cirdq\_cnt}} = \begin{bmatrix} I_{cirdq\_cnt,0} & I_{cirdq\_cnt,1} & I_{cirdq\_cnt,2} & \dots & I_{cirdq\_cnt,h} \end{bmatrix}^T$, and both $\mathbf{V_{mcir,d}}$ and $\mathbf{V_{mcir,q}}$ have the same number of harmonic orders as I_{cirdq_cnt}. (9.71) should be transformed into the $\alpha\beta$-frame in order to establish a mismatch equation between the solutions of the Circulating Current Control Block and Block 1. A new connection matrix, C_{con2}, defined in (9.72) is applied to (9.71) to obtain the circulating current in the $\alpha\beta$-frame.

$$\mathbf{C_{con2}} = \frac{1}{2} \begin{bmatrix} \mathbf{0} & \mathbf{0} & \mathbf{M_j} & \mathbf{0} & \mathbf{0} \\ \mathbf{0} & \mathbf{0} & \mathbf{0} & \ddots & \mathbf{0} \\ \mathbf{N_j} & \mathbf{0} & \mathbf{0} & \mathbf{0} & \mathbf{M_j} \\ \mathbf{0} & \ddots & \mathbf{0} & \mathbf{0} & \mathbf{0} \\ \mathbf{0} & \mathbf{0} & \mathbf{N_j} & \mathbf{0} & \mathbf{0} \end{bmatrix} \tag{9.72}$$

Consequently, (9.72) is converted into (9.73).

$$\mathbf{I_{cir\alpha\beta\_cnt}} = \mathbf{C_{con2}} \mathbf{Y_{mcir,dq}} \mathbf{V_{mcir,dq}} + \mathbf{C_{con2}} \mathbf{K_{cirdqref}} \mathbf{I_{cirdqref}} \tag{9.73}$$

Table 9.3 Parameters of the controller.

		1st	2nd
Output current controller	Proportional gain	0.1	1
	Integral time constant	0.1	0.1
Circulating current controller	Proportional gain	5	5
	Integral time constant	0.001	0.001
Voltage controller	Proportional gain	5	1
	Integral time constant	0.05	0.1

Table 9.4 Setting of the controllers.

Reference of dc voltage	$v_{dcref} = 8.32$ kV
Reference of current	$i_{tqref} = -1000$A
	$i_{cirdref} = 0$A
	$i_{cirqref} = 0$A
Filter gain of output current	1
Filter time constant of output current	0.0019895
Filter gain of circulating current	1
Filter time constant of circulating current	0.0019895

(9.73) can be rewritten as (9.74) since the modulation signals $\mathbf{V_{mcir,\alpha\beta}}$ are in the $\alpha\beta$ frame as shown in Fig. 9.17.

$$\mathbf{I_{cir\alpha\beta\_cnt}} = \mathbf{C_{con2}Y_{mt,dq}C_{con2}^{-1}V_{mt,\alpha\beta}} + \mathbf{C_{con2}K_{cirdqref}I_{cirdqref}} \tag{9.74}$$

To demonstrate that different controllers can greatly affect the harmonic spectra of a power converter, two sets of controllers as listed in Table 9.3 are used in the simulation. The reference values and filter parameters are listed in Table 9.4. The PWM strategy used is PS-PWM and the modulation frequency index is 27.

Figs. 9.20 and 9.21 show the harmonic spectra, $i_{t\alpha}$, $i_{cir\alpha}$, and V_{dc}, respectively, for the two controllers. These figures show that different controllers may amplify or suppress different harmonics. For example, the 27th harmonic of $i_{t\alpha}$ for the 1st controller is about 0.4A while that for the 2nd controller is about 2.5A, which is about 8 times larger. Moreover, the dc components of $i_{cir\alpha}$ for the 1st and 2nd controllers are 32A and 12A, respectively. Their difference is about 160%. Finally, the fundamental of the V_{dc} for that the 1st controller is about 19V whereas for the 2nd controller it is about 7V.

It is to be noted that both sets of controllers successfully regulate the fundamental current component of $i_{t\alpha}$ at 1000A and the 0th component of V_{dc} at 8320V, as evident from Figs. 9.19 and 9.21.

Although the 2nd harmonic components of $i_{cir\alpha}$ for both controllers are not exactly zero due to the limit of the control bandwidth, they are quite close to zero (about 5A). This has also been captured by the PSCAD/EMTDC model. □

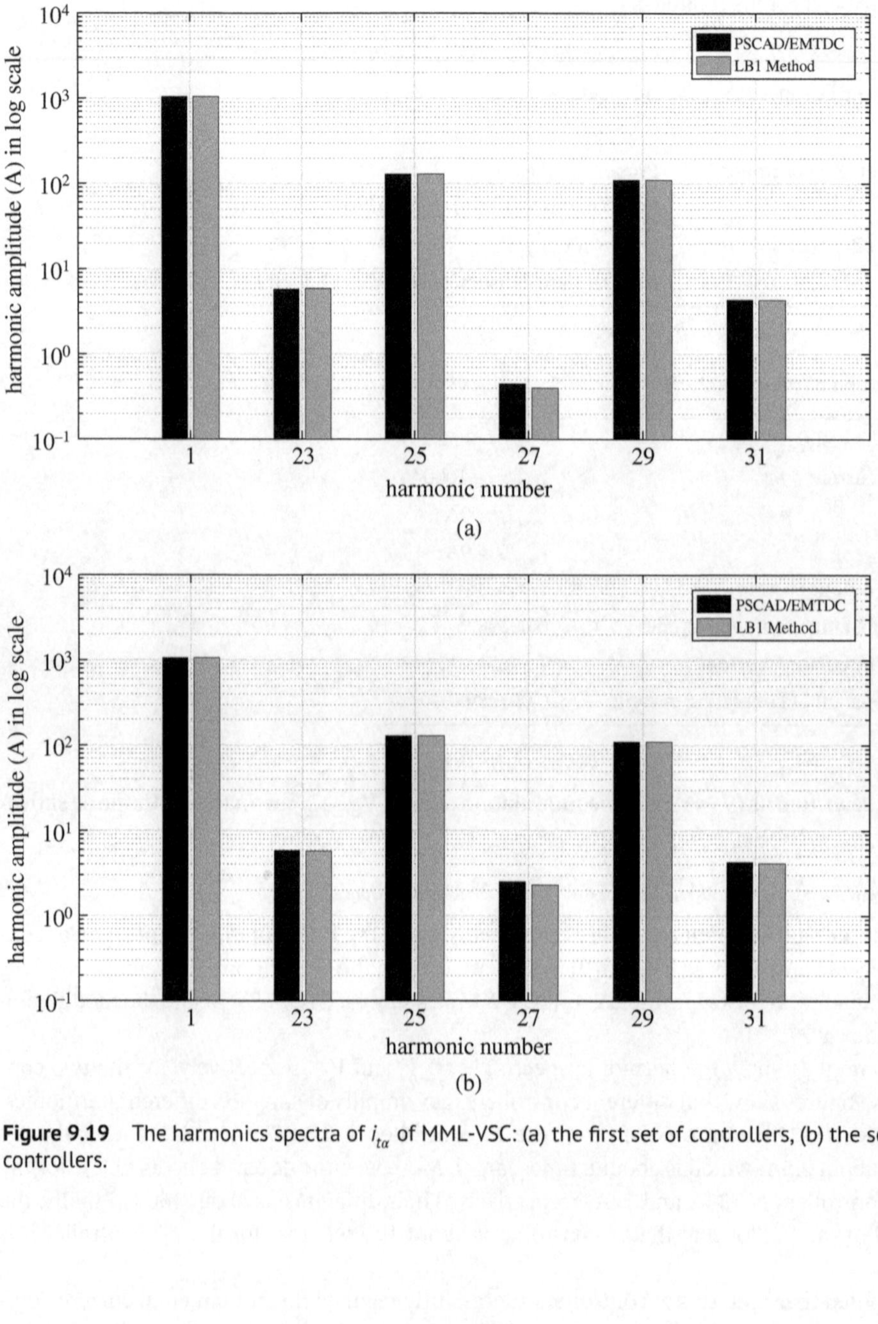

Figure 9.19 The harmonics spectra of $i_{t\alpha}$ of MML-VSC: (a) the first set of controllers, (b) the second set of controllers.

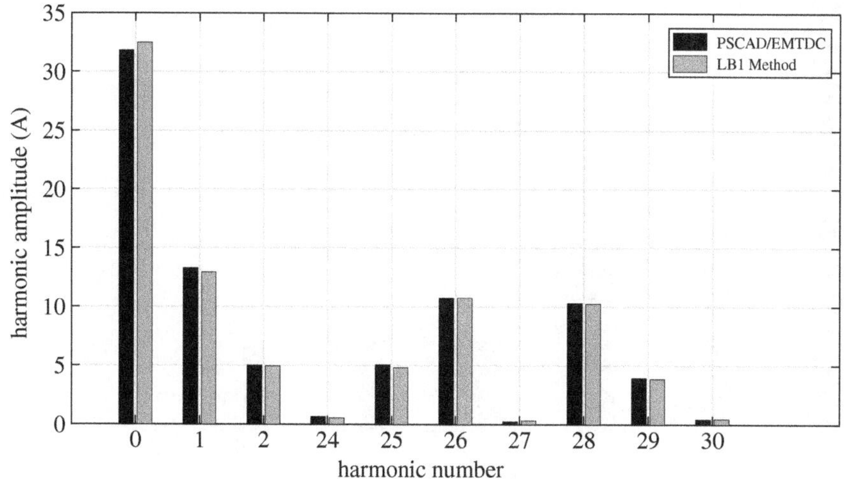

(a)

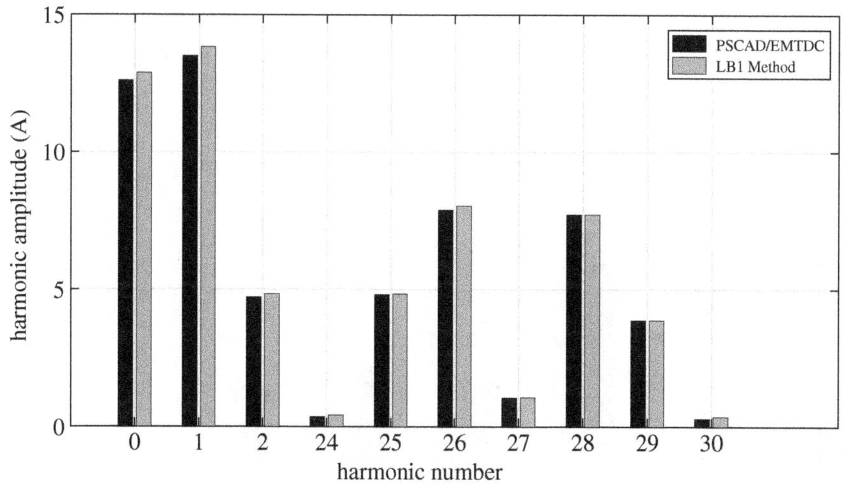

(a)

Figure 9.20 The harmonics spectra of $i_{cir\alpha}$ of MML-VSC: (a) the first set of controllers, (b) the second set of controllers.

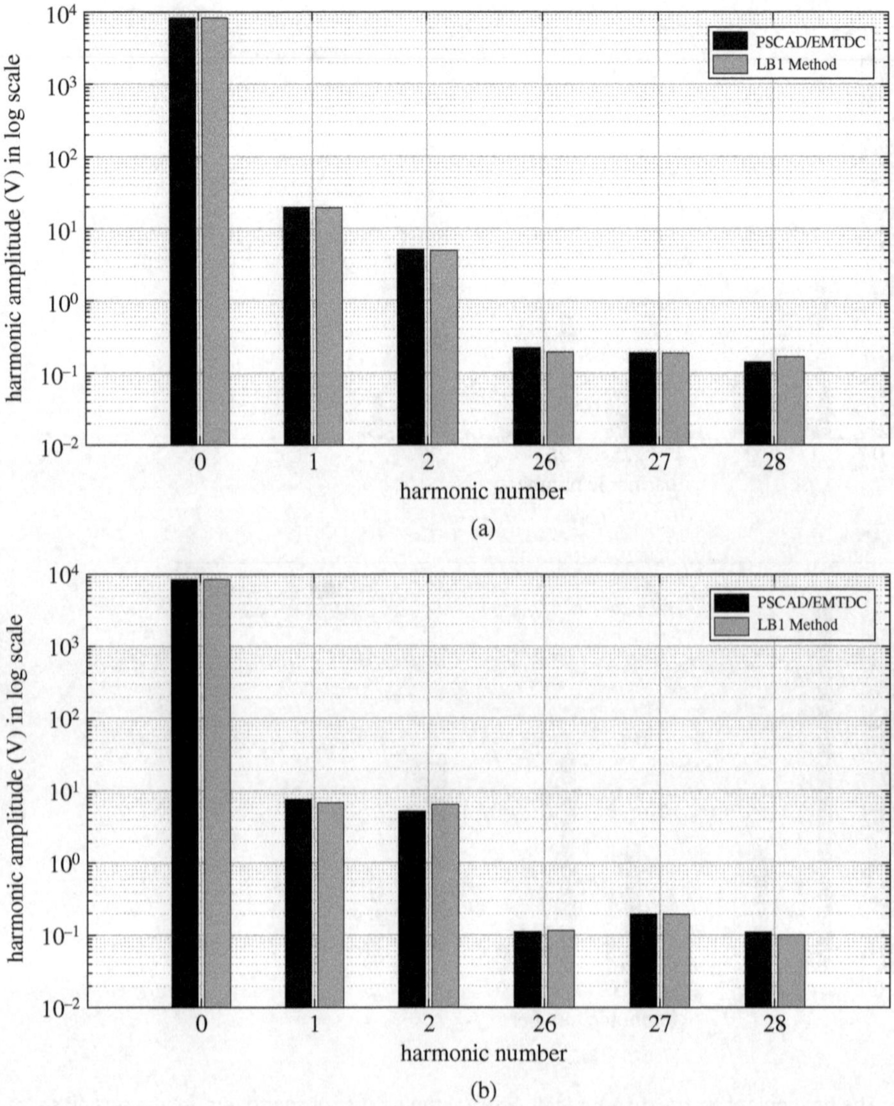

Figure 9.21 The harmonics spectra of V_{dc} of MML-VSC: (a) the first set of controllers, (b) the second set of controllers.

10

Loop-Breaking 2 Method for Solving a VSC

"Divide the fire and you will soon put it out."

– Greek Proverb

In this section, we will devote ourselves to using the loop-breaking 2 (LB 2) method to solve a two-level VSC. As mentioned in Chapter 8, the LB 2 method breaks within the individual blocks. It has been shown that such a method can be used to derive the generalized shooting method. In Section 10.1, an example of a dc-dc buck converter is used to demonstrate how augmented matrix methods can be easily incorporated into the LB 2 method, which avoids evaluating convolution integrals. In Section 10.2, it is demonstrated how a VSC, whose controllers are described in the *dq* frame, can also be modeled by the LB 2 method. Finally, we will show how a line-commutated power converter can be modeled by means of the LB 2 method. It is noted that for this type of converter, since the size of the state matrices may be different for different switching intervals, transition matrices are needed for harmonic evaluation.

10.1 Modeling for a Closed-Loop DC-DC Converter

Fig. 10.1 shows the schematic diagram of a closed-loop buck converter.[1] The parameter values are listed in Table 10.1. To simplify the analysis, $SW1$ and $D1$ are assumed to switch in a complementary manner. The output voltage $v_m(t)$ of the controller is compared with a sawtooth waveform to generate the correct turn-off time (see Fig. 10.2), t_{sw}, to achieve the desired reference voltage.

10.1.1 Model of the Buck Converter

When $SW1$ is on and $D1$ is off, the state space equation is given as

$$\frac{d}{dt}\mathbf{x_p} = \mathbf{A_{on}}\mathbf{x_p} + \mathbf{B_{on}}v_s. \tag{10.1}$$

When $D1$ is on and $SW1$ is off, the state space equation is given as

$$\frac{d}{dt}\mathbf{x_p} = \mathbf{A_{off}}\mathbf{x_p}, \tag{10.2}$$

1 The switches in Fig. 10.1 are ideal, which are different from the dc-dc converter presented in Fig. 3.9 of Section 3.4.5 whose diode is modeled as a continuous function

Harmonic Modeling of Voltage Source Converters using Basic Numerical Methods, First Edition.
Ryan Kuo-Lung Lian, Ramadhani Kurniawan Subroto, Victor Andrean, and Bing Hao Lin.

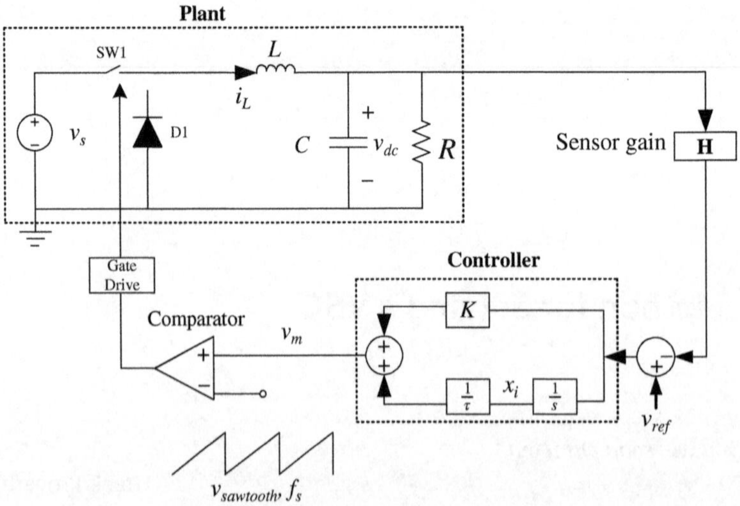

Figure 10.1 A regulated dc-dc buck converter.

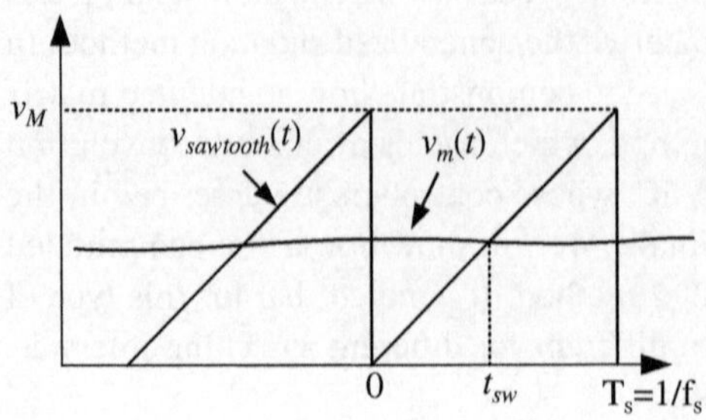

Figure 10.2 The output voltage of the controller of the buck converter intersects with the PWM sawtooth waveform.

where

$$\mathbf{x_p} = \begin{bmatrix} i_L \\ v_{dc} \end{bmatrix}, \mathbf{A_{on}} = \mathbf{A_{off}} = \begin{bmatrix} 0 & -\frac{1}{L} \\ \frac{1}{C} & -\frac{1}{RC} \end{bmatrix}, \mathbf{B_{on}} = \begin{bmatrix} \frac{1}{L} \\ 0 \end{bmatrix}.$$

The constant supply input voltage, v_s, can be cast as a state variable [101], and the following equations result:

When $SW1$ is on, and $D1$ is off,

$$\frac{d}{dt}\bar{\mathbf{x}} = \bar{\mathbf{A}}_{\mathbf{on}}\bar{\mathbf{x}}. \tag{10.3}$$

When $SW1$ is off, and $D1$ is on,

$$\frac{d}{dt}\bar{\mathbf{x}} = \bar{\mathbf{A}}_{\mathbf{off}}\bar{\mathbf{x}}, \tag{10.4}$$

where

$$\bar{\mathbf{x}} = \begin{bmatrix} \mathbf{x_p} \\ z \end{bmatrix}, z(0) = v_s, \bar{\mathbf{A}}_{\mathbf{on}} = \begin{bmatrix} \mathbf{A_{on}} & \mathbf{B_{on}} \\ \mathbf{0} & 0 \end{bmatrix}, \bar{\mathbf{A}}_{\mathbf{off}} = \begin{bmatrix} \mathbf{A_{off}} & 0 \\ \mathbf{0} & 0 \end{bmatrix}.$$

This makes the system autonomous and eliminates the need to evaluate convolution integrals within the state equations.

For the system of Fig. 10.1 to be in steady-state, we require (i) the plant and controller to be at steady-state, and (ii) the comparator must be in equilibrium (i.e. the sawtooth and modulating signal v_m must intersect at t_{sw} and thereby obey switching time constraints).

10.1.2 Constraints of Steady-State

Equation (10.5) describes the state transition over one period:

$$\bar{\mathbf{x}}(T) = e^{\bar{\mathbf{A}}_{\mathbf{off}}(T-t_{sw})} e^{\bar{\mathbf{A}}_{\mathbf{on}}(t_{sw})} \bar{\mathbf{x}}(0) \tag{10.5}$$

or

$$\begin{bmatrix} \mathbf{x_p}(T) \\ z(T) \end{bmatrix} = \bar{\mathbf{\Phi}} \begin{bmatrix} \mathbf{x_p}(0) \\ z(0) \end{bmatrix}. \tag{10.6}$$

According to [4.2], $\bar{\mathbf{\Phi}}$ takes the following form

$$\bar{\mathbf{\Phi}} = \begin{bmatrix} \mathbf{\Phi}(t_{sw}) & \mathbf{\Gamma}(t_{sw}) \\ \mathbf{0} & \mathbb{I} \end{bmatrix}, \tag{10.7}$$

where

$$\mathbf{\Phi}(t_{sw}) = e^{\mathbf{A}_{\mathbf{off}}(T-t_{sw})} e^{\mathbf{A}_{\mathbf{on}}(t_{sw})}, \; \mathbf{\Gamma}(t_{sw}) = e^{\mathbf{A}_{\mathbf{off}}(T-t_{sw})} \int_0^{t_{sw}} e^{\mathbf{A}_{\mathbf{on}}(t_{sw}-\tau)} \mathbf{B}_{\mathbf{on}} d\tau.$$

At steady-state, $\bar{\mathbf{x}}(T) = \bar{\mathbf{x}}(0)$ yielding,

$$\mathbf{x_p}(0) = \left\{ \mathbb{I} - \mathbf{\Phi}(t_{sw}) \right\}^{-1} \mathbf{\Gamma}(t_{sw}) z(0). \tag{10.8}$$

A steady-state constraint must also be imposed on the state variables of the control circuit. The output voltage of the PI controller in the frequency domain, $V_m(s)$, is given as

$$V_m(s) = \left(K + \frac{1}{\tau s} \right) \left(V_{ref} - H V_{dc}(s) \right). \tag{10.9}$$

As (10.9) indicates, only one additional state variable is introduced by the controller. Name this extra state variable as x_i, and define

$$\frac{dx_i}{dt} = v_{ref} - H v_{dc}. \tag{10.10}$$

At steady-state, the controller periodicity constraint $x_i(T) = x_i(0)$ requires

$$x_i(T) - x_i(0) = \begin{bmatrix} 0 & 0 & 0 & 1 & 0 \end{bmatrix} \left\{ \check{\mathbf{\Phi}} - \mathbb{I} \right\} \check{\mathbf{x}}(0) = 0, \tag{10.11}$$

where $\check{\mathbf{x}} = \begin{bmatrix} \mathbf{x_p} & z & x_i & r \end{bmatrix}^T$, $r(0) = v_{ref}$, $\check{\mathbf{\Phi}} = e^{\check{\mathbf{A}}_{\mathbf{off}}(T-t_{sw})} e^{\check{\mathbf{A}}_{\mathbf{on}} t_{sw}}$, and

$$\check{\mathbf{A}}_{\mathbf{on}} = \begin{bmatrix} 0 & -\frac{1}{L} & \frac{1}{L} & 0 & 0 \\ \frac{1}{C} & -\frac{1}{RC} & 0 & 0 & 0 \\ 0 & 0 & 0 & 0 & 0 \\ 0 & -H & 0 & 0 & 1 \\ 0 & 0 & 0 & 0 & 0 \end{bmatrix},$$

$$\check{\mathbf{A}}_{\mathbf{off}} = \begin{bmatrix} 0 & -\frac{1}{L} & 0 & 0 & 0 \\ \frac{1}{C} & -\frac{1}{RC} & 0 & 0 & 0 \\ 0 & 0 & 0 & 0 & 0 \\ 0 & -H & 0 & 0 & 1 \\ 0 & 0 & 0 & 0 & 0 \end{bmatrix}.$$

Note that in the above formulation, the constant controller reference input, v_{ref}, has been also cast as a state variable, leading to a 5th order system model.

10.1.3 Switching Time Constraints

The switching time constraint is formulated based on Fig. 10.2. *SW*1 turns off when v_m coincides with the sawtooth signal, $v_{sawtooth}$. Therefore,

$$v_{sawtooth}(t_{sw}) - v_m(t_{sw}) = \frac{v_M}{T_s} t_{sw} - \begin{bmatrix} 0 & -KH & 0 & \frac{1}{\tau} & K \end{bmatrix} e^{\check{\mathbf{A}}_{on} t_{sw}} \check{\mathbf{x}}(0) = 0, \tag{10.12}$$

where $T_s = \frac{1}{f_s}$, and f_s is the switching frequency.

Note that $v_m(t)$ can also be evaluated by convolving the impulse response of the controller with $v_{dc}(t)$ as is done by [102]. Although the convolution integral approach has the advantage of "not enlarging" the system matrix [102], it is limited to controllers whose transfer functions have a simple form such as $K + \frac{1}{\tau s}$. When it comes to VSC controllers whose transfer function takes the form of $K_1 + \frac{K_2 s}{s^2 + w^2}$ [103] in the *abc*-frame, the convolution integral is difficult to evaluate. In addition, the Jacobian matrix of the mismatch equations is also difficult to obtain [104].

10.1.4 Solution Flow Diagram

Fig. 10.3 shows the solution flow diagram to solve for the closed-loop buck converter. First, the two unknowns ($x_i(0)$ and t_{sw}) are initialized. t_{sw} is passed through (10.8) to obtain the steady-state initial conditions of the state variables of the plant. Once $\mathbf{x_p}(0)$ is obtained, the mismatch equations, based on (i) the steady-state constraint of controller's state variable and (ii) the switching time constraint, can be evaluated by (10.13) and (10.14), respectively:

$$M_1 = \begin{bmatrix} 0 & 0 & 0 & 1 & 0 \end{bmatrix} \left\{ \check{\mathbf{\Phi}} - \mathbb{I} \right\} \check{\mathbf{x}}(0) \tag{10.13}$$

$$M_2 = \frac{v_M}{T_s} t_{sw} - \begin{bmatrix} 0 & -KH & 0 & \frac{1}{\tau} & K \end{bmatrix} e^{\check{\mathbf{A}}_{\mathbf{on}} \mathbf{t}_{\mathbf{sw}}} \check{\mathbf{x}}(0). \tag{10.14}$$

Equation (10.13) is needed to ensure the controller is also in steady-state, whereas (10.14) is required to ensure that the loop is closed. More importantly, the mismatch equations set the stage for a Newton-type iterative method, which starts with an initial seed value ζ, and generates the sequence, $\{\zeta\}_{n=0}^{\infty}$ by (10.15):

$$\zeta^{(k+1)} = \zeta^{(k)} - \mathbf{J}^{-1}\mathbf{M}, \tag{10.15}$$

where $\mathbf{M} = [M_1 \;\; M_2]^T$, $\mathbf{J} = \begin{bmatrix} J_{11} & J_{12} \\ J_{21} & J_{22} \end{bmatrix} = \begin{bmatrix} \frac{\partial M_1}{\partial x_i(0)} & \frac{\partial M_1}{\partial t_{sw}} \\ \frac{\partial M_2}{\partial x_i(0)} & \frac{\partial M_2}{\partial t_{sw}} \end{bmatrix}$.

The analytical expressions for J_{11}, J_{12}, J_{21}, and J_{22}, are listed as follows:

$$J_{11} = \begin{bmatrix} 0 & 0 & 0 & 1 & 0 \end{bmatrix} \left\{ \check{\mathbf{\Phi}} - \mathbb{I} \right\} \begin{bmatrix} 0 \\ 0 \\ 0 \\ 1 \\ 0 \end{bmatrix}, \tag{10.16}$$

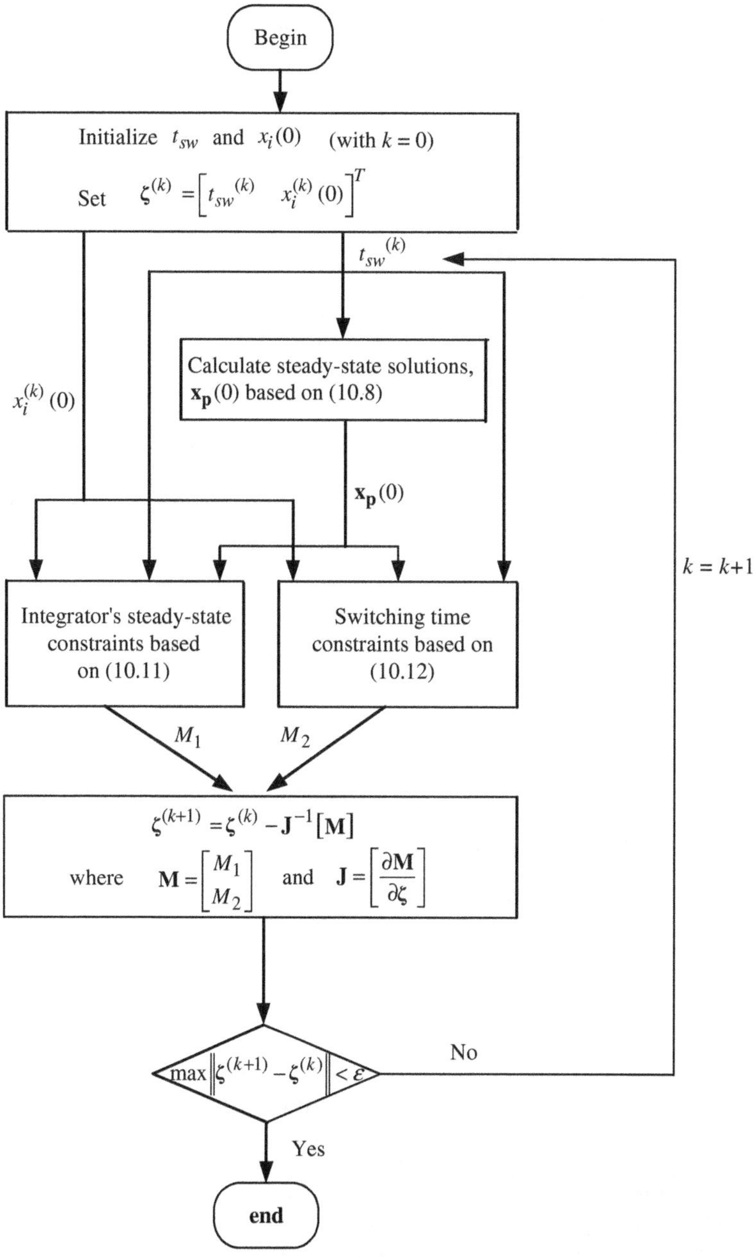

Figure 10.3 Solution flow chart of the shooting method for the buck converter.

$$J_{12} = \begin{bmatrix} 0 & 0 & 0 & 1 & 0 \end{bmatrix} \left((-\check{\mathbf{A}}_{\mathbf{off}})\check{\mathbf{\Phi}}\check{\mathbf{x}}(0) + \check{\mathbf{\Phi}}\check{\mathbf{A}}_{\mathbf{on}}\check{\mathbf{x}}(0) + \check{\mathbf{\Phi}} \begin{bmatrix} \dfrac{\partial \mathbf{x_p}(0)}{\partial t_{sw}} \\ 0 \\ 0 \\ 0 \end{bmatrix} \right), \tag{10.17}$$

Table 10.1 The parameter values of Fig. 10.1.

Source voltage	$v_s = 28V$
Source inductance	$L = 50\mu H$
DC capacitance	$C = 500\mu F$
Load resistance	$R = 3\Omega$
DC voltage reference	$v_{ref} = 5V$
PWM strategy	Natural-sampled PWM
Switching frequency	$f_s = 10kHz$
Sensor gain	$H = \frac{1}{3}$
Proportional gain	$K = 5.0512 \times 10^{-5}$
Integral constant	$\tau = 0.5543$

$$J_{21} = -\begin{bmatrix} 0 & \frac{-K}{3} & 0 & \frac{1}{\tau} & K \end{bmatrix} e^{\check{\mathbf{A}}_{\mathbf{on}} t_{sw}} \begin{bmatrix} 0 \\ 0 \\ 0 \\ 1 \\ 0 \end{bmatrix}, \tag{10.18}$$

$$J_{22} = \frac{v_M}{T_s} - \begin{bmatrix} 0 & \frac{-K}{3} & 0 & \frac{1}{\tau} & K \end{bmatrix} \left(e^{\check{\mathbf{A}}_{\mathbf{on}} t_{sw}} \check{\mathbf{A}}_{\mathbf{on}} \check{\mathbf{x}}(0) + e^{\check{\mathbf{A}}_{\mathbf{on}} t_{sw}} \begin{bmatrix} \frac{\partial \mathbf{x_p}(0)}{\partial t_{sw}} \\ 0 \\ 0 \\ 0 \end{bmatrix} \right), \tag{10.19}$$

where

$$\frac{\partial \mathbf{x_p}(0)}{\partial t_{sw}} = \left\{ -(\mathbb{I} - \mathbf{\Phi})^{-2}(-\mathbf{\Phi_d})\mathbf{\Gamma} + (\mathbb{I} - \mathbf{\Phi})^{-1}\mathbf{\Gamma_d} \right\} \mathbf{z(0)}. \tag{10.20}$$

Note that $\mathbf{\Phi_d}$ has the same size as $\mathbf{\Phi}$, whereas $\mathbf{\Gamma_d}$ has the same size as $\mathbf{\Gamma}$. Their values are extracted from the following matrices:

$$\frac{d\bar{\mathbf{\Phi}}}{dt_{sw}} = -\bar{\mathbf{A}}_{\mathbf{off}}\bar{\mathbf{\Phi}} + \bar{\mathbf{\Phi}}\bar{\mathbf{A}}_{\mathbf{on}} = \left[\begin{array}{c|c} \mathbf{\Phi_d} & \mathbf{\Gamma_d} \\ \hline \mathbf{0} & 0 \end{array}\right].$$

Equation (10.20) is found to be particularly useful when the number of switching actions is very high, as in the case of a VSC.

Example 10.1 Given that $V_s = 28V$, and $v_{ref} = 5V$ (refer to Fig. 10.1 and Table 10.1), a duty cycle of 0.5357, and $x_i(0) = 0.0$ are initialized. Only two iterations are required to converge to the true solution, $t_{sw} = 5.3571 \times 10^{-5}$ sec, and $x_i = 1.1878$.

Fig. 10.4 shows the waveform of the inductor current, $i_L(t)$, and the capacitor voltage, $v_{dc}(t)$. The dc value of the capacitor voltage is regulated at the desired reference value, 15 V (note that the sensor gain is $\frac{1}{3}$).

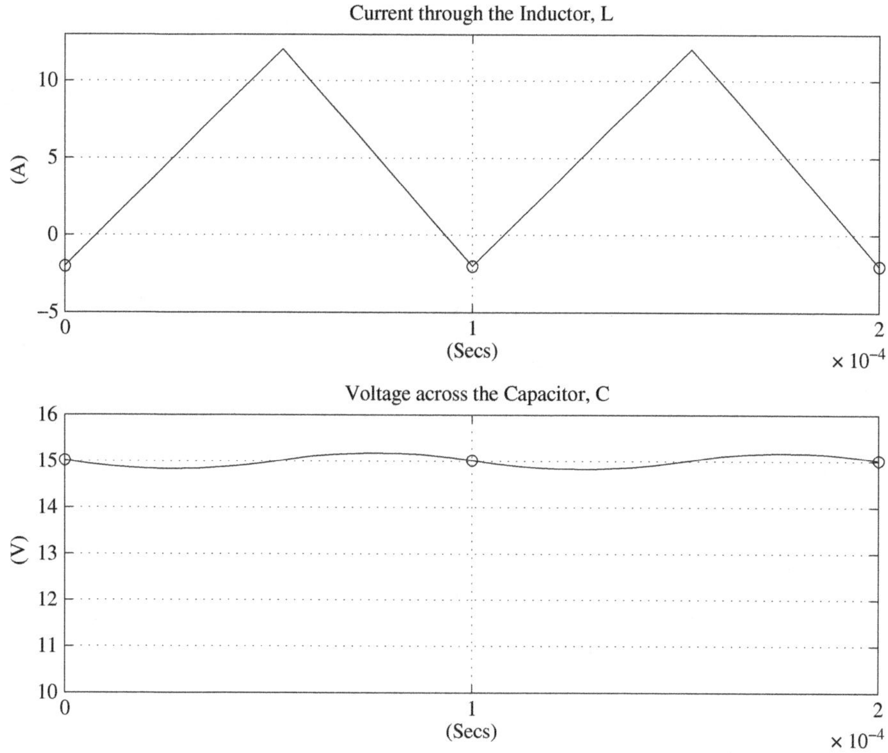

Figure 10.4 The inductor current and capacitor voltage of the regulated buck converter.

```

Vs=28.00;
L=50e-6;
C=500e-6;

R=3;
fsw=10e3;
T=1/fsw;

Gtf=tf(0.536,[L*C L/R 1]);
Vref=5;
P=5.0512e-4;
Ti=0.5543;

Aon=[0 -1/L; 1/C -1/(R*C)];
Aoff=[0 -1/L; 1/(C) -1/(R*C)];
B=[1/L;0];

%treat the input as one of the states
Aon2=[0 -1/L 1/L; 1/C -1/(R*C) 0;0 0 0];
Aoff2=[0 -1/L 0; 1/C -1/(R*C) 0;0 0 0];


%incorporate the control and input states
AonB=[0 -1/L 1/L 0 0; 1/C -1/(R*C) 0 0 0;0 0 0 0 0;0 -1/3 0 0 1;0 0 0 0 0];
AoffB=[0 -1/L 0 0 0; 1/C -1/(R*C) 0 0 0;0 0 0 0 0;0 -1/3 0 0 1;0 0 0 0 0];



```

```
k=0;
t1=T*0.5357;  %initialization
xi=0.0; %initialization
error=1;
counter=0;

while (error>1e-6)

% nominal values
phi2=expm(Aoff*(T-t1))*expm(Aon*t1);
phiB=expm(AoffB*(T-t1))*expm(AonB*t1);

phi=eye(3)-expm(Aoff2*(T-t1))*expm(Aon2*t1);
E=phi(1:2,1:2);
F=phi(1:2,3);
xss=-inv(E)*F*[Vs];


x1=expm(AonB*t1)*[xss;Vs;xi;Vref];

xss2=phiB*[xss;Vs;xi;Vref];

BB=(-((eye(2)-phi2)^-2)*(Aoff*phi2-phi2*Aon)*(-expm(Aoff*(T-t1))*...
     inv(Aon)*B+expm(Aoff*(T-t1))*inv(Aon)*expm(Aon*t1)*B)+...
     inv(eye(2)-phi2)*(expm(Aoff*(T-t1))*Aoff*inv(Aon)*B-...
     expm(Aoff*(T-t1))*Aoff*inv(Aon)*expm(Aon*t1)*B+ ...
     expm(Aoff*(T-t1))*inv(Aon)*expm(Aon*t1)*(Aon)*B))*Vs;


Jx=phiB*[0;0;0;1;0]; %phiB(4)-1
J11=Jx(4)-1;
J12B=-AoffB*phiB*[xss;Vs;xi;Vref]+phiB*AonB*[xss;Vs;xi;Vref]+...
     phiB*[BB;0;0;0];
J12=J12B(4);

J2A=-1/Ti*expm(AonB*t1)*[0;0;0;1;0];
J21=J2A(4);

J22=4/T-1*[0 -P/3 0 1/Ti P]*(expm(AonB*t1)*AonB*[xss;Vs;xi;Vref]+...
     expm(AonB*t1)*[BB;0;0;0]);

f1=xss2(4)-xi;
f2=4/T*t1-1*[0 -P/3 0 1/Ti P]*x1;


J=[J11 J12;J21 J22];


dt=inv(J)*[f1;f2];
xi=xi-dt(1);
t1=t1-dt(2);
counter=counter+1;
error=max(abs(dt./[xi;t1]))
end
```

□

10.2 Two-Level VSC Modeling: Open-Loop Equations

This section employs the LB 2 method to model a two-level VSC. Similar to the case of a buck converter, the steady-state formulation of a two-level VSC is presented in this section. To formulate

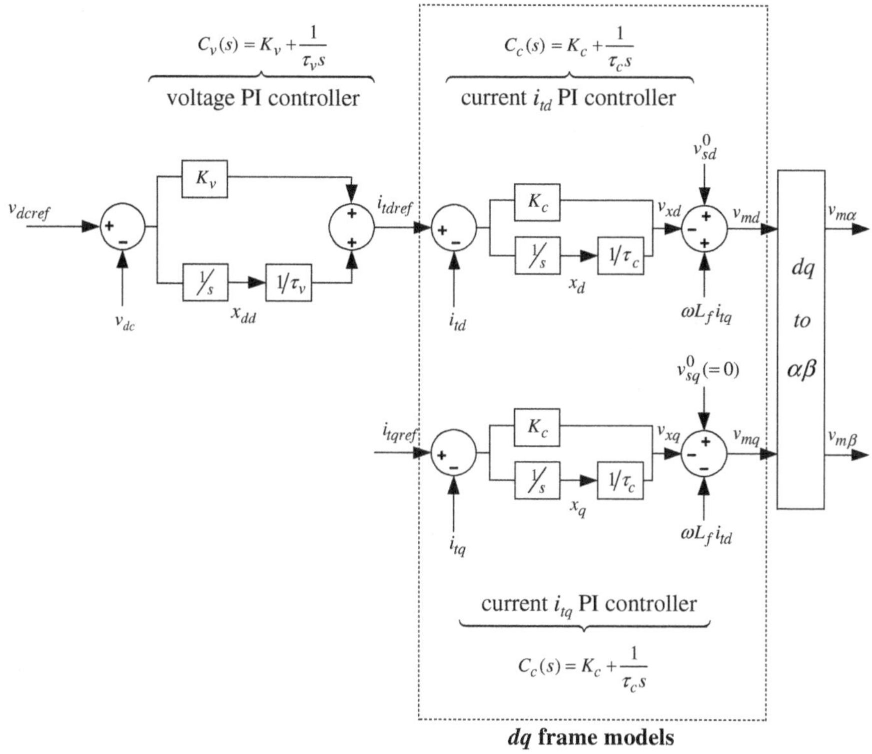

Figure 10.5 The overall controller of the D-STATCOM.

the switching time constraints, the expressions for the voltage modulating signals, $\mathbf{v_m}$, are needed. To formulate the steady-state constraints, the expressions for $\mathbf{x_p}(T)$ and $\mathbf{x_i}(T)$ are needed. To obtain these expressions for the VSC, the open-loop equations, consisting of both the power converter and the controllers for the VSC, need to be formulated. This requires the VSC converter circuit model and the controller's model to be in the same reference frame.

Fig. 10.5 shows the current controller of a VSC in the synchronous dq reference frame, together with a supervisory dc voltage controller [100]. Note that here we assume the phase-locked loop perfectly tracks the positive sequence component of the fundamental voltage. Hence, the signals v_{sd} and v_{sq} entering the controller are pure dc quantities, which are designated as v_{sd}^0 and v_{sq}^0, respectively.

As indicated in the figure, the three PI controllers introduce three state variables, x_d, x_q, and x_{dd}. To augment the VSC's differential equations with the controller's differential equations, one needs to convert the state variables in the dq-frame (i.e. x_d and x_q) to the equivalent $\alpha\beta$ state variables. This conversion process is based on the work of [105] and [106], and is described as follows:

The transfer function of the current controller is given as

$$C_c(s) = K_c + \frac{1}{\tau_c s}, \tag{10.21}$$

which can be converted into the time domain by introducing extra states, x_d and x_q, as shown in Fig. 10.6.

$$\frac{d}{dt}x_d = -i_{td} + i_{tdref} \tag{10.22}$$

$$\frac{d}{dt}x_q = -i_{tq} + i_{tqref}. \tag{10.23}$$

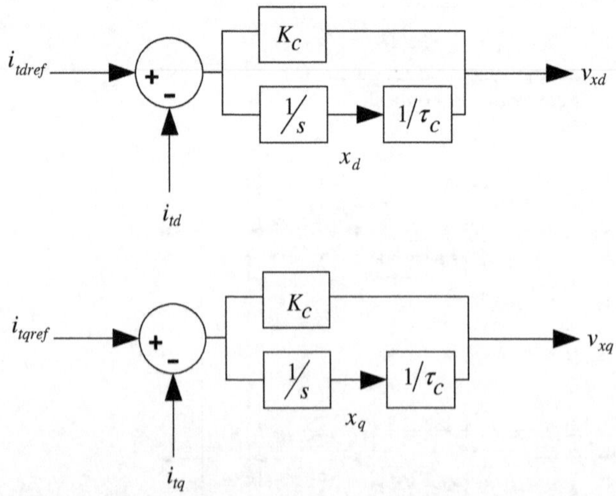

Figure 10.6 The current controllers in *dq*-frame.

Multiplying both sides of (10.23) by *j* and adding it to (10.22) defines the *dq*-frame space vectors:

$$\frac{d}{dt}(x_d + jx_q) = -(i_{td} + ji_{tq}) + (i_{tdref} + ji_{tqref}). \tag{10.24}$$

Assuming prefect synchronization to the positive sequence component, (10.24) can be rewritten as:

$$\frac{d}{dt}\left\{\left(x_\alpha + jx_\beta\right)e^{-j\omega t}\right\} = -\left\{\left(i_{t\alpha} + ji_{t\beta}\right)e^{-j\omega t}\right\} + \left\{\left(i_{t\alpha ref} + ji_{t\beta ref}\right)e^{-j\omega t}\right\}. \tag{10.25}$$

Let $x_{\alpha\beta} = x_\alpha + jx_\beta$, $i_{t\alpha\beta} = i_{t\alpha} + ji_{t\beta}$, and $i_{t\alpha\beta ref} = i_{t\alpha ref} + ji_{t\beta ref}$, and (10.26) results:

$$\left\{\frac{d}{dt}x_{\alpha\beta}\right\}e^{-j\omega t} + x_{\alpha\beta}e^{-j\omega t}(-j\omega) = -\left(i_{t\alpha\beta}e^{-j\omega t}\right) + \left(i_{t\alpha\beta ref}e^{-j\omega t}\right), \tag{10.26}$$

or

$$\left\{\frac{d}{dt}x_{\alpha\beta}\right\} + x_{\alpha\beta}(-j\omega) = -\left(i_{t\alpha\beta}\right) + \left(i_{t\alpha\beta ref}\right). \tag{10.27}$$

Express (10.27) in matrix form:

$$\frac{d}{dt}\begin{bmatrix} x_\alpha \\ x_\beta \end{bmatrix} = \begin{bmatrix} 0 & -\omega \\ \omega & 0 \end{bmatrix}\begin{bmatrix} x_\alpha \\ x_\beta \end{bmatrix} + \begin{bmatrix} i_{t\alpha ref} - i_{t\alpha} \\ i_{t\beta ref} - i_{t\beta} \end{bmatrix}. \tag{10.28}$$

Note that it is v_{dcref} and i_{tqref}, not $i_{t\alpha ref}$ and $i_{t\beta ref}$, which are specified by the users. Hence, $i_{t\alpha ref}$ and $i_{t\beta ref}$ of (10.28) need to be linked back to v_{dcref} and i_{tqref}. This is achieved by the dc voltage controller. Fig. 10.7 describes the transfer function of the dc voltage controller, $C_v(s)$:

$$C_v(s) = K_v + \frac{1}{\tau_v s} \tag{10.29}$$

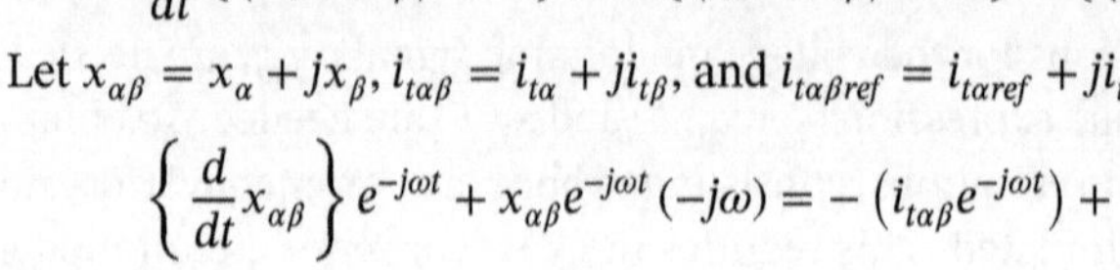

Figure 10.7 The DC voltage controller.

while its state space equations are

$$\frac{d}{dt}x_{dd} = v_{dcref} - v_{dc}, \tag{10.30}$$

and

$$i_{tdref} = K_v(v_{dcref} - v_{dc}) + \frac{1}{\tau_v}x_{dd}. \tag{10.31}$$

Converting i_{tdref} and i_{tqref} into $i_{t\alpha ref}$ and $i_{t\beta ref}$ via the Park transformation [67]:

$$\begin{bmatrix} i_{t\alpha ref} \\ i_{t\beta ref} \end{bmatrix} = \begin{bmatrix} \cos(\omega t) & -\sin(\omega t) \\ \sin(\omega t) & \cos(\omega t) \end{bmatrix} \begin{bmatrix} i_{tdref} \\ i_{tqref} \end{bmatrix}, \tag{10.32}$$

one obtains

$$i_{t\alpha ref} = K_v v_{dcref}\cos(\omega t) - K_v v_{dc}\cos(\omega t) + \frac{1}{\tau_v}x_{dd}\cos(\omega t) - i_{tqref}\sin(\omega t) \tag{10.33}$$

$$i_{t\beta ref} = K_v v_{dcref}\sin(\omega t) - K_v v_{dc}\sin(\omega t) + \frac{1}{\tau_v}x_{dd}\sin(\omega t) + i_{tqref}\cos(\omega t). \tag{10.34}$$

Substituting (10.33) and (10.34) into (10.28) allows the dc voltage reference and the reactive current reference to be incorporated into the $\alpha\beta$ form system model.

Finally, if combining all the differential equations, describing the controller's integrators (10.28) and (10.30), together with the VSC's differential equations, one will get (10.35):

$$\frac{d}{dt}\check{\mathbf{x}}(t) = \check{\mathbf{A}}_{\mathbf{tv}}(t)\check{\mathbf{x}}(t), \tag{10.35}$$

where

$$\check{\mathbf{x}} = \begin{bmatrix} \mathbf{x_p} & \mathbf{x_i} & \mathbf{z} & \mathbf{r} \end{bmatrix}^T,$$

$$\mathbf{x_p} = \begin{bmatrix} i_{t\alpha} & i_{t\beta} & v_{dc} \end{bmatrix}, \mathbf{x_i} = \begin{bmatrix} x_\alpha & x_\beta & x_{dd} \end{bmatrix}, \mathbf{z} = \begin{bmatrix} v_{s\alpha}^1 & v_{s\beta}^1 \end{bmatrix}, \mathbf{r} = \begin{bmatrix} i_{tqref} & v_{dcref} \end{bmatrix},$$

$$\check{\mathbf{A}}_{\mathbf{tv}}(t) = \begin{bmatrix} -\frac{R}{L} & 0 & -\frac{S_\alpha}{L} & 0 & 0 & 0 & \frac{1}{L} & 0 & 0 & 0 \\ 0 & -\frac{R}{L} & -\frac{S_\beta}{L} & 0 & 0 & 0 & 0 & \frac{1}{L} & 0 & 0 \\ \frac{3S_\alpha}{2C} & \frac{3S_\beta}{2C} & 0 & 0 & 0 & 0 & 0 & 0 & 0 & 0 \\ -1 & 0 & -K_v\cos\omega t & 0 & -\omega & \frac{1}{\tau_v}\cos\omega t & 0 & 0 & -\sin\omega t & K_v\cos\omega t \\ 0 & -1 & -K_v\sin\omega t & \omega & 0 & \frac{1}{\tau_v}\sin\omega t & 0 & 0 & \cos\omega t & K_v\sin\omega t \\ 0 & 0 & -1 & 0 & 0 & 0 & 0 & 0 & 0 & 1 \\ 0 & 0 & 0 & 0 & 0 & 0 & 0 & -\omega & 0 & 0 \\ 0 & 0 & 0 & 0 & 0 & 0 & \omega & 0 & 0 & 0 \\ 0 & 0 & 0 & 0 & 0 & 0 & 0 & 0 & 0 & 0 \\ 0 & 0 & 0 & 0 & 0 & 0 & 0 & 0 & 0 & 0 \end{bmatrix},$$

and $L = L_s + L_f$. Note that $v_{s\alpha}^1$ and $v_{s\beta}^1$ represent the positive-sequence fundamental component of $v_{s\alpha}^1$ and $vs\beta$.

Equation (10.35) is the overall open loop equation when the dq-frame state variables of the current controller, x_d and x_q, are converted into their equivalent $\alpha\beta$-frame state variables, x_α and x_β. Note that the input stimuli, $\mathbf{z}$, here only contains the positive sequence fundamental component to avoid the above system matrix size from becoming excessively large. In practice, $\mathbf{z}$ can hold as many harmonic oscillators as desired.

Equation (10.35) is a LTV system, and $\check{\mathbf{A}}_{\mathbf{tv}}$ has eight different forms [107] due to eight different combinations of the switching function, $\{S_a, S_b, S_c\}$.

The solution to (10.35) is

$$\check{\mathbf{x}}(t) = \check{\mathbf{\Phi}}_{\mathbf{tv}}(t,0)\check{\mathbf{x}}(0), \tag{10.36}$$

where $\check{\mathbf{\Phi}}_{\mathbf{tv}}(t,0)$ is the state transition matrix. Chapter 4 has discussed the technique of solving $\check{\mathbf{\Phi}}_{\mathbf{tv}}(t,0)$.

Further note that if i_{tdref} is set by the user rather than by the supervisory dc voltage controller, the open-loop equation will form a linear time-invariant system within every switching time interval. The resulting open-loop equation is given by (10.37):

$$\frac{d}{dt}\begin{bmatrix} i_\alpha \\ i_\beta \\ v_{dc} \\ x_\alpha \\ x_\beta \\ v_{s\alpha}^1 \\ v_{s\beta}^1 \\ i_{t\alpha ref} \\ i_{t\beta ref} \end{bmatrix} = \begin{bmatrix} \frac{-R}{L} & 0 & \frac{-S_\alpha}{L} & 0 & 0 & \frac{1}{L} & 0 & 0 & 0 \\ 0 & \frac{-R}{L} & \frac{-S_\beta}{L} & 0 & 0 & 0 & \frac{1}{L} & 0 & 0 \\ \frac{3S_\alpha}{2C} & \frac{3S_\beta}{2C} & 0 & 0 & 0 & 0 & 0 & 0 & 0 \\ -1 & 0 & 0 & 0 & -\omega & 0 & 0 & 1 & 0 \\ 0 & -1 & 0 & \omega & 0 & 0 & 0 & 0 & 1 \\ 0 & 0 & 0 & 0 & 0 & 0 & -\omega & 0 & 0 \\ 0 & 0 & 0 & 0 & 0 & \omega & 0 & 0 & 0 \\ 0 & 0 & 0 & 0 & 0 & 0 & 0 & 0 & -\omega \\ 0 & 0 & 0 & 0 & 0 & 0 & 0 & \omega & 0 \end{bmatrix}\begin{bmatrix} i_\alpha \\ i_\beta \\ v_{dc} \\ x_\alpha \\ x_\beta \\ v_{s\alpha}^1 \\ v_{s\beta}^1 \\ i_{t\alpha ref} \\ i_{t\beta ref} \end{bmatrix}. \tag{10.37}$$

Thus, for constant current references the system is LTI for each switching interval, just like the buck converter. The focus of the next section will be dealing with the more difficult case of the VSC with dc voltage control, which produces an LTV system for each switching time interval.

10.2.1 Steady-State Constraints

For the VSC, the steady-state initial conditions of the power circuit are developed in a manner analogous to that of the buck converter of Section 10.1.2:

$$\mathbf{x_p}(0) = \{\mathbb{I} - \mathbf{\Phi}(\mathbf{t_{sw}})\}^{-1}\mathbf{\Gamma}(\mathbf{t_{sw}})\mathbf{z}(0), \tag{10.38}$$

where

$$\mathbf{x_p} = \begin{bmatrix} i_\alpha & i_\beta & v_{dc} \end{bmatrix}^T,$$

$$\mathbf{z} = \begin{bmatrix} v_{s\alpha} & v_{s\beta} \end{bmatrix}^T,$$

$$\bar{\mathbf{\Phi}} = \prod_{k=n+1}^{1} e^{\bar{\mathbf{A}}_{\mathbf{k}}(t_k - t_{k-1})} = \begin{bmatrix} \mathbf{\Phi}(\mathbf{t_{sw}}) & \mathbf{\Gamma}(\mathbf{t_{sw}}) \\ \mathbf{0} & \mathbb{I} \end{bmatrix},$$

$$\bar{\mathbf{A}}_{\mathbf{k}} = \begin{bmatrix} \frac{-R}{L} & 0 & \frac{-S_\alpha}{L} & \frac{1}{L} & 0 \\ 0 & \frac{-R}{L} & \frac{-S_\beta}{L} & 0 & \frac{1}{L} \\ \frac{3S_\alpha}{2C} & \frac{3S_\beta}{2C} & 0 & 0 & 0 \\ 0 & 0 & 0 & 0 & -\omega \\ 0 & 0 & 0 & \omega & 0 \end{bmatrix},$$

and $t_{n+1} = T, t_0 = 0, t_1, \ldots, t_n$ are switching times like those obtained in Fig. 6.7.

The steady-state periodicity constraints on the controller's state variables are:

$$x_\alpha(T) - x_\alpha(0) = 0 \quad (10.39)$$

$$x_\beta(T) - x_\beta(0) = 0 \quad (10.40)$$

$$x_{dd}(T) - x_{dd}(0) = 0. \quad (10.41)$$

When the solution to (10.35) is found, (10.39), (10.40), and (10.41) can be expressed as (10.42), (10.43), and (10.44), respectively:

$$\begin{bmatrix}0 & 0 & 0 & 1 & 0 & 0 & 0 & 0 & 0 & 0\end{bmatrix}\left\{\check{\mathbf{\Phi}}_{\mathbf{tv}}(T,0) - \mathbb{I}\right\}\check{\mathbf{x}}(0) = 0 \quad (10.42)$$

$$\begin{bmatrix}0 & 0 & 0 & 0 & 1 & 0 & 0 & 0 & 0 & 0\end{bmatrix}\left\{\check{\mathbf{\Phi}}_{\mathbf{tv}}(T,0) - \mathbb{I}\right\}\check{\mathbf{x}}(0) = 0 \quad (10.43)$$

$$\begin{bmatrix}0 & 0 & 0 & 0 & 0 & 1 & 0 & 0 & 0 & 0\end{bmatrix}\left\{\check{\mathbf{\Phi}}_{\mathbf{tv}}(T,0) - \mathbb{I}\right\}\check{\mathbf{x}}(0) = 0, \quad (10.44)$$

where

$$\check{\mathbf{\Phi}}_{\mathbf{tv}}(T,0) = \prod_{k=n+1}^{k=1} \check{\mathbf{\Psi}}_{\mathbf{tv}}(t_k, t_{k-1}).$$

10.2.2 Switching Time Constraints

Switching time constraints are a function of the modulation technique employed. For illustration purposes, switching time constraints are formulated based on natural sampled PWM since this technique is most prominent in the literature.

Fig. 10.8 shows how switching times are generated. For example, t_7 is generated when the second positive slope of the triangular carrier signal intersects with the phase c modulating signal, v_{mc}, and the constraint equation is:

$$\frac{4}{T_{sw}}t_7 - 7 - v_{mc}(t_7) = 0, \quad (10.45)$$

where T_{sw} is the switching period.

t_9 is generated when the third negative slope of the triangular carrier signal intersects with the phase b modulating signal, v_{mb}, and the constraint equation is:

$$-\frac{4}{T_{sw}}t_9 + 9 - v_{mb}(t_9) = 0. \quad (10.46)$$

Therefore, the switching time constraints of the natural sampled PWM can be generally expressed using the form:

$$\pm\frac{4}{T_{sw}}\mathbf{t_a} \mp \mathbf{k_a} - v_{ma}(\mathbf{t_a}) = 0 \quad (10.47)$$

$$\pm\frac{4}{T_{sw}}\mathbf{t_b} \mp \mathbf{k_b} - v_{mb}(\mathbf{t_b}) = 0 \quad (10.48)$$

$$\pm\frac{4}{T_{sw}}\mathbf{t_c} \mp \mathbf{k_c} - v_{mc}(\mathbf{t_c}) = 0, \quad (10.49)$$

where $\mathbf{t_a}$, $\mathbf{t_b}$, and $\mathbf{t_c}$ represent the switching times for phases a, b, and c respectively, and constants, $\mathbf{k_a}$, $\mathbf{k_b}$, and $\mathbf{k_c}$ are the y-intercepts of the appropriate carrier wave segment.

Since the output of the controllers are $v_{m\alpha}$, and $v_{m\beta}$ (refer back Fig. 10.5), one needs to apply the Clarke transformation to them to obtain the modulating signals in the abc reference frame.

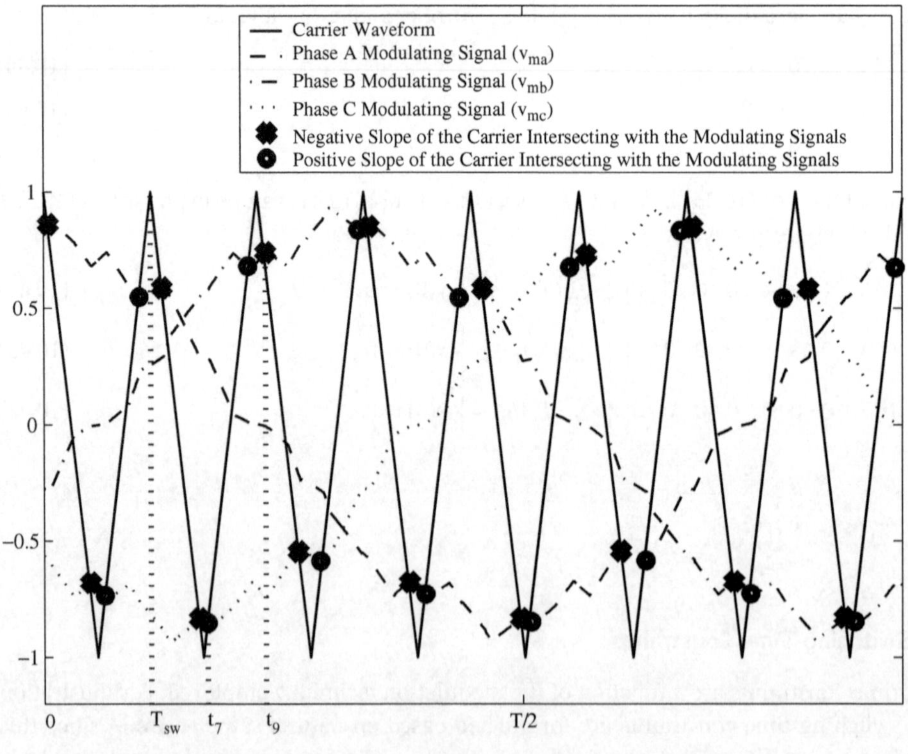

Figure 10.8 Triangular carrier waveform together with the modulating signals.

The resulting expressions are:

$$v_{ma} = v_{m\alpha} \tag{10.50}$$

$$v_{mb} = -\frac{1}{2}v_{m\alpha} + \frac{\sqrt{3}}{2}v_{m\beta} \tag{10.51}$$

$$v_{mc} = -\frac{1}{2}v_{m\alpha} - \frac{\sqrt{3}}{2}v_{m\beta}. \tag{10.52}$$

$v_{m\alpha}$ and $v_{m\beta}$ are, in turn, a linear combination of all the state variables, and their expressions are given in (10.53) and (10.54):

$$v_{m\alpha} = K_c i_{t\alpha} + \omega L_f i_{t\beta} + K_c K_v \cos(\omega t) v_{dc} - \frac{1}{\tau_c} x_\alpha - \frac{K_c}{\tau_v}\cos(\omega t) x_{dd} + v_{s\alpha}^1 + K_c \sin(\omega t) i_{tqref}$$
$$-K_c K_v \cos(\omega t) v_{dcref} \tag{10.53}$$

$$v_{m\beta} = -\omega L_f i_{t\alpha} + K_c i_{t\beta} + K_c K_v \sin(\omega t) v_{dc} - \frac{1}{\tau_c} x_\beta - \frac{K_c}{\tau_v}\sin(\omega t) x_{dd} + v_{s\beta}^1 - K_c \cos(\omega t) i_{tqref}$$
$$-K_c K_v \sin(\omega t) v_{dcref}. \tag{10.54}$$

Note that the derivations of (10.53) and (10.54) are described as follows:

To find out $v_{m\alpha}$ and $v_{m\beta}$, one needs to refer to Fig. 10.5. From Fig. 10.5, the following equations can be written:

$$v_{md} = v_{sd}^0 + \omega L_f i_{tq} - v_{xd} \tag{10.55}$$

$$v_{mq} = v_{sq}^0 - \omega L_f i_{td} - v_{xq}. \tag{10.56}$$

Multiplying both sides of (10.56) by j and adding it to (10.55) defines the dq-frame space vectors:

$$(v_{md} + jv_{mq}) = (v_{sd}^0 + jv_{sq}^0) + \omega L_f(i_{tq} - ji_{td}) - (v_{xd} + jv_{xq}). \tag{10.57}$$

Multiplying both sides of (10.57) by $e^{j\omega t}$ converts it into the $\alpha\beta$ reference frame:

$$(v_{m\alpha}+jv_{m\beta})=(v_{s\alpha}^{1}+jv_{s\beta}^{1})+\frac{\omega}{j}L_f(i_{t\alpha}+ji_{t\beta})-(v_{x\alpha}+jv_{x\beta}), \tag{10.58}$$

or equivalently,

$$v_{m\alpha}=v_{s\alpha}^{1}+\omega L_f i_{t\beta}-v_{x\alpha} \tag{10.59}$$

$$v_{m\beta}=v_{s\beta}^{1}-\omega L_f i_{t\alpha}-v_{x\beta}. \tag{10.60}$$

$v_{x\alpha}$ and $v_{x\beta}$ can be found by the following:

$$v_{xd}=K_c(i_{tdref}-i_{td})+\frac{1}{\tau_c}x_d \tag{10.61}$$

$$v_{xq}=K_c(i_{tqref}-i_{tq})+\frac{1}{\tau_c}x_q. \tag{10.62}$$

Converting (10.61) and (10.62) to the $\alpha\beta$ reference frame, one gets (10.63) and (10.64), respectively.

$$v_{x\alpha}=K_c(i_{t\alpha ref}-i_{t\alpha})+\frac{1}{\tau_c}x_\alpha \tag{10.63}$$

$$v_{x\beta}=K_c(i_{t\beta ref}-i_{t\beta})+\frac{1}{\tau_c}x_\beta. \tag{10.64}$$

Substituting (10.63) into (10.59), and (10.64) into (10.60), one obtains (10.53) and (10.54). Substituting (10.53) and (10.54) into (10.50), (10.51), and (10.52), one gets:

$$v_{ma}(t)=\mathbf{C_a}(t)\check{\mathbf{\Phi}}_{\mathbf{tv}}(t,0)\check{\mathbf{x}}(0), \tag{10.65}$$

where

$$\mathbf{C_a}(t)=\begin{bmatrix}K_c & \omega L_f & K_cK_v\cos(\omega t) & \frac{-1}{\tau_c} & 0 & \frac{-K_c}{\tau_c}\cos(\omega t) & 1 & 0 & K_c\sin(\omega t) & -K_cK_v\cos(\omega t)\end{bmatrix}. \tag{10.66}$$

$$v_{mb}(t)=\mathbf{C_b}(t)\check{\mathbf{\Phi}}_{\mathbf{tv}}(t,0)\check{\mathbf{x}}(0), \tag{10.67}$$

where

$$\begin{aligned}\mathbf{C_b}(t)=\Bigg[&\left(-\frac{\sqrt{3}}{2}\omega L_f-\frac{K_c}{2}\right)\ \left(\frac{\sqrt{3}}{2}K_c-\frac{1}{2}\omega L_f\right)\ \left(\frac{\sqrt{3}}{2}K_cK_v\sin\omega t-\frac{1}{2}K_cK_v\cos\omega t\right)\ \frac{1}{2\tau_c}\\ &-\frac{\sqrt{3}}{2\tau_c}\ \left(\frac{K_c}{2\tau_v}\cos\omega t-\frac{\sqrt{3}}{2\tau_v}K_c\sin\omega t\right)\ -\frac{1}{2}\ \frac{\sqrt{3}}{2}\ \left(-\frac{1}{2}K_c\sin\omega t-\frac{\sqrt{3}}{2}K_c\cos\omega t\right)\\ &\left(\frac{1}{2}K_cK_v\cos\omega t-\frac{\sqrt{3}}{2}K_cK_v\sin\omega t\right)\Bigg].\end{aligned} \tag{10.68}$$

$$v_{mc}(t)=\mathbf{C_c}(t)\check{\mathbf{\Phi}}_{\mathbf{tv}}(t,0)\check{\mathbf{x}}(0), \tag{10.69}$$

where

$$\begin{aligned}\mathbf{C_c}(t)=\Bigg[&\left(\frac{\sqrt{3}}{2}\omega L_f-\frac{K_c}{2}\right)\ \left(-\frac{\sqrt{3}}{2}K_c-\frac{1}{2}\omega L_f\right)\ \left(-\frac{\sqrt{3}}{2}K_cK_v\sin\omega t-\frac{1}{2}K_cK_v\cos\omega t\right)\ \frac{1}{2\tau_c}\\ &\frac{\sqrt{3}}{2\tau_c}\ \left(\frac{K_c}{2\tau_v}\cos\omega t+\frac{\sqrt{3}}{2\tau_v}K_c\sin\omega t\right)\ -\frac{1}{2}\ -\frac{\sqrt{3}}{2}\ \left(-\frac{1}{2}K_c\sin\omega t+\frac{\sqrt{3}}{2}K_c\cos\omega t\right)\\ &\left(\frac{1}{2}K_cK_v\cos\omega t+\frac{\sqrt{3}}{2}K_cK_v\sin\omega t\right)\Bigg].\end{aligned} \tag{10.70}$$

10.2.3 Solution Flow Diagram

The solution flow diagram of the VSC is similar to that of the dc-dc converter and is shown in Fig. 10.9. The mismatch equations of a VSC based on the steady-state and switching time constraints are as follows:

$$M_1 = \begin{bmatrix} 0 & 0 & 0 & 1 & 0 & 0 & 0 & 0 & 0 & 0 \end{bmatrix} \left\{ \check{\mathbf{\Phi}}_{\mathbf{tv}}(\mathbf{T},\mathbf{0}) - \mathbb{I} \right\} \check{\mathbf{x}}(\mathbf{0}) \tag{10.71}$$

$$M_2 = \begin{bmatrix} 0 & 0 & 0 & 0 & 1 & 0 & 0 & 0 & 0 & 0 \end{bmatrix} \left\{ \check{\mathbf{\Phi}}_{\mathbf{tv}}(\mathbf{T},\mathbf{0}) - \mathbb{I} \right\} \check{\mathbf{x}}(\mathbf{0}) \tag{10.72}$$

$$M_3 = \begin{bmatrix} 0 & 0 & 0 & 0 & 0 & 1 & 0 & 0 & 0 & 0 \end{bmatrix} \left\{ \check{\mathbf{\Phi}}_{\mathbf{tv}}(\mathbf{T},\mathbf{0}) - \mathbb{I} \right\} \check{\mathbf{x}}(\mathbf{0}) \tag{10.73}$$

$$\mathbf{M_4} = \pm\frac{4}{T_{sw}}\mathbf{t_a} \mp \mathbf{k_a} - \mathbf{C_a}(\mathbf{t_a})\check{\mathbf{\Phi}}_{\mathbf{tv}}(\mathbf{t_a}, 0)\check{\mathbf{x}}(0) \tag{10.74}$$

$$\mathbf{M_5} = \pm\frac{4}{T_{sw}}\mathbf{t_b} \mp \mathbf{k_b} - \mathbf{C_b}(\mathbf{t_b})\check{\mathbf{\Phi}}_{\mathbf{tv}}(\mathbf{t_b}, 0)\check{\mathbf{x}}(0) \tag{10.75}$$

$$\mathbf{M_6} = \pm\frac{4}{T_{sw}}\mathbf{t_c} \mp \mathbf{k_c} - \mathbf{C_c}(\mathbf{t_c})\check{\mathbf{\Phi}}_{\mathbf{tv}}(\mathbf{t_c}, 0)\check{\mathbf{x}}(0) \tag{10.76}$$

Similar to the case of buck converter, here M_1, M_2, and M_3 are used to ensure that the overall system is at steady-state whereas $\mathbf{M_4}$, $\mathbf{M_5}$, and $\mathbf{M_6}$ are vectors of constraint equations used to ensure that the overall system is in the closed-loop configuration.

10.2.4 Initialization

The circuit for obtaining initial values for starting up the iteration is the same as Fig. 9.8. The initialization model assumes infinite switching frequency and neglects both control action and converter losses. The phasor analysis yields modulation index (m_a) and firing angle (δ) associated with the fundamental frequency positive sequence component for a given value of $|e_s|$, θ_s, v_{dcref} and i_{tqref}. Once m_a and δ are obtained, they are passed to a PWM generator to generate the necessary switching times. Note that for a two-level VSC, the total number of switching times is $6m_f$. Since initial values of the controllers' integrators are not known, their values are initialized to 0.

10.2.5 Jacobian Matrix

Since the state transition matrix is LTV within each switching time interval, the following form of the derivations must be used [108] when one evaluates the sensitivity of the transition matrix with respect to the individual switching times:

$$\frac{d\check{\mathbf{\Psi}}(\tau_1, \tau_2)}{d\tau_1} = \check{\mathbf{A}}(\tau_1)\check{\mathbf{\Psi}}(\tau_1, \tau_2) \tag{10.77}$$

$$\frac{d\check{\mathbf{\Psi}}(\tau_1, \tau_2)}{d\tau_2} = -\check{\mathbf{\Psi}}(\tau_1, \tau_2)\check{\mathbf{A}}(\tau_2). \tag{10.78}$$

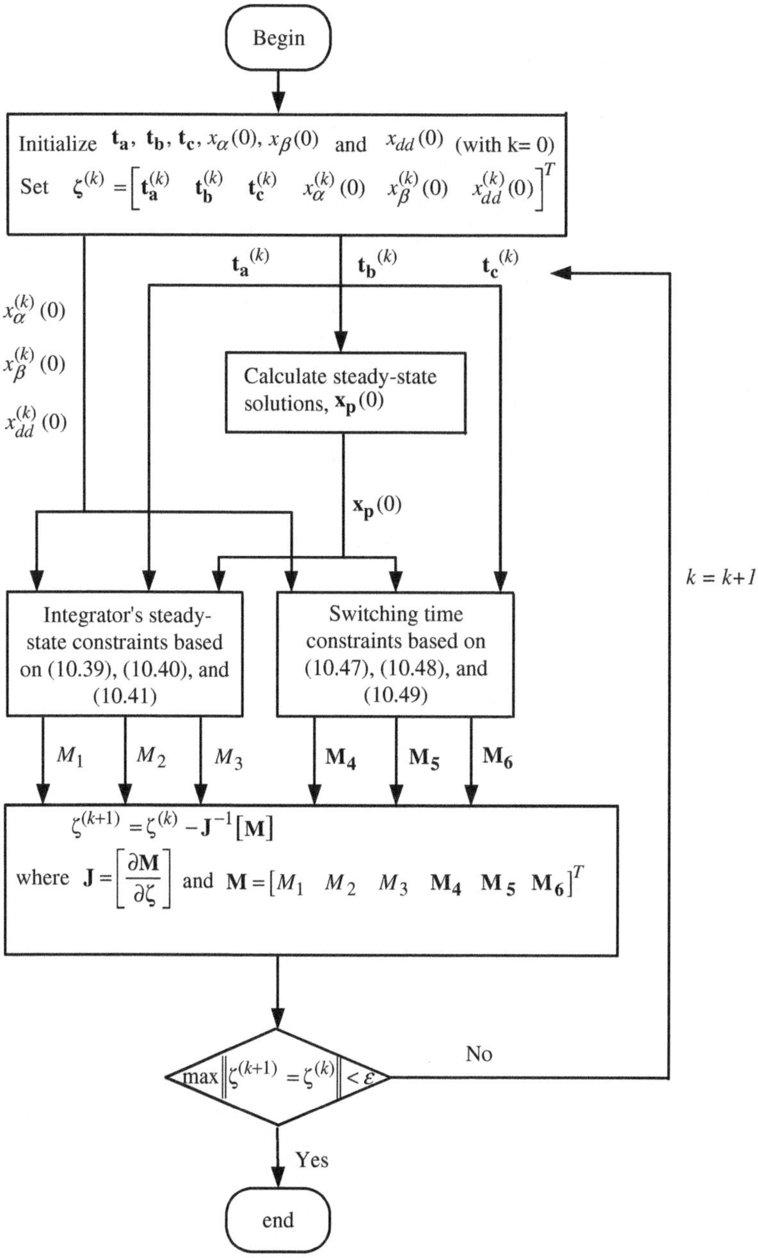

Figure 10.9 Solution flow chart of the shooting method for the VSC.

The Jacobian matrix is stated in (10.79):

$$\mathbf{J} = \left[\begin{array}{c|c} \mathbf{J}_{11} & \mathbf{J}_{12} \\ \hline \mathbf{J}_{21} & \mathbf{J}_{22} \end{array}\right] = \left[\begin{array}{ccc|cccc} \frac{\partial M_1}{\partial x_\alpha} & \frac{\partial M_1}{\partial x_\beta} & \frac{\partial M_1}{\partial x_{dd}} & \frac{\partial M_1}{\partial t_1} & \frac{\partial M_1}{\partial t_2} & \cdots & \frac{\partial M_1}{\partial t_n} \\ \frac{\partial M_2}{\partial x_\alpha} & \frac{\partial M_2}{\partial x_\beta} & \frac{\partial M_2}{\partial x_{dd}} & \frac{\partial M_2}{\partial t_1} & \frac{\partial M_2}{\partial t_2} & \cdots & \frac{\partial M_2}{\partial t_n} \\ \frac{\partial M_3}{\partial x_\alpha} & \frac{\partial M_3}{\partial x_\beta} & \frac{\partial M_3}{\partial x_{dd}} & \frac{\partial M_3}{\partial t_1} & \frac{\partial M_3}{\partial t_2} & \cdots & \frac{\partial M_3}{\partial t_n} \\ \hline \frac{\partial M_4}{\partial x_\alpha} & \frac{\partial M_4}{\partial x_\beta} & \frac{\partial M_4}{\partial x_{dd}} & \frac{\partial M_4}{\partial t_1} & \frac{\partial M_4}{\partial t_2} & \cdots & \frac{\partial M_4}{\partial t_n} \\ \frac{\partial M_5}{\partial x_\alpha} & \frac{\partial M_5}{\partial x_\beta} & \frac{\partial M_5}{\partial x_{dd}} & \frac{\partial M_5}{\partial t_1} & \frac{\partial M_5}{\partial t_2} & \cdots & \frac{\partial M_5}{\partial t_n} \\ \vdots & \vdots & \vdots & \vdots & \vdots & \vdots & \vdots \\ \frac{\partial M_{n+3}}{\partial x_\alpha} & \frac{\partial M_{n+3}}{\partial x_\beta} & \frac{\partial M_{n+3}}{\partial x_{dd}} & \frac{\partial M_{n+3}}{\partial t_1} & \frac{\partial M_{n+3}}{\partial t_2} & \cdots & \frac{\partial M_{n+3}}{\partial t_n} \end{array}\right] \tag{10.79}$$

The derivations of the individual elements in (10.79) are listed below:

Derivations of $\mathbf{J}_{11}$

$$\mathbf{J}_{11} = \begin{bmatrix} \frac{\partial M_1}{\partial x_\alpha} & \frac{\partial M_1}{\partial x_\beta} & \frac{\partial M_1}{\partial x_{dd}} \\ \frac{\partial M_2}{\partial x_\alpha} & \frac{\partial M_2}{\partial x_\beta} & \frac{\partial M_2}{\partial x_{dd}} \\ \frac{\partial M_3}{\partial x_\alpha} & \frac{\partial M_3}{\partial x_\beta} & \frac{\partial M_3}{\partial x_{dd}} \end{bmatrix}, \tag{10.80}$$

where

$$\frac{\partial M_1}{\partial x_\alpha} = \mathbf{C}_1 \check{\boldsymbol{\Phi}}_{\mathbf{tv}}(T,0)\mathbf{P}_1 - 1 \tag{10.81}$$

$$\frac{\partial M_1}{\partial x_\beta} = \mathbf{C}_1 \check{\boldsymbol{\Phi}}_{\mathbf{tv}}(T,0)\mathbf{P}_2 \tag{10.82}$$

$$\frac{\partial M_1}{\partial x_{dd}} = \mathbf{C}_1 \check{\boldsymbol{\Phi}}_{\mathbf{tv}}(T,0)\mathbf{P}_3 \tag{10.83}$$

$$\frac{\partial M_2}{\partial x_\alpha} = \mathbf{C}_2 \check{\boldsymbol{\Phi}}_{\mathbf{tv}}(T,0)\mathbf{P}_1 \tag{10.84}$$

$$\frac{\partial M_2}{\partial x_\beta} = \mathbf{C}_2 \check{\boldsymbol{\Phi}}_{\mathbf{tv}}(T,0)\mathbf{P}_2 - 1 \tag{10.85}$$

$$\frac{\partial M_2}{\partial x_{dd}} = \mathbf{C}_2 \check{\boldsymbol{\Phi}}_{\mathbf{tv}}(T,0)\mathbf{P}_3 \tag{10.86}$$

$$\frac{\partial M_3}{\partial x_\alpha} = \mathbf{C}_3 \check{\boldsymbol{\Phi}}_{\mathbf{tv}}(T,0)\mathbf{P}_1 \tag{10.87}$$

$$\frac{\partial M_3}{\partial x_\beta} = \mathbf{C}_3 \check{\boldsymbol{\Phi}}_{\mathbf{tv}}(T,0)\mathbf{P}_2 \tag{10.88}$$

$$\frac{\partial M_3}{\partial x_{dd}} = \mathbf{C}_3 \check{\boldsymbol{\Phi}}_{\mathbf{tv}}(T,0)\mathbf{P}_3 - 1, \tag{10.89}$$

and

$$\mathbf{C}_1 = \begin{bmatrix} 0 & 0 & 0 & 1 & 0 & 0 & 0 & 0 & 0 & 0 \end{bmatrix} \tag{10.90}$$

$$\mathbf{C}_2 = \begin{bmatrix} 0 & 0 & 0 & 0 & 1 & 0 & 0 & 0 & 0 & 0 \end{bmatrix} \tag{10.91}$$

$$\mathbf{C}_3 = \begin{bmatrix} 0 & 0 & 0 & 0 & 0 & 1 & 0 & 0 & 0 & 0 \end{bmatrix} \tag{10.92}$$

$$\mathbf{P}_3 = \begin{bmatrix} 0 & 0 & 0 & 1 & 0 & 0 & 0 & 0 & 0 & 0 \end{bmatrix}^T \tag{10.93}$$

$$\mathbf{P}_3 = \begin{bmatrix} 0 & 0 & 0 & 0 & 1 & 0 & 0 & 0 & 0 & 0 \end{bmatrix}^T \tag{10.94}$$

$$\mathbf{P}_3 = \begin{bmatrix} 0 & 0 & 0 & 0 & 0 & 1 & 0 & 0 & 0 & 0 \end{bmatrix}^T. \tag{10.95}$$

Derivations of $\mathbf{J}_{12}$

The components in $\mathbf{J}_{12}$ are listed in the following:

$$\frac{\partial M_1}{\partial t_1} = \mathbf{C}_1 \frac{\partial \check{\mathbf{\Phi}}_{\mathbf{tv}}}{\partial t_1} \check{\mathbf{x}}(0) + \mathbf{C}_1 \check{\mathbf{\Phi}}_{\mathbf{tv}} \frac{\partial \check{\mathbf{x}}(0)}{\partial t_1} \tag{10.96}$$

$$\frac{\partial M_1}{\partial t_2} = \mathbf{C}_1 \frac{\partial \check{\mathbf{\Phi}}_{\mathbf{tv}}}{\partial t_2} \check{\mathbf{x}}(0) + \mathbf{C}_1 \check{\mathbf{\Phi}}_{\mathbf{tv}} \frac{\partial \check{\mathbf{x}}(0)}{\partial t_2} \tag{10.97}$$

$$\vdots$$

$$\frac{\partial M_1}{\partial t_n} = \mathbf{C}_1 \frac{\partial \check{\mathbf{\Phi}}_{\mathbf{tv}}}{\partial t_n} \check{\mathbf{x}}(0) + \mathbf{C}_1 \check{\mathbf{\Phi}}_{\mathbf{tv}} \frac{\partial \check{\mathbf{x}}(0)}{\partial t_n} \tag{10.98}$$

$$\frac{\partial M_2}{\partial t_1} = \mathbf{C}_2 \frac{\partial \check{\mathbf{\Phi}}_{\mathbf{tv}}}{\partial t_1} \check{\mathbf{x}}(0) + \mathbf{C}_2 \check{\mathbf{\Phi}}_{\mathbf{tv}} \frac{\partial \check{\mathbf{x}}(0)}{\partial t_1} \tag{10.99}$$

$$\frac{\partial M_2}{\partial t_2} = \mathbf{C}_2 \frac{\partial \check{\mathbf{\Phi}}_{\mathbf{tv}}}{\partial t_2} \check{\mathbf{x}}(0) + \mathbf{C}_2 \check{\mathbf{\Phi}}_{\mathbf{tv}} \frac{\partial \check{\mathbf{x}}(0)}{\partial t_2} \tag{10.100}$$

$$\vdots$$

$$\frac{\partial M_2}{\partial t_n} = \mathbf{C}_2 \frac{\partial \check{\mathbf{\Phi}}_{\mathbf{tv}}}{\partial t_n} \check{\mathbf{x}}(0) + \mathbf{C}_2 \check{\mathbf{\Phi}}_{\mathbf{tv}} \frac{\partial \check{\mathbf{x}}(0)}{\partial t_n} \tag{10.101}$$

$$\frac{\partial M_3}{\partial t_1} = \mathbf{C}_3 \frac{\partial \check{\mathbf{\Phi}}_{\mathbf{tv}}}{\partial t_1} \check{\mathbf{x}}(0) + \mathbf{C}_3 \check{\mathbf{\Phi}}_{\mathbf{tv}} \frac{\partial \check{\mathbf{x}}(0)}{\partial t_1} \tag{10.102}$$

$$\frac{\partial M_3}{\partial t_2} = \mathbf{C}_3 \frac{\partial \check{\mathbf{\Phi}}_{\mathbf{tv}}}{\partial t_2} \check{\mathbf{x}}(0) + \mathbf{C}_3 \check{\mathbf{\Phi}}_{\mathbf{tv}} \frac{\partial \check{\mathbf{x}}(0)}{\partial t_2} \tag{10.103}$$

$$\vdots$$

$$\frac{\partial M_3}{\partial t_n} = \mathbf{C}_3 \frac{\partial \check{\mathbf{\Phi}}_{\mathbf{tv}}}{\partial t_n} \check{\mathbf{x}}(0) + \mathbf{C}_3 \check{\mathbf{\Phi}}_{\mathbf{tv}} \frac{\partial \check{\mathbf{x}}(0)}{\partial t_n}. \tag{10.104}$$

Derivations of $\mathbf{J}_{21}$

For the purpose of illustration, let us assume the following (refer to Fig. 10.8)

$$M_4 = -\frac{4}{T_{sw}} t_1 + 1 - v_{ma}(t_1) \tag{10.105}$$

$$M_5 = -\frac{4}{T_{sw}} t_2 + 2 - v_{mb}(t_2) \tag{10.106}$$

$$M_6 = \frac{4}{T_{sw}} t_3 - 3 - v_{mc}(t_3) \tag{10.107}$$

$$\vdots$$

$$M_{n+3} = \frac{4}{T_{sw}} t_n - (n) - v_{ma}(t_n). \tag{10.108}$$

Therefore,

$$\frac{\partial M_4}{\partial x_\alpha} = -\mathbf{C_a}(t_1)\check{\mathbf{\Phi}}_{\mathbf{tv}}(t_1, 0)\mathbf{P_1} \tag{10.109}$$

$$\frac{\partial M_5}{\partial x_\alpha} = -\mathbf{C_b}(t_2)\check{\mathbf{\Phi}}_{\mathbf{tv}}(t_2, 0)\mathbf{P_1} \tag{10.110}$$

$$\frac{\partial M_6}{\partial x_\alpha} = -\mathbf{C_c}(t_3)\check{\mathbf{\Phi}}_{\mathbf{tv}}(t_3, 0)\mathbf{P_1} \tag{10.111}$$

$$\vdots$$

$$\frac{\partial M_{n+3}}{\partial x_\alpha} = -\mathbf{C_a}(t_n)\check{\mathbf{\Phi}}_{\mathbf{tv}}(t_n, 0)\mathbf{P_1} \tag{10.112}$$

$$\frac{\partial M_4}{\partial x_\beta} = -\mathbf{C_a}(t_1)\check{\mathbf{\Phi}}_{\mathbf{tv}}(t_1, 0)\mathbf{P_2} \tag{10.113}$$

$$\frac{\partial M_5}{\partial x_\beta} = -\mathbf{C_b}(t_2)\check{\mathbf{\Phi}}_{\mathbf{tv}}(t_2, 0)\mathbf{P_2} \tag{10.114}$$

$$\frac{\partial M_6}{\partial x_\beta} = -\mathbf{C_c}(t_3)\check{\mathbf{\Phi}}_{\mathbf{tv}}(t_3, 0)\mathbf{P_2} \tag{10.115}$$

$$\vdots$$

$$\frac{\partial M_{n+3}}{\partial x_\beta} = -\mathbf{C_a}(t_n)\check{\mathbf{\Phi}}_{\mathbf{tv}}(t_n, 0)\mathbf{P_2} \tag{10.116}$$

$$\frac{\partial M_4}{\partial x_{dd}} = -\mathbf{C_a}(t_1)\check{\mathbf{\Phi}}_{\mathbf{tv}}(t_1, 0)\mathbf{P_3} \tag{10.117}$$

$$\frac{\partial M_5}{\partial x_{dd}} = -\mathbf{C_b}(t_2)\check{\mathbf{\Phi}}_{\mathbf{tv}}(t_2, 0)\mathbf{P_3} \tag{10.118}$$

$$\frac{\partial M_6}{\partial x_{dd}} = -\mathbf{C_c}(t_3)\check{\mathbf{\Phi}}_{\mathbf{tv}}(t_3, 0)\mathbf{P_3} \tag{10.119}$$

$$\vdots$$

$$\frac{\partial M_{n+3}}{\partial x_{dd}} = -\mathbf{C_a}(t_n)\check{\mathbf{\Phi}}_{\mathbf{tv}}(t_n, 0)\mathbf{P_3}. \tag{10.120}$$

Derivations of $\mathbf{J_{22}}$

Based on equation (10.105) to (10.108), the individual components of $\mathbf{J_{22}}$ are listed as follows:

$$\frac{\partial M_4}{\partial t_1} = \frac{-4}{T_{sw}} - \left[\frac{\partial \mathbf{C_a}(t_1)}{\partial t_1}\check{\mathbf{\Phi}}_{\mathbf{tv}}(t_1, 0)\check{\mathbf{x}}(0) + \mathbf{C_a}(t_1)\frac{\check{\mathbf{\Phi}}_{\mathbf{tv}}(t_1, 0)}{\partial t_1}\check{\mathbf{x}}(0) + \mathbf{C_a}(t_1)\check{\mathbf{\Phi}}_{\mathbf{tv}}(t_1, 0)\frac{\partial \check{x}(0)}{\partial t_1}\right] \tag{10.121}$$

$$\frac{\partial M_5}{\partial t_1} = -\mathbf{C_b}(t_2)\check{\mathbf{\Phi}}_{\mathbf{tv}}(t_2, 0)\frac{\partial \check{\mathbf{x}}(0)}{\partial t_1} \tag{10.122}$$

$$\frac{\partial M_6}{\partial t_1} = \mathbf{C_c}(t_3)\check{\mathbf{\Phi}}_{\mathbf{tv}}(t_3, 0)\frac{\partial \check{\mathbf{x}}(0)}{\partial t_1} \tag{10.123}$$

$$\vdots \tag{10.124}$$

$$\frac{\partial M_{n+3}}{\partial t_1} = \mathbf{C_a}(t_n)\check{\mathbf{\Phi}}_{\mathbf{tv}}(t_n, 0)\frac{\partial \check{\mathbf{x}}(0)}{\partial t_1} \tag{10.125}$$

$$\frac{\partial M_4}{\partial t_2} = -\mathbf{C_a}(t_2)\check{\mathbf{\Phi}}_{\mathbf{tv}}(t_2, 0)\frac{\partial \check{\mathbf{x}}(0)}{\partial t_2} \tag{10.126}$$

$$\frac{\partial M_5}{\partial t_2} = \frac{-4}{T_{sw}} - \left[\frac{\partial \mathbf{C_b}(t_2)}{\partial t_2}\check{\mathbf{\Phi}}_{\mathbf{tv}}(t_2, 0)\check{\mathbf{x}}(0) + \mathbf{C_b}(t_2)\frac{\check{\mathbf{\Phi}}_{\mathbf{tv}}(t_2, 0)}{\partial t_2}\check{\mathbf{x}}(0) + \mathbf{C_b}(t_2)\check{\mathbf{\Phi}}_{\mathbf{tv}}(t_2, 0)\frac{\partial \check{\mathbf{x}}(0)}{\partial t_2}\right] \tag{10.127}$$

$$\frac{\partial M_6}{\partial t_2} = -\mathbf{C_c}(t_3)\check{\mathbf{\Phi}}_{\mathbf{tv}}(t_3, 0)\frac{\partial \check{\mathbf{x}}(0)}{\partial t_2} \tag{10.128}$$

$$\vdots$$

$$\frac{\partial M_{n+3}}{\partial t_2} = -\mathbf{C_a}(t_n)\check{\mathbf{\Phi}}_{\mathbf{tv}}(t_n, 0)\frac{\partial \check{x}(0)}{\partial t_1} \tag{10.129}$$

$$\frac{\partial M_4}{\partial t_3} = -\mathbf{C_a}(t_1)\check{\mathbf{\Phi}}_{\mathbf{tv}}(t_1, 0)\frac{\partial \check{\mathbf{x}}(0)}{\partial t_3} \tag{10.130}$$

$$\frac{\partial M_5}{\partial t_3} = -\mathbf{C_b}(t_2)\check{\mathbf{\Phi}}_{\mathbf{tv}}(t_2, 0)\frac{\partial \check{\mathbf{x}}(0)}{\partial t_3} \tag{10.131}$$

$$\frac{\partial M_6}{\partial t_3} = \frac{-4}{T_{sw}} - \left[\frac{\partial \mathbf{C_c}(t_3)}{\partial t_3}\check{\mathbf{\Phi}}_{\mathbf{tv}}(t_3, 0)\check{\mathbf{x}}(0) + \mathbf{C_c}(t_3)\frac{\check{\mathbf{\Phi}}_{\mathbf{tv}}(t_3, 0)}{\partial t_3}\check{\mathbf{x}}(0) + \mathbf{C_c}(t_3)\check{\mathbf{\Phi}}_{\mathbf{tv}}(t_3, 0)\frac{\partial \check{\mathbf{x}}(0)}{\partial t_3}\right] \tag{10.132}$$

$$\vdots$$

$$\frac{\partial M_{n+3}}{\partial t_3} = -\mathbf{C_a}(t_n)\check{\mathbf{\Phi}}_{\mathbf{tv}}(t_n, 0)\frac{\partial \check{\mathbf{x}}(0)}{\partial t_3} \tag{10.133}$$

$$\frac{\partial M_4}{\partial t_n} = -\mathbf{C_a}(t_1)\check{\mathbf{\Phi}}_{\mathbf{tv}}(t_1, 0)\frac{\partial \check{x}(0)}{\partial t_n} \tag{10.134}$$

$$\frac{\partial M_5}{\partial t_n} = -\mathbf{C_b}(t_2)\check{\mathbf{\Phi}}_{\mathbf{tv}}(t_2, 0)\frac{\partial \check{\mathbf{x}}(0)}{\partial t_n} \tag{10.135}$$

$$\frac{\partial M_6}{\partial t_n} = -\mathbf{C_c}(t_3)\check{\mathbf{\Phi}}_{\mathbf{tv}}(t_3, 0)\frac{\partial \check{\mathbf{x}}(0)}{\partial t_n} \tag{10.136}$$

$$\vdots$$

$$\frac{\partial M_{n+3}}{\partial t_n} = \frac{-4}{T_{sw}} - \left[\frac{\partial \mathbf{C_a}(t_n)}{\partial t_n}\check{\mathbf{\Phi}}_{\mathbf{tv}}(t_n, 0)\check{\mathbf{x}}(0) + \mathbf{C_a}(t_n)\frac{\check{\mathbf{\Phi}}_{\mathbf{tv}}(t_n, 0)}{\partial t_n}\check{\mathbf{x}}(0) + \mathbf{C_a}(t_n)\check{\mathbf{\Phi}}_{\mathbf{tv}}(t_n, 0)\frac{\partial \check{\mathbf{x}}(0)}{\partial t_n}\right]. \tag{10.137}$$

Example 10.2 A balanced case similar to that presented in **Example 9.2** is studied here by means of the LB 2 method. The only difference is that no current filter is employed in the current controller, as presented in Fig. 10.5. Note that similar analysis can be done if filters are included. For the system under study, $|e_s| = 112$ V, $v_{dcref} = 230$ V, and $i_{tqref} = 30$A, resulting in initial values of m_a and δ equal to 0.8733 and $-3.0051°$. Next, m_a and δ are passed to the PWM pattern generator to produce the seed values for the switching times to begin iterations. The initial values of $x_\alpha(0)$, $x_\beta(0)$, and $x_{dd}(0)$ are all set to zeros.

Once the convergence is reached, the resulting switching times are substituted into the FCM to obtain the output ac current and dc voltage harmonics. Fig. 10.10 shows the ac current space

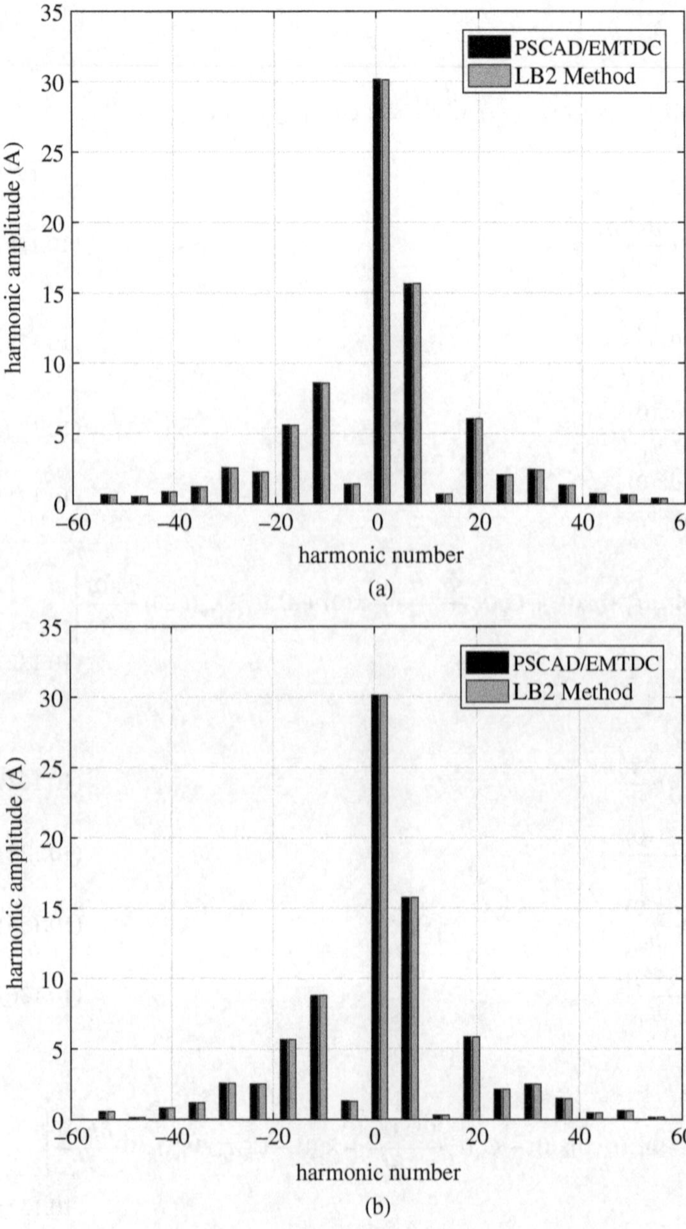

Figure 10.10 Balanced case, ac current harmonic spectra: (a) slow controller; (b) fast controller.

vector $(I_{t\alpha} + jI_{t\beta})$ harmonic spectrum due to the slow and fast current controllers under balanced operation.

The nonzero ac current harmonics shown in Fig. 10.11 (from left to right) are [−53 −47 −41 −35 −29 −23 −17 −11 −5 1 7 13 19 25 31 37 43 49 55]. It is to be noted that in **Example 9.2**, the VSC currents are first filtered with current filters prior to entering the current controllers whereas in this example, the currents are not filtered and directly fed to the controllers. Compared to Fig. 10.13, one can clearly see that the current filters suppress some of the harmonics. For instance, the -5th,

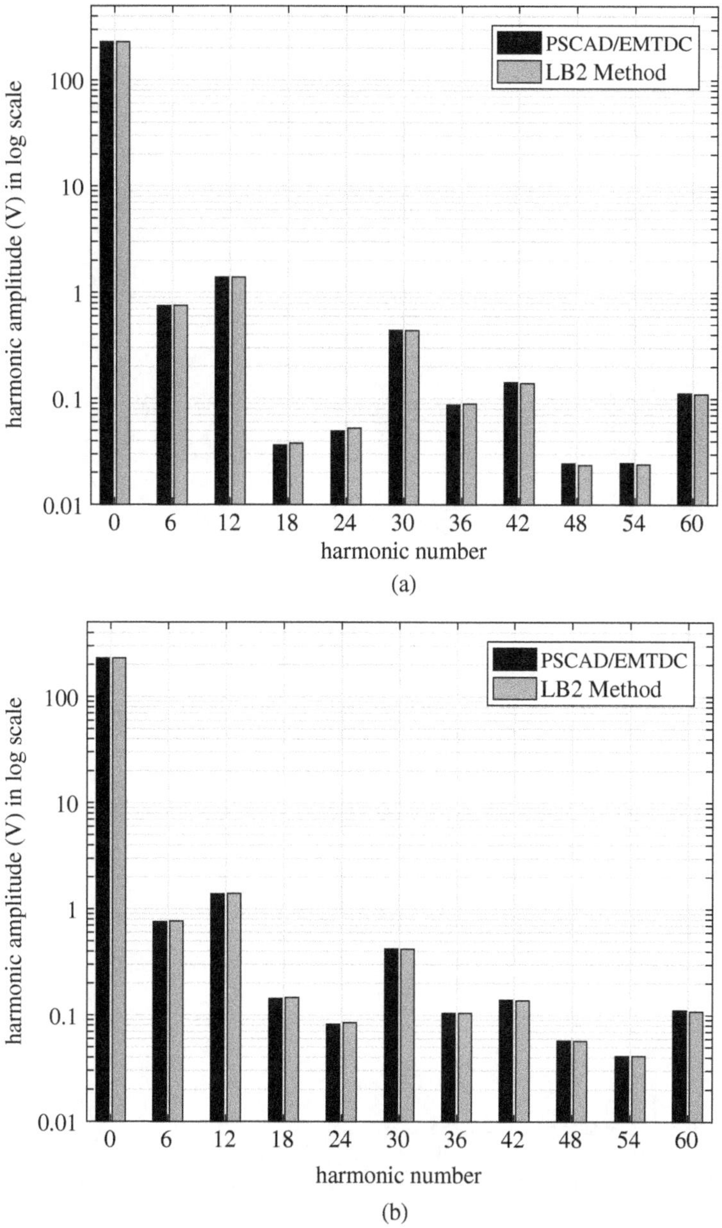

Figure 10.11 Balanced case, dc voltage harmonic spectra: (a) slow controller; (b) fast controller.

7th, and 13th harmonics are smaller, as compared to the case without the current filters. Moreover, the presence of current filters also affects the dc voltage spectra. This is evident by comparing Fig. 10.11 with Fig. 9.14. For instance, the 18th, 24th, and 48th harmonic components are effectively reduced. □

Example 10.3 Similar to **Example 9.3**, a 10% negative sequence voltage is superimposed on the system voltage to study operation under unbalanced conditions. Fig. 10.12 shows the ac current

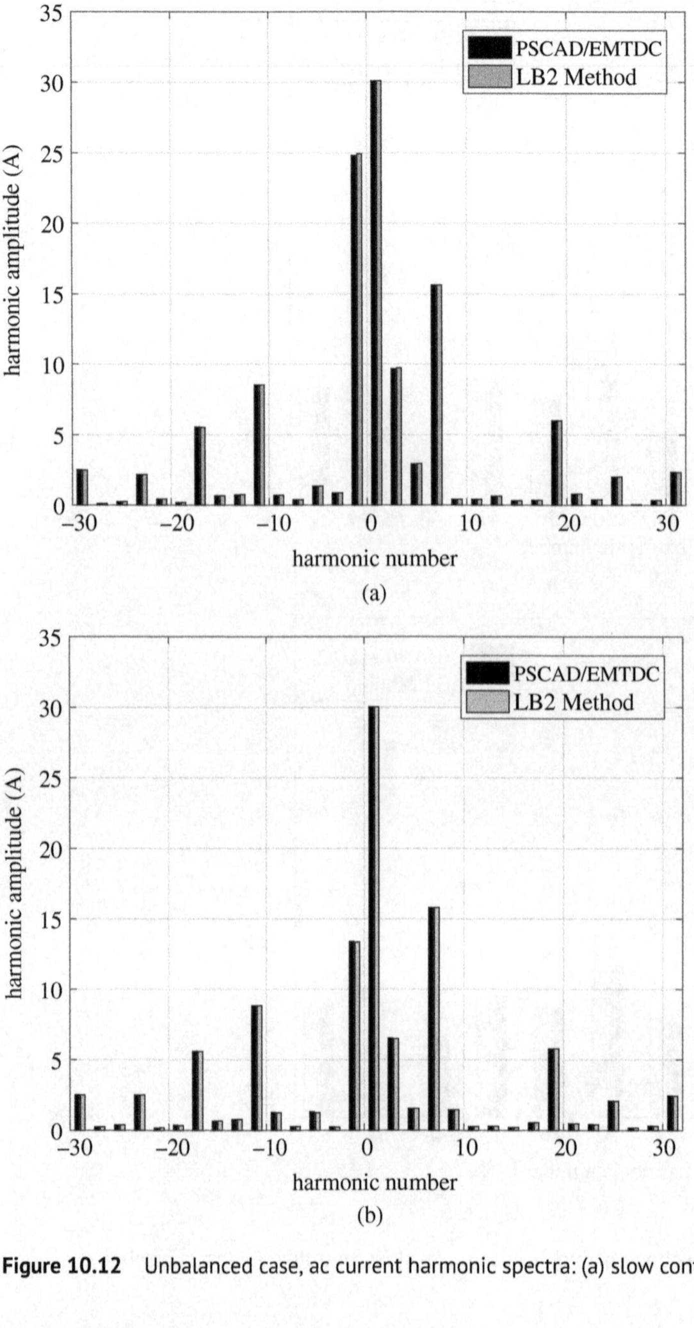

Figure 10.12 Unbalanced case, ac current harmonic spectra: (a) slow controller; (b) fast controller.

space vector ($I_{t\alpha} + jI_{t\beta}$) harmonic spectra due to the slow and fast current controllers under unbalanced operation. All odd harmonics are shown.

Fig. 10.13 shows the dc voltage harmonic spectrum for the slow and fast current controllers. All even harmonics are shown. As in the case of balanced operation, the magnitude of dc voltage harmonics have been plotted using a log scale to allow comparison of small, high-frequency components. □

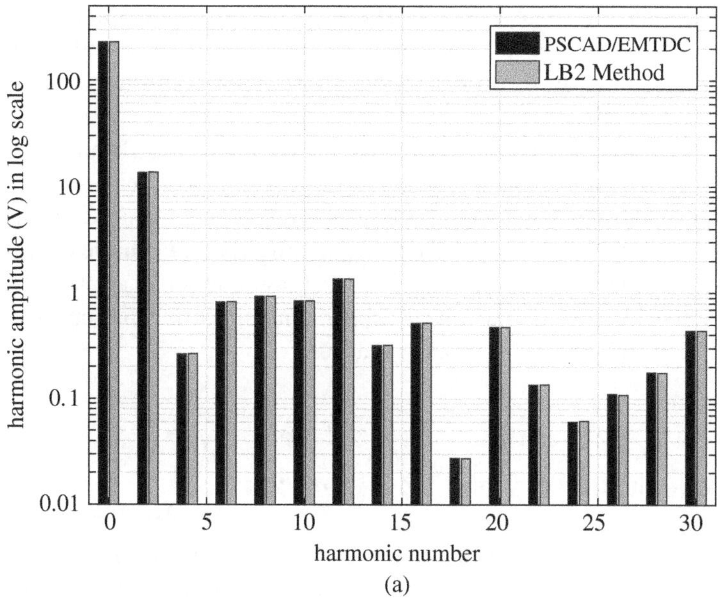

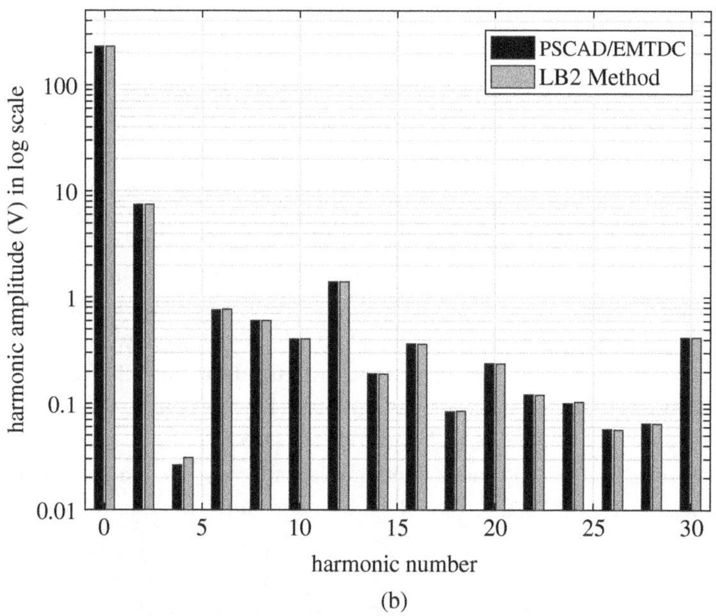

Figure 10.13 Unbalanced case, dc voltage harmonic spectra: (a) slow controller; (b) fast controller.

10.2.6 Discussions of Results

As can be noted from Figs. 10.10–10.13, excellent agreement exists between the LB 2 method and the transient time domain (PSCAD/EMTDC) simulation results.

The results allow us to see that ac current spectra for the controllers C_{c_slow} and C_{c_fast} differ at some frequencies, even for the case of balanced operation. Both controllers successfully regulate the positive sequence fundamental current to 30A in both balanced and unbalanced operation. However, comparing Fig. 10.10(a) and (b), one sees that harmonics at −47, +13, and +55 differ, and that

C_{c_fast} suppresses these three harmonics to almost zero, while C_{c_slow} does not. Under unbalanced operation, the ac current harmonic spectra for the two controllers differ very significantly, as is evident from Fig. 10.12(a) and (b). For instance, the negative sequence fundamental current component is about 24.8A (82.7% of the positive sequence of the fundamental component) for C_{c_slow} while it is only 13.4A (45% of the positive sequence of the fundamental component) when C_{c_fast} is employed.

Figs. 10.11 and 10.13 show that the dc component of the capacitor voltage V_{dc}^0 is successfully regulated to its reference value of 230 V in all cases. Similar to the ac current spectra, the dc voltage spectra also differ from one another. For example, under the balanced case, harmonics at 18, 24, 48, and 54 are different for different controllers. For unbalanced operation, noticeable discrepancies exist in the 2nd, 4th, 18th, 20th, 22nd, 26th, and 28th harmonic voltages.

10.3 Comparison Between the LB 1 and LB 2 Methods

In general, the higher the number of the unknowns, the longer the time required for iteration. This section presents three different case studies to compare the number of unknown variables required to simulate a two-level VSC based on the LB 1 and LB 2 methods, under different switching frequencies. The system parameters employed are the same as those used in Section 9.2, and the current control loop filters are included in both of the LB 1 and LB 2 methods.

10.3.1 Case #1: Balanced System

As stated in Section 9.3.3, only characteristic harmonics are non-zero under balanced operation. Therefore, only a small number of unknown variables need to be included in the LB 1 method. For the LB 2 method, only m_f switching times need to be included in the simulation because all the ac current waveforms exhibit sixth period symmetry. Therefore, a total of $m_f + 5$ unknown variables (i.e. three control variables and two filter variables) need to be initialized for the LB 2 method. Fig. 10.14 shows the number of unknown variables for the LB 1 and LB 2 methods when m_f is varied from 3 to 57. Note that m_f is chosen to be an odd multiple of 3 [11]. As can be noted from the figure, when C_{c_slow} is employed, the LB 1 method is more efficient for $m_f \geq 15$. For the case of C_{c_fast}, m_f must be equal to or greater than 27 before the LB 1 method is more efficient.

10.3.2 Case #2: Unbalanced System with AC Waveform Exhibiting Half-Wave Symmetry

If the ac current waveform caused by the harmonic distortion possesses half-wave symmetry, only odd harmonics of the voltage modulating signals and even harmonics of the active current reference need to be included in the simulation for the LB 1 method. A reduction in the number of unknown variables also exists in the LB 2 method when ac current waveforms exhibit half-wave symmetry because only half of the switching times need to be included. Thus, a total number of $3m_f + 5$ variables need to be initialized for the LB 2 method. As Fig. 10.15 indicates, the LB 1 method offers iteration time advantage over the LB 2 when $m_f \geq 9$ if C_{c_slow} is employed. For C_{c_fast}, the LB 1 method is more advantageous when $m_f \geq 33$.

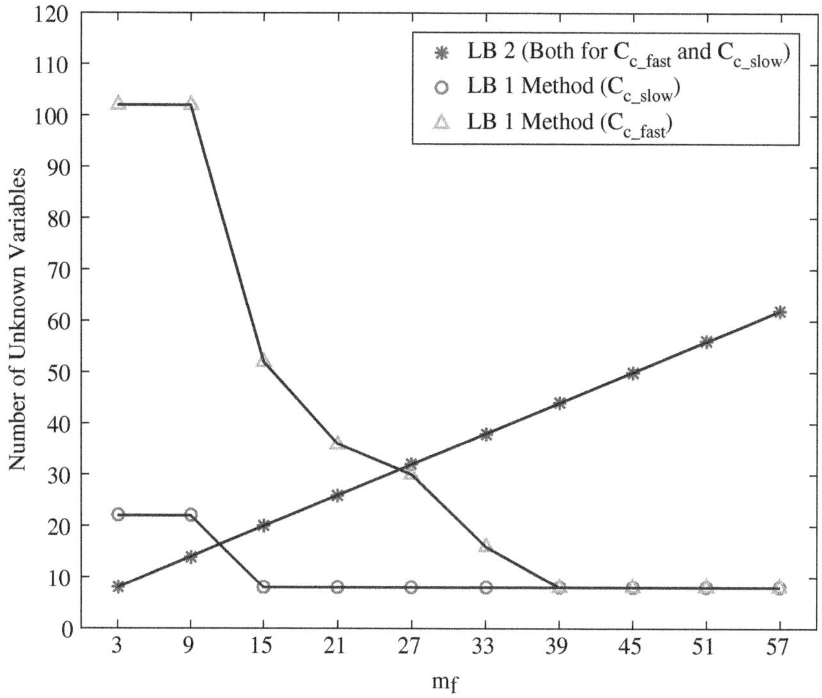

Figure 10.14 The number of unknown variables for the LB 1 and LB 2 methods under balanced operation.

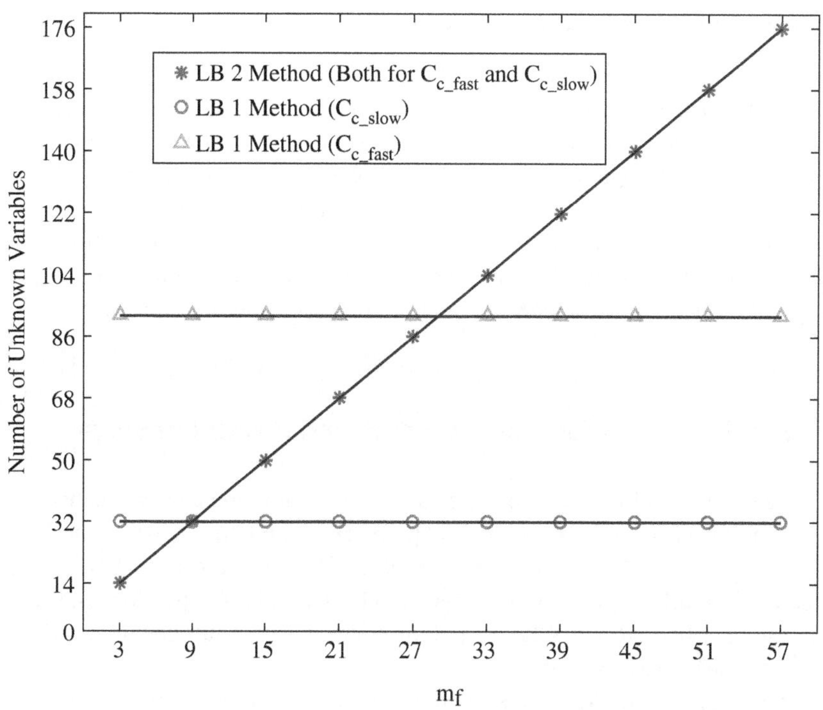

Figure 10.15 The number of unknown variables for the LB 1 and LB 2 methods under unbalanced operation whose AC current waveforms have half-wave symmetry.

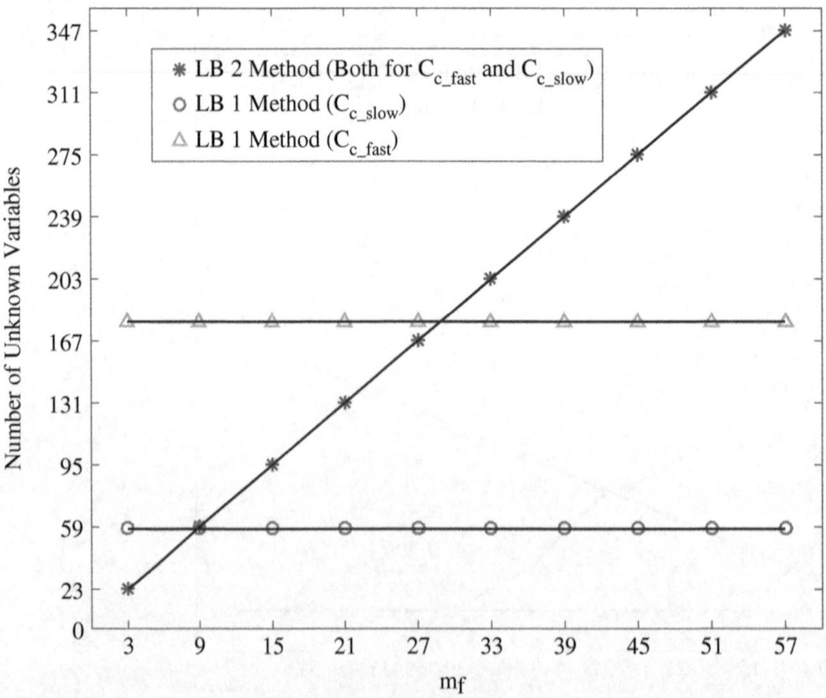

Figure 10.16 The number of unknown variables for the LB 1 and LB 2 methods under unbalanced operation whose ac current waveforms have no symmetry.

10.3.3 Case #3: Unbalanced System, No Waveform Symmetry

In many cases, a lack of knowledge about the harmonic distortions means no assumptions can be made regarding symmetry. It can also happen that the distortions in the system are so severe that there is no odd symmetry in the ac currents and no even symmetry in the dc voltage. In these cases, all the harmonics of the modulating signals and active current reference need to be included in the iteration for the LB 1 method. For the LB 2 method, all the switching times need to be included. Fig. 10.16 shows the number of unknown variables for the LB 1 and LB 2 methods. The situation is similar to when half-wave symmetry exists. The LB 1 method is more efficient for $m_f \geq 9$ if C_{c_slow} is employed. For C_{c_fast}, the LB 1 method is more efficient than the LB 2 method when $m_f \geq 33$.

10.4 Large-Signal Modeling for Line-Commutated Power Converter

In this section, we demonstrate how LB 2 can be used to model a line-commutated power converter. Fig. 10.17 shows a six-pulse capacitor-filtered diode bridge rectifier where the dc load is modeled as an equivalent resistance, R_l [11, 109]. The line harmonics are filtered by the ac chokes, L. This type of rectifier is frequently employed for battery charger application. It is also used for adjustable speed drives [8] because it has better drive isolation and lower dc current requirements [110, 111] than a conventional inductor-filtered rectifier.

To simplify the model development and discussion, only balanced operation of the converter is considered. However, the technique can well be extended to the unbalanced case.

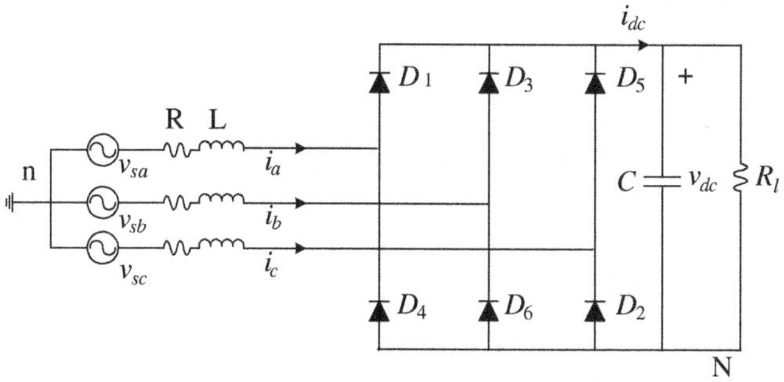

Figure 10.17 Six-pulse uncontrolled rectifier with capacitive dc smoothing.

10.4.1 Discontinuous Conduction Mode

For discontinued conduction mode (DCM), Fig. 10.18 shows a typical discontinuous phase current, together with the six line-to-line and dc voltages. As noted in the figure, two subintervals can be identified in every sixth of the period — conduction (η), and non-conduction ($\xi = T/6 - \eta$) intervals. This repetition pattern allows one to fully describe the behavior of the rectifier by only considering 1/6th of the period. The conduction interval with the diodes D_1 and D_6 on, commences at instant γ with respect to the zero reference (the intersection point between v_{sab} and v_{scb}). Fig. 10.19(a) shows the circuit involved in this duration, and (10.138) is the corresponding differential equation. In the nonconduction interval, $D1$ and $D6$ turn off, and the capacitor discharges through the load resistance (Fig. 10.19(b)). Its corresponding differential equation is given in (10.139).

$$\frac{d}{dt}\begin{bmatrix}\mathbf{x}_1\\ \mathbf{z}\end{bmatrix} = \mathbf{A}_{\mathbf{on}}\begin{bmatrix}\mathbf{x}_1\\ \mathbf{z}\end{bmatrix} = \begin{bmatrix}\mathbf{A}_{\mathbf{x1}} & \mathbf{N}_1\\ \mathbf{0} & \boldsymbol{\Omega}\end{bmatrix}\begin{bmatrix}\mathbf{x}_1\\ \mathbf{z}\end{bmatrix} \tag{10.138}$$

$$\frac{d}{dt}\begin{bmatrix}\mathbf{x}_2\\ \mathbf{z}\end{bmatrix} = \mathbf{A}_{\mathbf{off}}\begin{bmatrix}\mathbf{x}_2\\ \mathbf{z}\end{bmatrix} = \begin{bmatrix}\mathbf{A}_{\mathbf{x2}} & \mathbf{N}_2\\ \mathbf{0} & \boldsymbol{\Omega}\end{bmatrix}\begin{bmatrix}\mathbf{x}_2\\ \mathbf{z}\end{bmatrix} \tag{10.139}$$

where $\mathbf{x}_1 = \begin{bmatrix}i_d & v_{dc}\end{bmatrix}$, $\mathbf{x}_2 = [v_{dc}]$, $\mathbf{z} = \begin{bmatrix}v_{s\alpha} & v_{s\beta}\end{bmatrix}$,

$$\begin{bmatrix}v_{s\alpha}\\ v_{s\beta}\end{bmatrix} = \mathbf{C}_\mathbf{x}\begin{bmatrix}v_{sa}\\ v_{sb}\\ v_{sc}\end{bmatrix}, \mathbf{C}_\mathbf{x} = \frac{2}{3}\begin{bmatrix}1 & \frac{-1}{2} & \frac{-1}{2}\\ 0 & \frac{\sqrt{3}}{2} & \frac{-\sqrt{3}}{2}\end{bmatrix},$$

$$\mathbf{A}_{\mathbf{x1}} = \begin{bmatrix}0 & \frac{-1}{2L}\\ \frac{1}{C} & \frac{1}{R_lC}\end{bmatrix}, \mathbf{A}_{\mathbf{x2}} = \frac{-1}{R_lC}, \boldsymbol{\Omega} = \begin{bmatrix}0 & -\omega\\ \omega & 0\end{bmatrix},$$

$$\mathbf{N}_1 = \begin{bmatrix}\frac{3}{4L} & \frac{-\sqrt{3}}{4L}\\ 0 & 0\end{bmatrix}, \text{ and } \mathbf{N}_2 = \begin{bmatrix}0 & 0\end{bmatrix}.$$

Note that loop current (e.g. i_d in Fig. 10.19) and $v_{s\alpha}$, $v_{s\beta}$ are chosen as state variables to ensure minimum realization for the state space formulation.

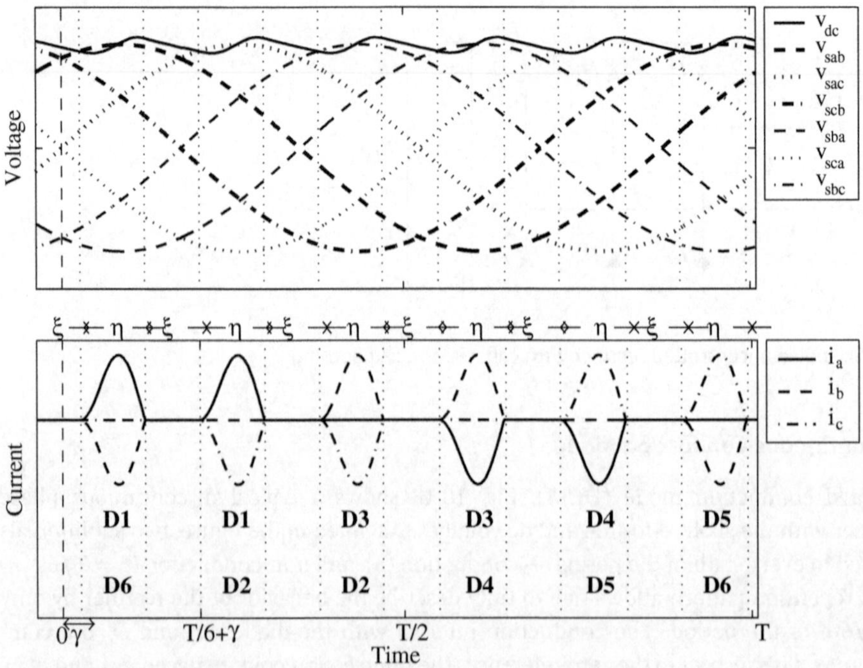

Figure 10.18 Rectifier waveforms in the DCM.

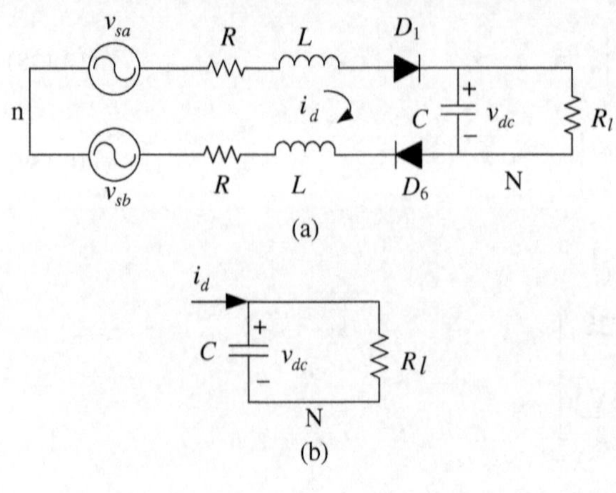

Figure 10.19 DCM models: (a) rectifier model during the conduction subinterval, (b) rectifier model during the non-conduction subinterval.

Under balanced operation, the states ($[\mathbf{x_2}(\gamma) \quad \mathbf{z}(\gamma)]^T$) at the end points of the sixth period interval can be linked by (10.140).

$$\begin{bmatrix} \mathbf{x_2}\left(\frac{T}{6}+\gamma\right) \\ \mathbf{z}\left(\frac{T}{6}+\gamma\right) \end{bmatrix} = \mathbf{\Phi_d} \begin{bmatrix} \mathbf{x_2}(\gamma) \\ \mathbf{z}(\gamma) \end{bmatrix}, \tag{10.140}$$

where $\mathbf{\Phi_d} = e^{\mathbf{A_{off}}\left(\frac{T}{6}-\eta\right)}\mathbf{P_d}e^{\mathbf{A_{on}}\eta}\mathbf{Q_d}$, and $\mathbf{Q_d} = \begin{bmatrix} 0 & 0 & 0 \\ 1 & 0 & 0 \\ 0 & 1 & 0 \\ 0 & 0 & 1 \end{bmatrix}$ is associated with the change of basis at the transition instant from the nonconduction to conduction interval, and $\mathbf{P_d} = \mathbf{Q_d}^T$, which is associated with the change of basis from conduction to nonconduction interval. Note the concept of injection and projection matrix can be traced back to as early as 1939, where Gabriel Kron in the book *Tensor Analysis of Networks* [112] introduced singular transformation matrices to transform from one network to the one with missing branches or meshes.

In addition, the sixth period symmetry that dc voltage (Fig. 10.18) and ac voltage space vector possesses yields

$$\begin{bmatrix} \mathbf{x_2}\left(\frac{T}{6}+\gamma\right) \\ \mathbf{z}\left(\frac{T}{6}+\gamma\right) \end{bmatrix} = \mathbf{I_{1/6,d}} \begin{bmatrix} \mathbf{x_2}(\gamma) \\ \mathbf{z}(\gamma) \end{bmatrix}, \tag{10.141}$$

where

$$\mathbf{I_{1/6,d}} = \begin{bmatrix} 1 & 0 & 0 \\ 0 & \cos\left(\frac{\pi}{3}\right) & -\sin\left(\frac{\pi}{3}\right) \\ 0 & \sin\left(\frac{\pi}{3}\right) & \cos\left(\frac{\pi}{3}\right) \end{bmatrix}.$$

Switching Times Determination

Since the state space equations of the diode bridge rectifier are different for different switching intervals, it is important to know the switching instants before evaluating the FCM. For a forced commutated converter like a VSC, the switching times are determined by control constraints whereas for a line-commutated converter, the switching times are determined by the circuit constraints. As shown in Fig. 10.17, the switching times of the diode bridge are found from the intersection points of circuit voltages or currents. Referring to Fig. 10.17, formulation of appropriate diode constraint equations is required to solve for γ and η:

Diode D_1 turn-on occurs when

$$v_{sab}(\gamma) - v_{dc}(\gamma) = \begin{bmatrix} -1 & \frac{3}{2} & \frac{-\sqrt{3}}{2} \end{bmatrix} \begin{bmatrix} \mathbf{x_2}(\gamma) \\ \mathbf{z}(\gamma) \end{bmatrix} = 0. \tag{10.142}$$

Diode D_1 turn-off occurs when

$$i_a(\gamma+\eta) = \begin{bmatrix} 1 & 0 & 0 & 0 \end{bmatrix} e^{\mathbf{A_{on}}\eta}\mathbf{Q_d} \begin{bmatrix} \mathbf{x_2}(\gamma) \\ \mathbf{z}(\gamma) \end{bmatrix} = 0. \tag{10.143}$$

To ensure that the rectifier is at steady-state, steady-state constraints also need to be included. This is to ensure that the switching times during steady-state, rather than transients, are taken into account. Thus, (10.142), (10.143), together with (10.140) should be simultaneously solved.

Nevertheless, similar to the analysis for the VSC, the steady-state constraints can be eliminated, reducing one of the unknowns. This can be reasoned as follows:

Combining (10.140) and (10.141), one gets

$$(\mathbf{I_{1/6,d}} - \mathbf{\Phi}) \begin{bmatrix} \mathbf{x_2}(\gamma) \\ \mathbf{z}(\gamma) \end{bmatrix} = \begin{bmatrix} \mathbf{D_d} & \mathbf{E_d} \\ \mathbf{F_d} & \mathbf{K_d} \end{bmatrix} \begin{bmatrix} \mathbf{x_2}(\gamma) \\ \mathbf{z}(\gamma) \end{bmatrix} = 0. \tag{10.144}$$

Therefore,

$$\mathbf{x_2}(\gamma) = -\mathbf{D_d^{-1}E_d z}(\gamma) = -\mathbf{D_d^{-1}E_d} e^{\Omega\gamma}\mathbf{z}(0). \tag{10.145}$$

Hence, $\boldsymbol{\zeta} = \begin{bmatrix}\gamma & \eta\end{bmatrix}^T$ is first initialized, allowing determination of $\mathbf{x_2}(\gamma)$ based on (10.145). Finally $\mathbf{x_2}(\gamma)$ is substituted into the diode constraint mismatch equations (10.142) and (10.143).

$$\zeta^{(k+1)} = \zeta^{(k)} - \mathbf{J^{-1}M}, \tag{10.146}$$

where $\mathbf{M} = \begin{bmatrix} M_1 \\ M_2 \end{bmatrix}$, and $\mathbf{J} = \begin{bmatrix} \dfrac{\partial M_{1,d}}{\partial \gamma} & \dfrac{\partial M_{1,d}}{\partial \eta} \\ \dfrac{\partial M_{2,d}}{\partial \gamma} & \dfrac{\partial M_{2,d}}{\partial \eta} \end{bmatrix}$

The iteration terminates when the difference between $\zeta^{(k+1)}$, and $\zeta^{(k)}$ reaches a required tolerance, ε.

To have quadratic convergence, an analytical Jacobian is constructed. The expression of each element in the Jacobian matrix is listed as follows:

$$\frac{\partial M_{1,d}}{\partial \gamma} = \begin{bmatrix} \dfrac{3}{2} & \dfrac{-\sqrt{3}}{2} \end{bmatrix} e^{\Omega\gamma}\boldsymbol{\Omega}\mathbf{z}(0) - \frac{\partial \mathbf{x_2}(\gamma)}{\partial \gamma} \tag{10.147}$$

where

$$\frac{\partial \mathbf{x_2}(\gamma)}{\partial \gamma} = (1 - \Phi_{xd})^{-1} e^{\mathbf{A}_{x2}\xi}\mathbf{P_{xd}}\boldsymbol{\Gamma_1} + \mathbf{N_1}e^{\Omega\eta}\mathbf{z}(\gamma) - e^{\mathbf{A}_{x1}\eta}\mathbf{N_1}\mathbf{z}(\gamma),$$

$$\mathbf{P}_{xd} = \begin{bmatrix}0 & 1\end{bmatrix},$$

$$\Phi_{x,d} = e^{\mathbf{A}_{x2}\left(\frac{T}{6} - \eta\right)}\mathbf{P_x}e^{\mathbf{A}_{x1}\eta}\mathbf{Q_{xd}},$$

$$\mathbf{Q_{xd}} = \mathbf{P_{xd}^T} = \begin{bmatrix}0\\1\end{bmatrix},$$

and

$$\boldsymbol{\Gamma_1} = \int_{\gamma}^{\gamma+\eta} e^{\mathbf{A}_{x1}(\gamma+\eta-\tau)}\mathbf{A_{x1}N_1z}(\tau)d\tau.$$

$$\frac{\partial M_{1,d}}{\partial \eta} = -\frac{\partial \mathbf{x_2}(\gamma)}{\partial \eta} \tag{10.148}$$

$$\begin{aligned}\frac{\partial \mathbf{x_2}(\gamma)}{\partial \eta} &= -(1 - \Phi_{xd})^{-2}(\mathbf{A_{x2}}\Phi_{xd} - e^{\mathbf{A}_{x2}\xi}\mathbf{P_{xd}}e^{\mathbf{A}_{x1}\eta}\mathbf{A_{x1}Q_{xd}})e^{\mathbf{A}_{x2}\xi}\mathbf{P_{xd}}\boldsymbol{\Gamma_2}\\ &-(1 - \Phi_{xd})^{-1}e^{\mathbf{A}_{x2}\xi}\mathbf{A_{x2}P_{xd}}\boldsymbol{\Gamma_2} + (1 - \Phi_{xd})^{-1}e^{\mathbf{A}_{x2}\xi}\mathbf{P_{xd}}\boldsymbol{\Gamma_1} + \mathbf{N_1}e^{\Omega\eta}\mathbf{z}(\gamma),\end{aligned}$$

where

$$\boldsymbol{\Gamma_2} = \int_{\gamma}^{\gamma+\eta} e^{\mathbf{A}_{x1}(\gamma+\eta-\tau)}\mathbf{N_1z}(\tau)d\tau.$$

$$\frac{\partial M_{2,d}}{\partial \gamma} = \begin{bmatrix}1 & 0\end{bmatrix}\{e^{\mathbf{A}_{x1}\eta}\mathbf{Q}\frac{\partial \mathbf{x_2}(\gamma)}{\partial \gamma} + \boldsymbol{\Gamma_1} + \mathbf{N_1}e^{\Omega\eta}\mathbf{z}(\gamma) - e^{\mathbf{A}_{x1}\eta}\mathbf{N_1z}(\gamma)\} \tag{10.149}$$

$$\frac{\partial M_{2,d}}{\partial \eta} = \begin{bmatrix}1 & 0\end{bmatrix}\{\mathbf{A_{x1}}e^{\mathbf{A}_{x1}\eta}\mathbf{Q}_{xd}\mathbf{x_2}(\gamma) + e^{\mathbf{A}_{x1}\eta}\mathbf{Q}_{xd}\frac{\partial \mathbf{x_2}(\gamma)}{\partial \eta} + \boldsymbol{\Gamma_1} + \mathbf{N_1}e^{\Omega\eta}\mathbf{z}(\gamma)\}. \tag{10.150}$$

Note that the two convolution integrals, $\boldsymbol{\Gamma}_1$ and $\boldsymbol{\Gamma}_2$, can be easily evaluated by using matrix augmentation technique:

$$\boldsymbol{\Gamma}_1 = \begin{bmatrix} 1 & 0 & 0 & 0 \\ 1 & 0 & 0 & 0 \end{bmatrix} e^{\mathbf{A}_{\mathbf{onG}}\eta} \begin{bmatrix} 0 \\ 0 \\ \mathbf{z}(\gamma) \end{bmatrix} \tag{10.151}$$

where $\mathbf{A}_{\mathbf{onG}} = \begin{bmatrix} \mathbf{A}_{\mathbf{x1}} & \mathbf{A}_{\mathbf{x1}}\mathbf{N}_1 \\ \mathbf{0} & \boldsymbol{\Omega} \end{bmatrix}$.

$$\boldsymbol{\Gamma}_2 = \begin{bmatrix} 1 & 0 & 0 & 0 \\ 1 & 0 & 0 & 0 \end{bmatrix} e^{\mathbf{A}_{\mathbf{on}}\eta} \begin{bmatrix} 0 \\ 0 \\ \mathbf{z}(\gamma) \end{bmatrix} \tag{10.152}$$

Construction of the FCM

Once γ and η are found, one can proceed to solve for the current and voltage harmonics. Since the sixth period symmetry also exists for ac current space vector (Fig. 10.20), the evaluation of the system harmonic can proceed as follows:

1) The system matrices are augmented with one additional equation [71] for each harmonic of interest, leading to the form:

$$\frac{d}{dt}\begin{bmatrix} \mathbf{x}_1 \\ \mathbf{z} \\ \mathbf{y}_1 \end{bmatrix} = \begin{bmatrix} \mathbf{A}_{\mathbf{x1}} & \mathbf{N}_1 & \mathbf{0} \\ \mathbf{0} & \boldsymbol{\Omega} & \mathbf{0} \\ \mathbf{G}_1 & \mathbf{0} & \mathbf{H}_1 \end{bmatrix}\begin{bmatrix} \mathbf{x}_1 \\ \mathbf{z} \\ \mathbf{y}_1 \end{bmatrix} = \mathbf{A}_{\mathbf{on,h}}\begin{bmatrix} \mathbf{x}_1 \\ \mathbf{z} \\ \mathbf{y}_1 \end{bmatrix} \tag{10.153}$$

$$\frac{d}{dt}\begin{bmatrix} \mathbf{x}_2 \\ \mathbf{z} \\ \mathbf{y}_2 \end{bmatrix} = \begin{bmatrix} \mathbf{A}_{\mathbf{x2}} & \mathbf{N}_2 & \mathbf{0} \\ \mathbf{0} & \boldsymbol{\Omega} & \mathbf{0} \\ \mathbf{G}_2 & \mathbf{0} & \mathbf{H}_2 \end{bmatrix}\begin{bmatrix} \mathbf{x}_2 \\ \mathbf{z} \\ \mathbf{y}_2 \end{bmatrix} = \mathbf{A}_{\mathbf{off,h}}\begin{bmatrix} \mathbf{x}_2 \\ \mathbf{z} \\ \mathbf{y}_2 \end{bmatrix}, \tag{10.154}$$

where $\mathbf{y}_1 = [\ I_{sp,h} \quad V_{dc,l}\ ]^T$, and $\mathbf{y}_2 = [V_{dc,l}]$. $I_{sp,h}$ is the h-th ac current space vector harmonic, which is given as

$$I_{sp,h} = I_{\alpha,h} + jI_{\beta,h}, \tag{10.155}$$

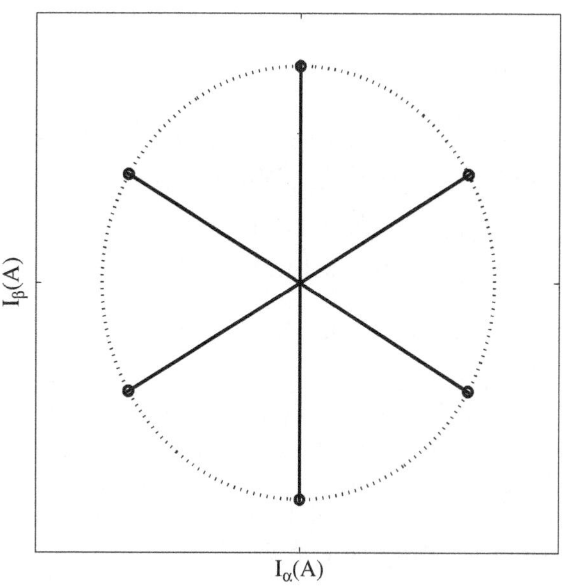

Figure 10.20 AC current space vector with one-sixth period symmetry in the DCM.

where

$$\begin{bmatrix} I_{\alpha,h} \\ I_{\beta,h} \end{bmatrix} = \mathbf{C_x} \begin{bmatrix} I_{a,h} \\ I_{b,h} \\ I_{c,h} \end{bmatrix},$$

and $I_{a,h}$, $I_{b,h}$, $I_{c,h}$ are the three-phase ac current harmonics. $V_{dc,l}$ represents the l-th dc voltage harmonics.

Since $i_a = i_d$, $i_b = -i_d$, $i_c = 0$ in the conduction subinterval (Fig. 10.19(a)), one can express $i_{sp} = \left(1 - j\frac{1}{\sqrt{3}}\right) i_d$.

Consequently, based on [71], expressions for G_1, G_2, H_1, and H_2 are as follows:

$$\mathbf{G_1} = \begin{bmatrix} \left(1 - j\frac{1}{\sqrt{3}}\right)\lambda_h & 0 \\ 0 & \lambda_l \end{bmatrix}, \mathbf{G_2} = \left[\lambda_l\right],$$

$$\lambda_h = \frac{6}{T} e^{-jh\omega\left(\frac{T}{6}+\gamma\right)}, \lambda_l = \frac{6}{T} e^{-jl\omega\left(\frac{T}{6}+\gamma\right)},$$

$$\mathbf{H_1} = \begin{bmatrix} jh\omega & 0 \\ 0 & jl\omega \end{bmatrix}, \mathbf{H_2} = \left[jl\omega\right], \omega = \frac{2\pi}{T}.$$

2) For characteristic harmonics (i.e. $h = 1, -5, 7, -11, \ldots$, and $l = 0, 6, 12, 18, \ldots$), the system harmonics can be obtained by evaluating (10.156):

$$\begin{bmatrix} I_{sp,h} \\ V_{dc,l} \end{bmatrix} = \mathbf{C_h} e^{\mathbf{A_{off,h}}\left(\frac{T}{6} - \eta\right)} \mathbf{P_{d,h}} e^{\mathbf{A_{on,h}}(\eta)} \mathbf{Q_{d,h}} \begin{bmatrix} \mathbf{x_1}(\gamma) \\ \mathbf{z}(\gamma) \\ 0 \\ 0 \end{bmatrix}, \quad (10.156)$$

where $\mathbf{C_h} = \left[\mathbf{0} \;\; \mathbf{0} \;\; \mathbb{I}_{2\times2}\right]$, $\mathbb{I} = \begin{bmatrix} 1 & 0 \\ 0 & 1 \end{bmatrix}$, $\mathbf{P_{d,h}} = \begin{bmatrix} \mathbf{P_d} & \mathbf{0} & \mathbf{0} \\ \mathbf{0} & \mathbf{0} & \mathbf{I} \end{bmatrix}$, and $\mathbf{Q_{d,h}} = \begin{bmatrix} \mathbf{Q_d} & \mathbf{0} \\ \mathbf{0} & \mathbf{0} \\ \mathbf{0} & \mathbb{I}_{2\times2} \end{bmatrix}$.

Note that the transition matrices $\mathbf{P_{d,h}}$ and $\mathbf{Q_{d,h}}$ are needed due to different basis of $\mathbf{A_{on,h}}$ and $\mathbf{A_{off,h}}$.

Example 10.4 To analyze the DCM, the following numerical values are used: $R = 0.001\Omega$, $L = 0.1mH$, $C = 1000\mu F$, $R_l = 25\Omega$, and source voltages:

$$v_{sa} = 120\sqrt{2}\sin(377t)$$
$$v_{sb} = 120\sqrt{2}\sin\left(377t - \frac{2\pi}{3}\right)$$
$$v_{sc} = 120\sqrt{2}\sin\left(377t + \frac{2\pi}{3}\right).$$

Fig. 10.21 shows ac currents, i_α and i_β, and dc capacitor voltage, v_{dc}, predicted by PSCAD/EMTDC. The values of diode conduction instant, γ, and conduction interval, η, and $\mathbf{x_2}(\gamma)$ (i.e. $v_{dc}(\gamma)$) in Fig. 10.21 are listed in Table 10.2 to compare with those predicted by the LB 2 method.

Figs. 10.22 and 10.23 compare ac current space vector and dc voltage harmonics obtained from PSCAD/EMTDC, and the LB 2 model method. Fig. 10.23 is shown in log scale because the dc voltage harmonic rolls off very rapidly.

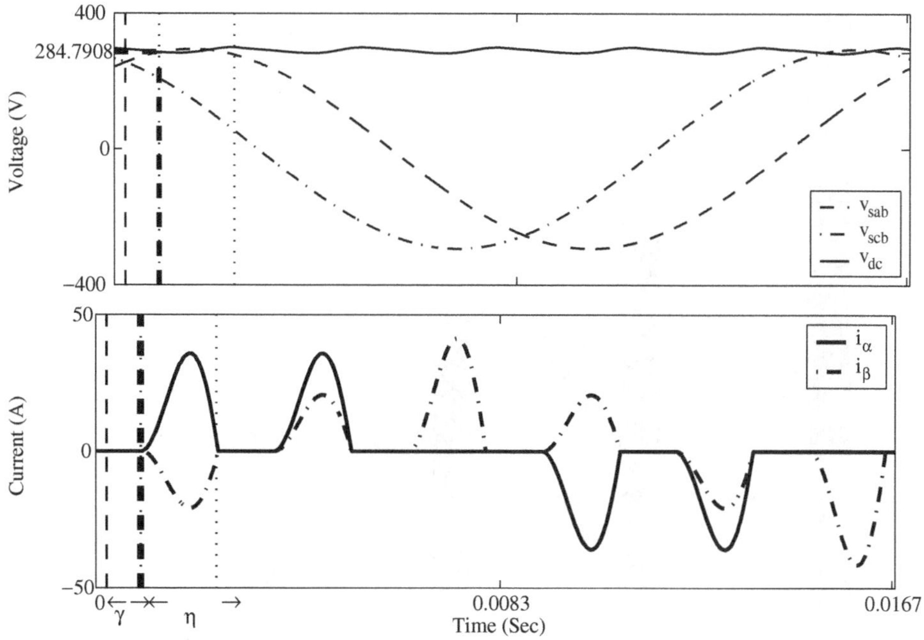

Figure 10.21 Voltage and current waveforms in the DCM produced by PSCAD/EMTDC.

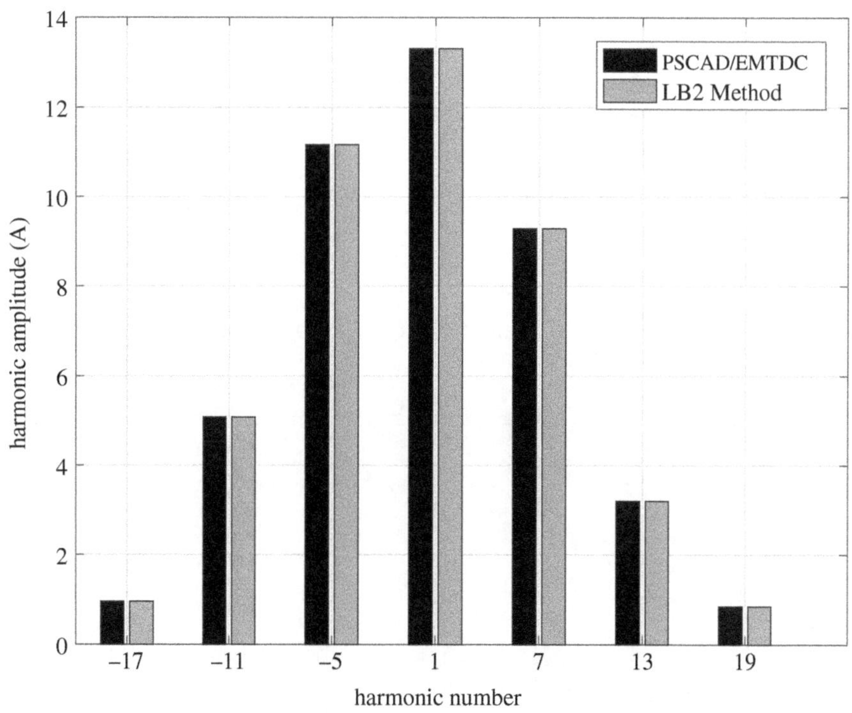

Figure 10.22 Complex harmonic spectrum of the capacitor filtered rectifier in the DCM.

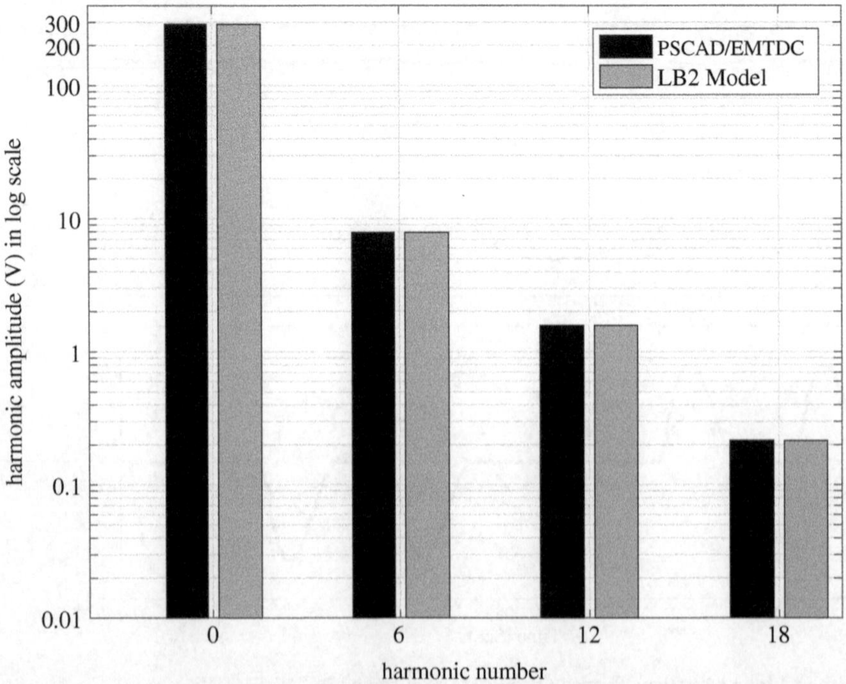

Figure 10.23 DC voltage harmonic spectrum of the capacitor filtered rectifier in the DCM.

Table 10.2 Results from PSCAD/EMTDC and the sampled-data model method in the DCM.

	PSCAD/EMTDC	Sampled-Data Model
γ	$7.3927 \times 10^{-4} s$	$7.2553 \times 10^{-4} s$
η	$1.6355 \times 10^{-3} s$	$1.6030 \times 10^{-3} s$
$v_{dc}(\gamma)$	$284.7908\ V$	$284.7951 V$

As can be seen from Table 10.2 and Figs. 10.22 and 10.23, excellent agreement exists between the two approaches.

```


freq=60;
w=2*pi*freq;
T=1/freq;
Ln=1e-4;
C=1000e-6;
Un=120*sqrt(2);

%R=10;
R=25;
%R1=0.0005;
R1=0.0005+0.0005;
R2=1e6;

```

```
A11=[-R1/Ln -1/(2*Ln) 3/(4*Ln) -sqrt(3)/(4*Ln);1/C -1/(R*C) 0 0;0 0 0 -w;
    0 0 w 0];
A22=[-1/(C*R) 0 0;0 0 -w; 0 w 0];
A3=[0 -w;w 0];

A1=[-R1/Ln -1/(2*Ln);1/C -1/(R*C)];
A2=[-1/(R*C)];

B1=[3/(4*Ln) -sqrt(3)/(4*Ln);0 0];


PP=[0 1 0 0;0 0 1 0;0 0 0 1];

P=[0 1];
I6=[1 0 0;0 cos(pi/3) -sin(pi/3);0 sin(pi/3) cos(pi/3)];

C1=[1 0 0 0];

ualph0=Un*sin(+pi/6);
ubeta0=-Un*cos(+pi/6);

gamma0=T/6/10; %initialization
mu0=T/6/3;  %initialization

gamma=gamma0;
mu=mu0;
error=1;
counter=0;

while (error>1e-10)
usbc=expm(A3*gamma)*[ualph0;ubeta0];
ualph_gamma=usbc(1);
ubeta_gamma=usbc(2);
zss=[ualph_gamma;ubeta_gamma];


phip=expm(A2*(T/6-mu))*P*expm(A1*(mu))*P';
phi=I6-expm(A22*(T/6-mu))*PP*expm(A11*(mu))*PP';
E=phi(1:1,1:1);
F=phi(1:1,2:length(A22));
uss=-inv(E)*F*[zss];

f1=[-1 3/2 -sqrt(3)/2]*[uss;zss];
f2=C1*expm(A11*(mu))*PP'*[uss;zss];

TT= expm([A1 A1*B1;zeros(2) A3]*mu)*[0;0;zss];
TT2=expm([A1 B1;zeros(2) A3]*mu)*[0;0;zss];


phi2=expm(A22*(T/6-mu))*PP*expm(A11*(mu))*PP';

E1=phi2(1:1,1:1);
F1=phi2(1:1,2:length(A22));

Run_phi2=-A22*phi2+expm(A22*(T/6-mu))*PP*expm(A11*mu)*A11*PP';

E2=Run_phi2(1:1,1:1);
F2=Run_phi2(1:1,2:length(A22));
```

```
xx_tst=inv(E)*(F1)*expm(A3*gamma)*A3*[ualph0;ubeta0];


yy_tst=-(E)^-2*(-E2)*F1*zss+inv(E)*F2*zss;

xx=inv(1-phip)*expm(A2*(T/6-mu))*P*(TT(1:2)+B1*expm(A3*mu)*zss-expm(A1*mu)*B1
    *zss);

J11=[3/2 -sqrt(3)/2]*expm(A3*gamma)*A3*[ualph0;ubeta0]-xx;
J11_tst=[-1 3/2 -sqrt(3)/2]*[xx_tst;expm(A3*gamma)*A3*[ualph0;ubeta0]];

J12x=-((1-phip)^-2)*(A2*phip-expm(A2*(T/6-mu))*P*expm(A1*(mu))
    *A1*P')*...
expm(A2*(T/6-mu))*P*TT2(1:2)+inv(1-phip)*expm(A2*(T/6-mu))*(-A2)*P*TT2(1:2)
    +...
inv(1-phip)*expm(A2*(T/6-mu))*P*(TT(1:2)+B1*expm(A3*mu)*[zss]);
J12=-J12x;




J21x=expm(A1*mu)*P'*xx+(TT(1:2)+B1*expm(A3*mu)*zss-expm(A1*mu)*B1*zss);
J21=J21x(1);


J22x=A1*expm(A1*mu)*P'*uss+expm(A1*mu)*P'*J12x+(TT(1:2)+B1*expm(A3*mu)
    *[zss]);
J22=J22x(1);



J=[J11 J12;J21 J22];
dt=inv(J)*[f1;f2];
gamma=gamma-dt(1);
mu=mu-dt(2);
counter=counter+1;
error=max(abs(dt));
end

counter;
```

□

10.4.2 Continuous Conduction Mode

Similar to the DCM, two subintervals can also be identified in every sixth of the period in the continuous conduction mode (CCM) (Fig. 10.24) — conduction (η) and commutation ($\mu = \frac{T}{6} - \eta$) intervals. However, unlike DCM, γ now refers to the time that the diode (D_5) turns off with respect to the zero reference.

The differential equations describing the conduction interval (Fig. 10.25(a)) are the same as (10.138). The differential equations describing the commutation interval (Fig. 10.25(b)) are given in (10.157):

$$\frac{d}{dt}\begin{bmatrix}\mathbf{x}_3\\ \mathbf{z}\end{bmatrix} = \mathbf{A}_{\mathbf{com}}\begin{bmatrix}\mathbf{x}_3\\ \mathbf{z}\end{bmatrix} = \begin{bmatrix}\mathbf{A}_{\mathbf{x3}} & \mathbf{N}_3\\ \mathbf{0} & \mathbf{\Omega}\end{bmatrix}\begin{bmatrix}\mathbf{x}_3\\ \mathbf{z}\end{bmatrix}, \tag{10.157}$$

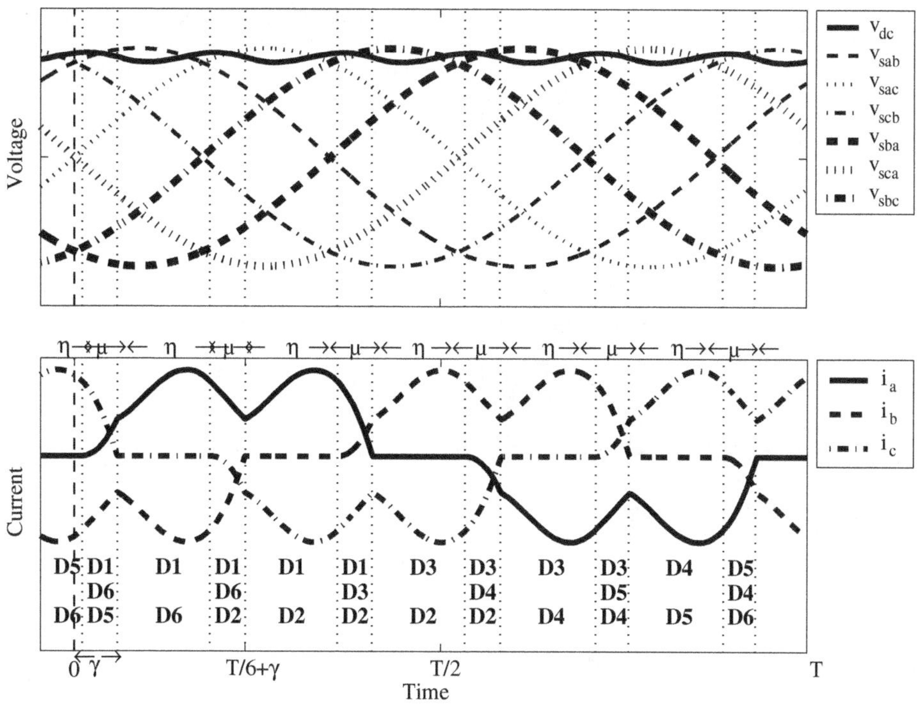

Figure 10.24 Rectifier waveforms in the CCM.

where $\mathbf{x_3} = \begin{bmatrix} i_e & i_d & v_{dc} \end{bmatrix}$, $\mathbf{N_3} = \begin{bmatrix} \frac{1}{2L} & \frac{-\sqrt{3}}{2L} \\ \frac{1}{L} & 0 \\ 0 & 0 \end{bmatrix}$, and $\mathbf{A_{x3}} = \begin{bmatrix} \frac{-R}{L} & 0 & \frac{-1}{3L} \\ 0 & \frac{-R}{L} & \frac{-2}{3L} \\ 0 & \frac{1}{C} & \frac{-1}{R_l C} \end{bmatrix}$.

Diode Constraint Equations

The diode constraint equations for the interval from γ to $\frac{T}{6} + \gamma$ in the CCM are now determined. Noting $v_{Nn} = -v_{dc}/3$ in Fig. 10.25(b), Diode D_2 turn-on occurs when

$$M_{1,c} = -3v_c(\gamma + \eta) - v_{dc}(\gamma + \eta) \tag{10.158}$$

$$= \begin{bmatrix} 0 & -1 & \frac{3}{2} & \frac{3\sqrt{3}}{2} \end{bmatrix} e^{\mathbf{A_{on}}\eta} \mathbf{P_c} \begin{bmatrix} \mathbf{x_3}(\gamma) \\ \mathbf{z}(\gamma) \end{bmatrix} = 0, \text{ where } \mathbf{P_c} = blkdiag(\mathbf{P_{xc}}, \mathbb{I}_{2\times 2}), \text{ and } \mathbf{P_{xc}} = \begin{bmatrix} \frac{-1}{2} & 1 & 0 \\ 0 & 0 & 1 \end{bmatrix}.$$

$\mathbf{P_c}$ is associated with change of basis at the diode transition instant from the commutation to the conduction interval in the CCM.

Diode D_5 turn-off (refer to Fig. 10.24) occurs when

$$M_{2,c} = -i_e(\gamma) = \begin{bmatrix} -1 & 0 & 0 & 0 & 0 \end{bmatrix} \begin{bmatrix} \mathbf{x_3}(\gamma) \\ \mathbf{z}(\gamma) \end{bmatrix} = 0 \tag{10.159}$$

Similar to the DCM, expressions linking $\mathbf{x_3}(\gamma)$ and $z(\gamma)$ to input $\mathbf{z}(0)$ are needed to solve for γ and η from (10.158), and (10.159).

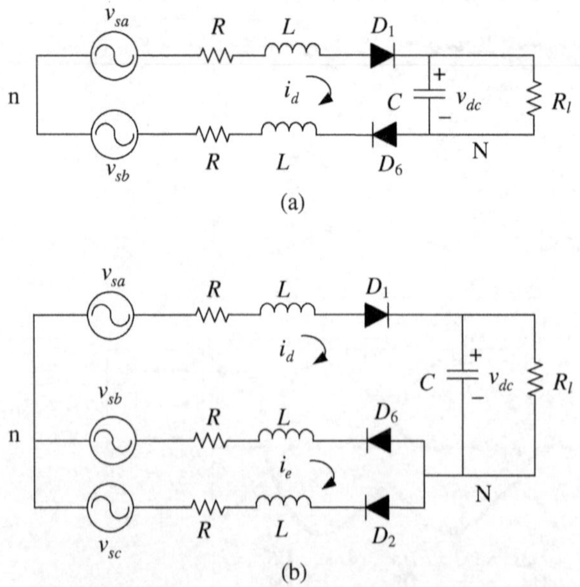

Figure 10.25 CCM Models: (a) rectifier model during the conduction subinterval, (b) rectifier model during the commutation subinterval.

10.4.3 Steady-State Constraint Equations

The steady-state constraint for CCM is expressed in (10.160):

$$\begin{bmatrix} \mathbf{x_3}\left(\frac{T}{6}+\gamma\right) \\ \mathbf{z}\left(\frac{T}{6}+\gamma\right) \end{bmatrix} = \mathbf{\Phi_c} \begin{bmatrix} \mathbf{x_3}(\gamma) \\ \mathbf{z}(\gamma) \end{bmatrix}, \tag{10.160}$$

where $\mathbf{\Phi_c} = e^{\mathbf{A_{com}}\left(\frac{T}{6}-\eta\right)} \mathbf{Q_c} e^{\mathbf{A_{on}}\eta} \mathbf{P_c}$, $\mathbf{Q_c} = blkdiag(\mathbf{Q_{xc}}, \mathbb{I}_{2\times 2})$, and $\mathbf{Q_{xc}} = \begin{bmatrix} 0 & 0 \\ 1 & 0 \\ 0 & 1 \end{bmatrix}$, which represents the transition matrix from conduction to commutation interval in the CCM.

Combining the sixth period mapping constraint, (10.161) can be obtained.

$$\begin{aligned} &(\mathbf{I_c^{1/6}} - \mathbf{\Phi_c}) \begin{bmatrix} \mathbf{x_3}(\gamma) \\ \mathbf{z}(\gamma) \end{bmatrix} = \begin{bmatrix} \mathbf{D_c} & \mathbf{E_c} \\ \mathbf{F_c} & \mathbf{K_c} \end{bmatrix} \begin{bmatrix} \mathbf{x_3}(\gamma) \\ \mathbf{z}(\gamma) \end{bmatrix} = 0 \\ &\Rightarrow \mathbf{x_3}(\gamma) = -\mathbf{D_c^{-1}} \mathbf{E_c} \mathbf{z}(\gamma) = -\mathbf{D_c^{-1}} \mathbf{E_c} e^{\mathbf{\Omega}\gamma} \mathbf{z}(0), \end{aligned} \tag{10.161}$$

where

$$\mathbf{I_{1/6,c}} = blkdiag(\mathbf{I_{ed}}, \mathbf{\Theta}), \text{ and } \mathbf{I_{ed}} = \begin{bmatrix} 0 & 1 & 0 \\ -1 & 1 & 0 \\ 0 & 0 & 1 \end{bmatrix}.$$

As in the DCM case, constraints (10.158) and (10.159) are again transcendental equations that must be solved numerically.

Numerical Iteration for Finding Unknowns

Similar to the DCM case, a Newton-type method can be formulated with (10.158) and (10.159) as mismatch equations, solving for γ and η. The elements of the required Jacobian matrix,

$\mathbf{J} = \begin{bmatrix} \frac{\partial M_{1,c}}{\partial \gamma} & \frac{\partial M_{1,c}}{\partial \eta} \\ \frac{\partial M_{2,c}}{\partial \gamma} & \frac{\partial M_{2,c}}{\partial \eta} \end{bmatrix}$, are listed as follows:

$$\frac{\partial M_{1,c}}{\partial \gamma} = \begin{bmatrix} \frac{3}{2} & \frac{3\sqrt{3}}{2} \end{bmatrix} e^{\boldsymbol{\Omega}(\gamma+\eta)} \boldsymbol{\Omega} z(0) - \begin{bmatrix} 0 & 1 \end{bmatrix} e^{\mathbf{A}_{x1}(\eta)} \mathbf{p}_{\mathbf{xc}} \frac{\partial \mathbf{x}_3(\gamma)}{\partial \gamma} \tag{10.162}$$

$$\frac{\partial M_{1,c}}{\partial \eta} = \begin{bmatrix} \frac{3}{2} & \frac{3\sqrt{3}}{2} \end{bmatrix} e^{\boldsymbol{\Omega}(\gamma+\eta)} \boldsymbol{\Omega} z(0) - \begin{bmatrix} 0 & 1 \end{bmatrix} \{ e^{\mathbf{A}_{x1}(\eta)} \mathbf{A}_{\mathbf{x}1} \mathbf{p}_{\mathbf{xc}} \mathbf{x}_3(\gamma) + e^{\mathbf{A}_{x1}(\eta)} \mathbf{p}_{\mathbf{xc}} \frac{\partial \mathbf{x}_3(\gamma)}{\partial \eta} \} \tag{10.163}$$

$$\frac{\partial M_{2,c}}{\partial \gamma} = - \begin{bmatrix} 1 & 0 & 0 \end{bmatrix} \frac{\partial \mathbf{x}_3(\gamma)}{\partial \gamma} \tag{10.164}$$

$$\frac{\partial M_{2,c}}{\partial \eta} = - \begin{bmatrix} 1 & 0 & 0 \end{bmatrix} \frac{\partial \mathbf{x}_3(\gamma)}{\partial \eta}, \tag{10.165}$$

where

$$\begin{aligned} \frac{\partial \mathbf{x}_3(\gamma)}{\partial \gamma} = {} & \left[\mathbf{I}_{ed} - \boldsymbol{\Phi}_{\mathbf{xc}}\right]^{-1} e^{\mathbf{A}_{x3}\mu} \mathbf{Q}_{\mathbf{xc}} \{ \boldsymbol{\Gamma}_1 + \mathbf{N}_1 e^{\boldsymbol{\Omega}\eta} \mathbf{z}(\gamma) \\ & - e^{\mathbf{A}_{x1}\eta} \mathbf{N}_1 \mathbf{z}(\gamma) \} + \boldsymbol{\Gamma}_3 + \mathbf{N}_3 e^{\boldsymbol{\Omega}(T/6)} \mathbf{z}(\gamma) \\ & - e^{\mathbf{A}_{x3}\mu} \mathbf{N}_3 e^{\boldsymbol{\Omega}\eta} \mathbf{z}(\gamma), \end{aligned}$$

$$\begin{aligned} \frac{\partial \mathbf{x}_3(\gamma)}{\partial \eta} = {} & - \left[\mathbf{I}_{\mathbf{e}d} - \boldsymbol{\Phi}_{\mathbf{xc}}\right]^{-2} (\mathbf{A}_{\mathbf{x}3} \boldsymbol{\Phi}_{\mathbf{xc}} - e^{\mathbf{A}_{x3}\mu} \mathbf{Q}_{\mathbf{xc}} e^{\mathbf{A}_{x1}\eta} \cdot \\ & \mathbf{A}_{\mathbf{x}1} \mathbf{P}_{\mathbf{xc}}) \cdot (e^{\mathbf{A}_{x3}\mu} \mathbf{Q}_{\mathbf{xc}} \boldsymbol{\Gamma}_2 + \boldsymbol{\Gamma}_4 + [\mathbf{I}_6 - \boldsymbol{\Phi}_{\mathbf{xc}}]^{-1} \cdot \\ & \{ -\mathbf{A}_{\mathbf{x}3} e^{\mathbf{A}_{x3}\mu} \mathbf{Q}_{\mathbf{xc}} \boldsymbol{\Gamma}_2 + e^{\mathbf{A}_{x3}\mu} \mathbf{Q}_{\mathbf{xc}} [\boldsymbol{\Gamma}_1 \\ & + \mathbf{N}_1 e^{\boldsymbol{\Omega}\mu} \mathbf{z}(\gamma)] - e^{\mathbf{A}_{x3}\mu} \mathbf{N}_3 e^{\boldsymbol{\Omega}\eta} \mathbf{z}(\gamma) \}, \end{aligned}$$

$$\boldsymbol{\Gamma}_3 = \int_{\gamma+\eta}^{T/6+\gamma} e^{\mathbf{A}_{x3}(T/6+\gamma-\tau)} \mathbf{A}_{\mathbf{x}3} \mathbf{N}_3 \mathbf{z}(\tau) d\tau,$$

$$\boldsymbol{\Gamma}_4 = \int_{\gamma+\eta}^{T/6+\gamma} e^{\mathbf{A}_{x3}(T/6+\gamma-\tau)} \mathbf{N}_3 \mathbf{z}(\tau) d\tau.$$

Note that both $\boldsymbol{\Gamma}_3$ and $\boldsymbol{\Gamma}_4$ can be found as follows:

$$\boldsymbol{\Gamma}_3 = \begin{bmatrix} 1 & 0 & 0 & 0 & 0 \\ 0 & 1 & 0 & 0 & 0 \\ 0 & 0 & 1 & 0 & 0 \end{bmatrix} e^{\mathbf{A}_{\mathbf{comG}}\mu} \begin{bmatrix} 0 \\ 0 \\ 0 \\ \mathbf{z}(\gamma+\mu) \end{bmatrix}, \tag{10.166}$$

where $\mathbf{A}_{\mathbf{comG}} = \begin{bmatrix} \mathbf{A}_{\mathbf{x}3} & \mathbf{N}_3 \\ \mathbf{0} & \boldsymbol{\Omega} \end{bmatrix}$.

$$\boldsymbol{\Gamma}_4 = \begin{bmatrix} 1 & 0 & 0 & 0 & 0 \\ 0 & 1 & 0 & 0 & 0 \\ 0 & 0 & 1 & 0 & 0 \end{bmatrix} e^{\mathbf{A}_{\mathbf{com}}\mu} \begin{bmatrix} 0 \\ 0 \\ 0 \\ \mathbf{z}(\gamma+\mu) \end{bmatrix} \tag{10.167}$$

Harmonic Analysis

As shown in Fig. 10.26, the current space vector of CCM also exhibits sixth period symmetry. Consequently, the harmonic analysis can proceed as follows:

1. Similar to the analysis of the DCM, the system matrices of CCM are augmented with one additional equation for each harmonic of interest. Consequently, (10.157) becomes (10.168).

$$\frac{d}{dt}\begin{bmatrix}\mathbf{x_3}\\ \mathbf{z}\\ \mathbf{y_1}\end{bmatrix} = \begin{bmatrix}\mathbf{A_{x3}} & \mathbf{N_3} & \mathbf{0}\\ \mathbf{0} & \mathbf{\Omega} & \mathbf{0}\\ \mathbf{G_3} & 0 & \mathbf{H_1}\end{bmatrix}\begin{bmatrix}\mathbf{x_3}\\ \mathbf{z}\\ \mathbf{y_1}\end{bmatrix} = \mathbf{A^h_{com}}\begin{bmatrix}\mathbf{x_3}\\ \mathbf{z}\\ \mathbf{y_1}\end{bmatrix}, \tag{10.168}$$

where

$$\mathbf{G_3} = \begin{bmatrix}\left(j\frac{2}{\sqrt{3}}\right)\lambda_h & \left(1-j\frac{1}{\sqrt{3}}\right)\lambda_h\\ 0 & \lambda_l\end{bmatrix}.$$

The expression of $\mathbf{G_3}$ is obtained based on the fact that $i = i_\alpha + ji_\beta$, and

$$\begin{bmatrix}i_\alpha\\ i_\beta\end{bmatrix} = \frac{2}{3}\begin{bmatrix}0 & \frac{3}{2}\\ \sqrt{3} & \frac{-\sqrt{3}}{2}\end{bmatrix}\begin{bmatrix}i_e\\ i_d\end{bmatrix} \tag{10.169}$$

during the commutation interval (Fig. 10.25(b)).

2. For characteristic harmonics, (i.e. $h = 1, -5, 7, -11, \ldots$, and $l = 0, 6, 12, 18, \ldots$), the system harmonics can be found by evaluating (10.170):

$$\begin{bmatrix}I^h_{sp}\\ V^l_{dc}\end{bmatrix} = \mathbf{C_h} e^{\mathbf{A_{com,h}}\left(\frac{T}{6}-\beta\right)} \mathbf{P_{c,h}} e^{\mathbf{A_{on,h}}(\beta)} \mathbf{Q_{c,h}} \begin{bmatrix}\mathbf{x_1}(\gamma)\\ \mathbf{z}(\gamma)\\ 0\\ 0\end{bmatrix}, \tag{10.170}$$

where $\mathbf{P_{c,h}} = \begin{bmatrix}\mathbf{P_c} & \mathbf{0}\\ \mathbf{0} & \mathbf{I}\end{bmatrix}$, and $\mathbf{Q_{c,h}} = \begin{bmatrix}\mathbf{Q_c} & \mathbf{0}\\ \mathbf{0} & \mathbf{I}\end{bmatrix}$.

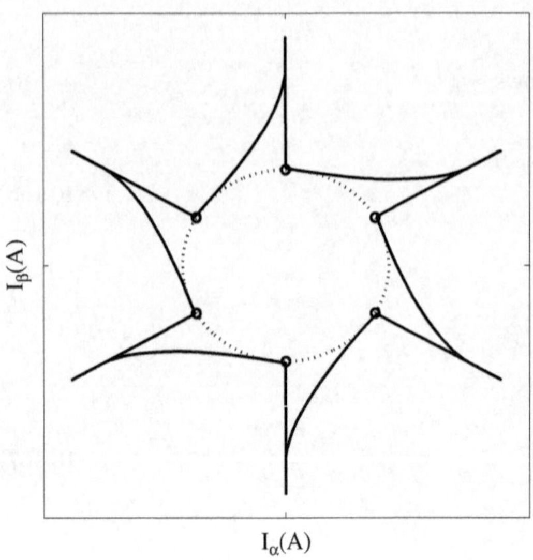

Figure 10.26 AC current space vector with one-sixth period symmetry in the CCM.

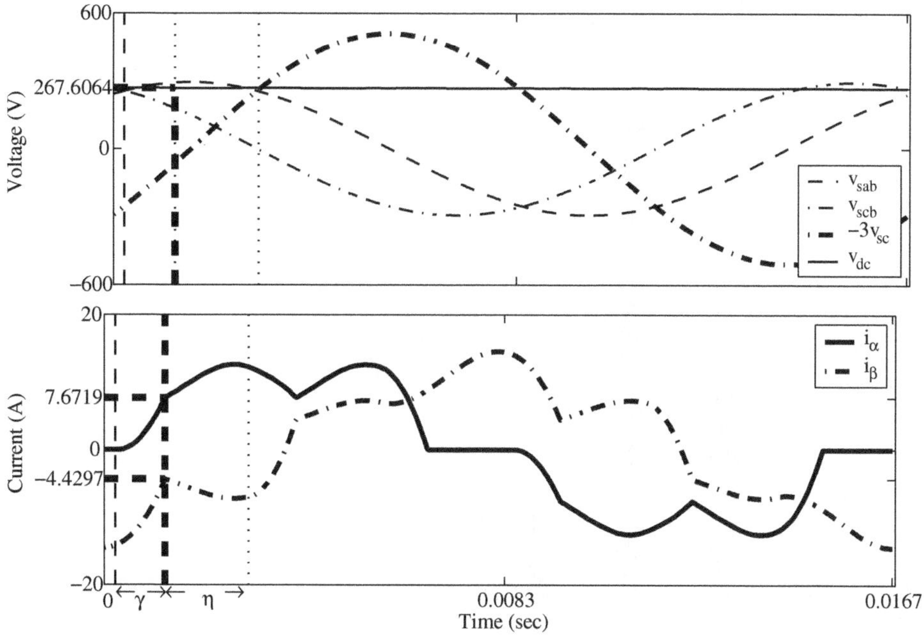

Figure 10.27 Voltage and current waveforms in the CCM produced by PSCAD/EMTDC.

Table 10.3 Results from PSCAD/EMTDC and the LB 2 model method in the CCM.

	PSCAD/EMTDC	LB 2 Model
γ	$1.0631 \times 10^{-3}s$	$1.0441 \times 10^{-3}s$
η	$1.8406 \times 10^{-3}s$	$1.8159 \times 10^{-3}s$
$i_\alpha(\gamma)$	7.6719A	7.6593A
$i_\beta(\gamma)$	−4.4297A	−4.4221A
$v_{dc}(\gamma)$	$267.6064V$	$267.6159V$

Example 10.5 As shown in [110], the rectifier operates in the CCM when the inductance of the ac choke in the DCM is changed to 3 mH and the rest of the parameters are kept intact.

Fig. 10.27 shows ac currents, i_α and i_β, and dc capacitor voltage, v_{dc}, predicted by PSCAD/EMTDC.

Table 10.3 lists the extinction time instant, γ, conduction interval, η, and the values of the steady-state ac currents and dc voltages obtained from Fig. 10.27, together with those obtained from the sampled-data model.

Note that $i_e(\gamma)$ and $i_d(\gamma)$ obtained from (10.161) has been converted to $i_\alpha(\gamma)$ and $i_\beta(\gamma)$ as shown in Table 10.3 via (10.169) so as to be comparable with PSCAD/EMTDC results.

The ac current space vector and dc voltage harmonics obtained from PSCAD/EMTDC and the LB 2 model method are shown in Figs. 10.28 and 10.29, respectively.

Similar to the case of the DCM, excellent agreement exists between PSCAD/EMTDC and the LB 2 model for the CCM.

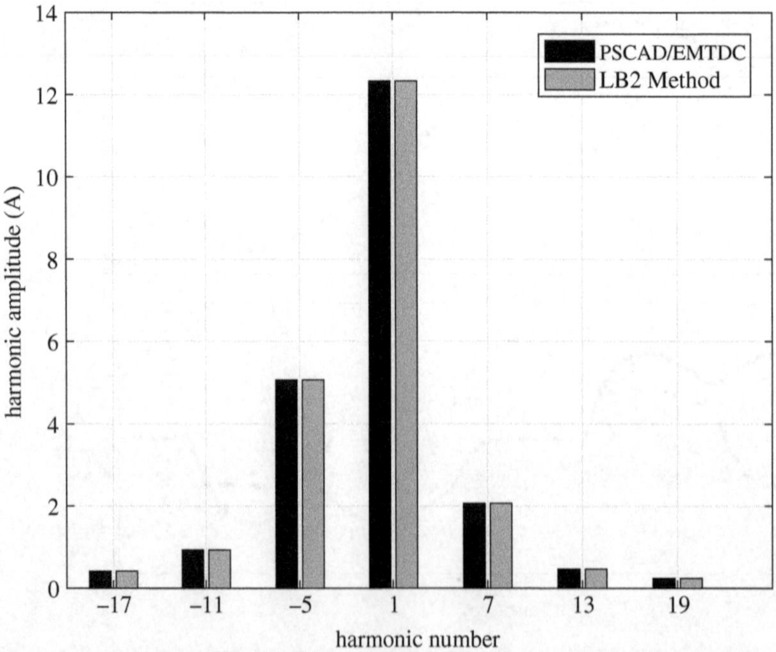

Figure 10.28 Complex harmonic spectrum of the capacitor filtered rectifier in the CCM.

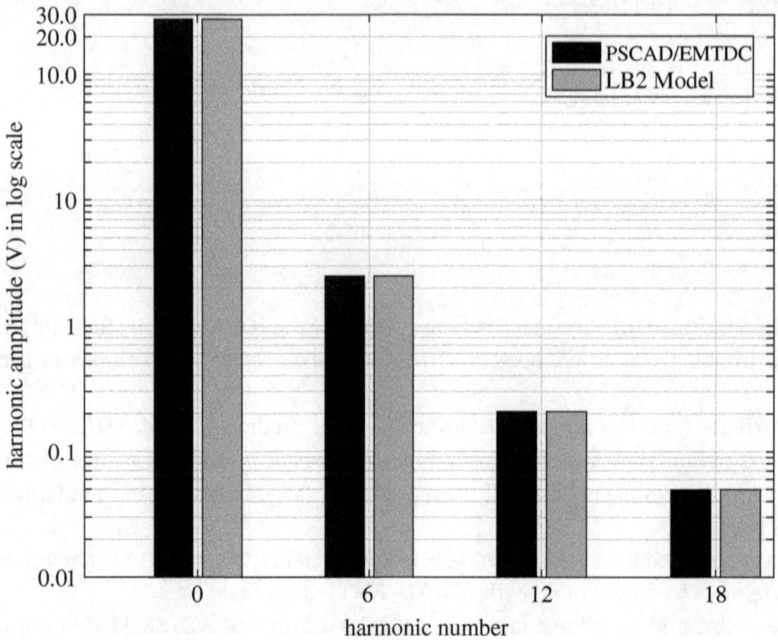

Figure 10.29 DC voltage harmonic spectrum of the capacitor filtered rectifier in the CCM.

```
%The size of A1 and A2 are the same

% Xn=0.18704;
% B0=0.377;
freq=60;
w=2*pi*freq;
T=1/freq;
Ln=3e-3;
C=1000e-6;
Un=120*sqrt(2);
%C_a=blkdiag([0 1;-sqrt(3)/2 1/2],1,eye(2))

%R=10;
R_dc=25;
R1=0.0005+0.0005;

P=[-1/2 1 0;0 0 1];
Q=[0 0;1 0;0 1];
P2=[-1/2 1 0 0 0;
    0    0 1 0 0;
    0    0 0 1 0;
    0    0 0 0 1];

Q2=[0 0 0 0;
   1 0 0 0;
   0 1 0 0;
   0 0 1 0;
   0 0 0 1];

A4=[-R1/Ln 0 -1/(3*Ln) 1/(2*Ln) sqrt(3)/(2*Ln);
    0 -R1/Ln -2/(3*Ln) 1/Ln 0;
    0 1/C -1/(C*R_dc) 0 0;
    0 0 0 0 -w;
    0 0 0 w 0];

A5=[-R1/(Ln) -1/(2*Ln) 3/(4*Ln) -sqrt(3)/(4*Ln);
    1/C -1/(C*R_dc) 0 0;
    0 0 0 -w;
    0 0 w 0];

Acomm=[-R1/Ln 0 -1/(3*Ln);
    0 -R1/Ln -2/(3*Ln);
    0 1/C -1/(C*R_dc)];
B3=[1/(2*Ln) sqrt(3)/(2*Ln);1/Ln 0;0 0];

Aon=[-R1/(Ln) -1/(2*Ln);
    1/C -1/(C*R_dc)];
B1=[3/(4*Ln) -sqrt(3)/(4*Ln);0 0];

A6=[-(R1)/(Ln) 0 -1/(3*Ln) 1/(2*Ln) -sqrt(3)/(2*Ln);
        0 -R1/Ln -2/(3*Ln) 1/(Ln) 0;
     0 1/C -1/(R_dc*C) 0 0;
    0 0 0 0 -w;
    0 0 0 w 0];


A9=[0 -w;w 0];



I6_2=[0 1 0 0 0;-1 1 0 0 0;0 0 1 0 0;0 0 0 cos(pi/3) -sin(pi/3);...
```

```
0 0 0 sin(pi/3) cos(pi/3)];

I6=[0 1 0;-1 1 0;0 0 1];


C1=[1 0 0 0 0 0 0 0 0 0];
C3=[0 0 1 0 0 0 0 0 0 0];

usax0=Un*sin(+pi/6);
usay0=-Un*cos(+pi/6);



alpha0=T/6/3; %guess
mu0=2*T/6/3; %guess



mu=mu0;
alpha=alpha0;

error=1;
counter=0;

while (error>1e-12)

usbc=expm(A9*(alpha))*[usax0;usay0];
uax_gamma=usbc(1);
uay_gamma=usbc(2);
zss=[uax_gamma;uay_gamma];

phip=expm(Acomm*(T/6-mu))*Q*expm(Aon*(mu))*P;
phi2=expm(A4*(T/6-mu))*Q2*expm(A5*(mu))*P2;
phi=I6_2-phi2;
E=phi(1:3,1:3);
F=phi(1:3,4:length(A6));
uss=-inv(E)*F*[zss];

f1=-uss(1);

f2=[0 -1 3/2 3*sqrt(3)/2]*expm(A5*mu)*P2*[uss;zss];




E1=phi2(1:3,1:3);
F1=phi2(1:3,4:length(A6));

Run_phi2=-A4*phi2+expm(A4*(T/6-mu))*Q2*expm(A5*mu)*A5*P2;

E2=Run_phi2(1:3,1:3);
F2=Run_phi2(1:3,4:length(A6));


TT= expm([Aon Aon*B1;zeros(2) A9]*mu)*[0;0;zss];
TT1=expm([Acomm Acomm*B3;zeros(2,3) A9]*(T/6-mu))*[0;0;0;expm(A9*mu)*zss];
TT2=expm([Aon B1;zeros(2) A9]*mu)*[0;0;zss];
TT3=expm([Acomm B3;zeros(2,3) A9]*(T/6-mu))*[0;0;0;expm(A9*mu)*zss];

xx=inv(I6-phip)*(expm(Acomm*(T/6-mu))*Q*(TT(1:2)+B1*expm(A9*mu)*zss-...
expm(Aon*mu)*B1*zss)+TT1(1:3)+B3*expm(A9*(T/6))*zss-...
```

```
expm(Acomm*(T/6-mu))*B3*expm(A9*mu)*zss);

yy=-((I6-phip)*\text{\couhat}*-2)*(Acomm*phip-expm(Acomm*(T/6-mu))*Q*expm
    (Aon*(mu))*Aon*P)*...
(expm(Acomm*(T/6-mu))*Q*TT2(1:2)+TT3(1:3))+inv(I6-phip)*...
(-Acomm*expm(Acomm*(T/6-mu))*Q*TT2(1:2)+expm(Acomm*(T/6-mu))*Q*(TT(1:2)+...
B1*expm(A9*mu)*zss)-expm(Acomm*(T/6-mu))*B3*expm(A9*mu)*zss);

J11=-xx(1);

J12=-yy(1);

J21=[3/2 3*sqrt(3)/2]*expm(A9*(alpha+mu))*A9*[usax0;usay0]-[0 1]
    *expm(Aon*mu)*P*xx;


J22=[3/2 3*sqrt(3)/2]*expm(A9*(alpha+mu))*A9*[usax0;usay0]-([0 1]
    *expm(Aon*mu)*...
Aon*P*uss+[0 1]*expm(Aon*mu)*P*yy);

J=[J11 J12;J21 J22];
dt=inv(J)*[f1;f2];
alpha=alpha-dt(1);
mu=mu-dt(2);
counter=counter+1;
error=max(abs(dt));
end

counter;
```

□

10.4.4 General Comments

From the above analysis, one can immediately note the strong similarity between the analysis of DCM and CCM.

In total, there are four different modes [113, 114] of converter operation: two discontinuous cases (modes 1 and 3) and two continuous cases (modes 2 and 4).

Mode 1 (also called 2/0 mode) happens when the intervals of conduction via two diodes alternate with intervals of zero conduction. Mode 2 (also called 3/2 mode) happens when conduction is via alternate 3- and 2-diode paths. Mode 3 (also called 2/3/2/0 mode) occurs when an interval of conduction via two diodes is followed by a 3-diode interval; this is followed by another 2-diode interval and then by a zero-conduction interval. In mode 4 (or 3/3 mode), conduction occurs via a sequence of 3-diode paths.

Only modes 1 and 2 are discussed because:

1. The operating range of mode 3 is very small [114]. Most of the existing literature [109, 115] only analyzes mode 1 for the DCM.
2. Operation at mode 4 is rare because it is very close to the short-circuit point [113, 114]. Most of the existing literature [109, 116], [117] only refers to mode 2 as the CCM.
3. The LB 2 model can be extended to mode 3 with slightly added complexity.
4. Constant voltage load can be assumed in mode 4 [118, 119] to have fairly accurate results, and this greatly simplifies harmonic analysis.

11

Loop-Breaking 3 Method

"Coming together is a beginning. Keeping together is progress. Working together is success."
Henry Ford

This chapter is devoted to the elaboration of the loop-breaking 3 (LB 3) method where the loop breaking is occurred within and between components. As mentioned in Chapter 8, the LB 3 method is suitable for systems involving a multi-converter system. Hence, we will show how the LB 3 method can be applied to solve for a power system containing multiple VSCs. The power system itself will be modeled in OpenDSS, which is a comprehensive electrical simulator, developed by Electric Power Research Institute (EPRI) for studies of distribution systems. OpenDSS allows one to interface with a variety of existing software such as MATLAB and Python. Thus, OpenDSS serves as an ideal platform for the LB 3 method. Section 11.1 will briefly describe OpenDSS. Section 11.2 shows how the MATLAB program can be interfaced with OpenDSS. Section 11.3 illustrates various examples on how developed VSC models can be interfaced with OpenDSS.

11.1 OpenDSS

OpenDSS is referred to as the open source Distribution System Simulator. The principal author of the software is Roger Dugan, who has been engaged in the development since 1997. The initial purpose for the development of the software was to serve the following objectives [120]:

- support distribution analysis applications
- provide a very flexible research platform
- fill gaps left by other distribution system analysis tools
- study new approaches to distribution analysis

OpenDSS has evolved over the years and is able to handle both deterministic and probabilistic calculations. Built-in solution capabilities include snapshot, daily, and yearly power flow, harmonics, fault current study, dynamics, Monte Carlo fault studies, etc. [121]. It is worth mentioning that OpenDSS evolved from programs designed for harmonic flow analysis of large, arbitrarily-meshed, multi-phase networks. Therefore, the circuit modeling and solution approach has more in common with harmonic analysis and electromagnetic transient tools than with typical distribution power flow programs, which usually employ the forward-backward sweeping method to solve radial circuits [120]. Hence, harmonic analysis is a native capability, not an add-on. Moreover, it is much

Harmonic Modeling of Voltage Source Converters using Basic Numerical Methods, First Edition.
Ryan Kuo-Lung Lian, Ramadhani Kurniawan Subroto, Victor Andrean, and Bing Hao Lin.

easier to make a harmonic analysis program to solve a power flow, while it is quite difficult to make a power flow program to perform harmonic analysis [120].

A salient feature of OpenDSS is that it allows the user to interface it with other external programs. A Component Object Model (COM) interface is implemented on the in-process server dynamic link library (DLL) version of the program to allow knowledgeable users to use the features of the program to perform new types of studies. Through the COM interface, the user is able to design and execute custom solution modes and features from an external program and perform the functions of the simulator, including definitions of the model data. Users commonly drive the OpenDSS with some of the familiar software programs such as MATLAB, Python, C, and other languages. This provides powerful external analytical capabilities as well as excellent graphics for displaying results.

11.2 Interfacing OpenDSS with MATLAB

To illustrate how the OpenDSS can be interfaced with MATLAB, we will present a simple example as illustrated in Fig. 11.1(a), which shows a simple RLC circuit. The circuit is divided into two subsystems, where subsystem A is modeled in OpenDSS, while subsystem B is modeled in MATLAB. Iteration techniques need to be employed to bridge both subsystems. Gauss–Seidel iteration is considered to be a very convenient technique in this regard as there is no need to derive the Jacobian matrix from the subsystems. Consequently, the two broken subsystems can be modeled as Fig. 11.1(b) where subsystem A is connected to a current source, while subsystem B is connected to a voltage source.

The OpenDSS code for the subsystem A is shown below:

```
Clear
New Circuit.SimpleTest
~ basekv=0.01 pu=1.0000 phases=1 bus1=Vs Angle=0
New Line.Simplebus1   Phases=1 Bus1=Vs.1   Bus2=Vx.1   R1=0.2 X1=0.377
~R0=0.2 X0=0.377 C1=0 C0=0 Length=1 Rg=0 Xg=0
```

The voltage source is modeled by the voltage source object, which is modeled as a voltage source behind a Thevenin equivalent impedance. The format of the voltage source in OpenDSS is described as:

```
New Circuit.<name> basekv=<value1> pu=<value2> angle=<value3>
Bus1=<name2> ~R1=<value4> X1=<value5> R0=<value6> X0=<value7>
```

New Circuit.<name> is used to define a voltage source with the <name>. The basekv represents the rated voltage base for the circuit with <value1>. The operating voltage and angle are specified by the pu and angle properties with <value2> and <value3>, respectively. The Bus1 specifies the positive node of the voltage source. The ground node does not need to be specified. The positive sequence reactance and resistance of the source are specified by X1 and R1 in ohms. The X0 and R0 properties represent the zero sequence reactance and resistance of the source respectively in ohms. Note that ~ means continuation of the previous line.

The resistor and inductor in subsystem A of Fig. 11.1(a) can be modeled using the equivalent line model offered in OpenDSS. The format of line definition in OpenDSS is described as:

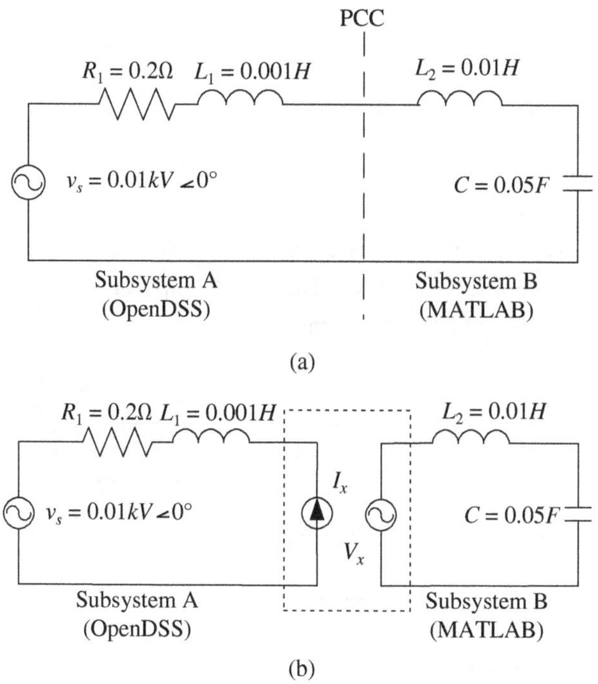

Figure 11.1 (a) Division between OpenDSS and MATLAB. (b) The interfacing circuits for linking OpenDSS and MATLAB.

```
New Line.
<line name> bus1=<bus1name>  bus2=<bus2name> length=<value>
~ units=<mi,kft,km,m,Ft,in,cm>geometry=<table 11.1>
(or linecode=<table 11.2>)
```

`New Line.<line name>` is used to define a line with the `<line name>`. `bus1` and `bus2` specify the two nodes that the line is connected to. The `length` of the line and its corresponding `units` are specified. The details of the line `geometry` and `linecode` are presented in Tables 11.1 and 11.2, respectively. Line geometry is the class of objects used to specify the position of the conductors, the type of conductors, and the Kron reduction settings [122]. The different line geometry properties are listed in Table 11.1. The line code objects are general library objects that contain impedance characteristics for the lines and cables. The impedance of a line is described by its series impedance and nodal capacitive admittance matrices. These matrices may be specified directly or they can be generated by specifying the symmetrical component data. The different line code properties are listed in Table 11.2.

To interface OpenDSS with MATLAB, we can run OpenDSS through MATLAB using the following command:

```
[DSSStartOK, DSSObj, DSSText]=DSSStartup;
```

Table 11.1 Line geometry properties.

Properties	Descriptions
Nconds	Number of conductors in the geometry
Nphases	Number of phases.
Cond	This number identifies each conductor in the geometry
Wire	Refers to the type of wire used and links to the Wire Data Class
X,H	X coordinate and the height of the conductor in mi—kft—km—m—Ft—in—cm
Reduce	Yes/No. Reduce out the neutrals using Kron Reduction formula.
Normamps	Normal ampacity, amperes
Emergamps	Emergency ampacity, amperes

Table 11.2 Line code properties.

Properties	Description
Nphases	Number of phases.
R1	Positive-sequence resistance in ohms per unit length
R0	Zero-sequence resistance in ohms per unit length
X1	Positive-sequence reactance in ohms per unit length
X0	Zero-sequence reactance in ohms per unit length
C1	Positive-sequence capacitance in nano-farads per unit length
C0	Zero-sequence capacitance in nano-farads per unit length
Normamps	Normal ampacity, amperes
Emergamps	Emergency ampacity, amperes
Base Freq	Base frequency at which the impedance values are specified

The `DSSStartup` function is as follows:

```
[DSSStartOK, DSSObj, DSSText]=DSSStartup;
% Function for starting up the DSS
% instantiate the DSS object
Obj= actxserver('OpenDSSEngine.DSS');

%Start the DSS
Start=Obj.Start(0);
% Define the text interface
Text=Obj.Text;
```

The circuit defined in OpenDSS (i.e. subsystem A) is then loaded.

```
DssText.Command = 'Compile "C:\Users\SimpleTest.dss"';
```

The interface variables are then prepared:

```
% assign a variable to the text interface for easier access
DssText = Dssobj.Text;
% get an interface to the active circuit called "DssCircuit"
DssCircuit = Dssobj.ActiveCircuit;
```

Note that the circuit and text interfaces are the primary tools for interacting with the COM server. The text interface can be used to pass command strings to OpenDSS. The text interface allows string commands to be passed to OpenDSS and run directly in OpenDSS.

To commence the iteration between subsystems A and B, we first solve for subsystem A for initialization:

```
DssText.Command = ['set DefaultBasefrequency=60'];
DssText.Command = ['set mode=snapshot'];
DssText.Command = ['solve'];
DssText.Command = ['export voltage'];
DssText.Command = ['export current'];
```

Finally, the overall loop can be formulated. Note that the text interface also allows the user to add and edit circuit elements of OpenDSS in MATLAB. Using the OpenDSS commands `new` and `edit`, different elements can be added, moved, and changed via MATLAB, as shown below:

```
Sys_converge = 0;
Pcc_Bus = 'Vx'; %
ite=0; % initialize the iteration number

% Iteration Loop
while Sys_converge < 1        % Convergence condition
    ite = ite + 1     % Iteration counting
    % Access System Bus Voltage
    BusV = DssCircuit.AllBusVolts;
    BusName = DssCircuit.AllBusNames;
    for n = 1:length(BusName)
        if strcmp(BusName(n), Pcc_Bus) == 1
            Vpcc_mag(ite)  = abs( BusV((n-1)*2+1) + j*BusV((n-1)*2+2));
            Vpcc_pha(ite)  = angle( BusV((n-1)*2+1) + j*BusV((n-1)*2+2));
        end
    end

    % Calculate Current across through L, C Component
    Ix = Vpcc_mag(ite)*sqrt(2)*exp(j*Vpcc_pha(ite))/(1/(j*w*Cap)+(j*w*Lmh));

    % Current Data Processing for Isource Component
    Ix_mag = abs(Ix)/sqrt(2);
    Ix_pha = angle(Ix)*180/pi+180;

    if ite == 1
        % Import Current of Matlab result to Isource
        DssText.Command = ...
            ['New Isource.Ix phases=1 bus1=',Pcc_Bus,'.1 amps=',...
            num2str(Ix_mag),' angle=',num2str(Ix_pha),' frequency=60'];
    else
        % Update Current value
```

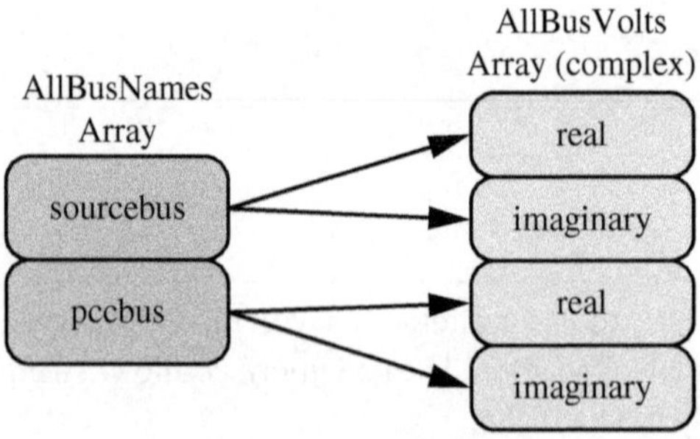

Figure 11.2 The structure of the array for AllBusVolts and AllBusNames.

Table 11.3 PCC voltage of simple case interface.

Platform	Voltage amplitude (V)	Angle (degree)
LB3	14.8671	−1.5690
PSCAD/EMTDC	15.0239	−1.5699

```
        DssText.Command = ...
            ['Edit Isource.Ix phases=1 bus1=',Pcc_Bus,'.1 amps=',...
            num2str(Ix_mag),' angle=',num2str(Ix_pha),' frequency=60'];
    end

    % Solve System Power Flow in Snapshot mode
    DssText.Command = ['set DefaultBasefrequency=60'];
    DssText.Command = ['set mode=snapshot'];
    DssText.Command = ['solve'];
    DssText.Command = ['export voltage'];
    DssText.Command = ['export current'];

    if ite ~=1
    %calculate error
        error = abs((Vpcc_mag(ite)-Vpcc_mag(ite-1))/Vpcc_mag(ite));
        if error < tol
            Sys_converge = Sys_converge + 1;
        else
            Sys_converge = 0;
        end
    end
end % end while
```

Note that the calculated results from OpenDSS can be retrieved directly through the COM interface as well as from an output file such as Microsoft Excel, written in comma-separated value (CSV) format. One of the properties of `DssCircuit` is `AllBusNames`, which returns as an array of strings giving the names of all buses in the system. The other is `AllBusVolts`, which returns the voltage (complex number) of each node in the circuit as a variant array of doubles. The order of the array of `AllBusVolts` is the same as `AllBusNames` property, as shown in Fig. 11.2. The structure of the array is bus-by-bus and then by node at each bus. Thus, all nodes from each bus are grouped together.

The final converged value of the PCC voltage (v_x) and that obtained from PSCAD/EMTDC are listed in Table 11.3. Their values are in agreement, justifying the LB 3 method.

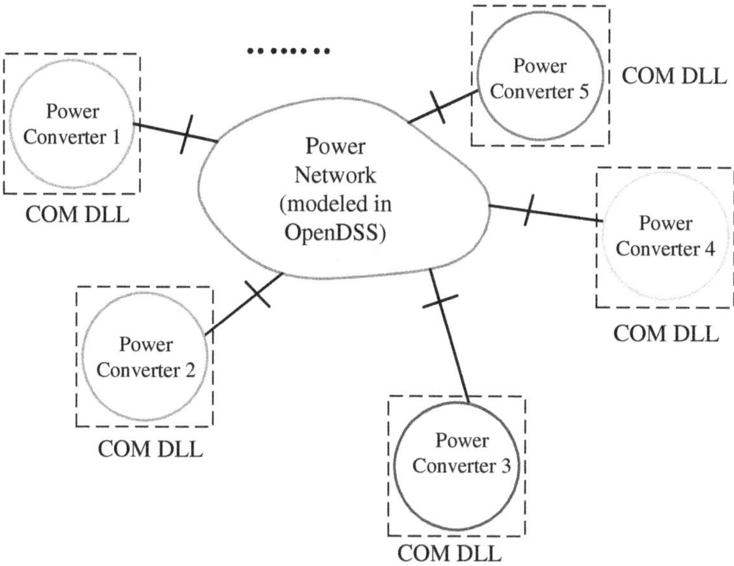

Figure 11.3 Interfacing OpenDss with harmonic models of VSCs.

11.3 Interfacing OpenDSS with Harmonic Models of VSCs

Since OpenDSS is inherently a harmonic analysis program, it is an excellent choice for interfacing with harmonic models of VSCs, derived in previous chapters. As shown in Fig. 11.3, the power network is modeled in OpenDSS, while the power converters are written as a COM DLL. LB 3 method is employed, where the VSCs are broken at the PCCs of the power network, as shown in Fig. 11.3. Again, the iteration technique employed between the OpenDSS and the individual converters is Gauss–Seidel. The advantage of using Gauss–Seidel in this situation is that there is no need to obtain the harmonic impedance of the power network and VSC to form the overall Jacobian matrix. (This will be further elaborated in Chapter 12.) Fig. 11.4 shows how the broken system of Fig. 11.3 looks when Gauss–Seidel is employed. In subsystem A, all the converters are modeled as current harmonic sources. On the other hand, in other subsystems (i.e. B, C, D, etc.), the power network is modeled as voltage harmonic sources. Through Gauss–Seidel iterations, the harmonic current sources in subsystem A and the voltage harmonic sources in subsystems B, C, D, etc. will converge to the current harmonics injected by the VSCs and the voltage harmonics produced by the power network, respectively. Finally, all the VSC models in subsystems B, C, D, etc. can be modeled by the LB 1 and 2 methods, as described in Chapters 9 and 10. The following four examples demonstrate four possible interfacings and they are (1) a power system interfaced with open-loop VSCs; (2) a power system interfaced with closed-loop VSCs; (3) a power system interfaced with open-loop VSC including power-flow constraints; and (4) a power system interfaced with closed-loop VSC including power-flow constraints. The presented cases are very simple but complete, and serve as templates for readers to turn in more complicated and practical cases.

Example 11.1 A modified version of the standard IEEE 13-bus distribution system is used for simulation study, as depicted in Fig. 11.5. The modifications to the distribution system were the shorting out of the regulator and the addition of three VSCs at the secondary sides of buses 650, 670, and 680. These VSCs are open-loop, three-level MML-VSC at bus 650; open-loop, seven-level

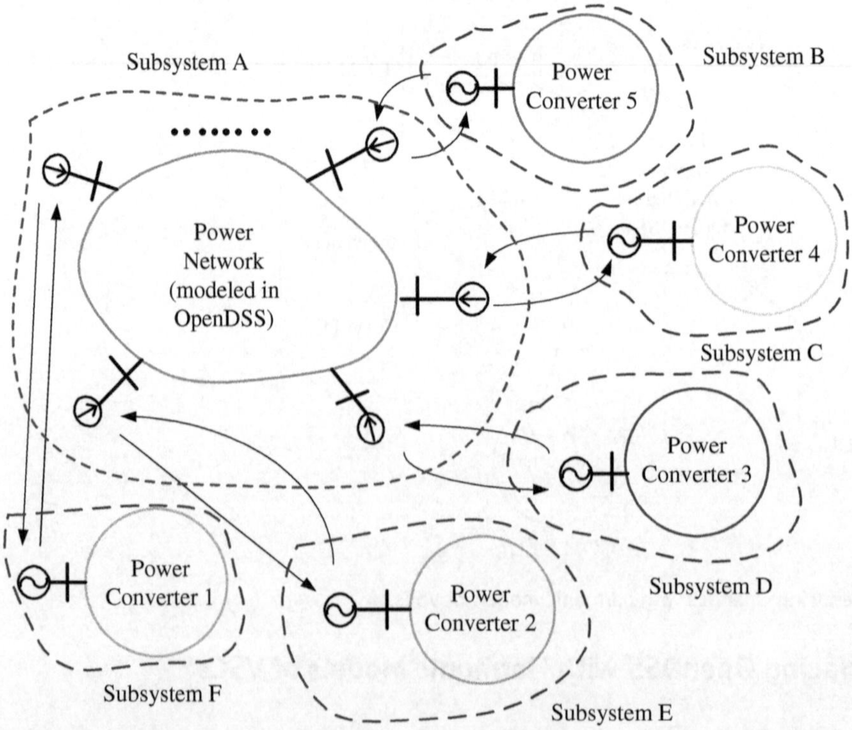

Figure 11.4 Iteration process between OpenDSS and COM DLLs.

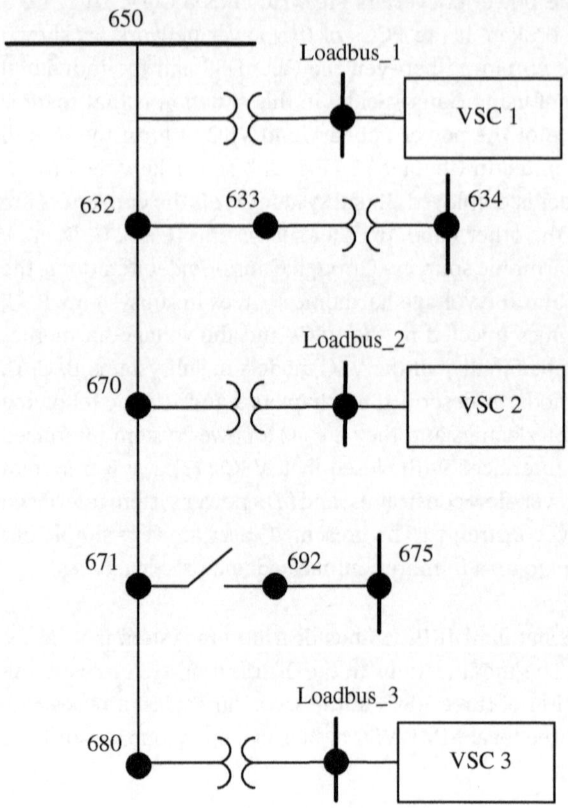

Figure 11.5 Iteration process between OpenDSS and COM DLLs.

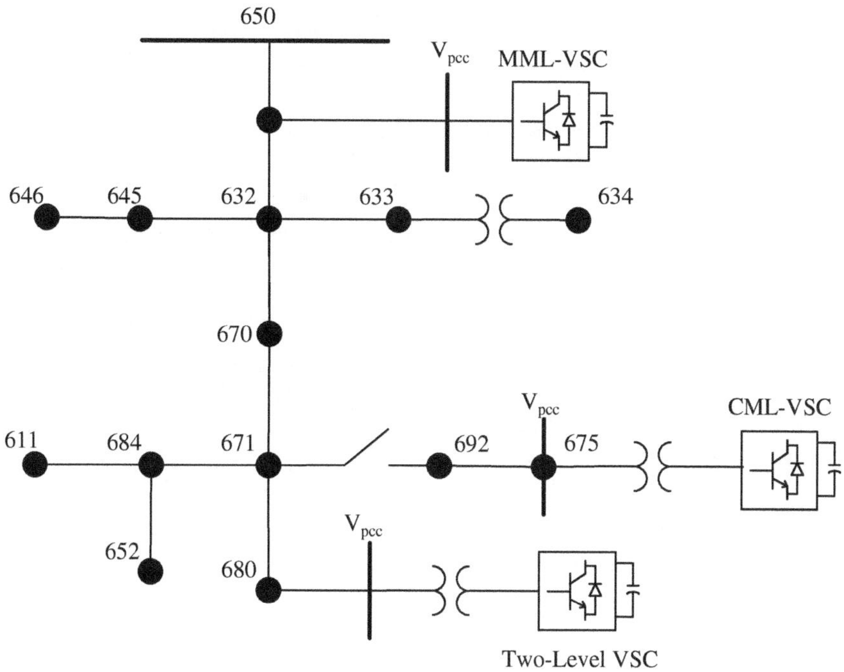

Figure 11.6 The system topology in Example 11.1.

Table 11.4 Modulation indices of VSCs.

	VSC model	m_f	m_a	δ
VSC 1	MML-VSC	27	0.95	−3.05°
VSC 2	CMC-VSC	3	0.8	2.19°
VSC 3	Two-level VSC	15	0.8816	−3.21°

CML-VSC at bus 675; and open-loop, two-level VSC at bus 680, as illustrated in Fig. 11.6. Table 11.4 shows the modulation indices (m_a) and firing angles (δ) of these VSC models.

Using the LB 3 technique presented, the system can be iteratively solved. The converter output current amplitudes of the three VSCs are shown in Fig. 11.7 and the voltages of the PCC are shown in Fig. 11.8. Their values are in great agreement with PSCAD/EMTDC. □

The codes for generating Figs. 11.7 and 11.8 are listed as follows:

First, set up the system environment and assign the parameters to the system.

```
DSSname ='IEEE13Nodeckt_mdf'; % Name of grid file in OpenDSS
ite = 0;  % Iteration times
freq = 60; % System frequency
Hn = 19; % Number of harmonics
% System harmonic order
spectrum = [1,5,7,11,13,17,19,23,25,29,31,35,37,41,43,47,49,53,55];
tol = 0.001; % Iteration tolerance
vsc_num = 3; % Number of VSC
% The PCC bus of OpenDSS Grid which interfaced with VSC model
bus_interfaced = {'PCCbus1', 'PCCbus2', 'PCCbus3'};
```

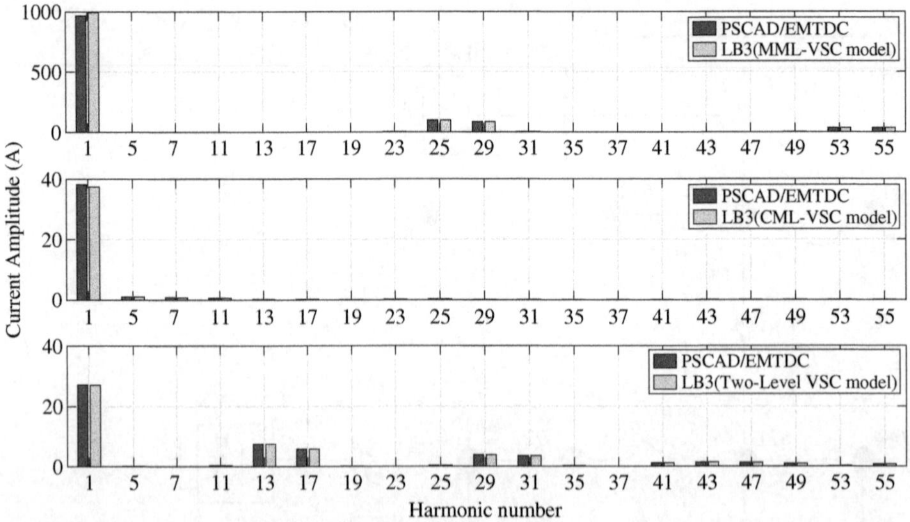

Figure 11.7 Current amplitudes of the converters in Example 11.1.

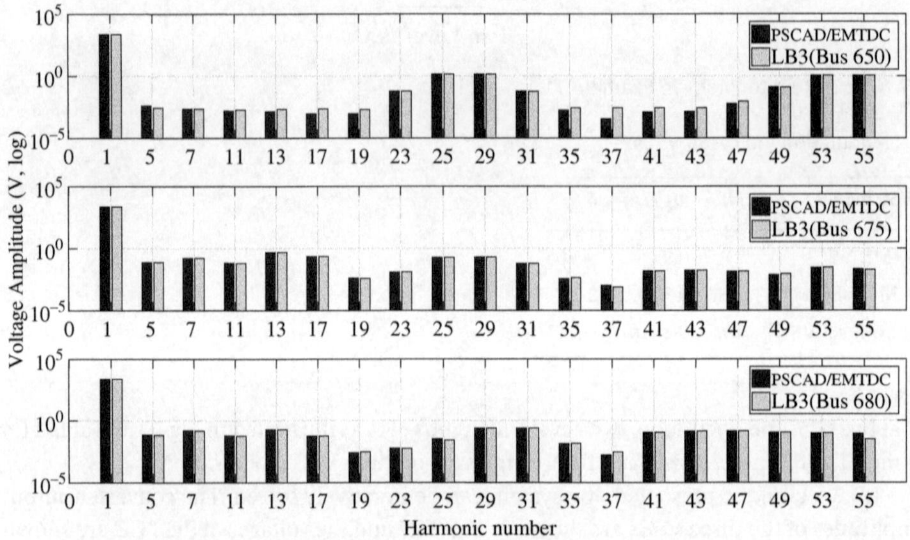

Figure 11.8 Bus voltages of the PCC in Example 11.1.

Then, activate the OpenDSS engine and initialize parameters in OpenDSS to obtain the initial PCC voltages. Set the class interfaces in OpenDSS as variables in MATLAB to facilitate the use of their properties. Table 11.5 shows the class interfaces in OpenDSS and the corresponding variables in MATLAB.

```
Dssobj = actxserver('OpenDSSEngine.DSS'); % Active engine of OpenDSS
if ~Dssobj.Start(0)
    disp('Unable to start the DSSEngine')
    return
end
```

Table 11.5 OpenDSS class interface and MATLAB variable correspondence.

OpenDSS	MATLAB
Dssobj.Text	DssText
Dssobj.ActiveCircuit	DssCircuit
DssCircuit.Monitors	Dssmonitor

```
% Set a variable for easy access to the text interface
DssText = Dssobj.Text;
% Set a variable for easy access to the circuit interface.
DssCircuit = Dssobj.ActiveCircuit;
% Set a variable for easy access to the monitors interface
Dssmonitor = DssCircuit.Monitors;
% Point to grid file path
DssText.Command = 'Compile "C:\Users\User\Desktop\IEEE13Nodeckt_mdf.dss"';
% Solve the grid in OpenDSS to generate the fundamental PCC voltage
DssText.Command = ['solve'];
% Export all node voltages on each bus as a CSV file
DssText.Command = ['export voltage'];
% Access variant array of bus names string in OpenDSS
BusName = DssCircuit.AllBusNames;
```

Unlike the simple case presented in Section 11.2, the VSC models access the PCC voltage values by reading the exports file for initialization. The default filename structure of the voltage output is 'DSSname' + 'EXP VOLTAGES.csv' and the output tables stretch out horizontally (as shown in the dashed box of Fig. 11.9) instead of vertically. The data type in the csv file are in polar forms. Fig. 11.9 shows the data correspondence between the 'AllBusVolts' array and voltages export table.

```
for c = 1:vsc_num % Access PCC voltage of each bus which interface with VSC model
 for n = 1:length(BusName) % To find the PCC bus array location
  if strcmp(BusName(n), bus_interfaced(c)) == 1
    va_mag(1,ite,c)=csvread([DSSname,'_EXP_VOLTAGES.csv'],n,3,[n,3,n,3]);
    va_ang(1,ite,c)=csvread([DSSname,'_EXP_VOLTAGES.csv'],n,4,[n,4,n,4])*pi/180;
    vb_mag(1,ite,c)=csvread([DSSname,'_EXP_VOLTAGES.csv'],n,7,[n,7,n,7]);
    vb_ang(1,ite,c)=csvread([DSSname,'_EXP_VOLTAGES.csv'],n,8,[n,8,n,8])*pi/180;
    vc_mag(1,ite,c)=csvread([DSSname,'_EXP_VOLTAGES.csv'],n,11,[n,11,n,11]);
    vc_ang(1,ite,c)=csvread([DSSname,'_EXP_VOLTAGES.csv'],n,12,[n,12,n,12])
       *pi/180;
     end
    end
 end

     for i = 2: Hn
         va_mag(i,ite,c) = 0;
         va_ang(i,ite,c) = 0;
         vb_mag(i,ite,c) = 0;
         vb_ang(i,ite,c) = 0;
         vc_mag(i,ite,c) = 0;
         vc_ang(i,ite,c) = 0;
     end
```

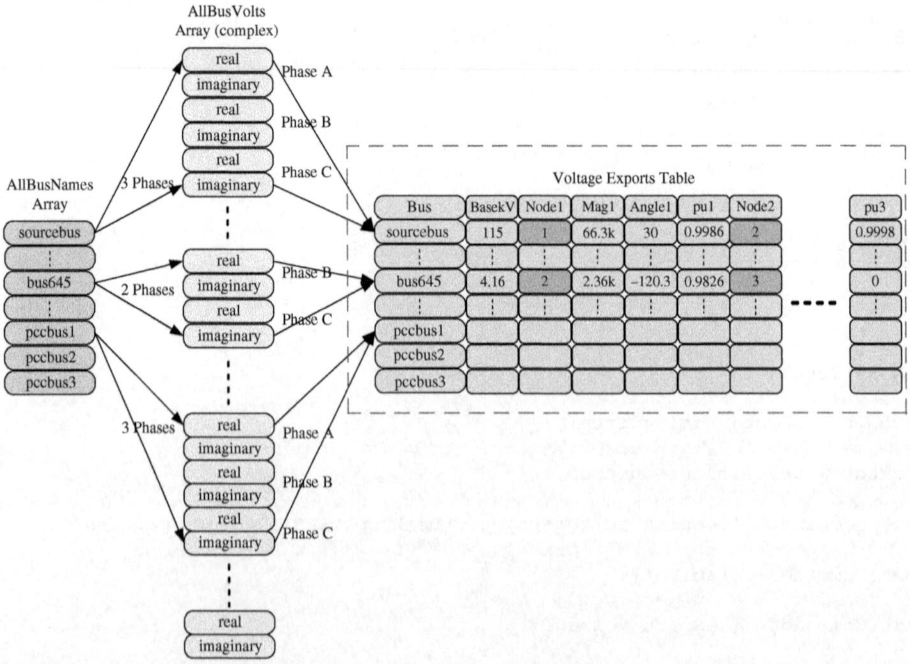

Figure 11.9 The array structure of the voltages export table.

Input the PCC voltage to VSC models and solve them for obtaining the output current harmonics. Then convert the output current harmonics type from complex to polar forms.

```
if c == 1 % Solve the model
    [ia, ib, ic] = MMC_model(va_mag(i,ite,c), va_ang(i,ite,c), ...
    vb_mag(i,ite,c), vb_ang(i,ite,c), vc_ang(i,ite,c), freq, spectrum);
elseif c == 2
    [ia, ib, ic] = CMC_model(va_mag(i,ite,c), va_ang(i,ite,c), ...
    vb_mag(i,ite,c), vb_ang(i,ite,c), vc_ang(i,ite,c), freq, spectrum);
else
    [ia, ib, ic] = VSC2Level_model(va_mag(i,ite,c), va_ang(i,ite,c), ...
    vb_mag(i,ite,c), vb_ang(i,ite,c), vc_ang(i,ite,c), freq, spectrum);
end

% Converter the current data form into OpenDSS input form
ia_mag(:,ite,c) = abs(ia)/sqrt(2);
ib_mag(:,ite,c) = abs(ib)/sqrt(2);
ic_mag(:,ite,c) = abs(ic)/sqrt(2);
ia_ang(:,ite,c) = angle(ia)*180/pi+180;
ib_ang(:,ite,c) = angle(ib)*180/pi+180;
ic_ang(:,ite,c) = angle(ic)*180/pi+180;
```

In order to represent the VSC models as harmonic current sources in OpenDSS, the current harmonics need to be expressed as the per unit value of the fundamental current harmonic and are represented as an inherited property, `spectrum`.

```
% Expressed the current harmonics in each phase as per unit value
ia_pu(:,ite,c) = ia_mag(:,ite,c)/ia_mag(1,ite,c)*100;
ib_pu(:,ite,c) = ib_mag(:,ite,c)/ib_mag(1,ite,c)*100;
ic_pu(:,ite,c) = ic_mag(:,ite,c)/ic_mag(1,ite,c)*100;
```

```

    % Set the spectrum property
    % Create new A phase of spectrum for VSC model(c)
    DssText.Command = ['New spectrum.iavsc_',num2str(c)];
    % Assign the spectrum property
    DssCircuit.SetActiveElement(['spectrum.iavsc_',num2str(c)]);
    % Number of harmonics
    DssCircuit.ActiveDSSElement.Properties('NumHarm').val = num2str(Hn);
    % Order of harmonics
    DssCircuit.ActiveDSSElement.Properties('harmonic').val = mat2str(spectrum);
    % Magnitude of harmonics
    DssCircuit.ActiveDSSElement.Properties('%mag').val = mat2str (ia_pu(:,ite,c)');
    % Angle of harmonics
    DssCircuit.ActiveDSSElement.Properties('angle').val = mat2str (ia_ang(:,ite,c)');

    % Create new B phase of spectrum for VSC model(c)
    DssText.Command = ['New spectrum.ibvsc_',num2str(c)];
    DssCircuit.SetActiveElement(['spectrum.ibvsc_',num2str(c)]);
    DssCircuit.ActiveDSSElement.Properties('NumHarm').val = num2str(Hn);
    DssCircuit.ActiveDSSElement.Properties('harmonic').val = mat2str(spectrum);
    DssCircuit.ActiveDSSElement.Properties('%mag').val = mat2str (ib_pu(:,ite,c)');
    DssCircuit.ActiveDSSElement.Properties('angle').val = mat2str (ib_ang(:,ite,c)');

    % Create new C phase of spectrum for VSC model(c)
    DssText.Command = ['New spectrum.icvsc_',num2str(c)];
    DssCircuit.SetActiveElement(['spectrum.icvsc_',num2str(c)]);
    DssCircuit.ActiveDSSElement.Properties('NumHarm').val = num2str(Hn);
    DssCircuit.ActiveDSSElement.Properties('harmonic').val = mat2str(spectrum);
    DssCircuit.ActiveDSSElement.Properties('%mag').val = mat2str (ic_pu(:,ite,c)');
    DssCircuit.ActiveDSSElement.Properties('angle').val = mat2str (ic_ang(:,ite,c)');
```

In the first iteration, new Isource objects are added to represent the VSC models in OpenDSS and spectrum are assigned to these sources. In order to access the updated voltage of the Isource, the monitor objects are added, which can record the complex values of voltage and current of the circuit elements. Hence, in the subsequent iterations, only the properties of Isource need to be updated.

```
if ite == 1
        % Add Isources which represent VSC models in OpenDSS
 DssText.Command = ['New Isource.ias_',num2str(c),' phases=1 bus1=', ...
 bus_interfaced{c},'.1 amps=',num2str(ia_mag(1,ite,c)),' angle=', ...
 num2str(ia_ang(1,ite,c)),' frequency=60 spectrum=iavsc_',num2str(c)];
 DssText.Command = ['New Isource.ibs_',num2str(c), ...
 ' phases=1 bus1=',bus_interfaced{c},'.2 amps=', ...
 num2str(ib_mag(1,ite,c)),' angle=',num2str(ib_ang(1,ite,c)),...
 ' frequency=60 spectrum=ibvsc_',num2str(c)];
 DssText.Command = ['New Isource.ics_',num2str(c),' phases=1 bus1=', ...
 bus_interfaced{c},'.3 amps=',num2str(ic_mag(1,ite,c)),' angle=', ...
 num2str(ic_ang(1,ite,c)),' frequency=60 spectrum=icvsc_',num2str(c)];

        % Add monitor for each Isource
 DssText.Command = ['New monitor.mon_ia_',num2str(c),...
 ' Element=Isource.ias_',num2str(c),' Mode=0'];
 DssText.Command = ['New monitor.mon_ib_',num2str(c),...
 ' Element=Isource.ibs_',num2str(c),' Mode=0'];
 DssText.Command = ['New monitor.mon_ic_',num2str(c),...
 ' Element=Isource.ics_',num2str(c),' Mode=0'];
else
        % Update the current harmonics of Isources
 DssText.Command = ['Edit Isource.ias_',num2str(c),' phases=1 bus1=', ...
```

```
bus_interfaced{c},'.1 amps=',num2str(ia_mag(1,ite,c)), ...
' angle=',num2str(ia_ang(1,ite,c)),' frequency=60 spectrum=iavsc_',num2str(c)];
DssText.Command = ['Edit Isource.ibs_',num2str(c),' phases=1 bus1=', ...
bus_interfaced{c},'.2 amps=',num2str(ib_mag(1,ite,c)),' angle=', ...
num2str(ib_ang(1,ite,c)),' frequency=60 spectrum=ibvsc_',num2str(c)];
DssText.Command = ['Edit Isource.ics_',num2str(c),' phases=1 bus1=', ...
bus_interfaced{c},'.3 amps=',num2str(ic_mag(1,ite,c)),' angle=',...
num2str(ic_ang(1,ite,c)),' frequency=60 spectrum=icvsc_',num2str(c)];
   end
```

A snapshot power flow is first performed for initialization. Then, the harmonic flow is performed. The SaveAll is executed so that all of the data are saved in the temporary monitor buffer.

```
    DssText.Command = ['set mode=snapshot'];
    DssText.Command = ['solve'];

    DssText.Command = ['set harmonic=',mat2str(spectrum)];
    DssText.Command = ['set mode=harmonics'];
    DssText.Command = ['solve'];
    Dssmonitor.SaveAll;
```

After the system harmonic flow is solved in the first iteration, the updated PCC voltage can be directly accessed from the Monitor object in the subsequent iterations and be fed into the VSC model to generate updated current harmonics.

```
    Dssmonitor.Name = ['mon_ia_',num2str(c)];
    va_mag(:,ite,c) = Dssmonitor.Channel(1)';
    va_ang(:,ite,c) = Dssmonitor.Channel(2)'*pi/180;
    Dssmonitor.Name = ['mon_ib_',num2str(c)];
    vb_mag(:,ite,c) = Dssmonitor.Channel(1)';
    vb_ang(:,ite,c) = Dssmonitor.Channel(2)'*pi/180;
    Dssmonitor.Name = ['mon_ic_',num2str(c)];
    vc_mag(:,ite,c) = Dssmonitor.Channel(1)';
    vc_ang(:,ite,c) = Dssmonitor.Channel(2)'*pi/180;
```

This process repeats until the difference of current harmonics between consecutive iterations is less than a tolerance value.

```
  if ite ~=1
        for s = 1:conv_num
            a_err(s,1) = abs(sum(ia_mag(:,ite,s)-ia_mag(:,ite-1,s)));
            b_err(s,1) = abs(sum(ib_mag(:,ite,s)-ib_mag(:,ite-1,s)));
            c_err(s,1) = abs(sum(ic_mag(:,ite,s)-ic_mag(:,ite-1,s)));
        end
        [error] = max(a_err(:,1),b_err(:,1));
        [error] = max(error,c_err(:,1))
        if error < tol
            Done = Done + 1;
        else
            Done = 0;
        end
    end
end
DssText.Command = ['BusCoords IEEE13Node_BusXY.csv'];
```

```
DssText.Command = ['Plot circuit Power max=1000 dots=y labels=n C1=$FF0000'];

for i = 1:conv_num
    ia_vsc(:,i) = ia_mag(:,ite,i)*sqrt(2).*exp(j*(ia_ang(:,ite,i)-180)*pi/180);
    ib_vsc(:,i) = ib_mag(:,ite,i)*sqrt(2).*exp(j*(ib_ang(:,ite,i)-180)*pi/180);
    ic_vsc(:,i) = ic_mag(:,ite,i)*sqrt(2).*exp(j*(ic_ang(:,ite,i)-180)*pi/180);
    [Ialpha_vsc(:,i), Ibeta_vsc(:,i)] = abc2alphabeta(ia_vsc(:,i), ib_vsc(:,i), ...
    ic_vsc(:,i));
end

for i = 1:conv_num
    Va_vsc(:,i) = Vbusa_mag(:,i).*exp(j*(Vbusa_ang(:,i)));
    Vb_vsc(:,i) = Vbusb_mag(:,i).*exp(j*(Vbusb_ang(:,i)));
    Vc_vsc(:,i) = Vbusc_mag(:,i).*exp(j*(Vbusc_ang(:,i)));
    [Valpha_vsc(:,i), Vbeta_vsc(:,i)] = abc2alphabeta(Va_vsc(:,i),
    Vb_vsc(:,i), ...
    Vc_vsc(:,i));
end
```

Example 11.2 The system to be studied is the same as **Example 11.1**. However, the VSCs are equipped with the controllers (closed-loop), as shown in Fig. 11.10. A closed-loop, three-level MML-VSC is connected to the system at bus 650, while a closed-loop, two-level VSC is connected to the system at bus 680. The output current amplitudes of the two VSCs are shown in Fig. 11.11 and the voltages of the PCC are shown in Fig. 11.12.

Note that in this case, the closed-loop models described in Chapters 9 and 10 replace the open-loop models while the rest of the codes remain the same as in **Example 11.1**. □

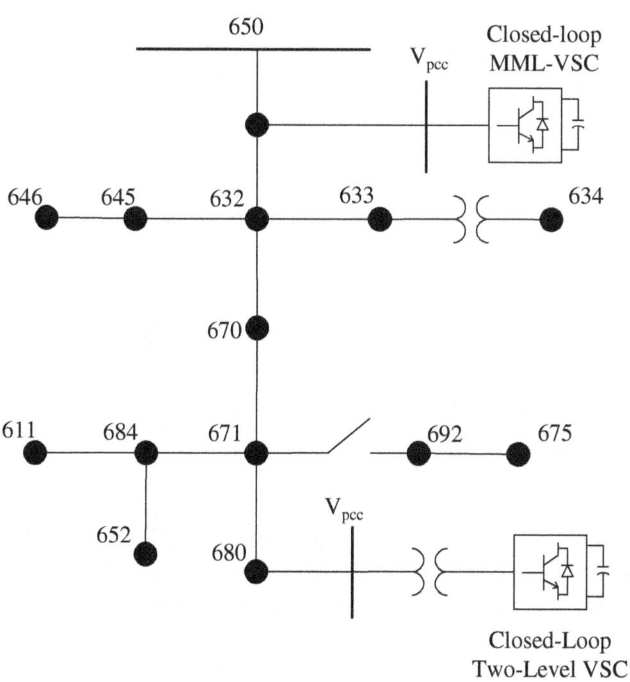

Figure 11.10 The system topology in Example 11.2.

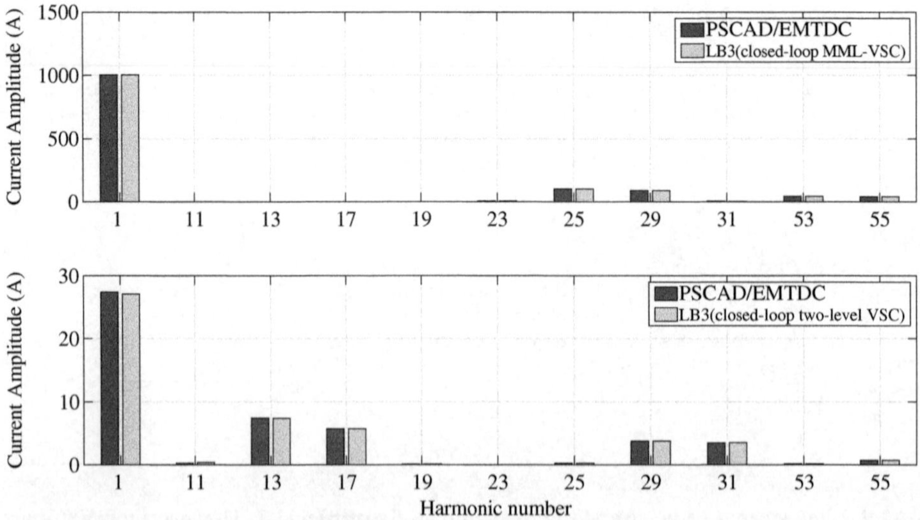

Figure 11.11 Current amplitude of converters in Example 11.2.

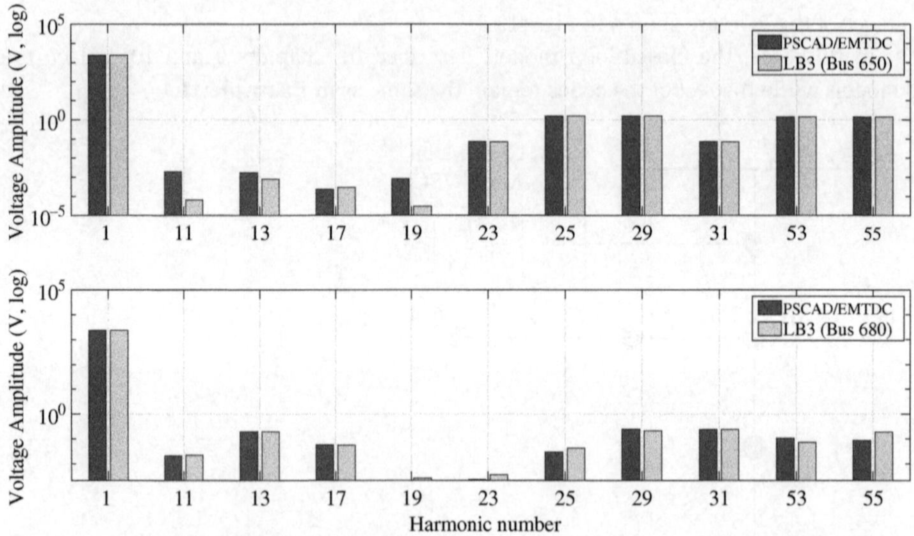

Figure 11.12 Bus voltages of PCC in Example 11.2.

Example 11.3 In this case, as shown in Fig. 11.13, an open-loop, two-level VSC is interfaced with the IEEE-13 bus system with the same modifications as in **Example 11.1**. To simplify the problem, power flow constraints are only imposed to the VSC, which are stated in (11.1) and (11.2).

$$V_{dc} - V_{dcref} = 0 \tag{11.1}$$

$$Q_{ac} - Q_{acref} = 0, \tag{11.2}$$

where V_{dc} and V_{dcref} represent the dc voltage and the reference of the dc voltage, while Q_{ac} and Q_{acref} are the injected reactive power and reference reactive power, respectively.

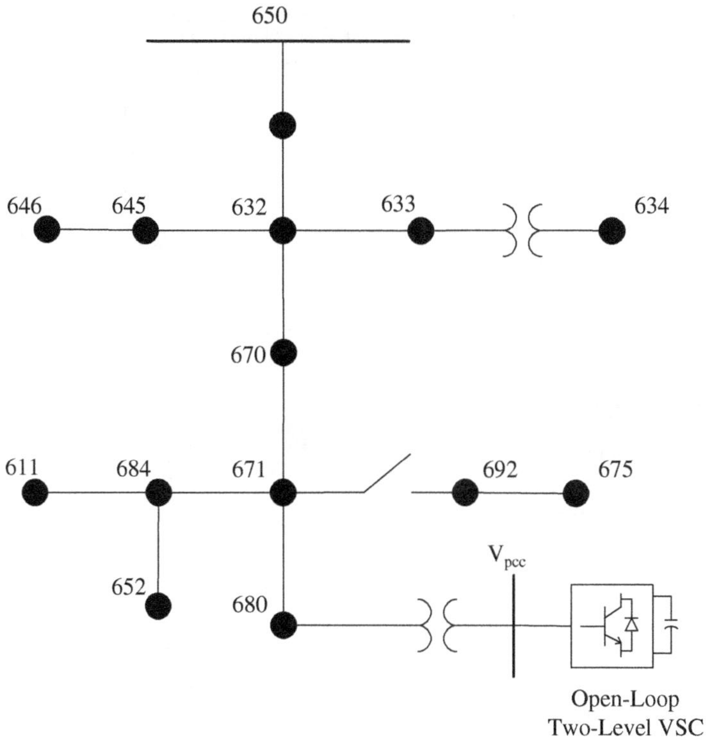

Figure 11.13 The system topology in Example 11.3.

V_{dc} can be directly obtained by the FCM of the two-level VSC model, and the reactive power can be calculated by (11.3).

$$Q_{ac} = imag\left(v_{ta1}i_{ta1}{}^{*} + v_{tb1}i_{tb1}{}^{*} + v_{tc1}i_{tc1}{}^{*}\right), \tag{11.3}$$

where v_{ta1}, v_{tb1}, and v_{tc1} are the fundamental components of the converter voltages of phases a, b, and c, respectively, while i_{ta1}, i_{tb1}, and i_{tc1} are the fundamental components of the converter currents of phases a, b, and c, respectively.

A Newton-type iterative method can be formulated to search for the correct modulation index (m_a) and firing angle (δ) to drive the mismatch equations imposed by the power flow constraints to close to zero.

$$\begin{bmatrix} m_a{}^{(k+1)} \\ \delta^{(k+1)} \end{bmatrix} = \begin{bmatrix} m_a^{(k)} \\ \delta^{(k)} \end{bmatrix} - \mathbf{J}^{-1} \begin{bmatrix} V_{dc} - V_{dcref} \\ Q_{ac} - Q_{acref} \end{bmatrix}, \tag{11.4}$$

where

$$\mathbf{J} = \begin{bmatrix} \dfrac{\partial V_{dc}}{\partial m_a} & \dfrac{\partial V_{dc}}{\partial \delta} \\ \dfrac{\partial Q_{ac}}{\partial m_a} & \dfrac{\partial Q_{ac}}{\partial \delta} \end{bmatrix} \tag{11.5}$$

Figs. 11.14 and 11.15 show the converged converter current and PCC voltage, respectively. The corresponding code for the Newton-type iteration is stated as below.

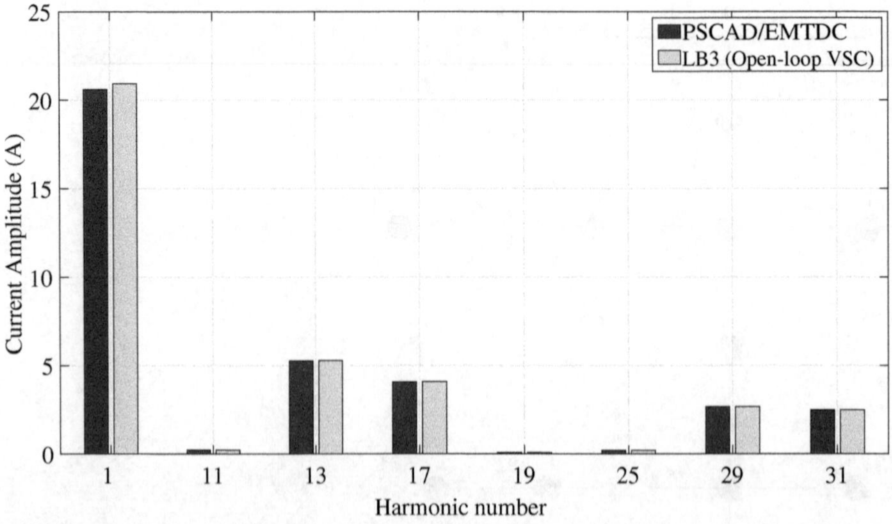

Figure 11.14 Current amplitude of converters in Example 11.3.

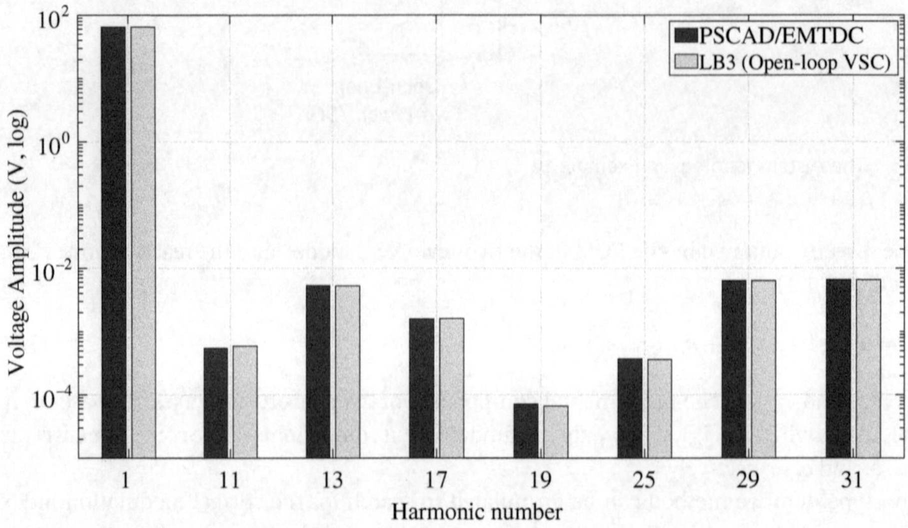

Figure 11.15 Bus voltages of PCC in Example 11.3.

```
% Set grid restriction reference
Vdcsch = 230;
Qsch = -3.6e3;

% Initial index value
Ma_initial = 0.8;
Delta_initial = -0.05;

% Index disturb for Jacobian
Ma_xxx = 1e-3;
Delta_xxx = 1e-3;

% Calculate the output of VSC model for mismatch equation
[ia_mis, ib_mis, ic_mis, Vdc_mis = open_loop(Vsa, Vsb, Vsc,...
Ma_initial, Delta_initial)];
```

```

% Per unit of restriction difference
Vdcpu = abs(Vdc_mis(1))/Vdcsch;

Qpu = (imag(Vsa(1).*conj(ia_mis(1)))+...
    imag(Vsb(1).*conj(ib_mis(1)))+...
    imag(Vsc(1).*conj(ic_mis(1)))) / Qsch;

% Mismatch equation
M=[(Vdcpu - 1); (Qpu - 1)];

% Calculate the output of VSC model for Jacobian matrix
Ma_M = Ma_initial + Ma_xxx;
Delta_M = Delta_initial + Delta_xxx;
[ia_Mx, ib_Mx, ic_Mx, Vdc_Mx] = open_loop_Vdc_abc(Vsa, Vsb, Vsc,...
    Ma_M, Delta_initial);
[ia_Dx, ib_Dx, ic_Dx, Vdc_Dx] = open_loop_Vdc_abc(Vsa, Vsb, Vsc,...
    Ma_initial, Delta_M);

% Per unit of restriction difference
Vdcpux = [abs(Vdc_Mx(1)),abs(Vdc_Dx(1))] / Vdcsch;

Qpux = [(imag(Vsa(1).*conj(ia_Mx(1)))+imag(Vsb(1).*conj(ib_Mx(1)))+...
     imag(Vsc(1).*conj(ic_Mx(1)))),...
     (imag(Vsa(1).*conj(ia_Dx(1)))+imag(Vsb(1).*conj(ib_Dx(1)))+...
     imag(Vsc(1).*conj(ic_Dx(1))))] / Qsch;

% ratio of change of the grid restriction when modulation index changed
pIpM_Vdc = (Vdcpux(:,1) - Vdcpu) / Ma_xxx;
pIpD_Vdc = (Vdcpux(:,2) - Vdcpu) / D_xxx;
pIpM_Q = (Qpux(:,1) - Qpu) / Ma_xxx;
pIpD_Q = (Qpux(:,2) - Qpu) / D_xxx;

% Jacobian matrix
J = [pIpM_Vdc, pIpD_Vdc;
    pIpM_Q, pIpD_Q];

% Newton equation for index updated
MJ = inv(J)*M;
Ma_next = Ma_initial - MJ(1);
Delta_next = Delta_initial - MJ(2);

% Matrix for Broyden method
JJ = inv(J);
M_old = M;
dt_ans = -MJ;

% Iteration loop
while Err_max > 1e-4 % Iteration tolerance contrast
    Ma_old = Ma_next;
    Delta_old = Delta_next;

    % Calculate the output of VSC model to updated mismatch equation
    [ia_mis, ib_mis, ic_mis, Vdc_mis] = open_loop_Vdc_abc(Vsa, Vsb, Vsc,...
        Ma_old, Delta_old);

    % Per unit of restriction difference
    Vdcpu = abs(Vdc_mis(1))/Vdcsch;

    Qpu = (imag(Vsa(1).*conj(ia_mis(1)))+imag(Vsb(1).*conj(ib_mis(1)))+...
        imag(Vsc(1).*conj(ic_mis(1))))/Qsch;
```

Table 11.6 Case 4 grid restriction.

Grid restriction	Reference	Result
V_{dc}	230 V	230 V
Q_{ac}	3.6 kW	3.6021 kW

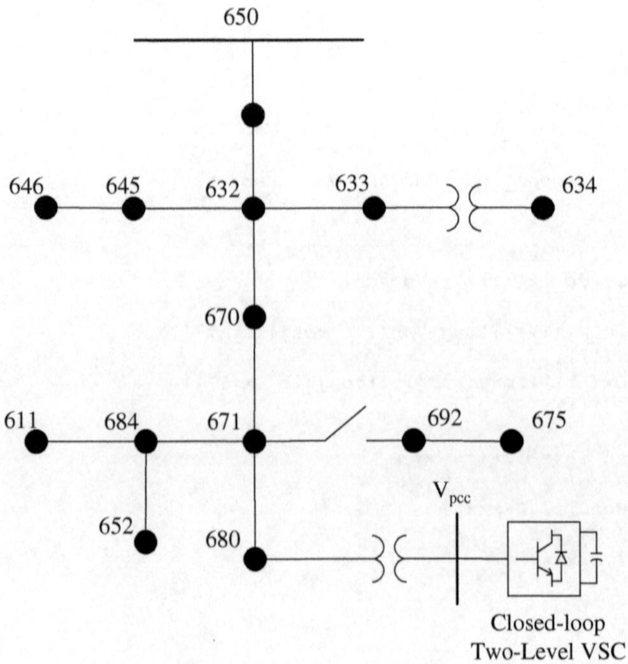

Figure 11.16 The system topology in Example 11.4.

```

    % Mismatch equation
    M=[(Vdcpu - 1); (Qpu - 1)];

    % Broyden method
    dt_M = M - M_old;
    JJ = JJ + ((dt_ans-JJ*dt_M)*dt_ans'*JJ) / (dt_ans'*JJ*dt_M);

    % index updated
    MJ = JJ*M;
    Ma_next = Ma_old - MJ(1);
    Delta_next = Delta_old - MJ(2);

    % Matrix for Broyden method
    M_old = M;
    dt_ans = - MJ;

    % Difference of index
    Err_Ma = abs(Ma_next - Ma_old);
    Err_Delta = abs(Delta_next - Delta_old);
    Err_max = max(Err_Ma,Err_Delta);
end
```

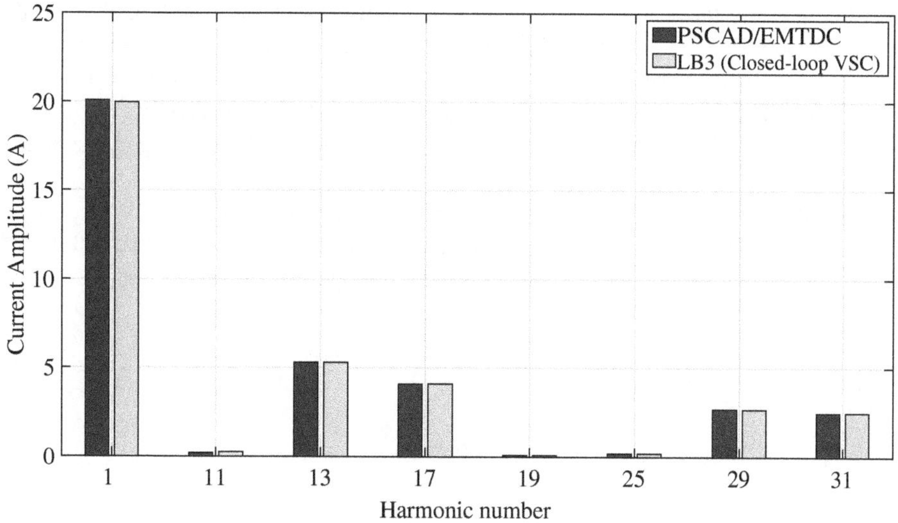

Figure 11.17 Current amplitude of converters in Example 11.4.

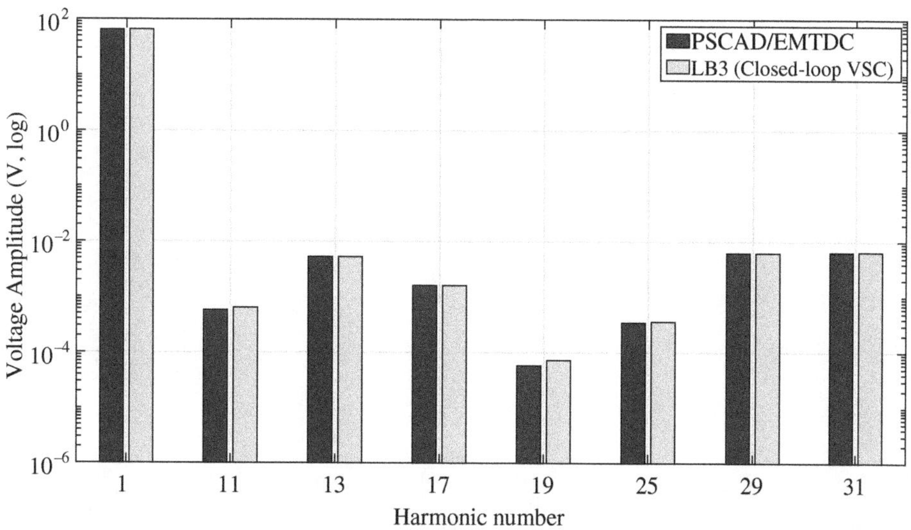

Figure 11.18 Bus voltages of PCC in Example 11.4.

Note that for ease of computation, the Jacobian matrix is obtained by the finite difference method. □

Example 11.4 The system under study is the same as **Example 11.3**, except that the VSC is in closed-loop and has controllers, as shown in Fig. 11.16. To satisfy the power constraints, the VSC needs to have the dc voltage and reactive power controls. The dc voltage controllers have been well explained in Chapters 9 and 10. It is the reactive power controller that needs to be added in the VSC. The reactive current reference can be obtained by (11.6), which is then used for the reactive current controller.

$$I_{tqref} = \frac{2}{3}\left(\frac{-Q_{ref}}{V_{sd}}\right) \tag{11.6}$$

Note that this case is similar to the cases presented in **Examples 11.1** and **11.2**, and one only needs to specify the references required for the controller in order to satisfy the power flow constraints. Thus, there is no need to formulate a Newton-type iterative loop for driving the constraints to zero or below a tolerance value, as done in **Example 11.3**. Figs. 11.17 and 11.18 show the converged converter current and PCC voltage, respectively. Table 11.6 lists the converged results for m_a and δ.

The output current amplitude of two-level VSC model with controller is shown in Fig. 11.17 and the voltages of PCC are shown in Fig. 11.18, which have similar harmonic spectra to those in **Example 11.3**. □

12

Small-Signal Harmonic Model of a VSC

"Be approximately right rather than exactly wrong."

– John W. Tukey

This chapter presents a time-domain SSM of a power electronic converter. It provides an analytical small signal VSC model for harmonic analysis. Also, such an SSM is required for constructing the Jacobian matrix for the LB 3 method if the interaction between power network and power converter is chosen to be Newton's method. The outline of this chapter is as follows: Section 12.1 explains the difference between the FCM presented in Chapter 5 and the harmonic SSM to be presented in this chapter. Then, Section 12.2 illustrates how a Gauss–Seidel-based LB 3 method is different from a Newton-based LB 3 method from an equivalent circuit point of view. The rest of the chapter closely follows that of Chapter 10. First, a model for a dc-dc converter is developed (in Section 12.3) to reveal some of the interesting patterns and characteristics of the proposed SSM. Then, in Section 12.4, a model for the VSC is developed, and its results are verified against the large-signal model.

12.1 Problem Statement

In Chapter 5, an FCM of a VSC is derived based on the assumption that the switching times are known in advance. However, in reality, the information about the switching times is not easily accessible. In addition, due to control action, any perturbation in the input sources will cause changes in the switching times, leading to a different FCM. The goal of this chapter is to find a small-signal FCM ($\mathbf{\Delta F}$), which relates the perturbed input harmonics ($\mathbf{\Delta U}$) to the perturbed output harmonics ($\mathbf{\Delta W}$ in Fig. 12.1).

The main difference between $\mathbf{\Delta F}$ and $\mathbf{F}$ is that $\mathbf{\Delta F}$ resembles more closely the concept of the admittance matrix for an LTI system. For an LTI system, an impedance or admittance matrix will not change as system currents or voltages change. It is merely a "proportionality" factor, relating terminal voltages and currents. When the perturbation from the nominal operating point is small, $\mathbf{\Delta F}$ acts as a proportionality factor, relating the perturbed input to the perturbed output. Specifically, for the case of VSC, the following form holds:

$$\begin{bmatrix} \mathbf{\Delta I_{ac}} \\ \mathbf{\Delta V_{dc}} \end{bmatrix} = \mathbf{\Delta F} \begin{bmatrix} \mathbf{\Delta V_{ac}} \\ \mathbf{\Delta I_{dc}} \end{bmatrix}. \tag{12.1}$$

One major difference, though, between $\mathbf{\Delta F}$ and an LTI system's impedance/admittance matrix (or generally speaking, transfer functions) is that $\mathbf{\Delta F}$ still preserves the coupling between different

Harmonic Modeling of Voltage Source Converters using Basic Numerical Methods, First Edition.
Ryan Kuo-Lung Lian, Ramadhani Kurniawan Subroto, Victor Andrean, and Bing Hao Lin.

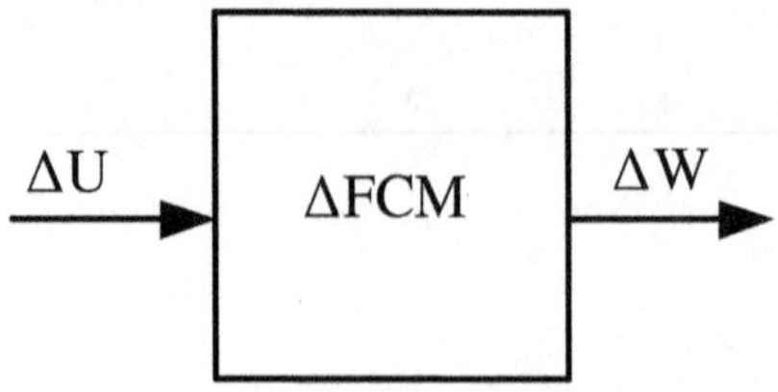

Figure 12.1 Purpose of SSM: find the resulting perturbed output harmonics due to the perturbed input harmonics.

orders of harmonics [123], and has a structure similar to those of Figs. 6.10 and 6.13. That is, $\mathbf{\Delta F}$, which is obtained from linearizing a nonlinear time-varying system, will in general be LTV, not LTI. LTI transfer functions have no coupling between different orders of harmonics and consequently only diagonal terms exist in the harmonic matrices.

The main objective of this chapter is to obtain a small-signal FCM such that (1) it preserves the structure of an LTV system, (2) it acts as a proportional factor between the perturbed input and output when the perturbation is small, and (3) it includes the effects of control action.

12.2 Gauss–Seidel LB 3 and Newton LB 3

This section illustrates that the Gauss–Seidel-based LB 3 method presented in Chapter 11 essentially regards a power converter as a voltage-dependent harmonic current source (VDHCS). On the other hand, if Newton-type iteration is chosen, then the power converter is regarded as a Norton equivalent harmonic current source (NEHCS). This can be illustrated by Fig. 12.2, which depicts a single line diagram of a power system interacting with two VSCs. The power system is assumed to contain only LTI components, and is represented by a highly simplified equivalent circuit whose source voltage is v_s and whose impedance is $Z_s = R_s + jh\omega L_s$. The two VSCs (VSC1 and VSC2) are connected in parallel at the PCC (point **A**) of the power system. The ac impedances of the two VSCs are $R_f + jh\omega L_f$ ($R_{f1} + jh\omega L_{f1}$ and $R_{f2} + jh\omega L_{f2}$, which represents the resistances and inductance of the coupling reactors. On the dc side of the VSCs, Z_{dc1} and Z_{dc2} represent the equivalent impedances of the dc loads and the switching losses of the converters.

12.2.1 Current Injection Method

The current injection method is the simplest way to integrate the VSCs with external systems. The VSC is treated as a VDHCS. Fig. 12.3 describes the equivalent circuit. $\mathbf{I_{c1}}$ and $\mathbf{I_{c2}}$ are VDHCS — their values change as $\mathbf{V_t}$ changes.

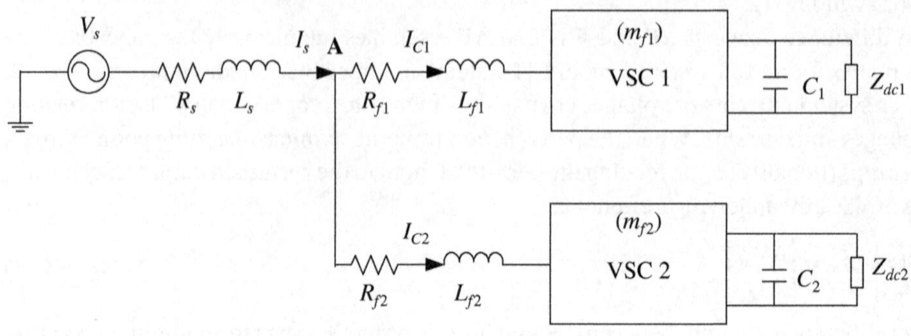

Figure 12.2 Two VSCs connected to a simple power network.

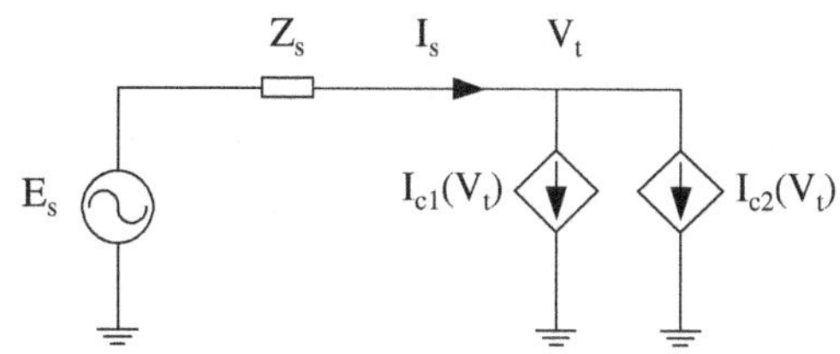

Figure 12.3 Current injection method.

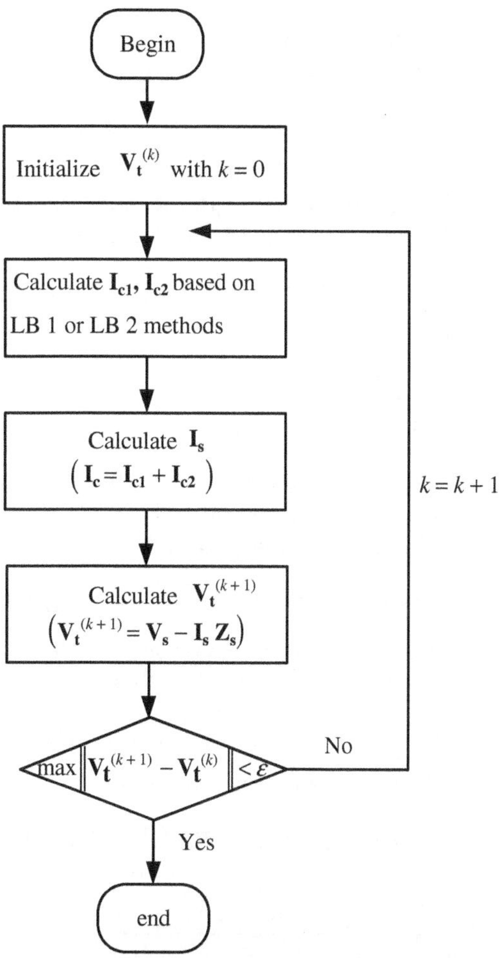

Figure 12.4 Current injection method – iteration based on the Gauss–Seidel method.

Note that Chapters 9 and 10 already described how to solve for the current harmonics for a given set of the input voltage harmonics. Basically, the switching times are first solved iteratively (either by the LB 1 or LB 2 methods), then their values are substituted into the FCM to get the output current harmonics. As delineated in Fig. 12.4, the voltage-dependent harmonic current source naturally leads to the Gauss–Seidel iteration, which echoes the contents discussed in Chapter 11.

12.2.2 Norton Circuit Method

The only disadvantage of using the current injection method is that it requires a significant number of iterations. For systems with high harmonic distortion, it is also possible that Gauss–Seidel methods may fail to converge [12]. The more elegant method is to use a Norton equivalent circuit method. This section will discuss how a VSC can be modeled as an NEHCS and how the NEHCS can be incorporated into an iterative harmonic power flow program.

Equation (12.2) represents the linearized model for the VSC (refer to Fig. 12.2):

$$\Delta \mathbf{I_c} = \Delta \mathbf{F} \cdot \Delta \mathbf{V_t}, \tag{12.2}$$

where

$$\Delta \mathbf{I_c} = \mathbf{I_c} - \mathbf{I_{c0}}, \tag{12.3}$$

$$\Delta \mathbf{V_t} = \mathbf{V_t} - \mathbf{V_{t0}}, \tag{12.4}$$

$$\Delta \mathbf{F} = \left. \frac{\partial \mathbf{F}}{\partial \mathbf{V_t}} \right|_{\mathbf{V_t}=\mathbf{V_{t0}}}, \tag{12.5}$$

and $\mathbf{I_{c0}}$, $\mathbf{V_{t0}}$ represent the current and voltage harmonics at the nominal operating point.

Substituting (12.3)–(12.5) into (12.2) produces the following expression:

$$\mathbf{I_c} = \mathbf{Y_N} \cdot \mathbf{V_t} + \mathbf{I_N}, \tag{12.6}$$

where

$$\mathbf{Y_N} = \Delta \mathbf{F} \tag{12.7}$$

$$\mathbf{I_N} = \mathbf{I_{c0}} - \mathbf{Y_N} \cdot \mathbf{V_{t0}}. \tag{12.8}$$

Equation (12.6) exactly represents the Norton equivalent circuit of the VSC. The incorporation of these equivalent circuits leads to Fig. 12.5. Alternatively, one can also convert the equivalent circuit of the linear power system to its Norton counterpart, which leads to Fig. 12.6 and (12.9).

$$\mathbf{V_t} = \left[\mathbf{Y_s} + \mathbf{Y_{N1}} + \mathbf{Y_{N2}}\right]^{-1} \left(\mathbf{Y_s E_s} - \mathbf{I_{N1}} - \mathbf{I_{N2}}\right) \tag{12.9}$$

However, (12.9) does not offer any insights on how an iterative algorithm can be formulated. To formulate an iterative algorithm based on (12.9), let us substitute (12.8) into (12.9):

$$\mathbf{V_t} = \left[\mathbf{Y_s} + \mathbf{Y_{N1}} + \mathbf{Y_{N2}}\right]^{-1} \left[\mathbf{Y_s E_s} - \left(\mathbf{I_{c01}} - \mathbf{Y_{N1} V_{t0}}\right) - \left(\mathbf{I_{c02}} - \mathbf{Y_{N2} V_{t0}}\right)\right]. \tag{12.10}$$

Rearranging (12.10) leads to (12.11):

$$\left[\mathbf{Y_s} + \mathbf{Y_{N1}} + \mathbf{Y_{N2}}\right] \mathbf{V_t} = \left(\mathbf{Y_s E_s} - \mathbf{I_{c01}} - \mathbf{I_{c02}}\right) + \left(\mathbf{Y_{N1}} + \mathbf{Y_{N2}}\right) \mathbf{V_{t0}}. \tag{12.11}$$

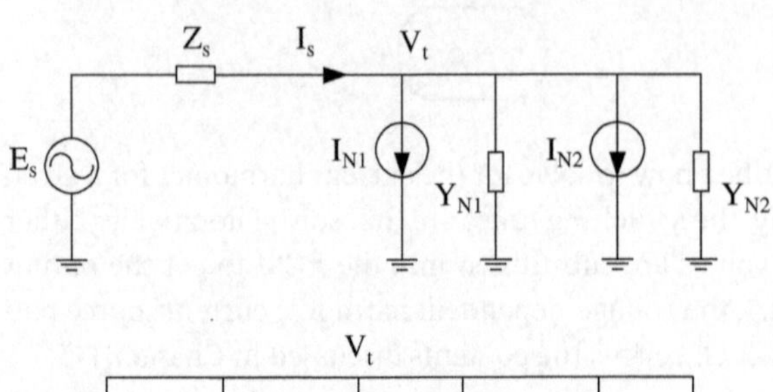

Figure 12.5 The VSCs are replaced with Norton equivalent circuit.

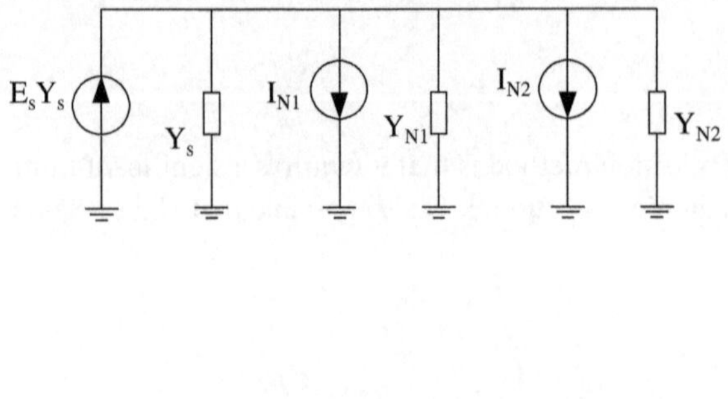

Figure 12.6 The equivalent circuit of the linear power system is replaced with the Norton circuit.

Adding $\mathbf{Y_s V_{t0}}$ to and subtracting $\mathbf{Y_s V_{t0}}$ from the right-hand side of (12.11) results in (12.12):

$$\left[\mathbf{Y_s} + \mathbf{Y_{N1}} + \mathbf{Y_{N2}}\right]\mathbf{V_t} = \left(\mathbf{Y_s E_s} - \mathbf{I_{c01}} - \mathbf{I_{c02}}\right) + \left(\mathbf{Y_s} + \mathbf{Y_{N1}} + \mathbf{Y_{N2}}\right)\mathbf{V_{t0}} - \mathbf{Y_s V_{t0}}. \tag{12.12}$$

Finally, one gets the following equation:

$$\mathbf{V_t} = \mathbf{V_{t0}} + \left[\mathbf{Y_s} + \mathbf{Y_{N1}} + \mathbf{Y_{N2}}\right]^{-1}\left[\mathbf{Y_s}\left(\mathbf{E_s} - \mathbf{V_{t0}}\right) - \mathbf{I_{c01}} - \mathbf{I_{c02}}\right]. \tag{12.13}$$

Comparing (12.13) with (9.12), one can see that $-\left[\mathbf{Y_s} + \mathbf{Y_{N1}} + \mathbf{Y_{N2}}\right]$ is the Jacobian matrix, $\left[\mathbf{Y_s}\left(\mathbf{E_s} - \mathbf{V_{t0}}\right) - \mathbf{I_{c01}} - \mathbf{I_{c02}}\right]$ is the mismatch equation, and $\mathbf{V_t}$ is the unknown variable to be solved. Hence, the Norton circuit model naturally leads to Newton iteration, and Fig. 12.7 shows the resulting iteration flow chart. Clearly, the Jacobian matrix requires the evaluation of the SSM of the converter, which is described in the following subsections.

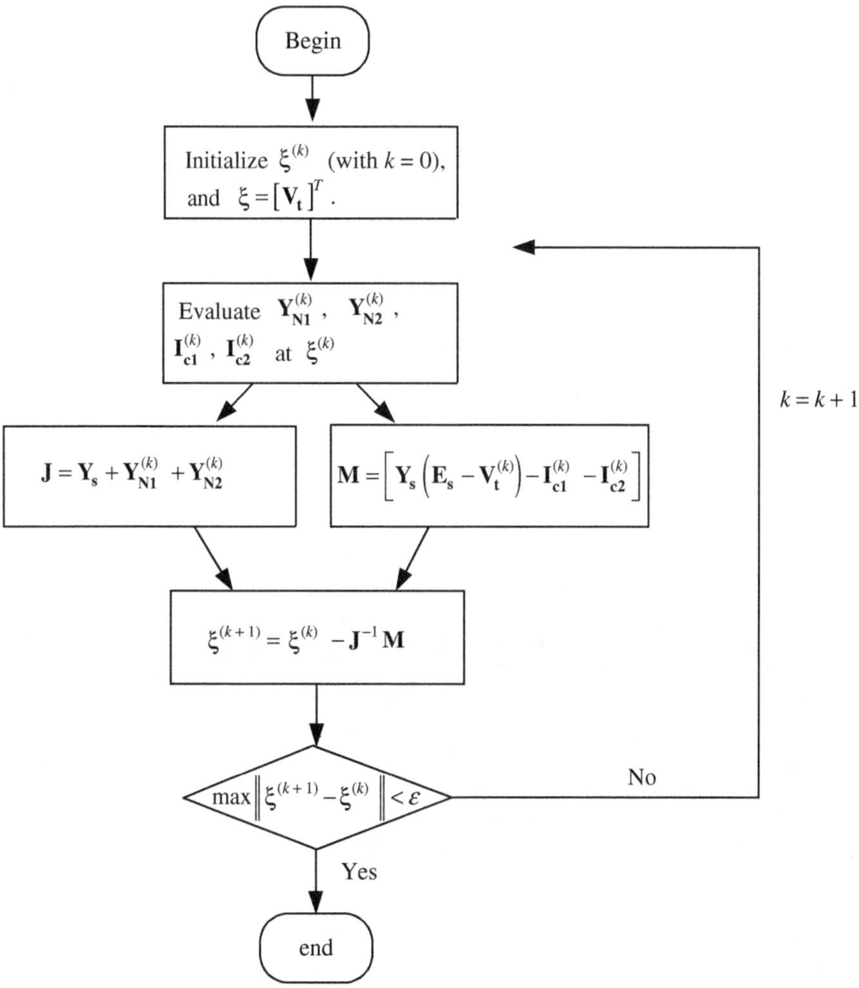

Figure 12.7 Norton equivalent circuit – iteration based on the Newton–Raphson method.

12.3 Small-Signal Analysis of DC-DC Converter

From the analysis of Chapters 5 and 10, the harmonics of interest and the sampled data value of the state variables at time $t = T$ for a buck converter can all be evaluated via (12.14). Note that here $x_h(T)$ represents the inductor current harmonic of interest.

$$\begin{bmatrix} \mathbf{x_p}(T) \\ z(T) \\ x_i(T) \\ v_{ref}(0) \\ x_h(T) \end{bmatrix} = \breve{\mathbf{\Phi}} \begin{bmatrix} \mathbf{x_p}(0) \\ z(0) \\ x_i(0) \\ v_{ref}(0) \\ 0 \end{bmatrix}, \tag{12.14}$$

where $\mathbf{x_p} = \begin{bmatrix} i_L & v_{dc} \end{bmatrix}^T$, $z(0) = \begin{bmatrix} v_s \end{bmatrix}$, $\breve{\mathbf{\Phi}} = e^{\breve{\mathbf{A}}_{\mathbf{off}}(T-t_{sw})} e^{\breve{\mathbf{A}}_{\mathbf{on}} t_{sw}}$,

$$\breve{\mathbf{A}}_{\mathbf{on}} = \begin{bmatrix} 0 & -\frac{1}{L} & \frac{1}{L} & 0 & 0 & 0 \\ \frac{1}{C} & -\frac{1}{RC} & 0 & 0 & 0 & 0 \\ 0 & 0 & 0 & 0 & 0 & 0 \\ 0 & -H & 0 & 0 & 1 & 0 \\ 0 & 0 & 0 & 0 & 0 & 0 \\ \frac{e^{-jh\omega T}}{T} & 0 & 0 & 0 & 0 & j\omega h \end{bmatrix}, \text{ and } \breve{\mathbf{A}}_{\mathbf{off}} = \begin{bmatrix} 0 & -\frac{1}{L} & 0 & 0 & 0 & 0 \\ \frac{1}{C} & -\frac{1}{RC} & 0 & 0 & 0 & 0 \\ 0 & 0 & 0 & 0 & 0 & 0 \\ 0 & -H & 0 & 0 & 1 & 0 \\ 0 & 0 & 0 & 0 & 0 & 0 \\ \frac{e^{-jh\omega T}}{T} & 0 & 0 & 0 & 0 & j\omega h \end{bmatrix}.$$

Without knowing the matrix structure of $\breve{\mathbf{\Phi}}$, one can write $x_h(T)$ as a linear combination of $\mathbf{x_p}(0)$, $z(0)$, $x_i(0)$, and v_{ref} as stated in (12.15):

$$x_h(T) = \mathbf{a}(t_{sw})\mathbf{x_p}(0) + b(t_{sw})z(0) + c(t_{sw})x_i(0) + d(t_{sw})v_{ref}, \tag{12.15}$$

where a, b, c, and d are coefficients corresponding to a function of switching time, t_{sw}.

Substituting (10.8) into (12.15) results in (12.16):

$$x_h(T) = m(t_{sw})z(0) + c(t_{sw})x_i(0) + d(t_{sw})v_{ref}, \tag{12.16}$$

where $m(t_{sw}) = \mathbf{a}(t_{sw})\left\{\mathbb{I} - \mathbf{\Phi}(t_{sw})\right\}^{-1}\mathbf{\Gamma}(t_{sw}) + b(t_{sw})$.

Equation (12.16) holds true for both open-loop and closed-loop cases.

For the open-loop case (see Fig. 12.8), $\Delta x_h(T)$ is obtained by

$$\Delta x_h(T) = \frac{\partial x_h(T)}{\partial t_{sw}}\Delta t_{sw} + \frac{\partial x_h(T)}{\partial z(0)}\Delta z(0) + \frac{\partial x_h(T)}{\partial x_i(0)}\Delta x_i(0) + \frac{\partial x_h(T)}{\partial v_{ref}}\Delta v_{ref} \tag{12.17}$$

$$= \left(\frac{\partial m(t_{sw})}{\partial t_{sw}}z(0) + \frac{\partial c(t_{sw})}{\partial t_{sw}}x_i(0) + \frac{\partial d(t_{sw})}{\partial t_{sw}}v_{ref}\right)\Delta t_{sw} + m(t_{sw})\Delta z(0) + c(t_{sw})\Delta x_i(0) + d(t_{sw})\Delta v_{ref}(0) \tag{12.18}$$

However, once the loop is closed, both integrator state, $x_i(0)$, and switching times, t_{sw}, are a sole function of $z(0)$ because changes in $z(0)$ produce associated changes in $x_i(0)$ and t_{sw}, provided that $v_{ref}(0)$ stays constant. This can be seen from the following.

From (12.15), one can see that

$$x_h(T) = f(\mathbf{x_p}(0), x_i(0), t_{sw}, z(0), v_{ref}). \tag{12.19}$$

Also, when the loop is closed, the following equation is true:

$$t_{sw} = g(\mathbf{x_p}(0), x_i(0), z(0)). \tag{12.20}$$

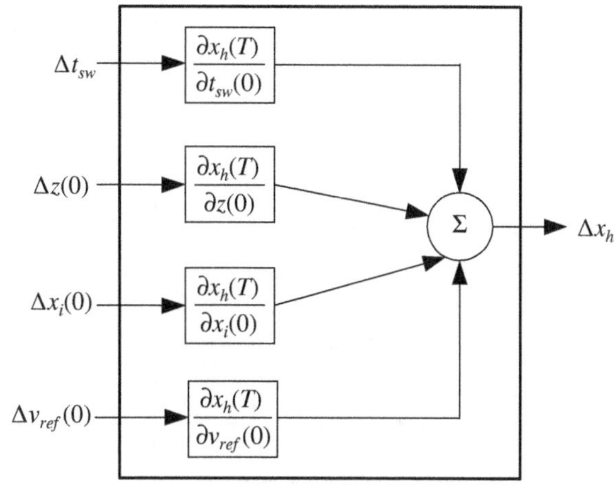

Figure 12.8 The input-output relation of the open-loop system at steady state.

Moreover, when the system is at steady state, the following two equations also hold true:

$$\mathbf{x_p}(0) = \mathbf{f_p}(z(0)) \tag{12.21}$$

$$x_i(0) = f_c(z(0)), \tag{12.22}$$

where $\mathbf{f_p}$ and f_c are all functions of $z(0)$.

Linearizing (12.19), (12.20), (12.21), and (12.22), one gets the following:

$$\Delta x_h(T) = \frac{\partial f}{\partial \mathbf{x_p}(0)}\Delta \mathbf{x_p}(0) + \frac{\partial f}{\partial x_i(0)}\Delta x_i(0) + \frac{\partial f}{\partial t_{sw}}\Delta t_{sw} + \frac{\partial f}{\partial z(0)}\Delta z(0) \tag{12.23}$$

$$\Delta t_{sw} = \frac{\partial g}{\partial \mathbf{x_p}(0)}\Delta \mathbf{x_p}(0) + \frac{\partial g}{\partial x_i(0)}\Delta x_i(0) + \frac{\partial g}{\partial z(0)}\Delta z(0) \tag{12.24}$$

$$\Delta \mathbf{x_p}(0) = \frac{\partial \mathbf{f_p}}{\partial z(0)}\Delta z(0) \tag{12.25}$$

$$\Delta x_i(0) = \frac{\partial f_c}{\partial z(0)}\Delta z(0) \tag{12.26}$$

Substituting (12.25) and (12.26) into (12.24) yields

$$\Delta t_{sw} = \frac{\partial g}{\partial \mathbf{x_p}(0)}\frac{\partial \mathbf{f_p}}{\partial z(0)}\Delta z(0) + \frac{\partial g}{\partial x_i(0)}\frac{\partial f_c}{\partial z(0)}\Delta z(0) + \frac{\partial g}{\partial z(0)}\Delta z(0) \tag{12.27}$$

$$= \left(\frac{\partial g}{\partial \mathbf{x_p}(0)}\frac{\partial \mathbf{f_p}}{\partial z(0)} + \frac{\partial g}{\partial x_i(0)}\frac{\partial f_c}{\partial z(0)} + \frac{\partial g}{\partial z(0)}\right)\Delta z(0) \tag{12.28}$$

$$= \frac{\partial t_{sw}}{\partial z(0)}\Delta z(0). \tag{12.29}$$

Therefore, $\mathbf{x_p}(0)$, $x_i(0)$, and t_{sw} are a function of $z(0)$.

Hence, $x_h(T)$ can be written as a function of $z(0)$ for the closed-loop configuration:

$$x_h(T) = f(z(0)). \tag{12.30}$$

Therefore, the sensitivity analysis graph (Fig. 12.8) for the open-loop case is changed to Fig. 12.9, and the expression for $\Delta x_h(T)$ is

$$\frac{\partial x_h(T)}{\partial z(0)} = \frac{\partial f}{\partial t_{sw}}\frac{\partial t_{sw}}{\partial z(0)} + \frac{\partial f}{\partial z(0)} + \frac{\partial f}{\partial x_i(0)}\frac{\partial x_i(0)}{\partial z(0)} + \frac{\partial f}{\partial v_{ref}(0)}\frac{\partial v_{ref}(0)}{\partial z(0)}. \tag{12.31}$$

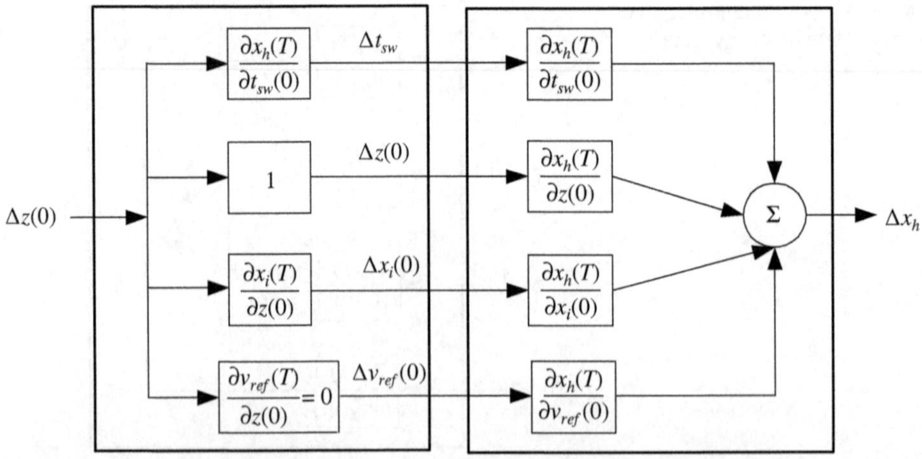

Figure 12.9 The input-output relation of the closed-loop system at steady state.

Alternatively, (12.31) may be expressed in a matrix form:

$$\frac{\partial x_h(T)}{\partial z(0)} = \begin{bmatrix} \frac{\partial f}{\partial t_{sw}} & \frac{\partial f}{\partial z(0)} & \frac{\partial f}{\partial x_i(0)} & \frac{\partial f}{\partial v_{ref}(0)} \end{bmatrix} \begin{bmatrix} \frac{\partial t_{sw}}{\partial z(0)} \\ 1 \\ \frac{\partial x_i(0)}{\partial z(0)} \\ \frac{\partial v_{ref}(0)}{\partial z(0)} \end{bmatrix}. \tag{12.32}$$

The terms, $\frac{\partial f}{\partial t_{sw}}$, $\frac{\partial f}{\partial z(0)}$, $\frac{\partial f}{\partial x_i(0)}$, and $\frac{\partial f}{\partial v_{ref}(0)}$, shown in the first block of Fig. 12.9, can be easily obtained, and their analytical expressions are stated in (12.33), (12.34), (12.35), and (12.36).

$$\frac{\partial f}{\partial t_{sw}} = \mathbf{C_1}\left(-\breve{\mathbf{A}}_{\mathbf{off}}\breve{\mathbf{\Phi}} + \breve{\mathbf{\Phi}}\breve{\mathbf{A}}_{\mathbf{on}}\right) \begin{bmatrix} \mathbf{x_p}(0) \\ z(0) \\ x_i(0) \\ v_{ref}(0) \\ 0 \end{bmatrix} + \mathbf{C_1}\breve{\mathbf{\Phi}} \begin{bmatrix} \frac{\partial \mathbf{x_p}(0)}{\partial t_{sw}} \\ 0 \\ 0 \\ 0 \\ 0 \end{bmatrix} \tag{12.33}$$

$$\frac{\partial f}{\partial z(0)} = \mathbf{C_1}\breve{\mathbf{\Phi}} \begin{bmatrix} \frac{\partial \mathbf{x_p}(0)}{\partial \mathbf{z}(0)} \\ 1 \\ 0 \\ 0 \\ 0 \end{bmatrix} \tag{12.34}$$

$$\frac{\partial f}{\partial x_i(0)} = \mathbf{C_1}\breve{\mathbf{\Phi}} \begin{bmatrix} 0 \\ 0 \\ 0 \\ 1 \\ 0 \\ 0 \end{bmatrix} \tag{12.35}$$

$$\frac{\partial f}{\partial v_{ref}(0)} = \mathbf{C_1}\check{\boldsymbol{\Phi}}\begin{bmatrix}0\\0\\0\\0\\1\\0\end{bmatrix}, \tag{12.36}$$

where $\mathbf{C_1} = \begin{bmatrix}0 & 0 & 0 & 0 & 0 & 1\end{bmatrix}$.

The terms, $\frac{\partial t_{sw}}{\partial z(0)}$, $\frac{\partial x_i(0)}{\partial z(0)}$, and $\frac{\partial v_{ref}(0)}{\partial z(0)}$, shown in the second block of Fig. 12.9, are obtained as follows.

$\frac{\partial v_{ref}(0)}{\partial z(0)}$ is simply zero, as the voltage reference is assumed to be constant.

To obtain $\frac{\partial t_{sw}}{\partial z(0)}$ and $\frac{\partial x_i(0)}{\partial z(0)}$, one needs to impose the steady-state and switching time constraints, as stated in (10.13) and (10.14). These constraints are restated here for convenience.

$$x_i(T) - x_i(0) = \begin{bmatrix}0 & 0 & 0 & 1 & 0\end{bmatrix}\left\{\check{\boldsymbol{\Phi}} - \mathbb{I}\right\}\check{\mathbf{x}}(0) = 0 \tag{12.37}$$

$$\frac{V_M}{T}t_{sw} - v_{out}(t_{sw}) = \frac{V_M}{T}t_{sw} - \begin{bmatrix}0 & -KH & 0 & \frac{1}{\tau} & K\end{bmatrix}e^{\check{\mathbf{A}}_{\mathbf{on}}t_{sw}}\check{\mathbf{x}}(0) = 0 \tag{12.38}$$

Differentiating (12.37) with respect to $z(0)$ yields

$$\begin{bmatrix}0 & 0 & 0 & 1 & 0\end{bmatrix}\left(-\check{\mathbf{A}}_{\mathbf{off}}\check{\boldsymbol{\Phi}}\check{x}(0) + \check{\boldsymbol{\Phi}}\check{A}_{on}\check{\mathbf{x}}(0) + \check{\boldsymbol{\Phi}}\begin{bmatrix}\frac{\partial \mathbf{x_p}(0)}{\partial t_{sw}}\\0\\0\\0\end{bmatrix}\right)\frac{\partial t_{sw}}{\partial z(0)}$$

$$+ \begin{bmatrix}0 & 0 & 0 & 1 & 0\end{bmatrix}\left(\check{\boldsymbol{\Phi}}\begin{bmatrix}0\\0\\0\\1\\0\end{bmatrix} - 1\right)\frac{\partial x_i(0)}{\partial z(0)} + \begin{bmatrix}0 & 0 & 0 & 1 & 0\end{bmatrix}\cdot$$

$$\check{\boldsymbol{\Phi}}\begin{bmatrix}\frac{\partial \mathbf{x_p}}{\partial z(0)}\\1\\0\\0\end{bmatrix} = 0. \tag{12.39}$$

Differentiating (12.38) with respect to $z(0)$ yields

$$\left\{\frac{V_M}{T} - \begin{bmatrix}0 & -\frac{K}{3} & 0 & \frac{1}{\tau} & K\end{bmatrix}\left(e^{\check{\mathbf{A}}_{\mathbf{on}}t_{sw}}\check{\mathbf{A}}_{\mathbf{on}}\check{\mathbf{x}} + e^{\check{\mathbf{A}}_{\mathbf{on}}t_{sw}}\begin{bmatrix}\frac{\partial \mathbf{x_p}(0)}{\partial t_{sw}}\\0\\0\\0\end{bmatrix}\right)\right\}\frac{\partial t_{sw}}{\partial z(0)}$$

$$- \begin{bmatrix}0 & -KH & 0 & \frac{1}{\tau} & K\end{bmatrix}e^{\check{\mathbf{A}}_{\mathbf{on}}t_{sw}}\begin{bmatrix}0\\0\\0\\1\\0\end{bmatrix}\frac{\partial x_i(0)}{\partial z(0)} - \begin{bmatrix}0 & -KH & 0 & \frac{1}{\tau} & K\end{bmatrix}e^{\check{\mathbf{A}}_{\mathbf{on}}t_{sw}}.$$

$$\begin{bmatrix} \dfrac{\partial \mathbf{x_p}(0)}{\partial z(0)} \\ 1 \\ 0 \\ 0 \end{bmatrix} = 0. \tag{12.40}$$

Rearranging (12.39) and (12.40) leads to the form

$$\begin{bmatrix} a_{11} & a_{12} \\ a_{21} & a_{22} \end{bmatrix} \begin{bmatrix} \dfrac{\partial x_i(0)}{\partial z(0)} \\ \dfrac{\partial t_{sw}}{\partial z(0)} \end{bmatrix} = \begin{bmatrix} d_1 \\ d_2 \end{bmatrix}. \tag{12.41}$$

Comparing (12.41) with (10.15), one can conclude that $a_{11} = J_{11}$, $a_{12} = J_{12}$, $a_{21} = J_{21}$, and $a_{22} = J_{22}$.

Therefore,

$$\begin{bmatrix} \dfrac{\partial x_i(0)}{\partial z(0)} \\ \dfrac{\partial t_{sw}}{\partial z(0)} \end{bmatrix} = \begin{bmatrix} J_{11} & J_{12} \\ J_{21} & J_{22} \end{bmatrix}^{-1} \begin{bmatrix} d_1 \\ d_2 \end{bmatrix} \tag{12.42}$$

where

$$d_1 = -\begin{bmatrix} 0 & 0 & 0 & 1 & 0 \end{bmatrix} \breve{\mathbf{\Phi}} \left(\begin{bmatrix} \dfrac{\partial \mathbf{x_p}(0)}{\partial z(0)} \\ 1 \\ 0 \\ 0 \end{bmatrix} \right) \tag{12.43}$$

$$d_2 = \begin{bmatrix} 0 & -\dfrac{K}{3} & 0 & \dfrac{1}{\tau} & K \end{bmatrix} e^{\breve{\mathbf{A}}_{\mathbf{on}} t_{sw}} \left(\begin{bmatrix} \dfrac{\partial \mathbf{x_p}(0)}{\partial z(0)} \\ 1 \\ 0 \\ 0 \end{bmatrix} \right). \tag{12.44}$$

Two important points can be observed from the above derivations:

1. d_1 and d_2 can be easily obtained by differentiating the negative of the mismatch equations, $-M_1$ and $-M_2$ of (10.15) with respect to $z(0)$ while treating $x_i(0)$ and t_{sw} as constants.
2. To obtain $\dfrac{\partial x_i(0)}{\partial z(0)}$ and $\dfrac{\partial t_{sw}}{\partial z(0)}$, one can simply multiply the inverse of the Jacobian matrix of (10.15) with $\begin{bmatrix} d_1 & d_2 \end{bmatrix}^T$.

Example 12.1 To validate the SSM of the derived dc-dc converter, the closed-loop dc-dc converter of **Example 10.1** is simulated. The original operating point or nominal operating point is defined when $v_s = 28V$ and $v_{ref} = 5V$, as presented in Section 10.1. The switching time and the steady-state initial condition of the controller's integrator are $5.3571 \times 10^{-5} sec$ and 1.1878V-sec, respectively. These values are then substituted in (12.32) to obtain the small-signal FCM, $\mathbf{\Delta F}$.

Assume that the input v_s is changed from $28V$ to $28.84V$ (a 3% change in v_s), and that the reference v_{ref} remains $5V$. The estimated perturbed inductor current harmonics ($\mathbf{\Delta I_L}$) can be obtained by

$$\mathbf{\Delta I_L} = \mathbf{\Delta F \Delta V_s}, \tag{12.45}$$

where $\mathbf{\Delta V_s} = \begin{bmatrix} \Delta V_s^0 & \Delta V_s^1 & \Delta V_s^1 & \cdots & \Delta V_s^n \end{bmatrix}^T = \begin{bmatrix} \Delta v_s & 0 & 0 & \cdots & 0 \end{bmatrix}^T$.

Table 12.1 DC current of the closed-loop buck converter, with $v_{ref} = 5V$, and base frequency is 10 kHz.

Harmonic number	Original operating point ($v_s = 28V$)	Large-signal harmonic model ($v_s = 28.84V$)	Small-signal harmonic model
0	5 A	5 A	5 A
1	−2.8301 + j0.31857 A	−2.9403 + j0.18569 A	−2.9472 + j0.18704 A
3	−0.28115 + j0.098378i A	−0.31351 + j0.060143 A	−0.31939 + j0.063958 A
5	−0.081391 + j0.051142i A	−0.10565 + j0.03453 A	−0.10965 + j0.040241 A

Then, $\Delta\mathbf{I}_{\mathbf{dc}}$ is added to the current harmonics of the original operating point to produce an estimate of the dc current harmonics. That is how one finds the estimated output harmonics from the small-signal analysis. Alternately, the dc current can be found by iterative large-signal analysis when v_s is changed to $28.84V$. Table 12.1 shows the values of the dc current obtained by this large-signal approach and by the SSM when a 3% increase in the dc voltage occurs.

As can be noted from the table, the SSM prediction of the real and imaginary parts of the harmonic is reasonably close to the large-signal model result. □

12.4 Small-Signal Analysis of a Two-Level VSC

12.4.1 Approach from Section 12.3

The SSM of a two-level VSC is derived in a manner similar to the analysis of the dc-dc converter in the previous section. The harmonics of interest can be solved by (12.46):

$$\check{\mathbf{x}} = \check{\mathbf{\Phi}}_{\mathbf{tv}}(T,0)\check{x}(0), \tag{12.46}$$

where

$$\check{\Phi}_{\mathbf{tv}}(T,0) = \prod_{k=n+1}^{k=1} \check{\mathbf{\Phi}}_{\mathbf{tv}}(t_k, t_{k-1}),$$

and $\check{\mathbf{\Phi}}_{\mathbf{tv}}$ is the solution to (12.47).

$$\frac{d}{dt}\hat{\mathbf{x}}(t) = \check{\mathbf{A}}_{\mathbf{tv}}(t)\hat{\mathbf{x}}(t), \tag{12.47}$$

where

$$\check{\mathbf{x}} = \begin{bmatrix}\check{\mathbf{x}} & \mathbf{x_h}\end{bmatrix}^T$$

Note that $\mathbf{x_h}$ represents the harmonic state vector of the interest, and $\check{\mathbf{x}}$ contains the states defined in (10.35).

(12.46) can be written as a linear combination of $\mathbf{x_i}(0)$, $\mathbf{z}(0)$, and i_{tqref}, v_{dcref}:

$$\mathbf{x_h}(T) = \mathbf{m}(\mathbf{t_{sw}})\mathbf{z}(0) + \mathbf{c}(\mathbf{t_{sw}})\mathbf{x_i}(0) + \mathbf{d}(\mathbf{t_{sw}})i_{tqref} + \mathbf{e}(\mathbf{t_{sw}})v_{dcref}, \tag{12.48}$$

where $\mathbf{t_{sw}} = [t_1, t_2, \ldots, t_n]$ and $\mathbf{x_p}$ has been replaced with $\{\mathbb{I} - \mathbf{\Phi}(\mathbf{t_{sw}})\}^{-1}\mathbf{\Gamma}(\mathbf{t_{sw}})\mathbf{z}(0)$.

As in the case of a dc-dc converter, the harmonic state variable of a closed-loop VSC is a sole function of the input's initial conditions:

$$\mathbf{x_h} = \mathbf{f}(\mathbf{z}(\mathbf{0})). \tag{12.49}$$

Differentiating (12.49) with respect to $\mathbf{z}(0)$ yields

$$\frac{\partial \mathbf{x}}{\partial \mathbf{z}(0)} = \begin{bmatrix} \frac{\partial \mathbf{f}}{\partial \mathbf{x_i}(0)} & \frac{\partial \mathbf{f}}{\partial \mathbf{t_{sw}}} & \frac{\partial \mathbf{f}}{\partial \mathbf{z}(0)} & \frac{\partial \mathbf{f}}{\partial i_{tqref}} & \frac{\partial \mathbf{f}}{\partial v_{tqref}} \end{bmatrix} \begin{bmatrix} \frac{\partial \mathbf{x_i}(0)}{\partial \mathbf{z}(0)} \\ \frac{\partial \mathbf{t_{sw}}}{\partial \mathbf{z}(0)} \\ \mathbb{I} \\ \mathbf{0} \\ \mathbf{0} \end{bmatrix} \tag{12.50}$$

To obtain the values of $\frac{\partial \mathbf{x_i}(0)}{\partial \mathbf{z}(0)}$ and $\frac{\partial \mathbf{t_{sw}}}{\partial \mathbf{z}(0)}$, one needs to impose the steady-state and switching time constraints. Following the same procedures presented for the case of the buck converter results in

$$\begin{bmatrix} \frac{\partial \mathbf{x_i}(0)}{\partial \mathbf{z}(0)} \\ \frac{\partial \mathbf{t_{sw}}}{\partial \mathbf{z}(0)} \end{bmatrix} = \mathbf{J}^{-1}\mathbf{M_z}, \tag{12.51}$$

where $\mathbf{J}$ is the Jacobian matrix from (10.79), and $\mathbf{M_z}$ is obtained via

$$\begin{bmatrix} -\frac{\partial M_1}{\partial \mathbf{z}(0)} \\ -\frac{\partial M_2}{\partial \mathbf{z}(0)} \\ -\frac{\partial M_3}{\partial \mathbf{z}(0)} \\ -\frac{\partial \mathbf{M_4}}{\partial \mathbf{z}(0)} \\ -\frac{\partial \mathbf{M_5}}{\partial \mathbf{z}(0)} \\ -\frac{\partial \mathbf{M_6}}{\partial \mathbf{z}(0)} \end{bmatrix}$$

while treating $\mathbf{x_i}(0)$ and $\mathbf{t_{sw}}$ as independent of $\mathbf{z}(0)$. Note that M_1, M_2, M_3 $\mathbf{M_4}$, $\mathbf{M_5}$, $\mathbf{M_6}$ are obtained from (10.71) to (10.76).

12.4.2 Simpler Approach

In the previous sections of this chapter, the analysis was carried out without knowing the structure of $\breve{\mathbf{\Phi}}$. The matrix structure of $\breve{\mathbf{\Phi}}$ takes the following form:

$$\begin{bmatrix} \mathbf{x_p}(T) \\ \mathbf{x_i}(T) \\ \mathbf{z}(T) \\ \mathbf{r} \\ \mathbf{x_h}(T) \end{bmatrix} = \breve{\mathbf{\Phi}} \begin{bmatrix} \mathbf{x_p}(0) \\ \mathbf{x_i}(0) \\ \mathbf{z}(0) \\ \mathbf{r} \\ \mathbf{x_h}(0) \end{bmatrix}, \tag{12.52}$$

where $\breve{\mathbf{\Phi}} = \begin{bmatrix} \mathbf{\Phi} & \mathbf{0} & \mathbf{\Gamma} & \mathbf{0} & \mathbf{0} \\ \mathbf{F_p} & \mathbf{G_p} & \mathbf{K_p} & \mathbf{L_p} & \mathbf{0} \\ \mathbf{0} & \mathbf{0} & \mathbf{\Omega_p} & \mathbf{0} & \mathbf{0} \\ \mathbf{0} & \mathbf{0} & \mathbf{0} & \mathbf{0} & \mathbf{0} \\ \mathbf{H_p} & \mathbf{0} & \mathbf{Q_p} & \mathbf{0} & \mathbf{M_p} \end{bmatrix}$, and $\mathbf{r}$ represents the reference values.

Note that the structure of (12.52) also holds true for $\breve{\mathbf{\Phi}}_{\mathbf{tv}}$ for the case of the VSC.

From (12.52), $\mathbf{x_h}(T)$ can be written as

$$\mathbf{x_h}(T) = \mathbf{H_p}\mathbf{x_p}(0) + \mathbf{Q_p}\mathbf{z}(0) \tag{12.53}$$

and

$$\mathbf{x_p}(T) = \mathbf{\Phi}\mathbf{x_p}(0) + \mathbf{\Gamma}\mathbf{z}(0) \tag{12.54}$$

At steady state,

$$\mathbf{x_p}(0) = \{\mathbb{I} - \mathbf{\Phi}\}^{-1}\mathbf{\Gamma}\mathbf{z}(0). \tag{12.55}$$

Substituting (12.55) into (12.53) leads to the form:

$$\mathbf{x_h}(T) = \{\mathbf{H_p}[(\mathbb{I} - \mathbf{\Phi})^{-1}\mathbf{\Gamma}] + \mathbf{Q_p}\}\mathbf{z}(0). \tag{12.56}$$

(12.56) is none other than the FCM derived in Chapter 5.

Moreover, comparing (12.56) with (12.48), one can see that

$$\mathbf{m}(\mathbf{t_{sw}}) = \{\mathbf{H_p}[(\mathbb{I} - \mathbf{\Phi})^{-1}\mathbf{\Gamma}] + \mathbf{Q_p}\} = \mathbf{F}(\mathbf{t_{sw}}) \tag{12.57}$$

$$\mathbf{c}(\mathbf{t_{sw}}) = \mathbf{0} \tag{12.58}$$

$$\mathbf{d}(\mathbf{t_{sw}}) = \mathbf{0} \tag{12.59}$$

$$\mathbf{e}(\mathbf{t_{sw}}) = \mathbf{0}. \tag{12.60}$$

Therefore, the SSM of the closed-loop power converter, $\Delta\mathbf{F}$, can be found by directly differentiating the FCM, $\mathbf{F}$.

$$\Delta\mathbf{F} = \frac{\partial\mathbf{x_h}(T)}{\partial\mathbf{z}(0)} = \mathbf{F}(\mathbf{t_{sw}}) + \frac{\partial\mathbf{F}(\mathbf{t_{sw}})}{\partial\mathbf{t_{sw}}}\frac{\partial\mathbf{t_{sw}}}{\partial\mathbf{z}(0)}\mathbf{z}(0) \tag{12.61}$$

A sample code for $\frac{\partial\mathbf{F}(\mathbf{t_{sw}})}{\partial\mathbf{t_{sw}}}$ is as follows:

```

PP=[];
% Calculate partial derivatives of the overall product


%when n=1;
pp_xx=eye(length(PHI3));
for i=2:length(tau)
    pp_xx=expm(Axxx_II(:,:,Sdseq_In(i))*(tau(i)-tau(i-1)))*pp_xx;
end
pp_xx=pp_xx*(-Axxx_II(:,:,Sdseq_In(2))+Axxx_II(:,:,Sdseq_In(1)))* ...
expm(Axxx_II(:,:,Sdseq_In(1))*tau(1));
PP=[PP,pp_xx];

counter=0;
%when n>= 2
for n=2:length(tau)-1
    PP2=expm(Axxx_II(:,:,Sdseq_In(1))*tau(1));
    phi22=eye(length(PHI3));
    for k=2:n
        PP2=expm(Axxx_II(:,:,Sdseq_In(k))*(tau(k)-tau(k-1)))*PP2;
    end
    phi=(-Axxx_II(:,:,Sdseq_In(n+1))+Axxx_II(:,:,Sdseq_In(n)))*PP2;
    for p=n+1:length(tau)
        phi22=expm(Axxx_II(:,:,Sdseq_In(p))*(tau(p)-tau(p-1)))*phi22;
    end
```

```
        PP=[PP,phi22*phi];
end
%%%%%%%%%%%%%%%%%%%%%%%%%%%%%%%%%%%%%%%%%%%%%%%%%%%%%%%%%%%%%%%%%%%%%%%%%%%%%%%%%%%%


%Remember the first one is with respect to tn;
counter=0;
for n=1:length(PHI3):length(PP)
    counter=counter+1;
    A_partial=PP(:,n:n+length(PHI3)-1);

% Assume that it is a two-level VSC so the system state is 3 (i_alpha, i_beta, ...
% and v_dc)
    Adp=A_partial(1:3,1:3);
%The input is assumed to be balanced and is modeled in the alpha-beta frame so ...
% needs two states each.
    Ndp=A_partial(1:3,4:7);
    H1dp=A_partial(8:size(PHI3),1:3);
    H2dp=A_partial(8:size(PHI3),4:7);

    FCM_part_tau=H1dp*(inv(II-Ap))*Np+H1p*(II-Ap)$\Hat$(-2)*(Adp)*Np+H1p*...
    (inv(II-Ap))*Ndp+H2dp;
end
```

Note that based on the analysis of Section 6.2 of Chapter 5, the input initial conditions are equal to the input harmonic phasors (broken into real and imaginary parts). Therefore,

$$\Delta\mathbf{F} = \mathbf{F}(\mathbf{t_{sw}}) + \frac{\partial\mathbf{F}(\mathbf{t_{sw}})}{\partial\mathbf{t_{sw}}}\frac{\partial\mathbf{t_{sw}}}{\partial\mathbf{z}(0)}\mathbf{V}. \tag{12.62}$$

To obtain $\frac{\partial\mathbf{t_{sw}}}{\partial\mathbf{z}(0)}$, again one needs to impose the steady state and switching time constraint and solve for (12.51). Note that the analyses in Sections 12.3 to 12.4.1 are exactly the same as those of this section. They have been included to bring out the important role of $\frac{\partial\mathbf{x_i}(0)}{\partial\mathbf{z}(0)}$. As (12.61) shows, only the term $\frac{\partial\mathbf{t_{sw}}}{\partial\mathbf{z}(0)}$ is needed to evaluate $\Delta\mathbf{F}$. One might naively choose to only consider the switching time constraints and treat $\mathbf{x_i}(0)$ as independent of $\mathbf{z}(0)$, but this would lead to erroneous results. Although $\frac{\partial\mathbf{x_i}(0)}{\partial\mathbf{z}(0)}$ is not explicitly required to evaluate (12.61), the term $\frac{\partial\mathbf{t_{sw}}}{\partial\mathbf{z}(0)}$ is affected by it via (12.51). At steady-state, both the switching times and integrator's steady-state constraints need to be simultaneously satisfied.

To validate the SSM of the VSC, two types of perturbation are carried out — positive sequence perturbation and negative sequence perturbation. $\Delta\mathbf{F}$ is first evaluated at the nominal operating point. Then it multiplies with the perturbed inputs to obtain the perturbed outputs (i.e. ac current and dc voltage harmonics).

The nominal operating points for both cases are the balanced cases studied in Chapter 10. That is, $|e_s| = 112\frac{\sqrt{2}}{\sqrt{3}}V$, $\theta = 0°$, $v_{dcref} = 230V$, and $i_{tqref} = 30$A. The controllers considered in both cases are the same as those in Chapter 10.

Example 12.2 Positive Sequence Perturbation

For this case, the only nonzero perturbed input is the positive sequence of the fundamental ac voltage, which is $112\frac{\sqrt{2}}{\sqrt{3}} \times 3\%V$. The resulting perturbed ac current ($\Delta\mathbf{I_{ac}}$) and dc voltage ($\Delta\mathbf{V_{dc}}$) are added to the nominal output ac current ($\mathbf{I_{aco}}$) and dc voltage ($\mathbf{V_{dco}}$) to obtain the estimated ac current and dc voltage harmonics based on the small-signal analysis.

Fig. 12.10 shows the nominal ac current harmonics predicted by the large-signal analysis and the SSM. Note that the nonzero ac current harmonic shown in the figure (from left to right) are [−53 −47 −41 −35 −29 −23 −17 −11 −5 1 7 13 19 25 31 37 43 49 55]. As can be seen from the figure, the SSM predict accurately which harmonic should be increased from the nominal operating point (such as the 7th and −11th harmonics), and which harmonic should be decreased from the nominal values (such as the −17th and 19th harmonics).

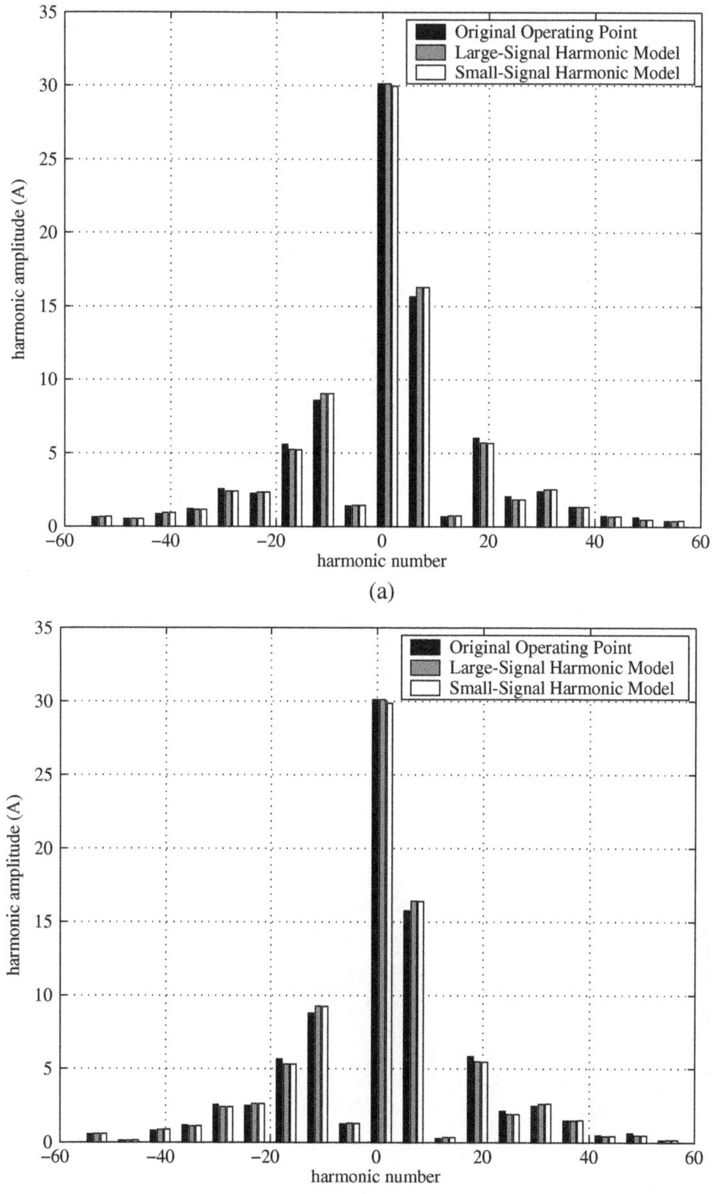

Figure 12.10 Positive sequence perturbation, ac current harmonics: (a) slow controller; (b) fast controller.

Fig. 12.11 shows the nominal dc voltage harmonics predicted by the large-signal analysis and the SSM. Note that the plot is shown in the log scale and the nonzero harmonic numbers shown are [0 6 12 18 24 30]. As can be noted, the SSM also predicts well which dc voltage harmonic should be increased from the nominal value (such as the 24th harmonic) and which should be decreased from the nominal value (such as the 18th harmonic). Their values are almost the same as those predicted by the large-signal model. □

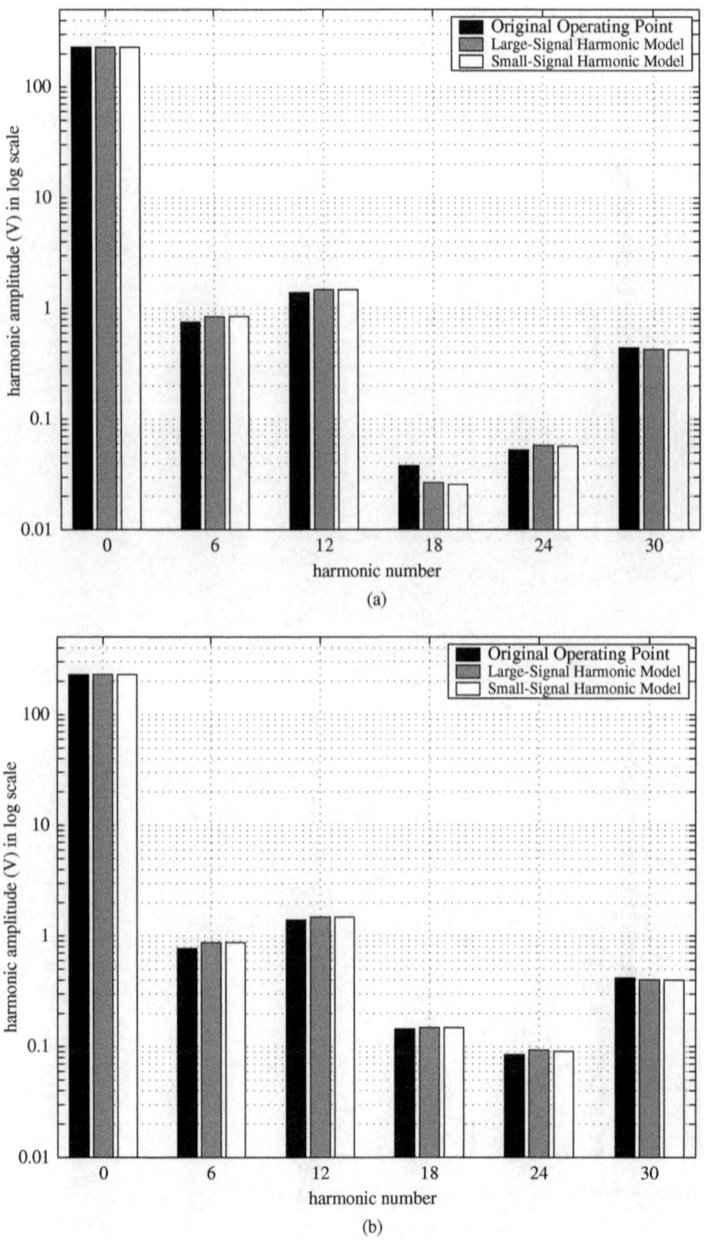

Figure 12.11 Positive sequence perturbation, dc voltage harmonics: (a) slow controller; (b) fast controller.

Example 12.3 Negative Sequence Perturbation

In this case, the nonzero perturbed input is the negative sequence of the fundamental voltage and its value is also $112\frac{\sqrt{2}}{\sqrt{3}} \times 3\%V$. The perturbed output ac current and dc voltage harmonics are obtained via multiplying $\Delta\mathbf{F}$ with the perturbed input. These values are then added back to the nominal ac current and dc voltage harmonics.

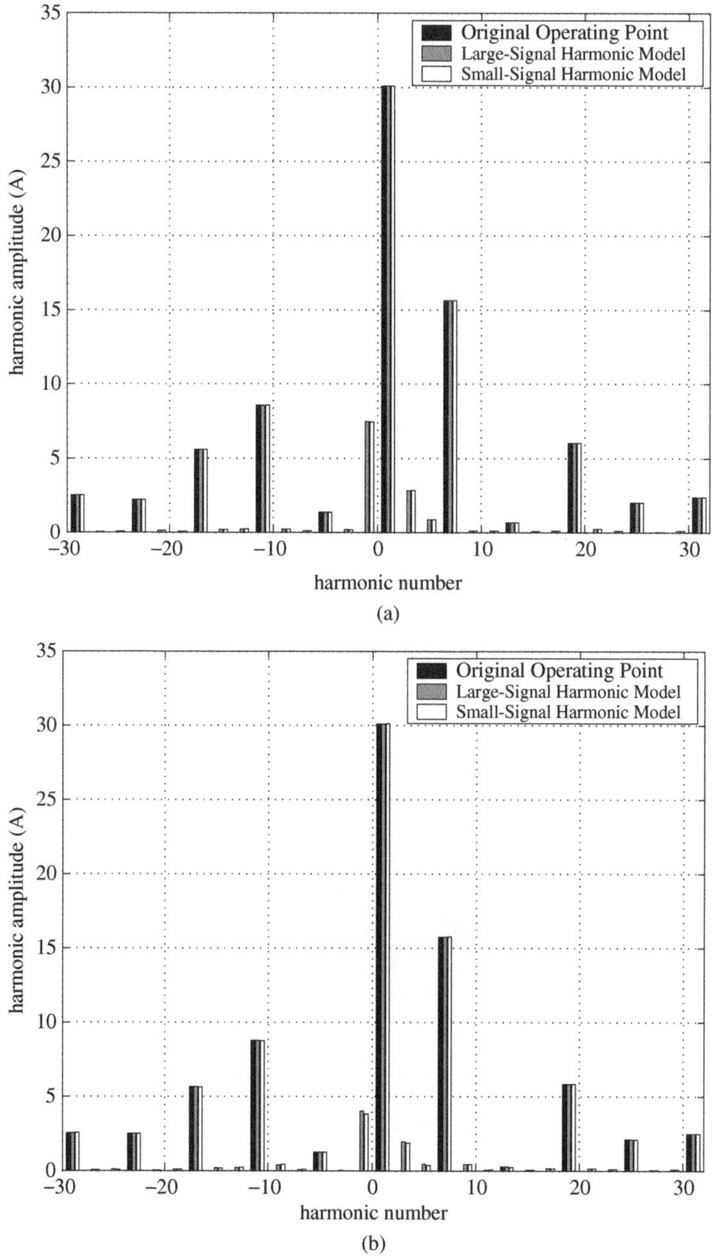

Figure 12.12 Negative sequence perturbation, ac current harmonics: (a) slow controller; (b) fast controller.

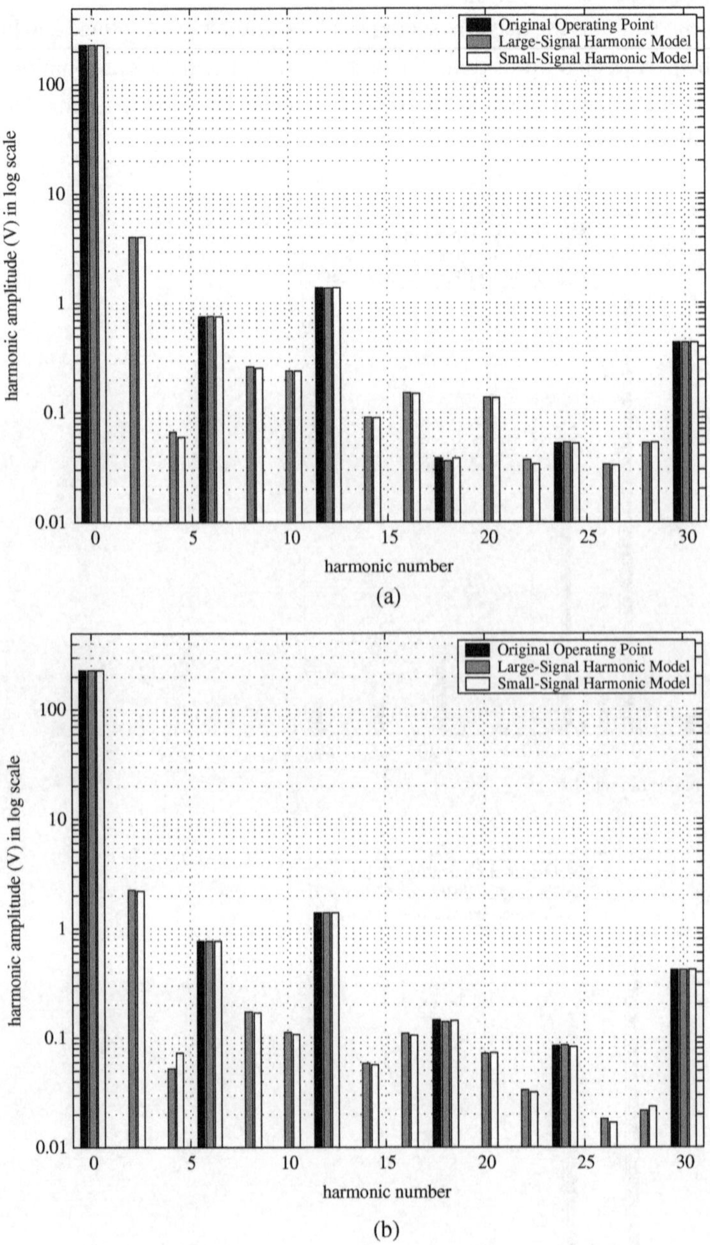

Figure 12.13 Negative sequence perturbation, dc voltage harmonics: (a) slow controller; (b) fast controller.

Fig. 12.12 shows the nominal ac current harmonics predicted by the large-signal analysis and the SSM. All the odd harmonic numbers are shown in the figure. Since the nominal operating points are evaluated under balanced conditions, only characteristic harmonics are nonzero for the nominal ac current harmonics. Note that even though this is the case, the SSM can still predict accurately the nonzero uncharacteristic harmonics accurately. For example, the negative sequence fundamental component, positive third, and positive fifth harmonic are all accurately predicted by the SSM as compared with the large-signal model.

Similarly, all the non-characteristic dc voltage harmonics are all well predicted by the SSM, which are otherwise zero at the nominal operating point, as clearly shown in Fig. 12.13. □

13

Parameter Estimation for a Single VSC

"It takes both sides to build a bridge."

- Frederik Nael

As mentioned in the previous chapters, VSCs have been widely used in modern power systems. Consequently, reliable operations of VSCs are essential for these systems to operate properly. A parameter estimator for a VSC acts as a useful tool for problem diagnosis and preventive maintenance. For instance, it has been reported in [124] that capacitors cause 30% of the total failures in converter circuits, and capacitance value is a good indication of the health conditions of the capacitor. Moreover, a parameter estimator also allows one to construct an accurate simulation model of the system for component or control design [125], stability analysis [126, 127], fault detection [128, 129], and so on. For a typical two-level VSC (refer to Fig. 13.1) at least nine parameters need to be identified [130]. They are the grid equivalent resistance (R_s) and inductance (L_s), filter resistance (R_f) and inductance (L_f), dc capacitance (C_{dc}), and the proportional (K_c and K_v) and integral time constants (τ_c and τ_v) of current and voltage controllers. Since we have derived closed-loop VSC models, it is possible to use these models for parameter estimation (PE).

13.1 Background on Parameter Estimation

Generally speaking, depending on how the measurement data are obtained, a PE method can be classified into invasive and non-invasive. An invasive method usually requires the use of external excitation signals to capture the responses of these excitations. For example, Huang, *et. al.* [126] proposed to inject a single-frequency excitation signal and estimate the magnitude and phase of the impedance of a VSC at that frequency. Neshvad, *et. al.* [131] proposed to inject a pseudo-random-binary-sequence signal to estimate the harmonic impedance of a VSC. On the other hand, Chang, *et. al.* [132] proposed to add sinusoidal sweep signals to the dc reference signals to obtain the time-domain response and estimated the proportional and integral time constants of converter's controllers. However, other circuit parameters of the converters are assumed to be known and no estimation is performed. Asiminoaei, *et. al.* [133] proposed to inject a noncharacteristic harmonic current and measured the grid voltage response to estimate the grid impedance of a grid-connected photovoltaic inverter. Sumner, *et. al.* [134, 135] proposed to inject a short duration current spike into the converter system and derived the impedance from the measured current and voltage. Liu, *et. al.* [136] proposed to inject a very short period of current

Harmonic Modeling of Voltage Source Converters using Basic Numerical Methods, First Edition.
Ryan Kuo-Lung Lian, Ramadhani Kurniawan Subroto, Victor Andrean, and Bing Hao Lin.

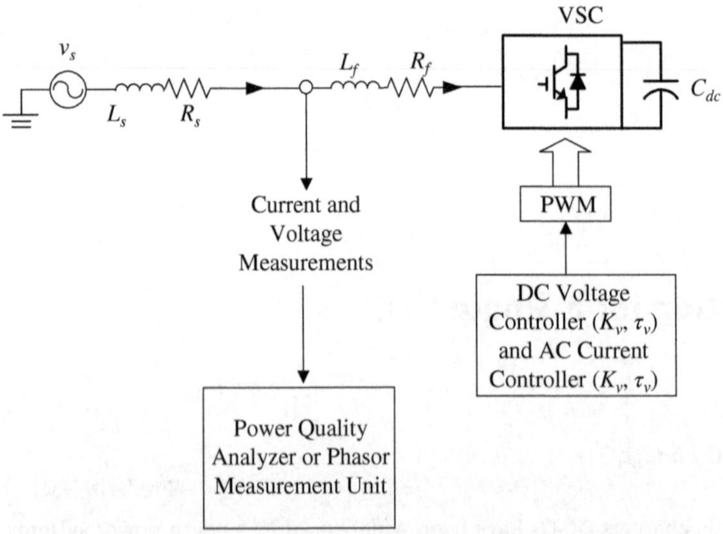

Figure 13.1 The schematic diagram of a typical voltage source converter.

with negative value to a permanent magnet synchronous machine (PMSM) drive system. Then they used dynamic PSO to simultaneously estimate the parameters of PMSM and the distorted voltage caused by a VSC.

The signal injections generally should not be too large to disturb normal system operation, and at the same time, should not be too small to suffer from measurement noise [137]. The invasive methods are difficult to be applied widely due to practical complexity [138]. Consequently, non-invasive methods, which do not require any external excitation, are generally preferred for PE. Moreover, with the advancement of phasor measurement units (PMU) and power quality analyzers (PQA), it is possible to have accurate PE via a non-invasive method. For example, the LOw COst (LOCO) PMU proposed in [139] is able to provide synchrophasors, frequency, and rate of change of frequency with an accuracy and maximum delay as specified in the IEEE Std C37.118.1-2011. The analog-to-digital converter (ADC) employed by LOCO PMU has a sampling rate of 12.5 kHz per channel. Moreover, the LOCO PMU can limit the total vector errors (TVEs) [140] to only 0.57%. With this high sampling rate and low value of TVE, it is possible to estimate the converter parameters via measured voltage and current harmonic data of PMU or PQA. Over the past few years, there have been various non-invasive methods proposed. Hoffmann [141] proposed an extended Kalman filter to estimate the grid impedance using the disturbances that are already present in the network. Esparza, *et. al.* [142] have constructed the harmonic model of a VSC using the extended harmonic domain method and extracted the resistances and inductances of the VSC by minimizing the error between the measured and calculated values via a power quality analyzer. However, the VSC is operated in the open-loop fashion so the control parameters are not identified. Li, *et. al.* [138] employed voltage and current harmonic measurements to estimate the equivalent Norton circuit of an ultra-high voltage direct current system via derived harmonic admittance matrix. Similarly, no control parameters are identified. This chapter demonstrates a non-invasive PE method, which is able to predict simultaneously the circuit and control parameters of a VSC by means of the derived closed-loop steady-state models. These models are used to generate the training dataset to obtain an artificial neural network (ANN) model to account for various uncertainties in a practical circuit. Then experimental measurements are fed to the ANN model to estimate the parameters of a VSC. It is to be emphasized that no experimental data are used for training the ANN models. All the training data are generated by the derived VSC models, which are constructed in MATLAB and run offline. Hence, an accurate mathematical model (i.e. white-box model) together with an ANN or deep neural network (i.e. black-box model) is a viable, time-saving, and cost-effective method for PE.

13.2 Parameter Estimator Based on White-Box-and-Black-Box Models

Since the models presented in Chapters 9 and 10 are very accurate and the results match well with the experimental results as shown in [97], one can use these detailed models (i.e. white-box model) for generating a large amount of the training dataset for constructing an ANN model (i.e. black-box model). This neural network model, which is aware of every combination of the parameters, due to the training dataset containing every combination of parameters, can take the measured harmonic currents or/and voltages and identify what combination of the parameters corresponds to these measurements. This section discusses how a neural network is constructed for parameter estimation of a VSC. The neural network we used is composed of an input layer, four hidden layers, and an output layer, as shown in Fig. 13.2. As can be seen, the input layer takes the current harmonics generated by the VSC and the voltage harmonics at PCC. These harmonics are broken into magnitudes and phase angles. Hence, for N number of harmonics, the input layer consists of $2N$ neurons, as listed in Table 13.1. The output layer, on the other hand, yields the estimated parameter values. Hence, if there are M numbers of parameters to be estimated, then the output layer will consist of M numbers of neurons. The hidden layers are to connect the input and output layers, and the numbers of neurons for each hidden layer are listed in Table 13.1. The activation functions (AF) used for the input and hidden layers are rectified linear (ReLu) unit, and is stated as

$$f(\alpha) = Max(0, \alpha), \tag{13.1}$$

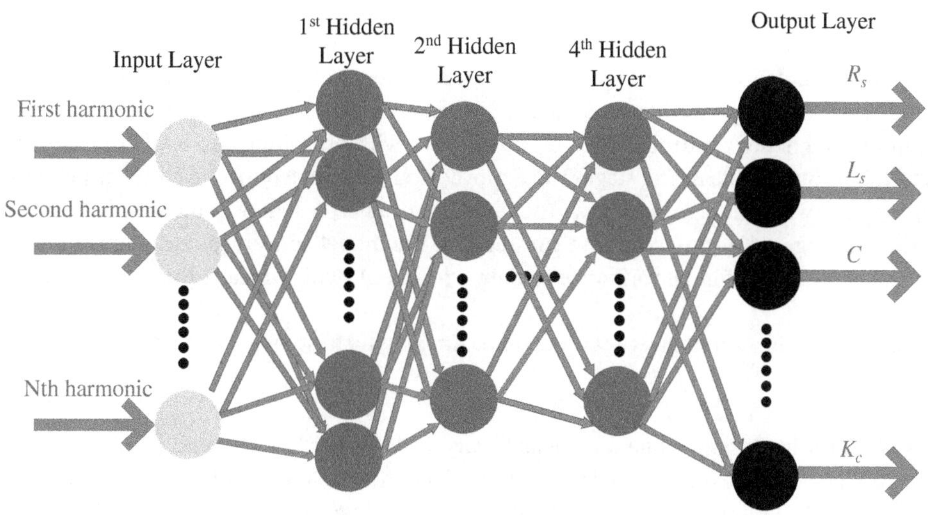

Figure 13.2 Proposed artificial neural network.

Table 13.1 ANN Architecture.

Layer type	Format	Activation function
Input	Units = $2 * N$	ReLU
1st fully connected	Units = $M * 16$	ReLU
2nd fully connected	Units = $M * 8$	ReLU
3rd fully connected	Units = $M * 4$	ReLU
4th fully connected	Units = $M * 2$	ReLU
Dropout	dropout rate = 0.5	
Output	Units = M	Linear

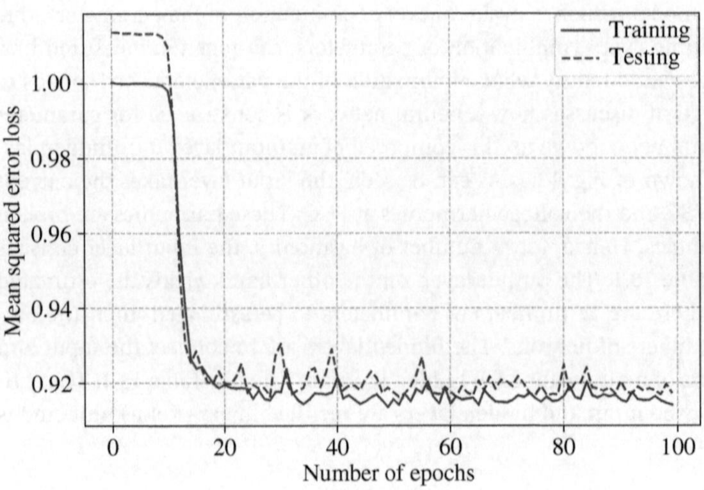

Figure 13.3 The mean squared error loss over training epochs.

where α is the input to a neuron. Note that ReLU can cause the output of some neurons to be 0, which can make the neural network sparse and alleviate the problem of overfitting. In addition, dropout regularization is applied between the last hidden layer and output layer to prevent overfitting. The individual nodes are dropped out of the network with probability, 0.5. Only the reduced network is trained on the data in that stage. The removed nodes are then reinserted into the network with their original weights. Fig. 13.3 shows the mean squared error loss over the training epochs for both the training and testing sets. One can see that the model converged reasonably quickly and both training and testing performance remain equivalent, which means that no overfitting occurred.

Finally, the AF for the output layer is chosen to be linear, which is

$$f(\alpha) = \alpha. \tag{13.2}$$

The ANN is built in Python, using the Keras library with Tensorflow backend (Please also see Section 14.10.). The training dataset used is generated from the VSC models (in MATLAB), derived in Chapter 9 and 10. With the help of these models, we are able to generate a large training set by varying each parameter from −10% to +10% of its nominal values. In this work, we have a total number of 2000 samples for training the neural network. Each sample set contains voltage and

Table 13.2 Hyperparameters of the proposed ANN.

Parameter	Specifications
Number of epochs	100
Batch size	128
Cost function	Mean Square Error (MSE)
Optimizer	Adaptive Moment Estimation (ADAM)

current harmonics. Note that we have divided the training dataset into several batches, and 128 samples constitute one batch size. The individual nodes are dropped out of the network with probability, 0.5. Only the reduced network is trained on the data in that stage. The removed nodes are then reinserted into the network with their original weights. Table 13.2 lists the hyperparameters used.

The Python code of the ANN model is listed as follows, which uses Keras as an application programming interface (API) with Tensorflow as the backend.

```
from keras.models import Sequential
from keras.layers import Dense
from keras.layers import Dropout

model=Sequential()
model.add(Dense(units=para_num*16, init = 'uniform' , activation = 'relu',
 input_dim = har_num*2))
model.add(Dense(units=para_num*8, init = 'uniform' , activation = 'relu' ))
model.add(Dense(units=para_num*4, init = 'uniform' , activation = 'relu' ))
model.add(Dense(units=para_num*2, init = 'uniform' , activation = 'relu' ))
model.add(Dropout(0.5))
model.add(Dense(output_dim=para_num,init='uniform',activation='linear'))
model.compile(optimizer='adam' , loss='mse')
model.summary()

history = model.fit(X_train,y_train, epochs=epochs_, batch_size=batch_size_,
                        validation_split=0.1,callbacks=[cb])
sd.plot_history(history)
```

The main type of model is a sequence of layers called `Sequential`, which is a linear stack of layers. `Dense` is a function for fully connected layers. The argument `units` is a positive integer, representing the dimensionality of the output space. `para_num` is the number of parameter. The argument `init` is the initializer, defining the way to set the initial random weights of the network layer. The shape of the input of the first layer is defined by `input_dim`, which is equal to `har_num`*2, where `har_num` is the number of harmonics. Once the model is created, we can configure the model with losses and metrics with `model.compile`, and train the model with `model.fit`. Note that $|x_{train}|$ and $|y_{train}|$ are the input and output of the ANN that is in training.

13.3 Estimation Validations

In this section, the parameter estimator will be validated in experiment and with an offline simulator such as PSCAD/EMTDC. Moreover, the commonly used optimization technique for parameter estimation, PSO, is also implemented to contrast the performance of the presented white-box-and-black-box model. To solve such a problem using PSO, an objective function is required, which is defined in (13.4).

13.3.1 Experimental Validation

The experimental data of the IEEE benchmark system for the D-STATCOM [100] are used to test the parameter estimator. Fig. 13.1 shows the circuit diagram and Figs. 9.5 and 9.6 are the control structures for the inner current and outer dc voltage controls. Fig. 13.4 shows the experimental PCC voltage, v_{pcc}, and converter current waveforms, i_t of phase a.

The parameters in this case to be estimated are L_s, L_f, R_s, R_f, C_{dc}, K_c, τ_c, K_v, τ_v, and τ_f. Table 13.3 lists the predicted values of these ten parameters. As the table shows, all the predicted values are

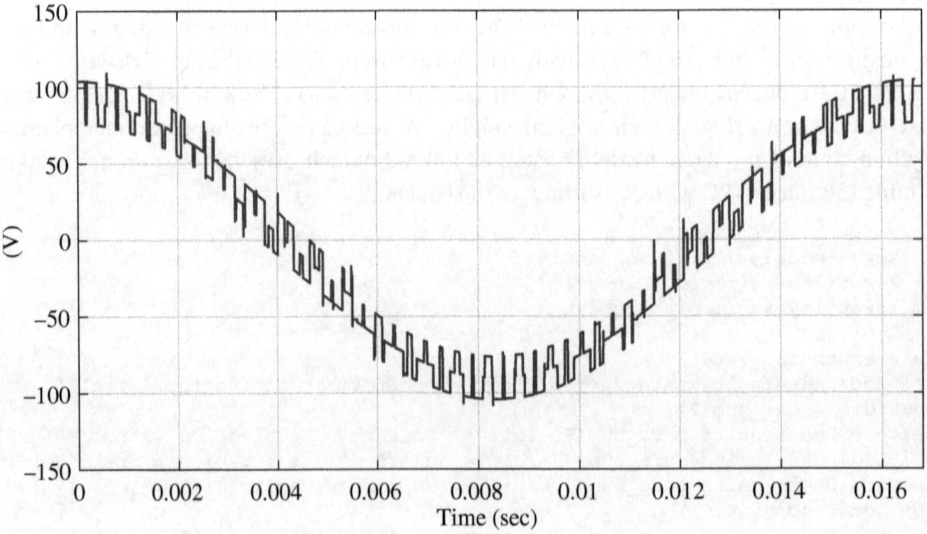

(a)

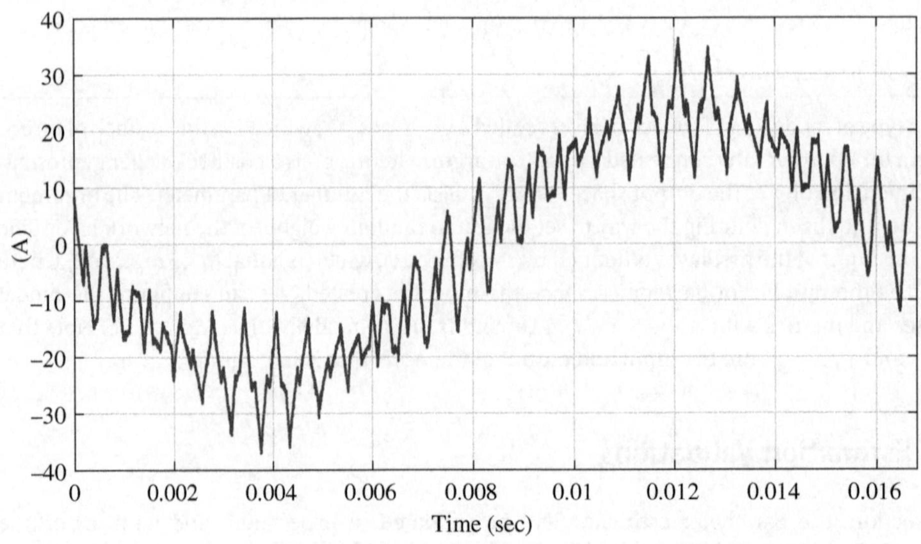

(b)

Figure 13.4 Experimental steady-state waveforms for the VSC: (a) PCC voltage; (b) converter current.

Table 13.3 Parameter estimation by the white-box-and-black-box method (experimental data).

Parameter	Estimated values	Actual values	Error rates, *E* (%)
L_s	1.5101×10^{-4}	1.4500×10^{-4}	4.1471
L_f	5.8928×10^{-4}	6.3700×10^{-4}	7.4906
R_s	0.0255	0.0255	0.0616
R_f	0.1502	0.1500	0.1565
C_{dc}	4.8144×10^{-3}	4.820×10^{-3}	0.1170
K_c	0.04256	0.04245	0.2119
τ_c	0.1003	0.1000	0.2862
K_v	0.500	0.5000	0.0000
τ_v	0.0398	0.0400	0.5323
τ_f	1.9963×10^{-3}	$1.9895 \times 10^{-}3$	0.3767
		Average error rate, $\bar{E}$	1.3380

quite close to the actual values. The error rate, E, is defined as (13.3).

$$E = \left| \frac{\xi' - \xi}{\xi} \right| \times 100\%, \tag{13.3}$$

where ξ and ξ' represent the actual and estimated values, respectively. As indicated in Table 13.3, the total average error, $\bar{E}$, is as small as 1.338%, demonstrating the validity of the white-box-and-black-box model.

These parameter values can also be found using PSO as suggested in [136]. To solve such a problem using PSO, an objective function is required, which is defined in (13.4) to minimize the cost function f_c.

$$Min\{f_c\} = Min\left\{ \sum_{n=1}^{N} w_n (X_{PMU,n} - X_{MDL,n})^2 \right\}, \tag{13.4}$$

where w_n is the weighting factor for the n-th harmonic, $X_{PMU,n}$ represents the measured voltage (V) or current (I) of the n-th harmonic recorded at PMU or PQA, and $X_{MDL,n}$ represents the voltage or current of the n-th harmonic obtained by the mathematical models. This objective function can be solved by PSO, whose code is listed as follows:

```
% Nominal value of Ls, Lf,  Rs, Rf, total R, total L, Cdc
Ls=145e-6;
Lf=0.637e-3;
Rs=0.0255;
Rf=0.15;
R_nominal = Rf+Rs;
L_nominal = Lf+Ls;
C_nominal = 4820e-6;
Pc_nominal=0.04247;
Tc_nominal=0.1;
Pv_nominal=0.5;
Tv_nominal=1/25;
swarm_size=50;               % Number of swarms
```

```
%initial value
for i=1:swarm_size
    swarm_x(1,i) = R_nominal*(0.9+0.2*rand);
    swarm_x(2,i) = L_nominal*(0.9+0.2*rand);
    swarm_x(3,i) = C_nominal*(0.9+0.2*rand);
    swarm_x(4,i) = Pc_nominal*(0.9+0.2*rand);
    swarm_x(5,i) = Tc_nominal*(0.9+0.2*rand);
    swarm_x(6,i) = Pv_nominal*(0.9+0.2*rand);
    swarm_x(7,i) = Tv_nominal*(0.9+0.2*rand);

    v(1,i) = R*unifrnd(0,0.2);
    v(2,i) = L*unifrnd(0,0.2);
    v(3,i) = C*unifrnd(0,0.2);
    v(4,i) = Pc*unifrnd(0,0.2);
    v(5,i) = Tc*unifrnd(0,0.2);
    v(6,i) = Pv*unifrnd(0,0.2);
    v(7,i) = Tv*unifrnd(0,0.2);

end
for i=1:swarm_size
    pbest_val(:,i)=[inf;inf;inf];
end
pbest=zeros(parameter_num,swarm_size);
% pso parameter
c1=1.5;c2=1.5;
w_max=0.9;w_min=0.4;
s=1;conver=0;
iteration=100;    % Max iteration number
S=iteration;
tol=1e-5;         % Tolerance
parameter_num=7; % Number of parameters
harnum=16;        % Number of harmonics to be considered

% The condition of iteration convergence
while converge<=converge_times && s<iteration

     w(s,1)=w_max-(w_max-w_min)*s/S;    % Linear decrease
     for i=1:swarm_size
          [Ialpha Ibeta] = Closed_loop_VSC_model(swarm_x(:,i));
          value(1,i) =  0;

% Calculate the sum of squares of the differences (Objective function)
                    for k=1:harnum
                         value(1,i)=(real(Ialpha_measurement(k)-Ialpha(k)))^2+...
                         (imag(Ialpha_measurement(k)-Ialpha(k)))^2+value(1,i);
                    end

          %if value i < pbest ,update pbest to this value
          if value(1,i)<pbest_val(1,i);
                    pbest(:,i)=swarm_x(:,i);      % update pbest position
                    pbest_val(:,i)=value(:,i);    % update pbest value
          end

     end
     if s==1
          [var Index]=min(value(1,:));
           gbest_val(:,s)=value(:,Index);
           % In the first iteration, gbest equal to minimum pbest
           gbest(:,s)=swarm_x(:,Index);
     else %s~=1
          % if value < gbest ,update gbest to this value
```

```
        [var Index]=min(value(1,:));
        if var<gbest_val(1,s-1)
                gbest_val(:,s) =value(:,Index);
                gbest(:,s)=swarm_x(:,Index);
        else
                gbest_val(:,s)=gbest_val(:,s-1);
                gbest(:,s)=gbest(:,s-1);
        end
    end
    vnew=zeros(parameter_num,swarm_size);
     % r1, r2 is random number
    r1=unifrnd(-1,1);r2=unifrnd(-1,1);
    for i=1:swarm_size
        for n=1:parameter_num
            vnew(n,i)=w(s,1)*v(n,i)+c1*r1*(pbest(n,i)-swarm_x(n,i))+...
            c2*r2*(gbest(n,s)-swarm_x(n,i));  update  formula
        end
    end
    swarm_x=swarm_x+vnew;
    v=vnew;
    if s==1
        converge=0;
    else
        if abs(sum(gbest(:,s))-sum(gbest(:,s-1)))<tol;
          % If the difference between the gbest of this iteration and
          % the previous gbest is less than tolerance number, increase
          % the number of convergence times
                converge=converge+1;
         else
                converge=0;
         end
    end
    s=s+1
end
```

Table 13.4 lists the estimated values obtained by PSO. As can be seen, most of these values deviate from the actual values and the average error rate is about 8.2836%. This is because the formulation based on (13.4) is difficult to account for the parameter variations due to the temperature, operating conditions, and so on. Hence, the deviations between the measured data and the mathematical model can cause the PSO to converge to values quite different from the actual ones.

13.3.2 PSCAD/EMTDC Validation

Fig. 13.5 shows a VSC with an LCL filter, which is simulated in PSCAD/EMTDC. Fig. 13.6 shows the waveforms for phase a of the PCC voltage (v_{pcca}), source current (i_{sa}), and converter current (i_{ta}) of phase a.

According to IEEE Std C37.118.1-2011, the TVE should be less than 1%. TVE is a vectorial difference between the measured and expected value of the phasors, which is mathematically defined as

$$TVE = \frac{|\tilde{y} - y|}{|y|} \times 100\%, \tag{13.5}$$

where $\tilde{y}$ and y represent measured and actual phase values, respectively.

The TVE combines errors from the magnitude and phase angle measurements. Their relationship can be stated in (13.6) [143]

$$TVE = \sqrt{2(1 \pm \lambda)(1 - cos(\gamma)) + \lambda^2}, \tag{13.6}$$

Table 13.4 Parameter estimation by PSO (experimental data).

Parameter	Estimated values	Actual values	Error rates, *E* (%)
L_s	1.3052×10^{-4}	1.4500×10^{-4}	9.9872
L_f	7.2892×10^{-4}	6.3700×10^{-4}	14.4304
R_s	0.0230	0.0255	9.8538
R_f	0.1350	0.1500	10.0062
C_{dc}	5.2519×10^{-3}	4.820×10^{-3}	8.9622
K_c	0.0460	0.04245	8.2711
τ_c	0.0912	0.1000	8.7794
K_v	0.5085	0.5000	1.6928
τ_v	0.0383	0.0400	4.2978
τ_f	2.1199×10^{-3}	1.9895×10^{-3}	6.5546
		Average error rate, $\bar{E}$	8.2836

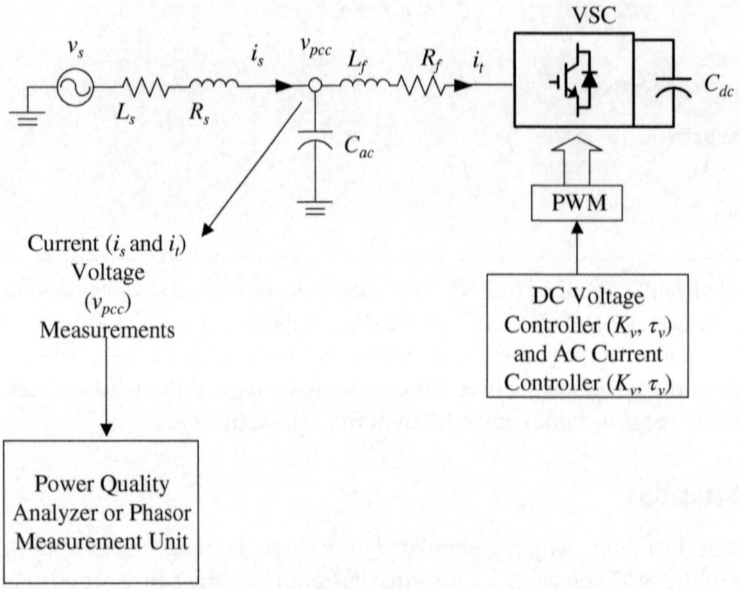

Figure 13.5 The schematic diagram of a VSC with an LCL filter.

where λ is the magnitude error and γ is the phase angle error. We have added a very severe condition of TVE of 0.99% to all the harmonic phasors of the voltage and currents of Fig. 13.6, and fed them into the trained ANN to see if the parameters can be accurately estimated. We have divided the case into three scenarios: (1) the amplitude error for the TVE (denoted as λ) is 0.99% while the phase angle error for the TVE (denoted as γ) is 0%; (2) $\gamma = 0.99\%$ (or 0.573°) while $\lambda = 0\%$; (3) $\lambda = \sqrt{0.499}\%$ or 0.706% while $\gamma = 0.706\%$.

The parameters to be estimated in this case are the ac capacitor, C_{ac}, L_s, L_f, R_s, R_f, C_{dc}, K_c, τ_c, K_v, τ_v, and τ_f, as shown in Fig. 13.5. Table 13.5 lists the predicted values of these 11 parameters

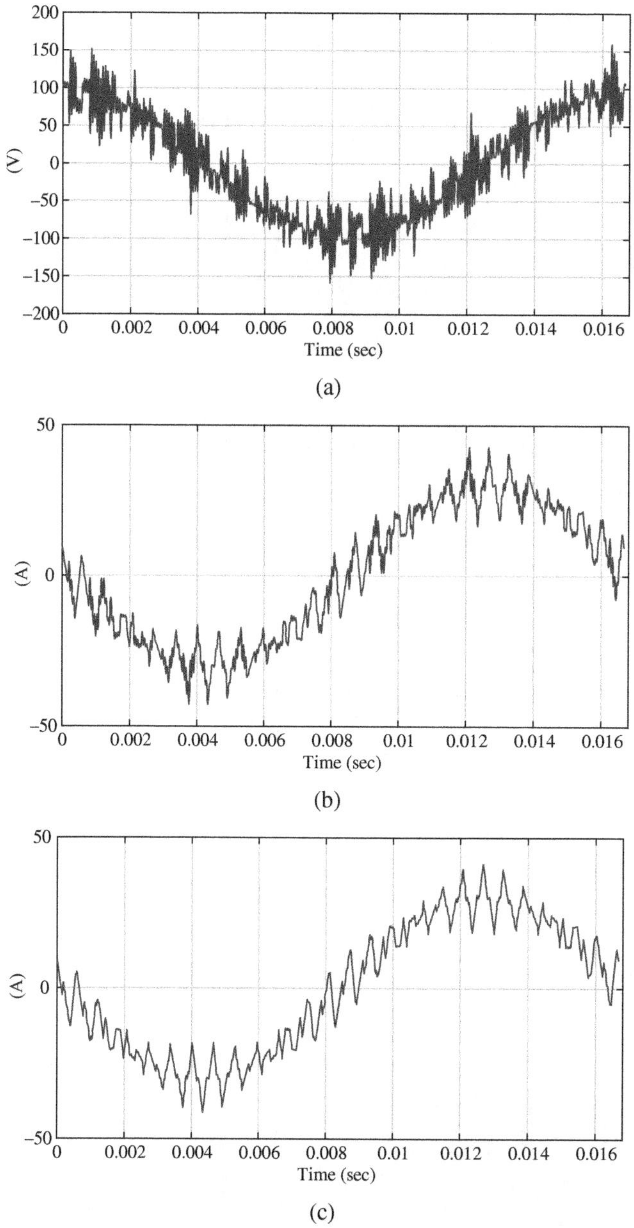

Figure 13.6 The steady-state voltage and current waveforms for phase a of a VSC with an LCL filter simulated by PSCAD/EMTDC: (a) PCC voltage (v_{pcca}); (b) source current (i_{sa}); (c) converter current (i_{ta}).

and their errors. It is to be noted that the largest value of these eleven parameters is τ_v, which is 2.1049 whereas the lowest one is C_{ac}, which is 1 μF. Hence, the largest parameter is about 2.1049 × 10^6 larger than the smallest one. Nevertheless, the proposed method can still yield satisfactory performance under such a challenging condition. As Table 13.5 indicates, the error rates on average are about 0.3%. On the other hand, as Table 13.6 indicates, the average error rates for PSO are about

Table 13.5 Parameter estimation by the white-box-and-black-box model for a two-level VSC with LCL filter (PSCAD/EMTDC data with various TVE conditions).

Parameter	Actual values	Estimated values ($\lambda = 0.99\%$; $\gamma = 0\%$)	Error rates, E_{λ} (%)	Estimated values ($\lambda = 0\%$; $\gamma = 0.99\%$)	Error rates, E_{γ} (%)	Estimated values ($\lambda = 0.706\%$; $\gamma = 0.706\%$)	Error rates, $E_{\lambda\gamma}$ (%)
L_s	1.4500×10^{-4}	1.4530×10^{-4}	0.2065	1.4525×10^{-4}	0.1728	1.4559×10^{-4}	0.4095
L_f	6.3700×10^{-4}	6.4356×10^{-4}	1.0297	6.4284×10^{-4}	0.9137	6.4607×10^{-4}	1.4237
R_s	0.0255	0.0255	0.0284	0.0255	0.0393	0.0255	0.0013
R_f	0.1500	0.1506	0.4315	0.1507	0.4451	0.1504	0.3596
C_{ac}	1.0000×10^{-6}	1.0006×10^{-6}	0.0577	1.0006×10^{-6}	0.0584	1.0008×10^{-6}	0.0835
C_{dc}	0.001	9.9683×10^{-4}	0.3169	9.9660×10^{-4}	0.3395	$9.9880e - 4$	0.1202
K_c	0.25	0.2497	0.0821	0.2499	0.0356	0.2496	0.1494
τ_c	0.0248	0.0247	0.4125	0.0247	0.3830	0.0247	0.2469
K_v	0.2766	0.2757	0.3125	0.2757	0.3187	0.2762	0.1513
τ_v	2.1049	2.1087	0.1783	2.1076	0.1285	2.1105	0.2681
τ_f	1.9895×10^{-3}	1.9826×10^{-3}	0.3443	1.9826×10^{-3}	0.3475	1.9837×10^{-3}	0.2883
		Average error rate, $\bar{E}_{\lambda}$ (%)	0.3020	Average error rate (%), $\bar{E}_{\gamma}$	0.2893	Average error rate, $\bar{E}_{\lambda\gamma}$ (%)	0.3183

Table 13.6 Parameter estimation by the PSO for a two-level VSC with LCL filter (PSCAD/EMTDC data with various TVE conditions).

Parameter	**Actual values**	**Estimated values ($\lambda = 0.99\%$; $\gamma = 0\%$)**	**Error rates, E_λ (%)**	**Estimated values ($\lambda = 0\%$; $\gamma = 0.99\%$)**	**Error rates E_γ (%)**	**Estimated values ($\lambda = 0.706\%$; $\gamma = 0.706\%$)**	**Error rates, $E_{\lambda\gamma}$ (%)**
L_s	1.4500×10^{-4}	1.5809×10^{-4}	9.0306	1.5864×10^{-4}	9.4049	1.5866×10^{-4}	9.4203
L_f	6.3700×10^{-4}	7.0059×10^{-4}	9.9826	7.0050×10^{-4}	9.9685	7.0024×10^{-4}	9.9285
R_s	0.0255	0.0264	3.3818	0.0280	9.7491	0.0251	1.4709
R_f	0.1500	0.1489	0.7362	0.1395	6.9983	0.1601	6.7254
C_{ac}	1.0000×10^{-6}	1.0095×10^{-6}	0.9539	1.0156×10^{-6}	1.5562	1.1000×10^{-6}	9.9998
C_{dc}	0.001	1.0437×10^{-3}	4.3762	9.5154×10^{-4}	4.4860	1.0260×10^{-3}	2.6039
K_c	0.25	0.2555	2.1811	0.2729	9.1696	0.2500	0.0047
τ_c	0.0248	0.0261	5.2902	0.025	0.6092	0.0259	4.6277
K_v	0.2766	0.2855	3.2241	0.2530	8.5183	0.2689	2.7994
τ_v	2.1049	2.2628	7.5036	2.1058	0.0422	2.0856	0.9158
τ_f	1.9895×10^{-3}	2.1340×10^{-3}	7.2627	2.1775×10^{-3}	9.4522	1.9443×10^{-3}	2.2722
		Average error rate, $\bar{E}_\lambda$ (%)	4.9021	Average error rate, $\bar{E}_\gamma$ (%)	6.3922	Average error rate, $\bar{E}_{\lambda\gamma}$ (%)	4.6153

4.6%–6.4%. This shows that the proposed method can accurately estimate both control and circuit parameters, whose values may vary greatly for a VSC with LCL filter, under various severe TVE conditions. Note that although the estimated values of K_v under the three TVE conditions listed in Table 13.5 are all 0.2757, their error rates differ because rounding occurred after the mathematical operations were completed.

14

Parameter Estimation for Multiple VSCs with Domain Adaptation

"Seek success, but prepare for vegetables."

– An AI-driven bot

Chapter 13 discusses how we can use simulation to generate a large number of training data for constructing an ANN model. We have shown that such a model still yields accurate parameter estimation (PE) even when the testing data fed to the model are the actual experimental measurement data. However, if the measurement data are contaminated with severe noises, we need to use domain adaptation (DA) techniques from deep learning (DL) to narrow the gap between simulation and measurement data. We believe that DA, which bridges the simulation and real-world domains, will find its useful roles in many power system applications. Parameter estimation PE is just one of them!

This chapter is to extend the work of Chapter 13 such that the following objectives can be achieved:

1. a wide range of parameters of a VSC can be estimated
2. the parameters of parallel-connected multiple VSCs can be estimated
3. the estimation can still be accurate even if severe noise is present in the measurement data.

It is to be noted that an ANN model would have difficulty in achieving these three objectives. Thus, this chapter uses DL methods for achieving them. The organization of this chapter is as follows: Section 14.1 provides a brief introduction on DL. Section 14.2 gives a brief introduction on DA. Section 14.3 describes the tasks of parameter estimations of multiple VSCs. Section 14.4 prepares the reader with notations commonly used for DA. Sections 14.5 and 14.6 describe the models addressing the tasks of regression and classification of DA. Section 14.7 describes how the training and test data are generated. Section 14.8 introduces metrics for evaluating the performance of the models. Section 14.9 discusses the results of various models for DA. Section 14.10 describes the software required to run the codes provided in Section 14.11. The code examples provided in this chapter are written in Python and MATLAB.

14.1 Introduction of Deep Learning

In general, artificial intelligence (AI), machine learning (ML), and DL are related as follows: DL is a subset of ML, which itself is a subset of AI [144], as shown in Fig. 14.1.

AI refers to the simulation of human intelligence in machines that are programmed to think like humans and mimic their actions. ML is a branch of AI that builds mathematical models based

Harmonic Modeling of Voltage Source Converters using Basic Numerical Methods, First Edition.
Ryan Kuo-Lung Lian, Ramadhani Kurniawan Subroto, Victor Andrean, and Bing Hao Lin.

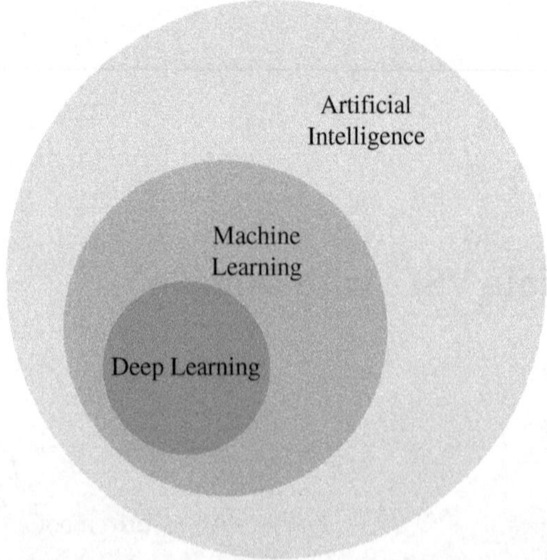

Figure 14.1 Relationship of DL, ML, and AI.

on sample data, known as training data, in order to make predictions or decisions without being explicitly programmed to do so. DL is an ML method that takes in an input X, and uses it to predict an output of Y. Given a large dataset of input and output pairs, a DL algorithm tries to minimize the difference between its prediction and expected output. By doing this, it tries to learn the association/pattern between given inputs and outputs, which allows a DL model to generalize the inputs that it has not seen before. Traditional ML models have always been very powerful to handle structured data and have been widely used in many applications. The success of these models is highly dependent on the performance of the feature engineering. The more we extract relevant knowledge from the structured data, the more powerful the model will be. When it comes to unstructured data (images, text, voice, etc.), hand-engineered features are time-consuming, brittle, and not scalable in practice. That is why DL is becoming more and more popular, thanks to its ability to automatically discover the representations needed for feature detection or classification from the raw data. This allows a machine to both learn the features and use them to perform a specific task. DL algorithms use ANN to find associations between a set of inputs and outputs. The "deep" part of DL refers to creating a deep neural network (DNN), which refers to a neural network with a large amount of layers, allowing it to approximate more complex functions.

DL generally can be categorized into three types: supervised, unsupervised, and reinforcement learning.

In a supervised learning model, the algorithm learns on a labeled dataset, providing an answer key that the algorithm can use to evaluate its accuracy on the training data. There are two main areas where supervised learning is useful: classification problems and regression problems. Classification problems ask the algorithm to predict a discrete value, identifying the input data as the member of a particular class or group. On the other hand, regression problems look at the continuous data.

In unsupervised learning, a DL model is handed a dataset without explicit instructions on what to do with it. The training dataset is a collection of examples without a specific desired outcome or correct answer. The neural network then attempts to automatically find structure in the data by extracting useful features and analyzing its structure. Depending on the problem at hand, the unsupervised learning model can organize the data in the following ways:

- clustering: the DL model looks for training data that are similar to each other and groups them together.
- anomaly detection: unlike clustering, unusual or dissimilar patterns can also be detected by the DL model.
- association: by looking at key attributes of a data point, an unsupervised learning model can predict the other attributes with which they are commonly associated.

In reinforcement learning, the learning agent works as a reward and action system. In this kind of ML, AI agents attempt to find the optimal way to accomplish a particular goal or improve performance on a specific task. As the agent takes action that goes toward the goal, it receives a reward. The overall aim is to predict the best next step to earn the biggest final reward.

14.2 Domain Adaptation

The accuracy of DL algorithms relies greatly on the amount of labeled training data. Collecting and annotating datasets for every new task can be an extremely expensive and time-consuming process. Moreover, sufficient training data may not always be available. Thus, one strategy is to make use of a pre-trained network on a large dataset for a similar problem, and we tune some of the parameters or layers using a small annotated dataset. This is called domain adaptation (DA). DA is a type of transfer learning in DNN [145] to leverage off-the-shelf-labeled data from a related domain (source domain) to improve the model for the domain of interest (target domain). The target domain may contain data collected by different sensors or under different conditions compared with the source domain, leading to a large domain gap. DA aims to minimize the domain gap and successfully transfer the model trained on the source domain to the target domain.

DA can be either homogeneous (the input feature spaces are the same, but have different data distributions) or heterogeneous (the feature spaces and dimensionalities may also differ). DA may also occur in one step (one-step DA), or through multiple steps, traversing one or more domains in the process (multi-step DA). Depending on the data availability from the target domain, DA can further be classified into supervised (i.e. one has labeled data from the target domain, but the amount is too small for training a whole model), semi-supervised (i.e. one has both labeled and unlabeled data), and unsupervised (i.e. one does not have any labeled data from the target domain). This chapter mainly addresses the problem of homogeneous supervised DA.

The following are several examples of DA or transfer learning across different fields.

- An example in a normal human life: A person who knows how to ride a bicycle can ride a motorcycle, after a few training sessions.
- A local health care center collects patient information in order to predict the probability of a patient developing a particular disease. After training a learning algorithm on this dataset, the resulting system is deployed in all hospitals in the whole nation. Researchers from [146] have suggested using similar approaches to deal with the COVID-19 pandemic.

- A spam filter is learned using a classification algorithm for a corporate mailbox of a given user. If this user installs this filter in his personal computer, DA is usually applied to narrow the difference between the contents of corporate and personal emails for the spam filter to work well for personal emails [147].
- A Photo-Sketch Face Recognition (PSFR) system may help law enforcement to quickly identify suspects [148]. A PSFR system utilizes a large database of face photos to train a DNN, which is then fine-tuned using a small database of sketches.

14.3 Parameter Estimation for Multiple VSCs

The parameter estimator presented in Chapter 13 is shown to estimate even up to a controller's proportional and integral gains. In practice, such an estimate may not be necessary or feasible if the structure of the controller is not known, and we can reduce the number of parameters by merely estimating the modulating signals, issued by the controller. Furthermore, in a microgrid system or a system with high renewable penetration, it is common to see multiple VSCs connected in parallel as shown in Fig. 14.2. Unlike a single VSC operating alone, the harmonics of different VSCs may interfere with each other and is more challenging to deal with for PE. In fact, the challenge lies in the fact that in most of the cases, only the aggregated rather than the individual current are available for a multiple VSC system, as indicated in Fig. 14.2. Thus, unlike Chapter 13, the capacitor voltages of individual VSCs need to be sensed in order for the DNN to tell the difference between these converters.

As mentioned in Chapter 13, to achieve high-performance PE in a time-saving and less expensive way, accurate steady-state models of VSCs are first derived (please refer to Chapters 8 to 11).

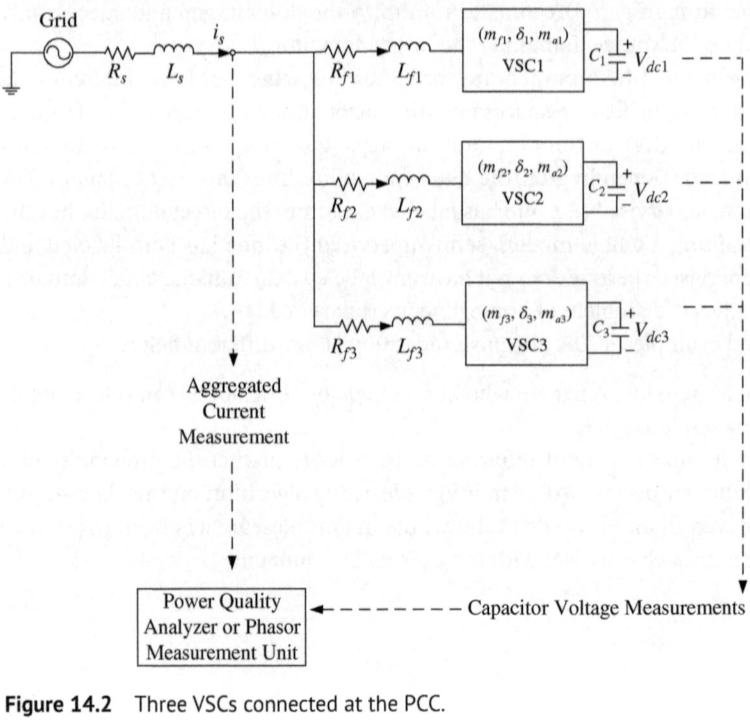

Figure 14.2 Three VSCs connected at the PCC.

This step plays an important role because this model is used to generate the training dataset as the source domain dataset. Then, only a few data are used as target domain training. To make the case more challenging, we assume the measurement data (i.e. the target domain data) to be quite different from the source domain data by adding aggressive noise to the simulation data while keeping their conditional probability distribution the same (this will be explained in Section 14.4).

The parameters of the VSC to be estimated consist of m_f, R (R_f and R_s), L (L_f and L_s), C, m_a, δ. Among them, m_f usually takes an integer. On the other hand, R, L, C, m_a, and δ take real values. Hence, we need two models for addressing regression and classification tasks. For the regression problem, a state-of-the-art method called Weighting Adversarial Neural Network (WANN) [149] is used. For the classification, a method proposed in [150] is employed.

14.4 Notations for DA

In this section, we introduce some of the commonly used notations in DA. Let a pair ($\boldsymbol{X}$,Y) define the input and the output spaces. In our case, $\boldsymbol{X}$ are real-valued vectors of finite dimension d, i.e. $\boldsymbol{X} \subseteq R^d$. Y in turn can be continuous or discrete. When Y is continuous, we are considering the regression problem. When Y is discrete, the classification problem is concerned. A hypothesis space $\mathcal{H}$ (also called hypothesis class) consists of functions that map each element of $\boldsymbol{X}$ to Y. A function or a hypothesis h is a classifier or regressor depending on the nature of Y. Given the training space, the goal of a learner is to find a good hypothesis $h \subseteq \mathcal{H}$ that captures the best relationship between $\boldsymbol{X}$ and Y [151]. A domain $\mathcal{D}$ is composed of a feature space with a marginal probability distribution $P(\boldsymbol{X})$. A task is defined by a label space Y with the conditional probability $P(Y|\boldsymbol{X})$. Table 14.1 shows different categories of deep learning strategies based on the types of domains and tasks. As the table indicates, when both the source and target domains are the same, and the tasks of the two domains are the same, this is traditional machine learning. If both the source and target domains are the same, but the tasks are different, this is called multitask learning [152], where multiple tasks are simultaneously learned by a shared model. If both the source and target domains are different but the tasks are the same, this is DA. Finally, if domains and tasks of the two domains are all different, this is the most general case of transfer learning. Note that DA and multitask learning are all special cases of transfer learning.

In order to measure how well the hypothesis reveals the relationship, we consider a loss function that gives a cost of $h(\boldsymbol{x})$ ($\boldsymbol{x} \in \boldsymbol{X}$), deviating from the output. For example, in regression, the squared error loss function as defined in (14.1) is often used:

$$l(y, \hat{y}) = (y - \hat{y})^2, \tag{14.1}$$

where $\hat{y}$ is an approximation of y.

Table 14.1 Types of training.

Source and target domains	Tasks	Categories
Same	Same	Traditional machine learning
Same	Different	Multitask learning
Different	Same	Domain adaptation
Different	Different	Most generalized cases of transfer learning

In binary classification, the 0/1 loss function, which is defined as (14.2), is often used:

$$y = \begin{cases} 1 & y \neq \hat{y} \\ 0 & y = \hat{y} \end{cases} \tag{14.2}$$

Then, to capture the notion of the loss over all samples coming from some probability distribution, we define the true risk as follows: Given a loss function, l, the true risk for a given hypothesis h on a distribution $\mathcal{D}$ over $\boldsymbol{X} \times Y$, $L_{\mathcal{D}}(h)$ is defined as

$$L_{\mathcal{D}}(h) = \mathop{\mathbb{E}}_{(\boldsymbol{x},y)\in \mathcal{D}} \left[\ell(h(\boldsymbol{x}), y)\right], \tag{14.3}$$

where $\mathbb{E}$ denotes statistical expectation. Note $L_{\mathcal{D}}(h)$ is used to measure the loss over a certain distribution (i.e. $\mathcal{D}$ can be either source domain ($\mathcal{D}_S$) or target domain ($\mathcal{D}_T$)). Nevertheless, for DA, which concerns the relationship between $\mathcal{D}_S$ and $\mathcal{D}_T$, we need another notion, which is called discrepancy. However, before we talk about discrepancy, an important concept of DA, covariate shift, needs to be discussed first. For an $\boldsymbol{X} \rightarrow Y$ problem, which means the class label is causally determined by the values of the instances, the joint probability distribution of samples of $\mathcal{D}_S$ and $\mathcal{D}_T$ are denoted as $P_s(\boldsymbol{X}, Y)$ and $P_t(\boldsymbol{X}, Y)$, respectively. Mathematically, $P_s(\boldsymbol{X}, Y)$ and $P_t(\boldsymbol{X}, Y)$ can be expressed as (14.4) and (14.5), respectively.

$$P_s(\boldsymbol{X}, Y) = P_s(\boldsymbol{X})P_s(Y|\boldsymbol{X}) \tag{14.4}$$

$$P_t(\boldsymbol{X}, Y) = P_t(\boldsymbol{X})P_t(Y|\boldsymbol{X}), \tag{14.5}$$

where $P_s(\boldsymbol{X})$ and $P_t(\boldsymbol{X})$ are the marginal probability distributions of the source and target, respectively, while $P_s(Y|\boldsymbol{X})$ and $P_t(Y|\boldsymbol{X})$ are the conditional probability distributions of the source and target, respectively. Covariate shift means that $P_s(\boldsymbol{X}) \neq P_t(\boldsymbol{X})$ while $P_s(Y|\boldsymbol{X}) = P_t(Y|\boldsymbol{X})$. In the case of the PE, we have an $\boldsymbol{X} \rightarrow Y$ problem with covariate shift because even though the marginal probability distribution of the harmonics in the source and target domains are different, their conditional probability distribution of the harmonics are assumed to be the same, given a particular set of R, L, C, m_a, δ, and m_f. Nevertheless, the source and target distributions should be somehow related; otherwise, the learning would be impossible.

Fig. 14.3 shows an intuitive explanation of DA [153]. Assume that squares and circles represent classes A and B, respectively. The grey-shaded markers represent the misclassified samples. In Fig. 14.3(a), hypothesis, h can distinguish classes A and B of the source domain very well, but not so well in the target domain. This is because the distributions of the source and target domains are dissimilar. After DA, h is relearned to be h'. As shown in Fig. 14.3(b), h' can perform well both on the source and target domains.

Looking at Fig. 14.3, we can see that a metric taking into account the hypothesis sets and loss functions is needed. Mansour et al. [154] defined the discrepancy distance as

$$disc(\mathcal{D}_S, \mathcal{D}_T) = \max_{h\in\mathcal{H}} \left|\mathcal{L}_{\mathcal{D}_T}(h, h') - \mathcal{L}_{\mathcal{D}_S}(h, h')\right|. \tag{14.6}$$

Ben-David et al. defined a similar concept, called d_A distance, which can be found in [155].

Fig. 14.4 provides an example of situations where *disc* is small or big. For illustration purposes, we assume that h and h' take the values as indicated in Fig. 14.4(a). The source and target distributions are represented by circles as shown in the figure. The shaded areas represent the situations where h and h' differ. Low values of *disc* are observed in Figs. 14.4(a) and (b) whereas Fig. 14.4(c) yields a large value of *disc*. It is interesting to note that *disc* is also small when both source and target domain are in the grey shaded areas as shown in Fig. 14.4(a). In the next section, we will see how the concept of discrepancy plays an essential role in DA.

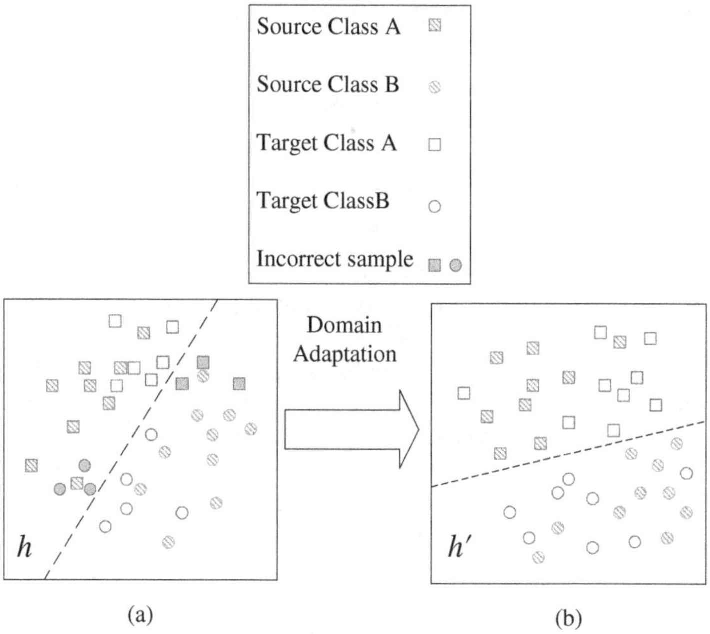

Figure 14.3 A simple illustration of DA.

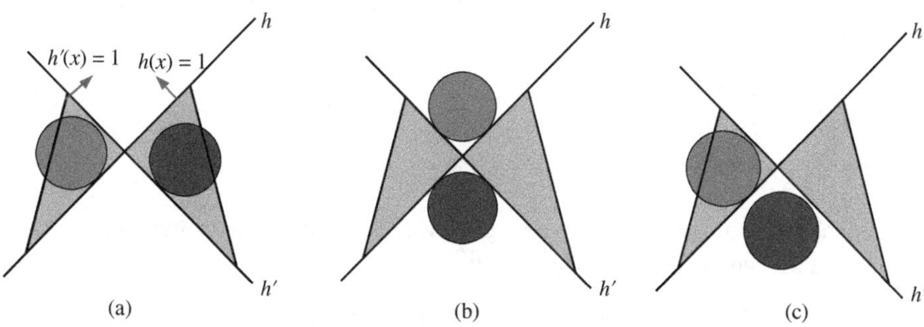

Figure 14.4 Discrepancy between source and target domains.

14.5 Supervised Domain Adaptation for Regression

Generally, there are two broad families of DA algorithms. The first is to map the source and target data into a new feature space where the difference between source and target distributions is much reduced. Such a method will be discussed in Section 14.6. The other algorithms are to use re-weighting techniques. These algorithms are essentially instance-based methods, which re-weight the loss of source instances in order to minimize a distance between source and target distributions. These re-weighted instances are directly used in the target domain for training. Using the re-weighted source samples helps the target learner to use only the relevant information from the source domain. A regression DA problem is usually tackled by the re-weighting technique [149]. WANN, proposed in [149], is employed for our regression task and is briefly reviewed. WANN is based on the notion that the task on the target domain can be learned efficiently by properly re-weighting the source instance during the training phase. To achieve it, the objective function stated in (14.7) is defined:

$$E(W, h_t, h_d) = \frac{1}{n} \sum_{(\boldsymbol{x}_j, y_j) \in \boldsymbol{D}_t} (h_t(\boldsymbol{x}_j) - y_j)^2 + \frac{1}{m} \sum_{(\boldsymbol{x}_i, y_i) \in \boldsymbol{D}_s} W(\boldsymbol{x}_i)(h_t(\boldsymbol{x}_i) - y_i)^2$$
$$+ \left| \frac{1}{m} \sum_{(\boldsymbol{x}_i, y_i) \in \boldsymbol{D}_s} W(\boldsymbol{x}_i)(h_d(\boldsymbol{x}_i) - y_i)^2 - \frac{1}{n} \sum_{(\boldsymbol{x}_j, y_j) \in \boldsymbol{D}_t} (h_d(\boldsymbol{x}_j) - y_j)^2 \right|. \tag{14.7}$$

As seen in (14.7), the objective function consists of three terms. The first is simply the target risk. The second term is a weighted source risk, which is the source risk multiplied with a weighting network, W. The purpose is to select the source instance that is more relevant to the target instance. h_t is the task hypothesis for the regression task. It can be computed by optimally weighting the loss of the source instances during the training phase. h_d is called the discrepancy hypothesis, which is trained to maximize the third term. Hence, we can view h_d as providing antagonist performance while h_t is the supporter. Such a concept is similar to the generative adversarial networks [144], which are algorithmic architectures that use two neural networks, pitting one against the other in order to generate new synthetic instances of data that can pass for real data. These types of problems are generally referred to as min-max problems [156], which can be solved by gradient descent ascent (GDA) methods [157]. One of the GDA methods is the alternating GDA based on the Gauss–Seidel iteration method [158], which is formulated as follows.

The unknown variables to be solved in (14.7) are the weights of the networks, h_t, h_d, and W, which are denoted as $\boldsymbol{\theta}_t$, $\boldsymbol{\theta}_d$, and $\boldsymbol{\theta}_w$, respectively. Essentially, (14.8) is solved first by treating $\boldsymbol{\theta}_d$ as a constant. Then, the solved $\boldsymbol{\theta}_t$ and $\boldsymbol{\theta}_w$ are substituted into (14.9) to update $\boldsymbol{\theta}_d$. The process is repeated until convergence is achieved.

$$(\theta_w, \theta_t) = \underset{\theta_w, \theta_m}{\operatorname{argmin}}\, E(\theta_w, \theta_t, \theta_d) \tag{14.8}$$

$$\theta_d = \underset{\theta_w}{\operatorname{argmax}}\, E(\theta_w, \theta_t, \theta_d) \tag{14.9}$$

It should be emphasized that the weighting network, W and the discrepancy hypothesis, h_d in (14.7), are important. Two degenerate cases can result:

- $W = 0$ and (14.7) does not have the third term, it is equivalent to having only target data for training, which can leads to over-fitting because the target data are scarce.
- $W = \mathbb{I}$ and (14.7) does not have the third term, this is equivalent to having both source and target data mixed together for training, which leads to under-fitting.

14.6 Supervised Domain Adaptation for Classification

For the classification task, we adopt the method from [150]. This method is called the classification and contrastive semantic alignment (CCSA) method and makes use of Siamese architecture [159] in Fig. 14.5, which consists of two identical neural networks sharing the same set of weights. Siamese network was first introduced in the early 1990s by Bromley et al. [160] to solve signature verification as an image matching problem. The Siamese network aims to compare a pair of feature vectors in terms of their semantic similarity or dissimilarity. It realizes a non-linear embedding of data with the objective to bring together similar samples and to move apart dissimilar samples. For DA, the main job of the Siamese network is to learn to map an input space into an embedding space where

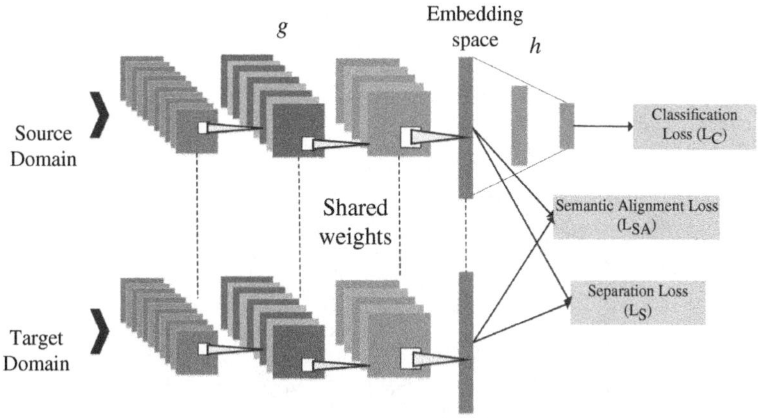

Figure 14.5 Siamese network for classification DA.

the distance between samples from different domains with the same class labels are "close" to each other and the distance between samples from different domains with different class labels are "far" from each other. This purpose can be achieved by defining a *semantic alignment loss* that minimizes the distance between samples from different domains with the same class labels and the *separation loss* that maximizes the distance between samples from different domains with different class labels. According to [150], this method has shown a remarkable speed of adaptation with only a few labeled data from target domain. As shown in Fig. 14.5, the classifier f is composed of two functions, i.e. $f = h \circ g$, where $g : \boldsymbol{X} \rightarrow \boldsymbol{Z}$ is a function, mapping from the input space, $\boldsymbol{X}$ to the embedding space, $\boldsymbol{Z}$, and $h : \boldsymbol{Z} \rightarrow Y$ is a function for mapping from the embedding space to the output label space.

The goal of function, g, is to align the distribution of the features from both domains in the embedding space. Therefore, we want samples from different domains with the same labels to be mapped in the embedding space as close as possible. Mathematically, a loss function can be defined in (14.10), which is referred to as *semantic alignment loss* [150]:

$$\mathcal{L}_{SA}(g) = \sum_{a=1}^{n} d((g(\boldsymbol{X}_a^s)), (g(\boldsymbol{X}_a^t))), \tag{14.10}$$

where n is the number of labels, $\boldsymbol{X}_a^s = \boldsymbol{X}^s | \{Y = a\}$, and $\boldsymbol{X}_a^t = \boldsymbol{X}^t | \{Y = a\}$. Moreover, d is a distance metric between the distributions of $\boldsymbol{X}^s$ and $\boldsymbol{X}^t$ in the embedding space as stated in (14.11).

$$d((g(\boldsymbol{X}_a^s)), (g(\boldsymbol{X}_a^t))) = \frac{1}{2} \sum_{i,j} \parallel g(\boldsymbol{x}_i^s) - g(\boldsymbol{x}_j^t) \parallel^2, \tag{14.11}$$

where it is assumed that $y_i^s = y_j^t = a$. Note that $y \in Y$.

To improve the performance, [150] suggested to encourage samples from different domains with different labels to be mapped as far apart as possible in the embedding space. This is referred to as *separation loss*, which is stated in (14.12):

$$\mathcal{L}_S(g) = \sum_{a,b|a \neq b} k((g(\boldsymbol{X}_a^s)), (g(\boldsymbol{X}_b^t))), \tag{14.12}$$

where k is a similarity metric that measures between the distributions of $\boldsymbol{X}_a^s$ and $\boldsymbol{X}_b^t$ in the embedding space as stated in (14.13), which adds a penalty when $g(\boldsymbol{X}_a^s)$ and $g(\boldsymbol{X}_b^s)$ (different labels) come close.

$$k((g(\boldsymbol{X}_a^s)), (g(\boldsymbol{X}_b^t))) = \frac{1}{2} \sum_{i,j} max(0, \alpha - \parallel g(\boldsymbol{x}_i^s) - g(\boldsymbol{x}_j^t) \parallel)^2, \tag{14.13}$$

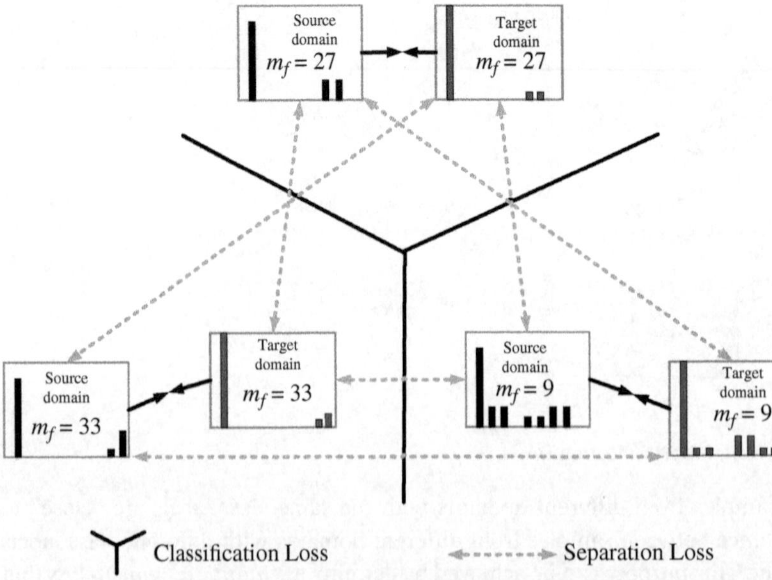

Figure 14.6 The illustration of semantic alignment loss, separation loss, and classification loss for m_f prediction.

where it is assumed that $y_i^s = a \neq y_j^t = b$ and α is the margin (in this case, we set to 1.) that specifies the separability in the embedding space.

Fig. 14.6 illustrates the different objectives of semantic and separation losses for the m_f prediction. Finally, the loss function (or the risk) for a classifier also needs to be satisfied, which is described by the cross-entropy [161] as stated in (14.14):

$$\mathcal{L}_C(f) = \frac{1}{N}\sum_{l=1}^{N}\left(-log(y_i{}^T\sigma(f(\boldsymbol{x}_i^s)))\right), \qquad (14.14)$$

where

$$\sigma(f(\boldsymbol{x}_i^s)) = \frac{e^{f(x_i^s)}}{\sum_{j=1}^{N} e^{x_j^s}}. \qquad (14.15)$$

Note that successful domain alignment can ensure that when L_C for $\boldsymbol{X}^s$ minimizes, L_C for $\boldsymbol{X}^t$ also minimizes.

To have a successful classifier, the total losses comprising the above three should be minimized. The total losses is referred to as *classification and contrastive semantic alignment* loss or *CCSA* loss [150], which is defined as (14.16):

$$\mathcal{L}_{CCSA}(f) = \mathcal{L}_C(h \circ g) + \mathcal{L}_{SA}(g) + \mathcal{L}_S(g). \qquad (14.16)$$

From now on, we will consistently refer to this method as the CCSA method.

14.7 Test Setup

In this section we will explain how the dataset is taken and how source and target domains are created for testing.

14.7.1 Data Generator

The multiple VSC system in Fig. 14.2 is modeled with OpenDSS interfaced with MATLAB as described in Chapter 11. By inputting numerous combinations of different parameters in certain ranges and obtaining the corresponding VSC harmonics, we can collect these data and create a dataset. Note that to simplify the problem, we assumed that L_s and R_s are negligible, and thereby, $L = L_f$ and $R = R_f$.

The labeled data of parameters consist of the following (for each VSC):

1. The values of R, L, and C with variations of $\pm 10\%$ of their nominal values.
2. The values of m_a whose ranges are between 0 and 1.
3. The value of δ whose ranges are between 0 and $\pm\frac{\pi}{2}$.
4. For simplicity, the values of m_f are assumed to take only 9, 15, 21, 27, and 33.

Total data collected are 10000 data. In order to create source and target domains, data are split into two groups (see Fig. 14.7); 5000 data each, in a random order. We add a 20% shift together with Gaussian noise (or random noise), whose probability density function is stated in (14.17), to the second group to form the target domain data. Fig. 14.8 shows three types of Gaussian noises with different standard deviations (STDs) of 1%, 5%, and 10%, respectively.

$$f(x) = \frac{1}{\sigma\sqrt{2\pi}} e^{-\frac{(x-\mu)^2}{2\sigma^2}}. \tag{14.17}$$

The contaminated target domain data are created with (14.18) and (14.19).

$$x_i^{shifted} = x_i \times (1 + shift) \tag{14.18}$$

$$x_i^{contaminated} = x_i^{shifted} \times (1 + noise(\%)), \tag{14.19}$$

where $noise(\%) \sim \mathcal{N}(0,\ \sigma^2)$.

Fig. 14.7 summarizes how the data are split. After having source domain data and target domain data, we then randomly sample 30 data from the target domain data. These 30 data are the training data of target domain. These are the data that will be used for training of DA method. The other 4970 data from the target domain are the test data, which are used to test the model performance. As Fig. 14.7 indicates, the training data consist of 5000 source domain data and 30 training data from the target domain.

14.7.2 Data Preprocessing

The purpose of data preprocessing is to filter out unnecessary features (or data cleaning), and to re-scale the input features so that every input feature has similar scale and variance.

When the scales of the input features are very different, the loss function will be skewed and the optimization process becomes difficult. In addition, data preprocessing in ML is a crucial step

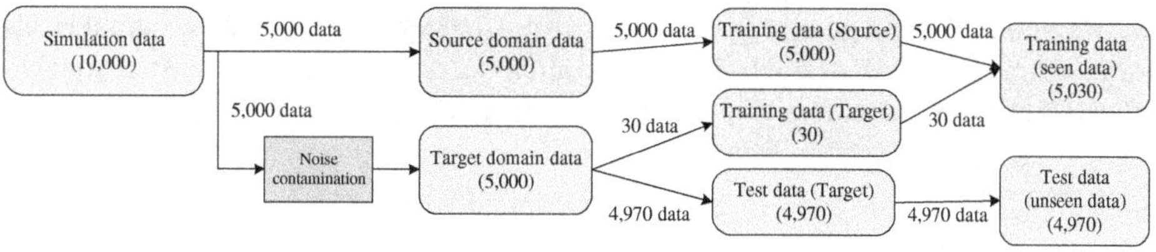

Figure 14.7 Data splitting.

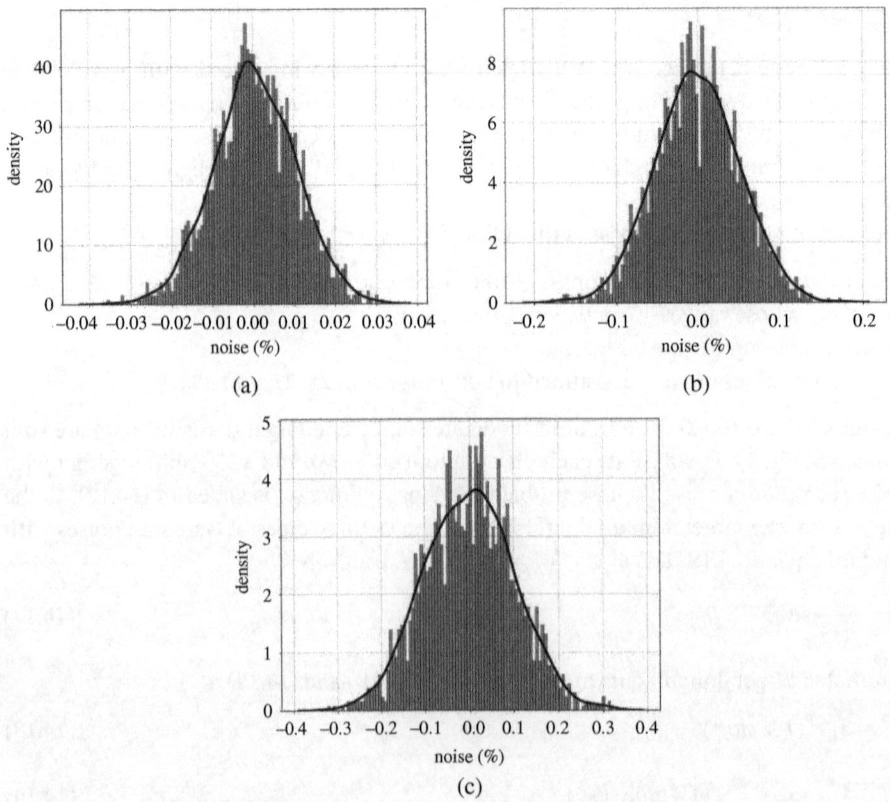

Figure 14.8 (a) Type A noise: noise with STD of 1%, (b) type B noise: noise with STD of 5%, (c) type C noise: noise with STD of 10%.

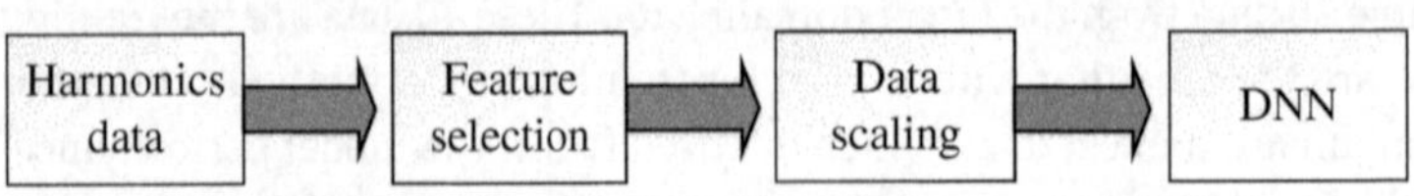

Figure 14.9 Data preprocessing.

that enhances the quality of data and helps the ML model to extract meaningful features from the data.

Fig. 14.9 describes the data preprocessing in this work. Feature selection in our case is to filter out some of the inputs that have no variations, which convey no information. These are the inputs that are regulated by the PI controllers and remain the same regardless of how the parameters are varied. After filtering out these inputs, data scaling is applied.

In general, there are two types of data preprocessing, namely normalization and standardization. In this task, we employ data preprocessing by standardizing the input feature and the output labels before feeding them into the DNN model. Standardization re-scales data to have a mean equal to zero and the STD of 1 (unit variance). The standardized data, (Z_i), are calculated by (14.20):

$$Z_i = \frac{x_i - mean(\boldsymbol{X})}{\sigma}, \tag{14.20}$$

where $\sigma = \sqrt{\frac{\sum_{i=1}^{T}(x_i - mean(\boldsymbol{X}))}{N}}$ and N is the number of sample points.

14.8 Performance Metrics

Performance metrics are used to compare the performance of different models. In this section, several performance metrics for regression and classification are discussed.

14.8.1 R square (Regression)

R^2, also known as coefficient of determination, is the proportion of the variation in the dependent variable that is predictable from the independent variable(s) [162]. Essentially, it is used to assess how good the model fits the data. (14.21) is the formula for R^2:

$$R^2 = 1 - \frac{SS_{res}}{SS_{tot}} = \frac{SS_{tot} - SS_{res}}{SS_{tot}}, \tag{14.21}$$

where SS_{res} is the sum of squares of errors between the true value, y_i, and the estimated value of a model, $\hat{y}_i$, which is defined as (14.22):

$$SS_{res} = \sum_{i=1}^{N} (y_i - \hat{y}_i)^2. \tag{14.22}$$

SS_{tot} is the total variation, which is stated in (14.23):

$$SS_{tot} = \sum_{i=1}^{N} (y_i - \bar{y})^2, \tag{14.23}$$

where

$$\bar{y} = \frac{1}{N} \sum_{i=1}^{N} y_i. \tag{14.24}$$

Note that SS_{tot} is a measure of variation or deviation from the mean. A model is expected to have residual sum of square (SS_{res}) smaller than SS_{tot}. Thus, R^2 is expected to have a value ranged between 0 and 1. $R^2 = 1$ represents a model that explains all the variation in the response variable around its mean. $R^2 = 0$ represents a model that does not explain any of the variation in the response variable around its mean. R^2 is negative when the chosen model has total errors that are bigger than the total variation of the data. This means that the chosen model is performing worse than a naive estimator, which is simply an averager.

14.8.2 Mean Absolute Percentage Error, MAPE (Regression)

Another well-known measurement is Mean Absolute Percentage Error (MAPE), which is defined in (14.25):

$$MAPE = \frac{1}{N} \sum_{i=1}^{n} \frac{|y_i - \hat{y}_i|}{|y_i|}. \tag{14.25}$$

MAPE is one of the easiest metrics and easy to infer and explain. Suppose the MAPE value of a particular model is 3%; this indicates that the average difference between the predicted value and the original value is 3%. Nevertheless, one drawback of MAPE is that it does not take into account the variance. Therefore, we do not know if our model performs better than a simple averager. Another drawback is that it produces infinite or undefined values when the actual values are zero or close to zero.

14.8.3 Accuracy (Classification)

Accuracy is a common measure for the classification task. It gives us a percentage of the data that are correctly classified against those that are wrongly classified. Mathematically, it can be expressed in (14.26):

$$\text{Accuracy} = \frac{\text{Number of correct predictions}}{\text{Total number of predictions}}, \tag{14.26}$$

For binary classification, accuracy can also be calculated in terms of positives and negatives from a confusion matrix [162] (see Table 14.2) as follows:

$$\text{Accuracy} = \frac{(TP + TN)}{(TP + TN + FP + FN)}, \tag{14.27}$$

where TP = True Positive, TN = True Negative, FP = False Positive, and FN = False Negative.

From the confusion matrix shown in Table 14.2, we conclude the following:

- If the model correctly predicts positive class, it is true positive, and if the model accurately predicts negative class, it is a true negative.
- If the model incorrectly (or falsely) predicts positive class, it is a false positive. If the model incorrectly (or falsely) predicts negative class, it is a false negative.

Accuracy gives us an intuitive measure toward the model. However, when dealing with imbalanced classes where the number of samples in one class outnumber the number of samples in an other class by a great deal, accuracy is not a good measure. In a case like an earthquake detector where an earthquake only happens a few times, a model that always predicts negative still yields a high score for accuracy.

14.8.4 F1 score (Classification)

Unlike Accuracy, F1 score is a better metric when there are imbalanced classes and when the false negatives and false positives are also important. This is because F1 is a weighted average of Precision and Recall.

Precision is the ratio of correctly predicted positive cases to the total predicted positive cases, and is defined in (14.28):

$$\text{Precision} = \frac{TP}{TP + FP}. \tag{14.28}$$

On the other hand, Recall is the ratio of correctly predicted positive cases to all the actual positive classes, and is defined in (14.29):

$$\text{Recall} = \frac{TP}{TP + FN}. \tag{14.29}$$

Table 14.2 A confusion matrix or a table of confusion.

		Actual Class	
		Positive	Negative
Predicted	Positive	True Positive	False Positive
Class	Negative	False Negative	True Negative

F1 is mathematically defined as (14.30):

$$\mathrm{F1} = 2 \cdot \frac{\text{Precision} \cdot \text{Recall}}{\text{Precision} + \text{Recall}}. \tag{14.30}$$

By substituting (14.28) and (14.29) into (14.30), (14.31) results:

$$\mathrm{F1} = \frac{TP}{TP + 1/2(FP + FN)}. \tag{14.31}$$

14.9 Test Results

Tests done in this work are divided into two cases, i.e. classification and regression cases. In the regression task, R, L, C, m_a, and δ are estimated. In the classification task, m_f is estimated. To better understand the advantage of using DA techniques compared to the conventional supervised learning, two models are created. The first model is trained only by data from the source domain (Source-only model). The second model is trained by the 30 training data from the target domain (Target-only model).

14.9.1 Classification Task on Multiple VSC

In this experiment, signal shift and the STD of the noises are set to be 0.2 and 0.01, respectively (i.e. type A noise). From the results shown in Table 14.3 we can conclude that DA and the Source-only model perform equally well. However, if we use the Target-only model, the accuracy and F_1 score area shown to be decreased, but not very significantly. These show that harmonic contents and m_f are strongly correlated and a conventional supervised learning alone can successfully extract feature vectors that are linearly separable in terms of label classes.

14.9.2 Regression Task on Multiple VSC

For the regression test, we compare the results of the WANN model with those of Source-only and Target-only models. Tables 14.4 and 14.5 summarize the performance of these methods. For

Table 14.3 Accuracy and F1 score for m_f.

Component	Converter	Method	Accuracy	F_1 score
m_f	VSC_1	Source-only	1	1
		Target-only	0.954	0.953
		CCSA	1	1
	VSC_2	Source-only	1	1
		Target-only	0.943	0.941
		CCSA	1	1
	VSC_3	Source-only	1	1
		Target-only	0.899	0.892
		CCSA	1	1

Table 14.4 MAPE and R^2 for R, L, and C.

Component	Converter	Method	MAPE	R^2
R	VSC_1	Source-only	12.690	−1.679
		Target-only	10.418	−1.583
		WANN	3.293	0.727
	VSC_2	Source-only	12.999	−1.756
		Target-only	7.759	−0.326
		WANN	3.530	0.711
	VSC_3	Source-only	13.729	−2.110
		Target-only	9.380	−1.016
		WANN	3.726	0.655
L	VSC_1	Source-only	9.695	−0.734
		Target-only	10.196	−1.334
		WANN	3.027	0.719
	VSC_2	Source-only	9.954	−0.789
		Target-only	9.420	−1.072
		WANN	3.193	0.733
	VSC_3	Source-only	10.257	−0.898
		Target-only	7.771	−0.330
		WANN	3.826	0.627
C	VSC_1	Source-only	9.363	−0.517
		Target-only	10.475	−1.587
		WANN	2.195	0.885
	VSC_2	Source-only	10.155	−0.694
		Target-only	10.556	−1.571
		WANN	2.590	0.845
	VSC_3	Source-only	9.292	−0.522
		Target-only	8.494	−0.699
		WANN	2.180	0.884

example, for the resistance (R) estimation of VSC_1, as shown in Table 14.4, the WANN model yields the lowest MAPE among the three models, and has a value of 0.727 for R^2, while the other two models yield negative values of R^2. Similarly, for estimation of the three δs, as shown in Table 14.5, the WANN model yields the lowest values of MAPE and an R^2 value closes to 1. For the overall average, the WANN, Source-only, and Target-only models yield MAPE values of 2.592, 9.224, and 7.66, respectively. For the averaged R^2, the WANN, Source-only, and Target-only models are 0.747, −1.020, and −0.939, respectively. This shows that WANN can adapt toward covariate shift easily with only a few samples from a new domain.

Table 14.5 MAPE and R^2 for m_a and δ.

Component	Converter	Method	MAPE	R^2
m_a	VSC_1	Source-only	0.893	−0.790
		Target-only	0.889	−1.303
		WANN	0.274	0.728
	VSC_2	Source-only	0.915	−0.864
		Target-only	0.819	−0.981
		WANN	0.274	0.754
	VSC_3	Source-only	0.941	−0.959
		Target-only	0.696	−0.340
		WANN	0.349	0.619
δ	VSC_1	Source-only	12.412	−1.029
		Target-only	9.389	−0.708
		WANN	3.002	0.820
	VSC_2	Source-only	11.880	−0.768
		Target-only	7.481	0.024
		WANN	3.521	0.779
	VSC_3	Source-only	13.188	−1.196
		Target-only	11.160	−1.263
		WANN	3.910	0.718
	Average	Source-only	9.224	−1.020
		Target-only	7.660	−0.939
		WANN	2.592	0.747

Besides performance metrics, visualizing the result is a good way to get an intuition of how the model performs. An actual-vs-predicted plot is a scatter plot that presents comprehensive information of how well predicted values compared to the actual values. Figs. 14.10 to 14.12 show these plots for Source-only, Target-only, and WANN models for the regression task of VSC_1, respectively. The x-axes of the figures represent predicted value and y-axes of the figures represent actual values. Perfect prediction is represented by the dashed line where all of the predicted values are equal to the actual values. Perfect prediction is when all of the dots fall on the dashed line. As Fig. 14.10 clearly shows, a trend line formed by the dots do not coincide with the dashed line. In addition, Fig. 14.11 shows no clear trend line is formed by the dots. On the other hand, as we can see from Fig. 14.12, the trend line formed by the dots coincides with the dashed line, showing the good performance of the WANN model.

Fig. 14.13 shows the relation between noise types, numbers of training data from the target domain, and the performance of the models, measured in R^2. As we increase the noise, the amount of training data from the target domain required to maintain the same level of performance also increases.

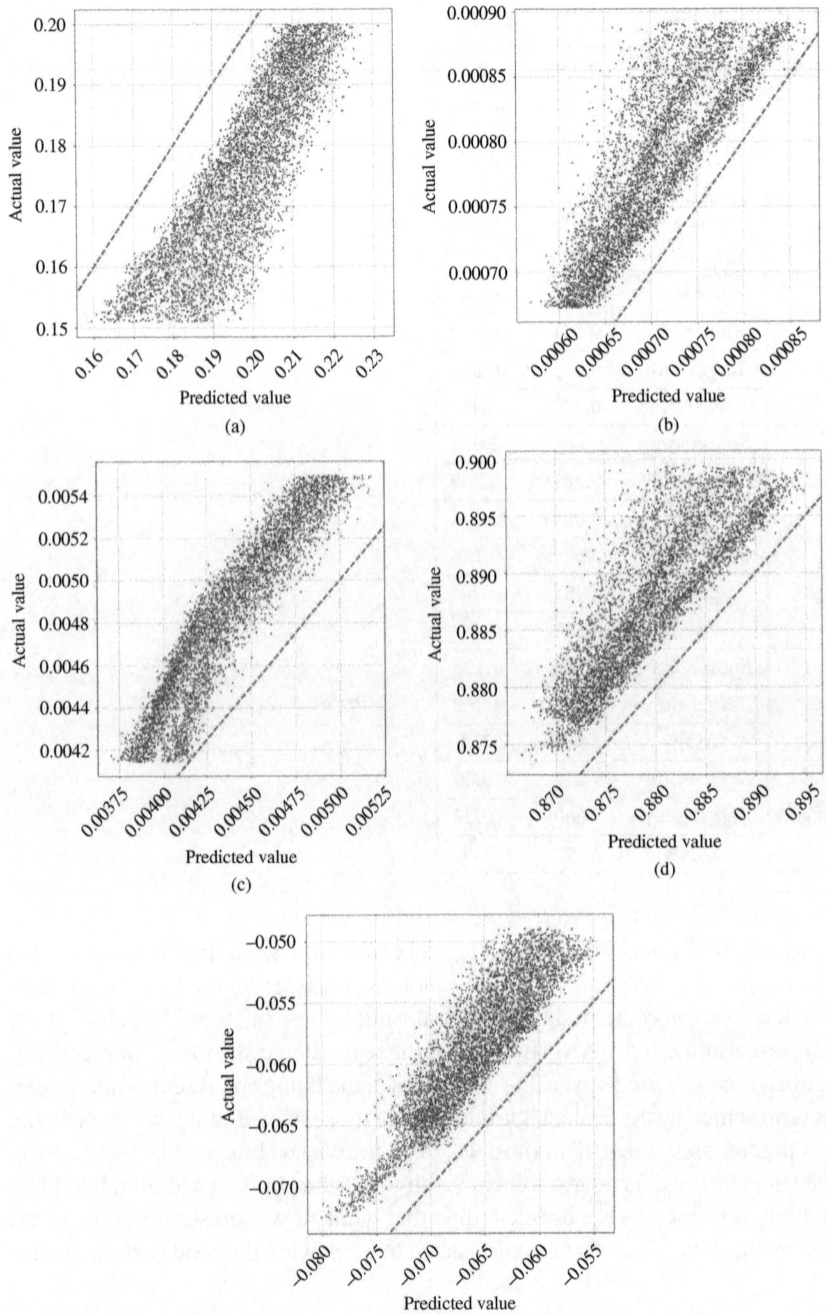

Figure 14.10 Actual vs. predicted plot of source-only model: (a) R of VSC_1, (b) L of VSC_1, (c) C of VSC_1, (d) m_a of VSC_1, (e) δ of VSC_1.

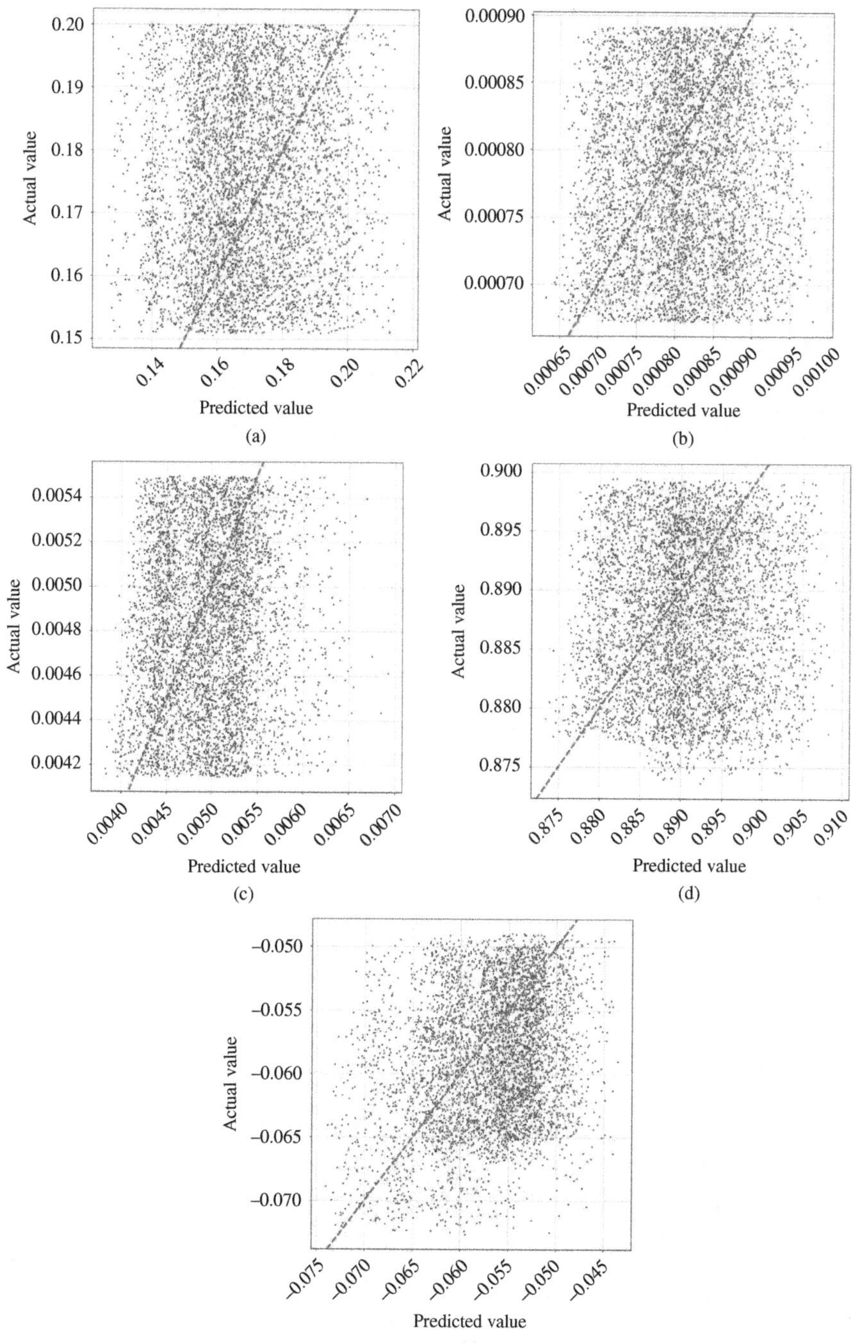

Figure 14.11 Actual vs. predicted plot of target-only model: (a) R of VSC_1, (b) L of VSC_1, (c) C of VSC_1, (d) m_a of VSC_1, (e) δ of VSC_1.

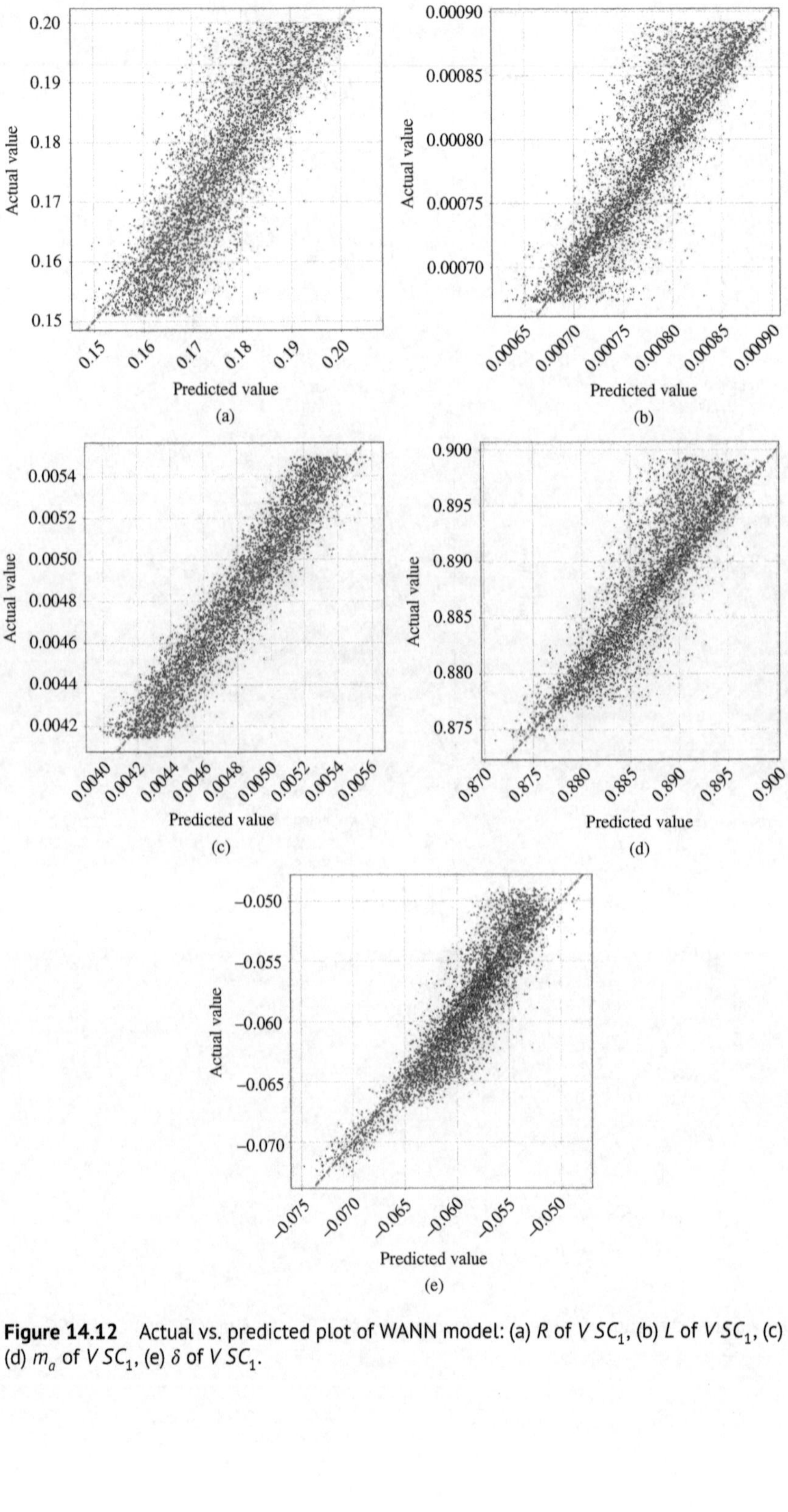

Figure 14.12 Actual vs. predicted plot of WANN model: (a) R of VSC_1, (b) L of VSC_1, (c) C of VSC_1, (d) m_a of VSC_1, (e) δ of VSC_1.

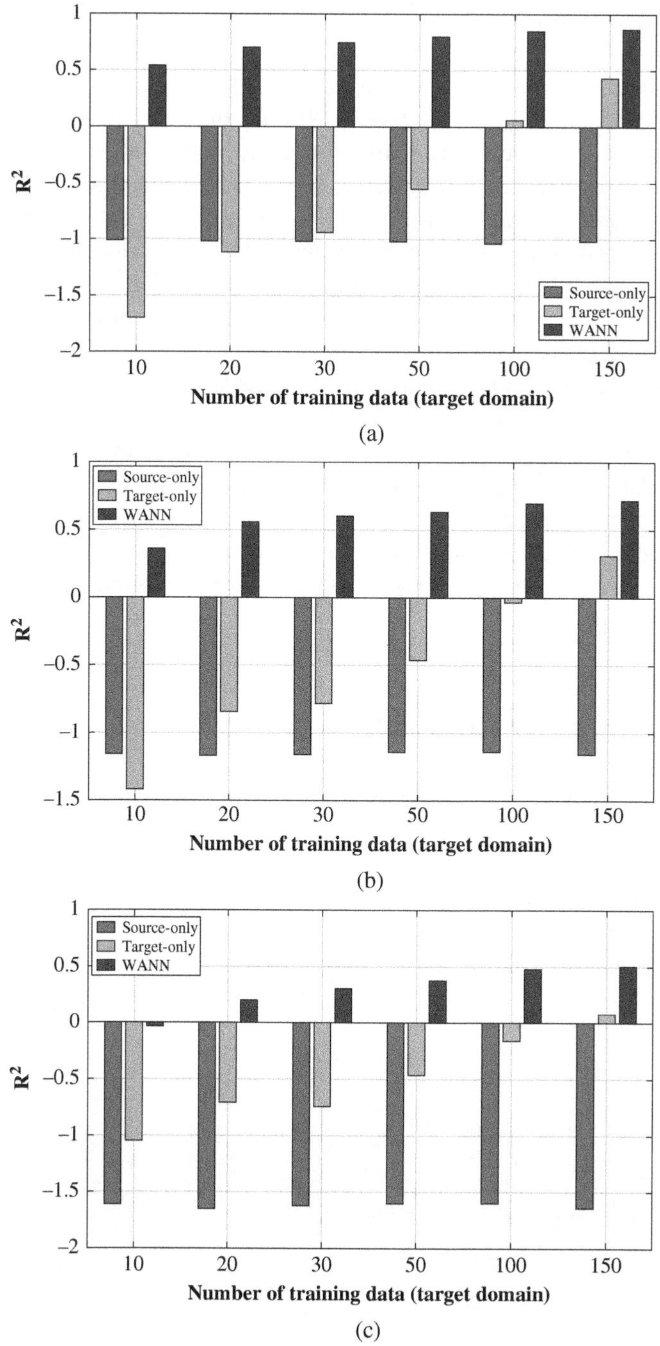

Figure 14.13 R^2 for (a) type A noise, (b) type B noise, (c) type C noise.

14.10 Software for Running the Codes

In addition to the MATLAB and OpenDss environment, which generates the simulation training data, we also use the Python programming language to develop NNs for the DA. Python is an open-source programming language that has a huge community, especially in the context of Deep Learning. The community has built many libraries such as Tensorflow and Keras that have made it possible to design and run complex neural networks with only several lines of code. Tutorials for using these libraries can be found easily on the internet or in books such as [163] and [164]. In addition to Tensorflow and Keras, several other libraries are also required to perform data processing and data visualization. Together they form a Python environment that can be used to run the code in this chapter. The Python community has been continuously growing, and it is very common to have a new version of library within a couple of weeks or months. To make sure the code can run without error, we need to make sure that the libraries installed inside the environment are the correct versions. The following are the list of the libraries we used to build the Python environment: matplotlib 3.1.1, numpy 1.17.4, pandas 0.25.3, scikit-learn 0.22, keras-lookahead 0.9.0, keras-radam 0.15.0, keras 2.2.4, h5py 2.10.0, tensorflow-gpu 2.0.0.

Note that the Python environment is built on top of a piece of software called Anaconda. Anaconda is designed to be able to manage various Python packages easily and to build separate environments for different projects.

14.11 Implementation of Domain Adaptation

This section provides the code for running regression and classification tasks, discussed previously for DA of PE.

14.11.1 Data Generation

The code for generating the training and test data is written in MATLAB as follows:

```
close all
clc
clear
addpath('../ieee13bus_pscad_opendss')
set_timeout=600; %10 menit
num_out=10;
% %The PI controller's gain and time constant
Pc=0.04247;
Tc=0.1;
Pv=0.5;
Tv=1/25;
freq=60;
%The parameter vales
Ls=0;
Lf=0.637e-3;
L=Ls+Lf;

Rs=0.0;
Rf=0.15;
R=Rs+Rf;
R_dc=inf;
```

```
%R=0.15;
C=4820e-6; %it was 4820uf
Ti = 0.0019895;
mf=[9 15 21 27 33];
% Hh=[1, 25, 29, 53, 55, 77, 79, 83, 85];
Hh=[1:2:67];
harNum=length(Hh);
i=0;
vdc_num=[1,3,6,8,11,14];
%  count =0
%% csv file setting
%ignore
for i=1:10000
    randata(i,:)=rand(1,9);
    mfrand(i,:)=[unidrnd(5),unidrnd(5),unidrnd(5)];
end
%+-14%from 1020 to

%% particles
for i=7700:8000
    tic
    for v=1:3
        part_R(v)=R*(0.86+0.28*randata(i,3*(v-1)+1));
        part_L(v)=L*(0.86+0.28*randata(i,3*(v-1)+2));
        part_C(v)=C*(0.86+0.28*randata(i,3*(v-1)+3));
        part_mf(v)=mf(mfrand(i,v));
    end
    %%
    fut= parfeval(@VSC_opdss_fund_only,num_out,part_R, part_L,...
                          part_C,Pc,Tc,Pv, Tv, Ti,part_mf);
    ok = wait(fut, 'finished', set_timeout);
    if ~ok
        disp(['sim: ',num2str(i),' DID NOT FINISH IN TIME!!!'])
    else
        disp(['sim: ',num2str(i),' Did finish in time.'])
        [Ialpha vdcc1 vdcc2 vdcc3 Ma1 Ma2 Ma3 Del1 Del2 Del3]=fetchOutputs(fut);
        Ma(1) = Ma1;
        Ma(2) = Ma2;
        Ma(3) = Ma3;
        Del(1) = Del1;
        Del(2) = Del2;
        Del(3) = Del3;
        for j=1:harNum
            Har(i,j*8-7:j*8) = [real(Ialpha(j)),imag(Ialpha(j)),real(vdcc1(j)),...
            imag(vdcc1(j)),real(vdcc2(j)),imag(vdcc2(j)),...
            real(vdcc3(j)),imag(vdcc3(j))];
        end
        for v=1:3
            Data(i,v*6-5:v*6)=[part_R(v), part_L(v), part_C(v),Ma(v),...
                                                     Del(v),part_mf(v)];
        end
    end
    clearvars fut
    timeElapsed = toc;
    print_=['timeElapsed= ',num2str(timeElapsed),' sec'];
    disp(print_)
end
writematrix(Data,'simulation_data/targets.csv')
writematrix(Har,'simulation_data/Inputs.csv')
```

The function, `VSC_opdss_fund_only.m` in line 50 contains the model of Fig. 14.2 whose code is similar to that of **Example 11.4** of Chapter 11. All of the data generated by this code will be stored in the following path, `simulation_data/Inputs.csv` and `simulation_data/Inputs.csv`. These data will later be used in the training and testing process.

14.11.2 Regression

For our regression task, we refer to the following listed python code. Note that the algorithm for WANN can be found in Github (https://github.com/antoinedemathelin/wann).

```
# =======================import necessary libraries=======================
import copy
import numpy as np
import pandas as pd
from config import *
from sklearn.preprocessing import StandardScaler
from sklearn.model_selection import train_test_split
from tensorflow.keras import backend as K
from lib.utils import BaggingModels,get_base_model,get_encoder,get_task
from lib.methods.WANN import WANN

# =======================import metric performance=======================
from sklearn.metrics import r2_score
def MAPE(Y_actual,Y_Predicted):
    mape = np.mean(np.abs((Y_actual - Y_Predicted)/Y_actual))*100
    return mape

def save_dataframe(X,y,y_pred,index,filename):
    df={'idx':index,
        'y_pred':y_pred[index],
         'y_true':y[index]}

    df=pd.DataFrame(df)
    df_X=pd.DataFrame(X[index])
    df=pd.concat([df_X,df],axis=1)
    PATH=os.path.join(*[dataframe_Path_3VSC, filename])
    df.to_csv(r'{}'.format(PATH))
    return df

# =======================main function=======================
def run_vsc_experiments(args):
    # =======================data loading=======================
    np.random.seed(args.random_state)
    ML_data = pd.read_csv(ML_3VSC_data_Path)
    ML_data = ML_data.iloc[:args.total_source]

    # ===================feature selection=============================
    inputs_data=ML_data.loc[:,'Re_I_1':'Im_vdcc3_67']
    inputs_summary = inputs_data.describe().transpose()
    inputs_list= inputs_summary.loc[inputs_summary['std']>1e-5].index.tolist()

    # =====split the data into source and target domain =================
    X_src=ML_data.loc[:args.num_source,inputs_list]
    y_src=ML_data.loc[:args.num_source,args.typeComp]

    X_tgt = ML_data.loc[args.num_source:, inputs_list]
    y_tgt = ML_data.loc[args.num_source:, args.typeComp]

```

```
    # ===================adding noise and signal shift================
    X_tgt = X_tgt.applymap(lambda x: x * (1 + args.signal_shift))
    X_tgt = X_tgt.applymap(lambda x: x * (1 + np.random.normal(args.noise_mean,
                                                  args.noise_std, 1)[0]))
    # ===================data concatenation============================
    y_all = pd.concat([y_src, y_tgt], axis=0, ignore_index=True)
    X_all = pd.concat([X_src, X_tgt], axis=0, ignore_index=True)

    shape = X_all.shape[1]
    # =======splitting the data into training and testing=================
    src_index = np.arange(0, X_src.shape[0])
    tgt_index = np.arange(X_src.shape[0], X_src.shape[0] + X_tgt.shape[0])

    tgt_train_index, tgt_test_index = train_test_split(tgt_index,
        train_size=args.n_target_labeled, random_state = args.random_state)
    train_index = np.concatenate((src_index, tgt_train_index))
    print('tgt_test_index: ', len(tgt_test_index))
    # ========================data normalization=========================
    X = X_all.values
    y = y_all.values

    std_X = StandardScaler()
    std_X.fit(X[train_index])
    X = std_X.transform(X)

    Scaler_y = StandardScaler()
    Scaler_y.fit(y[train_index].reshape(-1,1))
    y = Scaler_y.transform(y.reshape(-1,1))[:,0]
    # ==================defining model=================================
    base_estimator = BaggingModels(func=get_base_model,
                                   n_models=1,
                                   n_jobs=args.n_jobs,
                                   shape=shape,
                                   C=args.C,
                                   random_state=args.random_state)
    fit_params = dict(epochs=args.epochs, batch_size=args.batch_size, verbose=0)

    if args.method == "SrcOnly":
        model = copy.deepcopy(base_estimator)
        model.fit(X[src_index], y[src_index], **fit_params)

    if args.method == "TgtOnly":
        model = copy.deepcopy(base_estimator)
        model.fit(X[tgt_train_index], y[tgt_train_index], **fit_params)

    if args.method == "WANN":
        model = BaggingModels(WANN,
                              get_base_model=get_base_model,
                              C=args.C,
                              C_w=args.C_w,
                              n_models=args.n_models,
                              n_jobs=args.n_jobs,
                              random_state=args.random_state)
        model.fit(X, y, index=[src_index, tgt_train_index], **fit_params)

    #===================estimation process===================================
    y_pred = model.predict(X)
    y_pred = Scaler_y.inverse_transform(y_pred)[:,0]
    y = Scaler_y.inverse_transform(y.reshape(-1, 1))[:, 0]
    #================storing estimation result===============================
    filename=f'{args.method}_df_test_tgt_{args.typeComp}.csv'
```

```
110     save_dataframe(X,y,y_pred,tgt_test_index,filename)
111     filename=f'{args.method}_df_train_tgt_{args.typeComp}.csv'
112     save_dataframe(X,y,y_pred,tgt_train_index,filename)
113     filename=f'{args.method}_df_train_src_{args.typeComp}.csv'
114     save_dataframe(X,y,y_pred,src_index,filename)
115     #==================calculating metrics performance=========================
116     ''' R2 square'''
117     r2_src_train = r2_score(y[src_index], y_pred[src_index])
118     r2_tgt_train = r2_score(y[tgt_train_index], y_pred[tgt_train_index])
119     r2_tgt_test = r2_score(y[tgt_test_index], y_pred[tgt_test_index])
120     '''MAPE'''
121     mape_src_train=MAPE(y[src_index], y_pred[src_index])
122     mape_tgt_train = MAPE(y[tgt_train_index], y_pred[tgt_train_index])
123     mape_tgt_test = MAPE(y[tgt_test_index], y_pred[tgt_test_index])
124
125     df_measure={'r2_src_train':[r2_src_train],
126                 'r2_tgt_train':[r2_tgt_train],
127                 'r2_tgt_test':[r2_tgt_test],
128
129                 'mape_src_train':[mape_src_train],
130                 'mape_tgt_train': [mape_tgt_train],
131                 'mape_tgt_test': [mape_tgt_test]
132                 }
133
134     df_measure=pd.DataFrame(df_measure)
135     #==========================storing metrics performance=========================
136     PATH=os.path.join(*[Perf_Path_3VSC, f'{args.method}_Perf_{args.typeComp}.csv'])
137     df_measure.to_csv(r'{}'.format(PATH))
138     K.clear_session()
139
140 if __name__ == "__main__":
141     class ParamsConfig:
142         def __init__(self):
143             self.noise_std = 0.01 # noise standard deviation
144             self.noise_mean = 0.0
145             self.signal_shift = 0.2 # signal shift
146             self.n_target_labeled = 30 #data from target domain used for training
147             self.num_source = 5000 # data from source domain
148             self.total_source = 10000 # total data
149             self.random_state = 0 #numpy random seed.
150             self.C = 1 #Projecting constant of the weighting network
151             self.C_w = 0.1 #Projecting constant of the weighting network.
152             self.epochs = 700 #number of epoch
153             self.batch_size = 32 # batch size
154
155     args=ParamsConfig()
156     for method in ["WANN","TgtOnly", "SrcOnly"]:
157         args.method=method
158         for typeComp in ['R_1','L_1','C_1','Ma_1','Del_1',
159                          'R_2','L_2','C_2','Ma_2','Del_2',
160                          'R_3','L_3','C_3','Ma_3','Del_3']:
161             args.typeComp = typeComp
162             run_vsc_experiments(args)
163     print('finished!')
```

The code begins by importing the libraries required. In line 10 we are using code provided by [149]. Lines 13 and 14 import metrics performance, i.e. R^2 and MAPE. Line 18 is a function to store all the estimation data into a `.csv` format. `run_vsc_experiments` in line 31 is the main function that will be called in line 162. In lines 33–35, data generated from Section 14.11.1 are loaded. Lines 38–40 are the feature selection process. Line 40 means that we only consider features that have standard deviation more than 1×10^{-5}. Lines 43–47 are the process of splitting data into source and target domains. As shown by (14.18) and (14.19), lines 50–52 are the process of adding contamination to form the target domain data. The process is followed by splitting data, as described in lines 59–65, normalizing the data as described in lines 67–76. Lines 78–102 are the process of defining a model. We are comparing WANN with the two other models, as described in Section 14.9.2. Lines 105–107 are the estimation process. Lines 109–14 store estimation results together with the ground truth. Lines 117–123 calculate the metrics. Finally, lines 136–138 are storing these metric values. Lines 142–153 are the parameters and hyperparameters specified by the users. In line 155, we encapsulate these parameters inside an object, `args`. Lines 156–162 are the loop processes for regression, which execute the Source-only, Target-only, and WANN models together with their datasets.

14.11.3 Classification Network

The following is the Python code for the classification task. The algorithm for the CCSA method can be found at https://github.com/samotiian/CCSA.

```
# ========================import necessary libraries========================
from lib.model import Initialization
from keras.layers import Activation, Dropout, Dense
from keras.layers import Input, Lambda
from keras.models import Model
from sklearn.model_selection import train_test_split
from sklearn.preprocessing import StandardScaler
from sklearn.metrics import accuracy_score,precision_score,\
                            recall_score,f1_score
import numpy as np
import pandas as pd
from config import *

# ========================import metric performance========================
def calc_classification_perf(Y_actual, Y_pred):
    accuracy = accuracy_score(Y_actual, Y_pred)
    precision = precision_score(Y_actual, Y_pred, average='weighted')
    recall = recall_score(Y_actual, Y_pred, average='weighted')
    f1 = f1_score(Y_actual, Y_pred, average='weighted')
    perf_classification = pd.DataFrame({'accuracy': [accuracy],
                          'precision': [precision],
                          'recall': [recall],
                          'f1': [f1]})

    return perf_classification

def testing_the_model(model,data,name,Perf_Path,models_path,typeComp):
    filepath = os.path.join(*[models_path, f'{name}_model.h5'])
    model.load_weights(filepath)
    if name=='CCSA':
        y_pred = model.predict([data.X, data.X])[0]
    else:
        y_pred = model.predict(data.X)
    y_pred = np.argmax(y_pred, axis=1)

    perf_src_train = calc_classification_perf(data.y[data.src_index],
                                              y_pred[data.src_index])
    perf_tgt_train = calc_classification_perf(data.y[data.tgt_train_index],
                                          y_pred[data.tgt_train_index])
```

```
    perf_tgt_test = calc_classification_perf(data.y[data.tgt_test_index],
                                          y_pred[data.tgt_test_index])

    df_measures=pd.concat([perf_src_train,perf_tgt_train,perf_tgt_test],
                                      axis=0,ignore_index=False)
    df_measures.index=['src_train ','tgt_train','tgt_test']

    filepath = os.path.join(*[Perf_Path, f'{name}_Perf_{typeComp}.csv'])
    df_measures.to_csv(r'{}'.format(filepath))

# ======================main function=======================
def run_vsc_experiments(args):
    # =======================data loading=======================
    np.random.seed(args.random_state)
    ML_data = pd.read_csv(ML_3VSC_data_Path)
    ML_data = ML_data.iloc[:args.total_source]

    # ====================feature selection=============================
    inputs_data = ML_data.loc[:, 'Re_I_1':'Im_vdcc3_67']
    inputs_summary = inputs_data.describe().transpose()
    inputs_list = inputs_summary.loc[inputs_summary['std'] > 1e-5].index.tolist()

    # ====================data splitting==============================
    X_src = ML_data.loc[:args.num_source,inputs_list]
    y_src = ML_data.loc[:args.num_source, args.typeComp]

    X_tgt = ML_data.loc[args.num_source:,inputs_list]
    y_tgt = ML_data.loc[args.num_source:, args.typeComp]

    # ====================adding noise and signal shift=================
    X_tgt = X_tgt.applymap(lambda x: x * (1 + args.signal_shift))
    X_tgt= X_tgt.applymap(lambda x: x * (1 + np.random.normal(args.noise_mean,
                                          args.noise_std, 1)[0]))
    # ====================data concatenation============================
    X_all = pd.concat([X_src, X_tgt], axis=0, ignore_index=True)
    y_all = pd.concat([y_src, y_tgt], axis=0, ignore_index=True)
    # ====================converting class label into encoded values=====
    in_size=X_src.shape[1]
    out_size=len(y_src.unique())
    replace_table = {9: 0, 15: 1, 21: 2, 27: 3, 33: 4}
    y_all.replace(replace_table,inplace=True)
    # ==============================splitting============================
    src_index = np.arange(0, X_src.shape[0])
    tgt_index = np.arange(X_src.shape[0], X_src.shape[0] + X_tgt.shape[0])
    np.random.seed(0)
    tgt_train_index, tgt_test_index = train_test_split(tgt_index,
        train_size=args.n_target_labeled, random_state=args.random_state)
    train_index = np.concatenate((src_index, tgt_train_index))
    # ========================data normalization=========================
    X = X_all.values
    y = y_all.values

    std_X = StandardScaler()
    std_X.fit(X[train_index])
    X = std_X.transform(X)

    # =encapsulating information and data to be passed into training function===
    class DATA:
        def __init__(self):
            self.src_index = src_index
            self.tgt_index = tgt_index
```

```
            self.tgt_train_index = tgt_train_index
            self.tgt_test_index = tgt_test_index
            self.train_index=train_index
            self.X=X
            self.y=y
            self.std_X=std_X
            self.in_size=in_size
            self.out_size=out_size
            self.SrcOnly_epoch = args.SrcOnly_epoch
            self.TgtOnly_epoch = args.TgtOnly_epoch
            self.CCSA_epoch = args.CCSA_epoch
            self.batch_size=args.batch_size

    data=DATA()
    # ==================defining model====================================
    def getCCSA_model():
        '''constructing domain adaptation model'''
        model1=Initialization.Create_Model(in_size)
        input_shape = (in_size,)
        input_a = Input(shape=input_shape)
        input_b = Input(shape=input_shape)
        # Loss = (1-alpha)Classification_Loss + (alpha)CSA
        alpha =.25
        # Having two streams. One for source and one for target.
        processed_a = model1(input_a)
        processed_b = model1(input_b)
        # Creating the prediction function. This corresponds to h in the paper.
        out1 = Dropout(0.5)(processed_a)
        out1 = Dense(out_size)(out1)
        out1 = Activation('softmax', name='classification')(out1)
        distance = Lambda(Initialization.euclidean_distance,
                          output_shape=Initialization.eucl_dist_output_shape,
                          name='CSA')([processed_a, processed_b])
        model = Model(inputs=[input_a, input_b], outputs=[out1, distance])
        model.compile(loss={'classification': 'categorical_crossentropy',
                            'CSA': Initialization.contrastive_loss},
                      optimizer='adadelta',
                      loss_weights={'classification': 1 - alpha, 'CSA': alpha})
        return model

    def getSupervised_model():
        model1 = Initialization.Create_Model(in_size)
        input_shape = (in_size,)
        input_ = Input(shape=input_shape)
        processed_a = model1(input_)

        out1 = Dropout(0.5)(processed_a)
        out1 = Dense(out_size)(out1)
        out1 = Activation('softmax', name='classification')(out1)
        model = Model(inputs=input_, outputs=out1)
        model.compile(loss='categorical_crossentropy',optimizer='adam')
        return model

    # ==============================training process====================
    if args.method=='CCSA':
        model=getCCSA_model()
    else:
        model = getSupervised_model()

    if args.training_cmd:
        if args.method == 'CCSA':
```

```
162             Initialization.training_CCSA(model, data, models_path_3VSC)
163         elif args.method == 'SrcOnly':
164             Initialization.training_SrcOnly(model, data, models_path_3VSC)
165         elif args.method == 'TgtOnly':
166             Initialization.training_TgtOnly(model, data, models_path_3VSC)
167         else:
168             raise Exception('method is not available. check the typing!')
169 
170     # ===================testing and evaluation process=====================
171     testing_the_model(model,
172                       data,
173                       args.method,
174                       Perf_Path_3VSC,
175                       models_path_3VSC,
176                       args.typeComp)
177 
178 if __name__ == "__main__":
179     class ParamsConfig:
180         '''params config'''
181         def __init__(self):
182             self.training_cmd = True
183             self.noise_std = 0.01 # noise standard deviation
184             self.noise_mean = 0.0
185             self.signal_shift = 0.2 # signal shift
186             self.n_target_labeled = 30 # data from target domain used for training
187             self.total_source = 10000 # total data
188             self.num_source = 5000 # data from source domain
189             self.SrcOnly_epoch = 200 # number of epoch
190             self.TgtOnly_epoch = 200  # number of epoch
191             self.CCSA_epoch = 200 # number of epoch
192             self.random_state = 0  # numpy random seed.
193             self.batch_size = 32 # batch size
194 
195     args=ParamsConfig()
196     for method in ['CCSA', 'SrcOnly', 'TgtOnly']:
197         args.method=method
198         for typeComp in ['Mf_1','Mf_2','Mf_3']:
199             args.typeComp = typeComp
200             run_vsc_experiments(args)
201     print('finished!')
```

Lines 2–12 import all of the necessary libraries. Lines 15–25 define the metrics for classification. Lines 27–48 define the testing process and store the results in the form of a `.csv` file. The main code for this task starts from Line 51. Lines 53–55 import data generated by Section 14.11.1. Lines 58–60 perform the feature selection. Lines 63–67 split data into source and target domains. Lines 70–72 add noise contamination to form the target domain data. Lines 77–80 convert the output label data into binary encoded values, which is needed for calculating the *cross-entropy loss*. Lines 82–87 perform the data splitting process. Lines 89–94 are the normalization process. Lines 96–114 encapsulate the data and train information into one object (i.e `data`) to be passed for training in lines 162, 164, and 166. Lines 116–152 define the model used for training. Lines 171–176 define the function to perform the testing and store the result. Lines 178–193 are the parameters and hyperparameters specified by the users. In line 195, we encapsulate these parameters inside an object, `args`. Lines 196–201 are the loop processes for classification, which execute the Source-only, Target-only, and CCSA models together with their datasets.

References

1 F. Bedell, "History of a-c wave form, its determination and standardization," *Transactions of the American Institute of Electrical Engineers*, vol. 61, no. 12, pp. 864–868, Dec 1942.

2 E. L. Owen, "A history of harmonics in power systems," *IEEE Industry Applications Magazine*, vol. 4, no. 1, pp. 6–12, Jan 1998.

3 P. Mixon, "Technical origins of 60 Hz as the standard AC frequency in North America," *IEEE Power Engineering Review*, vol. 19, no. 3, pp. 35–37, March 1999.

4 M. Grady, *Understanding Power System Harmonics*. University of Texas at Austin, 2012.

5 "IEEE recommended practices and requirements for harmonic control in electrical power systems," *IEEE Std 519-1992*, pp. 1–112, 1993.

6 N. G. Hingorani and L. Gyugyi, *Understanding FACTS: Concepts and Technology of Flexible AC Transmission Systems*. New York: IEEE Press, 2001.

7 B. K. Bose, "Global energy scenario and impact of power electronics in 21st century," *IEEE Transactions on Industrial Electronics*, vol. 60, no. 7, pp. 2638–2651, July 2013.

8 A. E. Emanuel, *Power Definitions and The Physical Mechanism of Power Flow*, 1st ed. John Wiley & Sons, 2010.

9 M. F. McGranaghan, R. C. Dugan, and W. L. Sponsler, "Digital simulation of distribution system frequency-response characteristics," *IEEE Transactions on Power Apparatus and Systems*, vol. 100, no. 3, pp. 1362–1369, March 1981.

10 R. C. Dugan, M. F. McGranaghan, S. Santoso, and H. W. Beaty, *Electrical Power Systems Quality*. John & Sons: McGraw-Hill, 2002.

11 N. Mohan, T. Undeland, and W. Robins, *Power Electronics: Converters, Applications, and Design*, 2nd ed. John & Sons: Wiley, 1995.

12 E. Acha and M. Madrigal, *Power System Harmonics: Computer Modelling and Analysis*. Wiley, 01 2001.

13 A. Wood and C. M. Osauskas, "A linear frequency-domain model of a statcom," *IEEE Transactions on Power Delivery*, vol. 19, no. 3, pp. 1410–1418, July 2004.

14 A. McEachern, "Electric power definitions: a debate," in *2013 Power Engineering Society General Meeting*, 2013.

15 M. Bollen, M. Olofsson, A. Larsson, S. Ronnberg, and M. Lundmark, "Standards for supraharmonics (2 to 150 kHz)," *IEEE Electromagnetic Compatibility Magazine*, vol. 3, no. 1, pp. 114–119, 2014.

16 T. M. Mendes, C. A. Duque, L. R. Manso da Silva, D. D. Ferreira, J. Meyer, and P. F. Ribeiro, "Comparative analysis of the measurement methods for the supraharmonic range," *International Journal of Electrical Power and Energy Systems*, vol. 118, p. 105801, 2020. [Online]. Available: http://www.sciencedirect.com/science/article/pii/S014206151932455X

Harmonic Modeling of Voltage Source Converters using Basic Numerical Methods, First Edition.
Ryan Kuo-Lung Lian, Ramadhani Kurniawan Subroto, Victor Andrean, and Bing Hao Lin.

17 V. Khokhlov, J. Meyer, A. Grevener, T. Busatto, and S. Ronnberg, "Comparison of measurement methods for the frequency range 2–150 kHz (supraharmonics) based on the present standards framework," *IEEE Access*, vol. 8, pp. 77 618–77 630, 2020.

18 S. K. Ronnberg, M. H. J. Bollen, and M. Wahlberg, "Interaction between narrowband power-line communication and end-user equipment," *IEEE Transactions on Power Delivery*, vol. 26, no. 3, pp. 2034–2039, 2011.

19 V. Khokhlov, J. Meyer, and P. Schegner, "Thermal interactions in modern lighting equipment due to disturbances in the frequency range 2–150 kHz," in *2019 IEEE Milan PowerTech*, 2019, pp. 1–6.

20 C. Waniek, T. Wohlfahrt, J. M. A. Myrzik, J. Meyer, M. Klatt, and P. Schegner, "Supraharmonics: Root causes and interactions between multiple devices and the low voltage grid," in *2017 IEEE PES Innovative Smart Grid Technologies Conference Europe (ISGT-Europe)*, 2017, pp. 1–6.

21 T. Slangen, T. V. Wijk, V. Cuk, and S. Cobben, "The propagation and interaction of supraharmonics from electric vehicle chargers in a low-voltage grid," *Energies*, vol. 13, pp. 1–19, 2020.

22 S. K. Ronnberg, A. G. Castro, M. H. J. Bollen, A. Moreno-Munoz, and E. Romero-Cadaval, "Supraharmonics from power electronics converters," in *2015 9th International Conference on Compatibility and Power Electronics (CPE)*, 2015, pp.539–544.

23 R. N. Beres, X. Wang, F. Blaabjerg, M. Liserre, and C. L. Bak, "Optimal design of high-order passive-damped filters for grid-connected applications," *IEEE Transactions on Power Electronics*, vol. 31, no. 3, pp. 2083–2098, 2016.

24 E. L. Owen, "SCR is 50 years old," *IEEE Industry Applications Magazine*, vol. 13, no. 6, pp. 6–10, 2007.

25 D. Tiku, "DC power transmission mercury-arc to thyristor HVDC valves," *IEEE Power and Energy Magazine*, vol. 12, no. 2, pp. 76–96, 2014.

26 B. Ozpineci and L. M. Tolbert, "Comparison of wide-bandgap semiconductors for power electronics applications," Technical Report, Oak Ridge National Lab., USA, 2004.

27 A. Yoshikawa, *Development and Applications of Wide Bandgap Semiconductors*. Springer, 2007.

28 S. Bernet, "Recent developments of high power converters for industry and traction applications," *IEEE Transactions on Power Electronics*, vol. 15, no. 6, pp. 1102–1117, 2000.

29 I. Batarseh and A. Harb, *Power Electronics Circuit Analysis and Design*. Springer, 2018.

30 A. Yazdani and R. Iravani, Eds., *Voltage-Sourced Converters in Power Systems: Modeling, Control, and Applications*. Wiley-IEEE Press, 2010.

31 B. Wu, *Modular Multilevel Converters – Analysis, Control, and Applications*. IEEE Press, 2018.

32 B. Wu and M. Narimani, *High-power Converters and AC Drive*. IEEE Press, 2017.

33 V. K. Sood, Ed., *HVDC and FACTS Controllers: Applications of Static Converters in Power Systems*. Kluwer Academic Publishers, 2004.

34 P. Kim, *Kalman Filter for Beginners: with MATLAB Examples*. CreateSpace Independent Publishing Platform, 2011.

35 J. D. Glover and M. S. Sarma, *Power System Analysis and Design*, 3rd ed. Brooks/Cole, 2002.

36 J. J. Grainger and W. D. Stevenson, *Power System Analysis*, McGraw-Hill, 1994.

37 R. L. Burden and J. D. Faires, "Numerical analysis," *Brooks/Cole*, vol. 7, 2001.

38 T. J. Aprille and T. N. Trick, "Steady-state of nonlinear circuits with periodic inputs," *Proceedings of the IEEE*, vol. 60, no. 1, pp. 108–114, January 1972.

39 C. G. Broyden, "A class of methods for solving nonlinear simultaneous equations," *Mathematics of Computation*, vol. 19, no. 92, pp. 577–593, October 1965.

40 S. C. Chapra and R. P. Canale, *Numerical Method for Engineers with Software and Programming Applications*, 4th ed. McGraw-Hill, Toronto, 2002.

41 J. Kennedy and R. Eberhart, "Particle swarm optimization," in *Proceedings of ICNN'95 - International Conference on Neural Networks*, vol. 4, 1995, pp. 1942–1948.

42 X.-S. Yang, *Nature-Inspired Metaheuristic Algorithms*. Luniver Press, 2008.

43 C. F. V. Loan, "Computing integral involving the matrix exponential," *IEEE Transactions on Automatic Control*, vol. 23, no. 3, pp. 395–404, June 1978.

44 C. Moler and C. V. Loan, "Nineteen dubious ways to compute the exponential of a matrix, twenty-five years later," *SIAM Review*, vol. 45, no. 1, pp. 1–46, 2003.

45 ——, "Nineteen dubious ways to compute the exponential of a matrix," *SIAM Review*, vol. 20, no. 4, pp. 801–836, 1978.

46 G. H. Golub and C. F. V. Loan, *Matrix Computations*. The Johns Hopkins University Press, 2013.

47 W. H. Press, B. P. Flannery, S. A. Teukolsky, and W. T. Vetterling, *Numerical Recipes in C: The Art of Scientific Computing*. Cambridge University Press, 1992.

48 A. Semlyen, "Fundamental concepts of a Krylov subspace power flow methodology," in *IEEE Transactions on Power Systems*, vol. 11, no. 3, pp. 1528–1537, Aug. 1996.

49 M. Hochbruck and C. Lubich, "On Krylov subspace approximations to the matrix exponential operator," *SIAM Journal on Numerical Analysis*, vol. 34, no. 5, pp. 1911–1925, 1997.

50 M. A. Botchev and L. Knizhnerman, "Art: adaptive residual-time restarting for Krylov subspace matrix exponential evaluations," *J. Computational Applied Mathematics*, vol. 364, 2020.

51 R. Erickson and D. Maksimovic, Eds., *Fundamentals of Power Electronics*. Springer, 2001.

52 R. Tymerski, "Sampled-data modelling of switched circuits, revisited," in *Proceedings of IEEE Power Electronics Specialist Conference - PESC '93*, 1993, pp. 395–401.

53 M. Y. Wu, "A successive decomposition method for the solution of linear time-varying systems," *International Journal of Control*, vol. 33, no. 1, pp. 181–186, 1981.

54 ——, "Solution of certain classes of linear time-varying systems," *International Journal of Control*, vol. 31, no. 1, pp. 11–20, 1980.

55 R. Wu, S. B. Dewan, and G. R. Slemon, "A PWM AC-to-DC converter with fixed switching frequency," *IEEE Transactions on Industry Applications*, vol. 26, no. 5, pp. 880–885, Sept.-Oct. 1990.

56 J. G. Kassakian and T. M. Jahns, "Evolving and emerging applications of power electronics in systems," *IEEE Journal of Emerging and Selected Topics in Power Electronics*, vol. 1, no. 2, pp. 47–58, 2013.

57 S. Kouro, M. Malinowski, K. Gopakumar, J. Pou, L. G. Franquelo, B. Wu, J. Rodriguez, M. A. Perez, and J. I. Leon, "Recent advances and industrial applications of multilevel converters," *IEEE Transactions on Industrial Electronics*, vol. 57, no. 8, pp. 2553–2580, Aug 2010.

58 G. Carrara, S. Gardella, M. Marchesoni, R. Salutari, and G. Sciutto, "A new multilevel PWM method: a theoretical analysis," *Power Electronics, IEEE Transactions on*, vol. 7, no. 3, pp. 497–505, 1992.

59 M. Angulo, P. Lezana, S. Kouro, J. Rodriguez, and B. Wu, "Level-shifted PWM for cascaded multilevel inverters with even power distribution," in *Power Electronics Specialists Conference, 2007. PESC 2007. IEEE*, 2007, pp. 2373–2378.

60 A. Nabae, I. Takahashi, and H. Akagi, "A new neutral-point-clamped pwm inverter," *IEEE Transactions on Industry Applications*, vol. IA-17, no. 5, pp. 518–523, 1981.

61 S. A. Gonzalez, S. A. Verne, and M. I. Valla, *Multilevel Converters for Industrial Applications*. CRC Press, 2014.

62 M. Chaves, E. Margato, J. F. Silva, and S. F. Pinto, *Generalized State-Space Modeling for m Level Diode-Clamped Multilevel Converters*. Dordrecht: Springer Netherlands, 2014, pp. 67–85.

63 B. Wu, *High-Power Converters and AC Drives*. Wiley, 2007. [Online]. Available: https://books.google.com.tw/books?id=X1b1laKEtugC

64 A. Lesnicar and R. Marquardt, "An innovative modular multilevel converter topology suitable for a wide power range," in *2003 IEEE Bologna Power Tech Conference Proceedings,*, vol. 3, June 2003, 6 pp.

65 P. W. Lehn and K. L. Lian, "Frequency coupling matrix of a voltage-source converter derived from piecewise linear differential equations," *IEEE Transactions on Power Delivery*, vol. 22, no. 3, pp. 1603–1612, July 2007.

66 G. Strang, *Introduction to Applied Mathematics*. Wellesley-Cambridge Press, 1986.

67 D. Holmes and T. Lipo, *Pulse Width Modulation for Power Converters*. Hoboken, NJ: John Wiley, 2003.

68 W. R. Bennett, "New results in the calculation of modulation products," *The Bell System Technical Journal*, vol. 12, pp. 228–243, April 1993.

69 S. R. Bowes and B. M. Bird, "Novel approach to the analysis and synthesis of modulation processes in power convertors," *Proceedings of the Institution of Electrical Engineers*, vol. 122, no. 5, pp. 507–513, 1975.

70 I. Deslauriers, N. Avdiu, and B. T. Ooi, "Naturally sampled triangle carrier pwm band-width limit and output spectrum," *IEEE Transactions on Power Electronics*, vol. 20, no. 1, pp. 100–106, January 2005.

71 P. W. Lehn, "Direct harmonic analysis of the voltage source converter," *IEEE Transactions on Power Delivery*, vol. 18, no. 3, pp. 1304–1042, July 2003.

72 R. K. Subroto and K. L. Lian, "Modeling of a multilevel voltage source converter using the fast time-domain method," *IEEE Journal of Emerging and Selected Topics in Power Electronics*, vol. 2, no. 4, pp. 1117–1126, Dec 2014.

73 A. H. Al-Mohy and N. J. Higham, "Computing the action of the matrix exponential, with an application to exponential integrators," *SIAM Journal on Scientific Computing*, vol. 33, no. 2, pp. 488–511, 2011.

74 F. Blaabjerg, Ed., *Control of Power Electronic Converters and Systems*, vol. 2. Academic Press, 2018.

75 V. Kaura and V. Blasko, "Operation of a phase locked loop system under distorted utility conditions," *IEEE Transactions on Industry Applications*, vol. 33, no. 1, pp. 58–63, 1997.

76 M. Karimi-Ghartemani, *Enhanced phase-locked loop structures for power and energy applications*. John Wiley & Sons, 2014.

77 Y. Gui, X. Wang, F. Blaabjerg, and D. Pan, "Control of grid-connected voltage-source converters: The relationship between direct-power control and vector-current control," *IEEE Industrial Electronics Magazine*, vol. 13, no. 2, pp. 31–40, 2019.

78 J. Hu, L. Shang, Y. He, and Z. Q. Zhu, "Direct active and reactive power regulation of grid-connected DC/AC converters using sliding mode control approach," *IEEE Transactions on Power Electronics*, vol. 26, no. 1, pp. 210–222, 2011.

79 G. H. Lee, Y. Gui, C. Kim, and C. C. Chung, "Direct power control for three phase grid connected inverter via port-controlled Hamiltonian method," in *IECON 2015 - 41st Annual Conference of the IEEE Industrial Electronics Society*, 2015, pp. 002 312–002 317.

80 S. Buso and P. Mattavelli, *Digital Control in Power Electronics*. Morgan and Claypool, 2006.

81 K. Sharifabadi, L. Harnefors, H. P. Nee, S. Norrga, and R. Teodorescu, *Design, Control, and Application of Modular Multilevel Converters for HVDC Transmission Systems*. John Wiley & Sons, 2016.

82 B. Li, R. Li, B. W. Williams, and D. Xu, "Energy transfer analysis for capacitor voltage balancing of modular multilevel converters," in *2016 IEEE Transportation Electrification Conference and Expo*, 2016.

83 Q. Tu, Z. Xu, and L. Xu, "Reduced switching-frequency modulation and circulating current suppression for modular multilevel converters," *IEEE Transactions on Power Delivery*, vol. 26, no. 3, pp. 2009–2017, 2011.

84 J. Freytes, G. Bergna, J. A. Suul, S. DArco, F. Gruson, F. Colas, H. Saad, and X. Guillaud, "Improving small-signal stability of an mmc with ccsc by control of the internally stored energy," *IEEE Transactions on Power Delivery*, vol. 33, no. 1, pp. 429–439, 2018.

85 A. Medina, J. Segundo-Ramirez, P. Ribeiro, W. Xu, K. L. Lian, G. W. Chang, V. Dinavahi, and N. R. Watson, "Harmonic Analysis in Frequency and Time Domain," *IEEE Transactions on Power Delivery*, vol. 28, no. 3, pp. 1813–1821, July 2013.

86 K. L. Lian and P. W. Lehn, "Steady-state solution of a voltage-source converter with full closed-loop control," *IEEE Transactions on Power Delivery*, vol. 21, no. 4, pp. 2071–2081, Oct 2006.

87 J. Arrillaga, B. C. Smith, N. R. Watson, and A. R. Wood, *Power System Harmonic Analysis*. John Wiley & Sons, 1997.

88 B. K. H. Wong and H. Chung, "Dual-loop iteration algorithm for steady-state determination of current-programmed dc/dc switching converters," in *IEEE Transactions on Circuits and Systems I: Fundamental Theory and Applications*, vol. 46, no. 4, pp. 521–526, April 1999.

89 Y. Sun, G. Zhang, W. Xu, and J. G. Mayordomo, "A harmonically coupled admittance matrix model for AC/DC converters," *IEEE Transactions on Power Systems*, vol. 22, no. 4, pp. 1574–1582, 2007.

90 D. G. Bedrosian and J. Vlach, "An accelerated steady state method for networks with internally controlled switches," *IEEE Transactions on Circuits and Systems I: Fundamental Theory and Applications*, vol. 39, no. 7, pp. 520–530, July 1992.

91 B. K. H. Wong, H. S. H. Chung, and S. T. S. Lee, "Computation of the cycle state-variable sensitivity matrix of power DC/DC converters and its applications," *IEEE Transactions on Circuits and Systems – I: Fundamental theory and applications*, vol. 47, no. 10, pp. 1542–1548, October 2000.

92 R. C. Wong, "Accelerated convergence to the steady-state solution of closed-loop regulated switching-mode system as obtained through simulation," *IEEE Power Electronic Specialists Conference*, pp. 682–692, 1987.

93 M. Grötzbach and R. von Lutz, "Unified modeling of rectifier controlled DC power supplies," *IEEE Power Electronic Specialist Conference*, pp. 304–315, 1984.

94 C. Collins, N. Watson, and A. Wood, "UPFC modeling in the harmonic domain," *IEEE Transactions on Power Delivery*, vol. 21, no. 2, pp. 933–938, April 2006.

95 C. D. Collins, N. R. Watson, and A. R. Wood, "Unbalanced SSSC modelling in the harmonic domain," *The 7th International Power Engineering Conference*, vol. 29, pp. 705–710, Nov. 2005.

96 ——, "Unbalanced statcom analysis in the harmonic domain," in *11th International Conference on Harmonics and Quality of Power*, 2004, pp. 232–237.

97 K. L. Lian and P. W. Lehn, "Steady-state simulation methods of closed-loop power converter systems – a systematic solution procedure," *IEEE Transactions on Circuits and Systems I: Regular Papers*, vol. 59, no. 6, pp. 1299–1311, June 2012.

98 A. Semlyen, J. F. Eggleston, and A. Arrillaga, "Admittance matrix model of a synchronous machine for harmonic analysis," *IEEE Transactions on Power Systems*, vol. 2, no. 4, pp. 833–840, November 1987.

99 D. Shen, Z. Wang, J. Y. Chen, and Y. H. Song, "Harmonic resonance phenomena in statcom and relationship to parameters selection of passive components," *IEEE Transactions on Power Delivery*, vol. 16, no. 1, pp. 46–52, January 2001.

100 C. K. Sao, P. W. Lehn, M. R. Iravani, and J. A. Martinez, "A benchmark system for digital time-domain simulation of a pulse-width-modulated d-statcom," *IEEE Transactions on Power Delivery*, vol. 17, no. 4, pp. 1113–1120, October 2002.

101 H. Visser and P. van den Bosch, "Modeling of periodically switched networks," *IEEE Power Electronic Specialists Conference*, pp. 24–27, June 1991.

102 T. Suetsugu and S. Mori, "Steady-state analysis of closed-loop switching circuit by calculating convolution integral in the steady state," *Electronics and Communications in Japan*, vol. 77, no. 6, pp. 47–57, May 1994.

103 D. N. Zmood and D. G. Holmes, "Stationary frame current regulation of pwm inverters with zero steady-state error," *IEEE Transaction on Power Electronic*, vol. 18, no. 3, pp. 814–822, May 2003.

104 J. D. Turner, A. Messac, and J. L. Junkins, "Finite-time matrix convolution integral sensitivity calculations," *Journal of Guidance, Control, and Dynamics*, vol. 11, no. 5, pp. 473–475, 1988.

105 T. M. Rowan and R. J. Kerkman, "A new synchronous current regulator and an analysis of current regulated pwm inverters," *IEEE Transactions on Industrial Applications*, vol. 22, pp. 678–690, August 1986.

106 D. W. Novotny and T. A. Lipo, *Vector Control and Dynamics of AC Drives*. New York: Oxford Science, 1996.

107 P. W. Lehn, "Exact modeling of the voltage source converter," *IEEE Transactions on Power Delivery*, vol. 17, no. 1, pp. 217–222, January 2002.

108 H. D'Angelo, *Linear Time-Varying Systems: Analysis and Synthesis*. Boston: Allyn and Bacon, 1970.

109 G. Carpinelli, F. Iacovone, A. Russo, P. Varilone, and P. Verde, "Analytical modeling for harmonic analysis of line current of VSI-fed drives," *IEEE Transactions on Power Delivery*, vol. 19, no. 3, pp. 1212–1224, 2004.

110 B. Pilvelait, T. H. Ortmeyer, and M. Grizer, "Harmonic evaluation of inductor location in a variable speed drive," in *ICHPS V International Conference on Harmonics in Power Systems.*, 1992, pp. 267–271.

111 M. Grotzbach and R. Redmann, "Line current harmonics of VSI-fed adjustable-speed drives," *IEEE Transactions on Industry Applications*, vol. 36, no. 2, pp. 683–690, 2000.

112 G. Kron, *Tensor Analysis of Networks*. J. Wiley & Sons, 1939.

113 M. Hancock, "Rectifier action with constant load voltage: infinite-capacitance condition," *Proceedings of the Institution of Electrical Engineers*, vol. 120, no. 12, pp. 1529–1530, 1973.

114 W. F. Ray, "The effect of supply reactance on regulation and power factor for an uncontrolled 3-phase bridge rectifier with a capacitive load," in *International Conference on Power Electronics and Variable-Speed Drives*, May 1984, pp. 111–114.

115 M. Grotzbach and B. Draxler, "Line side behaviour of uncontrolled rectifier bridges with capacitive DC smoothing," in *Proceedings of 3rd EPE Conference (Aachen, Germany)*, 1989, pp. 761–764.

116 J. Schaefer, *Rectifier Circuits: Theory and Design*. John Wiley & Sons, 1965.

117 S. Hansen, L. Asiminoaei, and F. Blaabjerg, "Simple and advanced methods for calculating six-pulse diode rectifier line-side harmonics," in *38th IAS Annual Meeting on Conference Record of the Industry Applications Conference, 2003*, vol. 3, pp. 2056–2062, 2003.

118 J. A. M. Bleijs, "Continuous conduction mode operation of three-phase diode bridge rectifier with constant load voltage," *IEE Proceedings – Electric Power Applications*, vol. 152, no. 2, pp. 359–368, 2005.

119 J. G. Kassakian, M. F. Schlecht, and G. C. Verghese, *Principles of Power Electronics*. Addison-Wesley, 1991.

120 R. C. Dugan and T. E. McDermott, "An open source platform for collaborating on smart grid research," in *2011 IEEE Power and Energy Society General Meeting*, July 2011, pp. 1–7.

121 R. C. Dugan, *Reference Guide: The Open Distribution System Simulator*. Electric Power Research Institute, 2016.

122 V. Ramachandran, *Modeling of Utility Distribution Feeder in OpenDSS with Steady State Impact Analysis of Distributed Generation*. West Virginia University, 2011.

123 E. Möllerstedt and B. Bernhardsson, "Out of control because of harmonics – an analysis of the harmonic response of an inverter locomotive," *IEEE Control Systems Magazine*, pp. 70–81, August 2000.

124 S. Yang, D. Xiang, A. Bryant, P. Mawby, L. Ran, and P. Tavner, "Condition monitoring for device reliability in power electronic converters: A review," *IEEE Transactions on Power Electronics*, vol. 25, no. 11, pp. 2734–2752, Nov 2010.

125 B. Arif, L. Tarisciotti, P. Zanchetta, J. C. Clare, and M. Degano, "Grid parameter estimation using model predictive direct power control," *IEEE Transactions on Industry Applications*, vol. 51, no. 6, pp. 4614–4622, Nov 2015.

126 J. Huang, K. A. Corzine, and M. Belkhayat, "Small-signal impedance measurement of power-electronics-based AC power systems using line-to-line current injection," *IEEE Transactions on Power Electronics*, vol. 24, no. 2, pp. 445–455, Feb 2009.

127 A. N. R. L. Sirisha and A. K. Pradhan, "Parameter estimation of resonant fault current limiter for protection and stability analysis," *IEEE Transactions on Power Systems*, vol. 32, no. 3, pp. 2288–2295, May 2017.

128 E. M. Cimpoesu, B. D. Ciubotaru, and D. Stefanoiu, "Fault detection and diagnosis using parameter estimation with recursive least squares," in *2013 19th International Conference on Control Systems and Computer Science*, May 2013, pp. 18–23.

129 S. Zhai, W. Wang, and H. Ye, "Fault diagnosis based on parameter estimation in closed-loop systems," *IET Control Theory Applications*, vol. 9, no. 7, pp. 1146–1153, 2015.

130 B. H. Lin, J. T. Tsai, and K. L. Lian, "A non-invasive method for estimating circuit and control parameters of voltage source converters," *IEEE Transactions on Circuits and Systems I: Regular Papers*, vol. 66, no. 12, pp. 4911–4921, 2019.

131 S. Neshvad, S. Chatzinotas, and J. Sachau, "Wideband identification of power network parameters using pseudo-random binary sequences on power inverters," *IEEE Transactions on Smart Grid*, vol. 6, no. 5, pp. 2293–2301, Sep. 2015.

132 L. Chang, X. Jiang, M. Mao, and H. Zhang, "Parameter identification of controller for photovoltaic inverter based on l-m method," in *2018 IEEE International Power Electronics and Application Conference and Exposition (PEAC)*, Nov 2018, pp. 1–6.

133 L. Asiminoaei, R. Teodorescu, F. Blaabjerg, and U. Borup, "Implementation and test of an online embedded grid impedance estimation technique for pv inverters," *IEEE Transactions on Industrial Electronics*, vol. 52, no. 4, pp. 1136–1144, Aug 2005.

134 M. Sumner, B. Palethorpe, and D. W. P. Thomas, "Impedance measurement for improved power quality-part 1: the measurement technique," *IEEE Transactions on Power Delivery*, vol. 19, no. 3, pp. 1442–1448, July 2004.

135 ——, "Impedance measurement for improved power quality-part 2: a new technique for stand-alone active shunt filter control," *IEEE Transactions on Power Delivery*, vol. 19, no. 3, pp. 1457–1463, July 2004.

136 Z. Liu, H. Wei, Q. Zhong, K. Liu, X. Xiao, and L. Wu, "Parameter estimation for VSI-fed PMSM based on a dynamic pso with learning strategies," *IEEE Transactions on Power Electronics*, vol. 32, no. 4, pp. 3154–3165, April 2017.

137 B. Miao, R. Zane, and D. Maksimovic, "System identification of power converters with digital control through cross-correlation methods," *IEEE Transactions on Power Electronics*, vol. 20, no. 5, pp. 1093–1099, Sep. 2005.

138 Y. Li, Z. Deng, T. Wang, G. Zhao, and S. Zhou, "Coupled harmonic admittance identification based on least square estimation," *Energies*, vol. 11, no. 10, 2018. [Online]. Available: http://dx.doi.org/10.3390/en11102600

139 A. Angioni, G. Lipari, M. Pau, F. Ponci, and A. Monti, "A low cost PMU to monitor distribution grids," in *2017 IEEE International Workshop on Applied Measurements for Power Systems (AMPS)*, Sept 2017, pp. 1–6.

140 "IEEE standard for synchrophasor measurements for power systems," *IEEE Std C37.118.1-2011 (Revision of IEEE Std C37.118-2005)*, pp. 1–61, Dec 2011.

141 N. Hoffmann and F. W. Fuchs, "Minimal invasive equivalent grid impedance estimation in inductive–resistive power networks using extended Kalman filter," *IEEE Transactions on Power Electronics*, vol. 29, no. 2, pp. 631–641, Feb 2014.

142 M. Esparza, J. Segundo, C. Gurrola-Corral, N. Visairo-Cruz, E. Barcenas, and E. Barocio, "Parameter estimation of grid connected vsc using the extended harmonic domain," *IEEE Transactions on Industrial Electronics*, pp. 1–1, 2018.

143 D. R. Gurusinghe, A. D. Rajapakse, and K. Narendra, "Testing and enhancement of the dynamic performance of a phasor measurement unit," *IEEE Transactions on Power Delivery*, vol. 29, no. 4, pp. 1551–1560, 2014.

144 I. Goodfellow, Y. Bengio, and A. Courville, *Deep Learning*. The MIT Press, 2016.

145 D. Pardoe and P. Stone, "Boosting for regression transfer," in *ICML*, 2010.

146 Y. Hu, J. Jacob, G. J. M. Parker, D. J. Hawkes, J. R. Hurst, and D. Stoyanov, "The challenges of deploying artificial intelligence models in a rapidly evolving pandemic," *CoRR*, vol. abs/2005.12137, 2020. [Online]. Available: https://arxiv.org/abs/2005.12137

147 J. Jiang and C. Zhai, "Instance weighting for domain adaptation in NLP," in *Proceedings of the 45th Annual Meeting of the Association of Computational Linguistics*. Prague, Czech Republic: Association for Computational Linguistics, Jun. 2007, pp. 264–271. [Online]. Available: https://aclanthology.org/P07-1034

148 M. Wang and W. Deng, "Deep face recognition: A survey," *CoRR*, vol. abs/1804.06655, 2018. [Online]. Available: http://arxiv.org/abs/1804.06655

149 A. de Mathelin, G. Richard, M. Mougeot, and N. Vayatis, "Adversarial weighting for domain adaptation in regression," *CoRR*, vol. abs/2006.08251, 2020. [Online]. Available: https://arxiv.org/abs/2006.08251

150 S. Motiian, M. Piccirilli, D. A. Adjeroh, and G. Doretto, "Unified deep supervised domain adaptation and generalization," in *The IEEE International Conference on Computer Vision (ICCV)*, Oct 2017.

151 I. Redko, E. Morvant, A. Habrard, M. Sebban, and Y. Bennani, *Advances in Domain Adaptation Theory*. Elsevier, Aug. 2019. [Online]. Available: https://hal.archives-ouvertes.fr/hal-02286281

152 S. Ruder, "An overview of multi-task learning in deep neural networks," 2017.

153 X. Li, Y. Hu, J. Zheng, M. Li, and W. Ma, "Central moment discrepancy based domain adaptation for intelligent bearing fault diagnosis," *Neurocomputing*, vol. 429, pp. 12–24, 2021. [Online]. Available: https://www.sciencedirect.com/science/article/pii/S0925231220318579

154 Y. Mansour, M. Mohri, and A. Rostamizadeh, "Domain adaptation: Learning bounds and algorithms," CoRR, vol. abs/0902.3430, 2009.

155 S. Ben-David, J. Blitzer, K. Crammer, and F. Pereira, "Analysis of representations for domain adaptation," in *Advances in Neural Information Processing Systems*, B. Schölkopf, J. Platt, and T. Hoffman, Eds., vol. 19. MIT Press, 2007. [Online]. Available: https://proceedings.neurips.cc/paper/2006/file/b1b0432ceafb0ce714426e9114852ac7-Paper.pdf

156 D. Du and P. Pardalos, "Minimax and Applications." Springer, 1995.

157 L. Keke, Z. Ke, L. Qiang, and Y. Xinmin, "Training GANS with predictive projection centripetal acceleration," arXiv-CS-Machine Learning, 2020.

158 J. P. Bailey, G. Gidel, and G. Piliouras, "Finite regret and cycles with fixed step-size via alternating gradient descent-ascent," *Proceedings of Machine Learning Research*, vol. 125, pp. 1–17, 2020.

159 S. Roy, M. Harandi, R. Nock, and R. Hartley, "Siamese networks: The tale of two manifolds," in *2019 IEEE/CVF International Conference on Computer Vision (ICCV)*, 2019, pp. 3046–3055.

160 J. Bromley, J. W. Bentz, L. Bottou, I. Guyon, Y. LeCun, C. Moore, E. Säckinger, and R. Shah, "Signature verification using a 'siamese' time delay neural network," *International Journal of Pattern Recognition and Artificial Intelligence*, vol. 7, no. 4, pp. 669–688, 1993. [Online]. Available: http://oro.open.ac.uk/35662/

161 K. P. Murphy, *Machine Learning: A Probabilistic Perspective*. Cambridge, Mass. [u.a.]: MIT Press, 2013.

162 A. Thakur, "Approaching (almost) any machine learning problem." Abhishek Thakur, 2020.

163 A. Gulli and S. Pal, *Deep Learning with Keras*. Packt Publishing, 2017.

164 R. Atienza, *Advanced Deep Learning with TensorFlow 2 and Keras: Apply DL, GANs, VAEs, deep RL, unsupervised learning, object detection and segmentation, and more*, 2nd ed. Packt Publishing, 2020. [Online]. Available: https://books.google.com.tw/books?id=68rTDwAAQBAJ

Index

Harmonic Modeling of Voltage Source Converters using Basic Numerical Methods, First Edition.
Ryan Kuo-Lung Lian, Ramadhani Kurniawan Subroto, Victor Andrean, and Bing Hao Lin.